Watershed Management

Springer
New York
Berlin
Heidelberg
Barcelona
Budapest
Hong Kong
London
Milan
Paris
Tokyo

Robert J. Naiman
Editor

Watershed Management

Balancing Sustainability and Environmental Change

With 115 Illustrations

Robert J. Naiman
Center for Streamside Studies, AR-10
University of Washington
Seattle, WA 98195
USA

Library of Congress Cataloging-in-Publication Data
Watershed management: balancing sustainability and environmental change, / edited by Robert J. Naiman
p. cm.
"Symposium held in Seattle, Washington, USA, on new pwespectives for watershed management in the Pacific northwest" --CIP galley
Includes bibliographical references and index.
ISBN 0-387-94232-7 (pbk. printing)
ISBN 0-387-97790-2 (hbk. printing)
1. Watershed management--Northwest, Pacific--Congresses.
I. Naiman, Robert J.
TC423.7.W38 1992
333.91'009795--dc20 91-46693

Printed on acid-free paper.

Production managed by Natalie Johnson; manufacturing supervised by Robert Paella.
Typeset by Impressions, a Division of Edwards Brothers, Inc., Ann Arbor, MI.
Printed and bound by Edwards Brothers, Inc., Ann Arbor, MI.
Printed in the United States of America.

9 8 7 6 5 4 3 2

ISBN 0-387-94232-7 Springer-Verlag New York Berlin Heidelberg (pbk.)
ISBN 3-540-94232-7 Springer-Verlag Berlin Heidelberg New York

ISBN 0-387-97790-2 Springer-Verlag New York Berlin Heidelberg (hbk.)
ISBN 3-540-97790-2 Springer-Verlag Berlin Heidelberg New York

Contents

Contributors

TIMOTHY J. BEECHIE Center for Streamside Studies, AR-10, University of Washington, Seattle, WA 98195, USA

LEE E. BENDA Department of Geological Sciences, AJ-20, University of Washington, Seattle, WA 98195, USA

DEAN R. BERG Center for Streamside Studies, University of Washington, Seattle, WA 98195, USA

PETER A. BISSON Technology Center, Weyerhaeuser Company, Tacoma, WA 98477, USA

CAROLYN BLEDSOE Department of Land, Air and Water Resources, University of California, Davis, CA 95616, USA

PAUL CHANDLER International Resource Studies Program, Ball State University, Muncie, IN 47306, USA

COLLETTE DEFERRARI College of Forest Resources, AR-10, University of Washington, Seattle, WA 98195, USA

JAYNE DOLPH ManTech Environmental Technology, Inc., US EPA Environmental Research Laboratory, Corvallis, OR 97333, USA

WAYNE ELMORE Bureau of Land Management, Prineville, OR 97754, USA

RICHARD FLAMM Environmental Sciences Division, Oak Ridge National Laboratory, Oak Ridge, TN 37831, USA

JERRY F. FRANKLIN College of Forest Resources, AR-10, University of Washington, Seattle, WA 98195, USA

Robin Gottfried Department of Economics, The University of the South, Sewanee, TN 37375, USA

G.E. Grant USDA Forest Service, Forestry Sciences Laboratory, Corvallis, OR 97331, USA

Stanley V. Gregory Department of Fisheries and Wildlife, Oregon State University, Corvallis, OR 97331, USA

David A. Jay Geophysics Program, AK-50, University of Washington, Seattle, WA 98195, USA

K. Norman Johnson Department of Forest Resources, Oregon State University, Corvallis, OR 97331, USA

Kendall L. Johnson Department of Range Resources, University of Idaho, Moscow, ID 83843, USA

Carol A. Johnston Natural Resources Research Institute, University of Minnesota, Duluth, MN 55811, USA

George A. King ManTech Environmental Technology, Inc., US EPA Environmental Research Laboratory, Corvallis, OR 97333, USA

David R. Larsen College of Forest Resources, AR-10, University of Washington, Seattle, WA 98195, USA

Robert G. Lee College of Forest Resources, AR-10, University of Washington, Seattle, WA 98195, USA

Lee H. MacDonald Center for Streamside Studies, AR-10, University of Washington, Seattle, WA 98195, USA

Danny Marks ManTech Environmental Technology, Inc., US EPA Environmental Research Laboratory, Corvallis, OR 97333, USA

Edwin H. Marston Publisher, *High Country News*, Paonia, CO 81428, USA

Walter F. Megahan National Council of the Paper Industry for Air and Stream Improvement, Port Townsend, WA 98368, USA

Robert J. Naiman Center for Streamside Studies, AR-10, University of Washington, Seattle, WA 98195, USA

R.P. NEILSON Department of General Science, Oregon State University, Corvallis, OR 97331, USA

MATTHEW D. O'CONNOR Center for Streamside Studies, AR-10, University of Washington, Seattle, WA 98195, USA

KEVIN L. O'HARA School of Forestry, University of Montana, Missoula, MT 59812, USA

CHADWICK DEARING OLIVER College of Forest Resources, AR-10, University of Washington, Seattle, WA 98195 USA

PATRICIA L. OLSON Center for Streamside Studies, University of Washington, Seattle, WA 98195, USA

JOHN PASTOR Natural Resources Research Institute, AR-10, University of Minnesota, Duluth, MN 55811, USA

CLAY R. PATMONT Environmental Services Division, Hart Crowser, Inc., Seattle, WA 98102, USA

JOHN P. POTYONDY USDA Forest Service, Boise National Forest, Boise, ID 83702, USA

THOMAS P. QUINN School of Fisheries, WH-10, University of Washington, Seattle, WA 98195, USA

GORDON H. REEVES USDA Forest Service, Pacific Northwest Research Laboratory, Corvallis, OR 97331, USA

RAYMOND M. RICE USDA Forest Service, Redwood Sciences Laboratory, Arcata, CA 95521, USA

PAUL G. RISSER University of New Mexico, Albuquerque, NM 87131, USA

NATHAN SCHUMAKER College of Forest Resources, AR-10, University of Washington, Seattle, WA 98195, USA

KATHLEEN A. SEYEDBAGHERI USDA Forest Service, Intermountain Research Station, Boise, ID 83702, USA

CHRISTOPHER R. SHERWOOD School of Oceanography, WB-10, University of Washington, Seattle, WA 98195, USA

CHARLES A. SIMENSTAD Fisheries Research Institute, WH-10, University of Washington, Seattle, WA 98195, USA

LYNN R. SINGLETON Environmental Investigations and Laboratory Services, Washington State Department of Ecology, Olympia, WA 98504, USA

RAYMOND A. SOLTERO Department of Biology, Eastern Washington University, Cheney, WA 99004, USA

J.A. STANFORD Flathead Lake Biological Station, University of Montana, Polson, MT 59860, USA

E. ASHLEY STEEL Center for Streamside Studies, AR-10, University of Washington, Seattle, WA 98195, USA

F.J. SWANSON USDA Forest Service, Forestry Sciences Laboratory, Corvallis, OR 97331, USA

MONICA G. TURNER Environmental Sciences Division, Oak Ridge National Laboratory, Oak Ridge, TN 37831, USA

J.V. WARD Department of Biology, Colorado State University, Fort Collins, CO 80523, USA

DAVID WEAR USDA Forest Service, Forest Sciences Laboratory, Research Triangle Park, NC 27709, USA

Part 1
Global and National Perspectives

1

New Perspectives for Watershed Management: Balancing Long-Term Sustainability with Cumulative Environmental Change

ROBERT J. NAIMAN

Abstract

The background and rationale for the symposium held in Seattle, Washington, USA, on New Perspectives for Watershed Management in the Pacific Northwest are presented. As the region develops, natural resource availability is declining, increasing demands are being made on the remaining resources, and the cumulative impacts on the environmental and social systems are becoming severe. This has resulted in contentious debates within the growing population as global economic demands for forest, fish, water, and wildlife resources expand. Resolution of these conflicts requires new perspectives that combine social, economic, and environmental concerns with an approach to watershed management where forest, range, agricultural, wetland, and urban parcels are treated in an integrated manner.

Key words. Cumulative effects, environment, socioeconomic, sustainability, watershed management.

Introduction

The need for a new perspective in watershed management could not have been more apparent to the nearly 600 symposium participants at the University of Washington on November 27–29, 1990. Torrential rains engulfed western Washington and resulted in the worst flooding in over a century. The ecological, economic, and social consequences were on a scale that only a handful of people had ever witnessed. But few realized that the scale was small compared with those natural forces that had shaped the region in the last several hundred years. Predictably, the resource extraction methods used to obtain forest products and the changes in land use brought about by a growing urban population were blamed by the media and by special interest groups for the devastation. Yet the resulting debate served to highlight the need to understand the interactive processes at the watershed scale and to develop a new perspective for watershed management that balances long-term ecological, economic, and social stability with cumulative environmental change. As a result of increasingly severe environmental and social disturbances, the people of the Pacific Northwest are gradually realizing that a grand experiment involving the entire region inadvertently has been set in motion without the means to monitor the results or alter the outcome.

A major goal of the symposium was to explore how the Pacific Northwest, as an integrated regional society, could encourage economic strength through wise and efficient use of natural resources while maintaining (or improving) the quality of life and the long-term vitality of the environment. At best, this is a difficult challenge, especially for a region that has until recently enjoyed an abundance of renewable natural resources, including timber, fish, wildlife, and water. Historically, these resources have been used and managed independently. As a result, an inadequate understanding of their interrelations often has led to controversial resource management disputes. Resolving such disputes, and ultimately managing the natural resources wisely, will require integrated research and information at watershed and landscape scales. Understanding natural resource ecology and interactions at such scales will eventually provide the basis for developing comprehensive management guidelines; however, our understanding is still at a rudimentary stage.

Regional and global demands for renewable natural resources continue to grow. These resources are an important component of the global market system, and their value is often determined outside the region (Brown et al. 1987, 1991). As wood fiber, water, fish products, and recreational opportunities decline in other regions, the value and demand for these relatively abundant resources escalate in the Pacific Northwest. This can be seen in the declining stocks of anadromous salmonids over the last century, as habitat has been modified to support other resource demands (Figure 1.1) and in the hydrologic transformation of the streams and rivers (Benke 1990).

Growth within the region is also having a significant impact on the availability of renewable resources and their management. The urban population in northern California, Oregon, Washington, and British Columbia has grown rapidly since the 1930s, increasing demands on space and resources and changing expectations for resource management (Figures 1.2 and 1.3). Today, conflicts over renewable natural resources are contentious and will most likely intensify as access and availability of the resources become limiting (Lee, this volume). As a result, resource managers, scientists, and policy analysts must be able to state technical issues clearly and seek innovative solutions if conflicts are to be resolved in a positive manner, and they must be able to implement adaptive management strategies at the watershed scale (Lee and Lawrence 1986, Walters 1986).

It was against this background that scientists, resource managers, and policy analysts from government agencies, private industries, Indian tribes, environmental groups, and universities assembled to present and discuss cur-

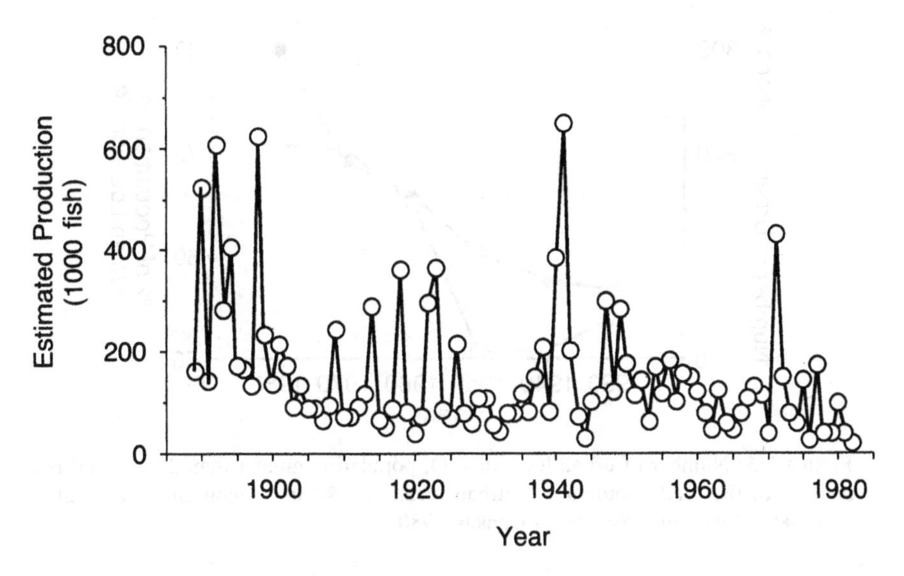

FIGURE 1.1. Estimated production of adult anadromous salmonids in the state of Washington, 1890–1985. Data from Shepard et al. (1985).

rent knowledge and future trends in watershed management and research. The specific objectives of the symposium were (1) to articulate the current status of watershed research and describe regional models, (2) to present indicators of environmental change, (3) to examine new technical tools and system models, and (4) to explore how new techniques and models could be related to integrated resource management. This symposium was the fourth in a series sponsored by the Center for Streamside Studies devoted to watershed issues in the Pacific Northwest. The first symposium addressed for-

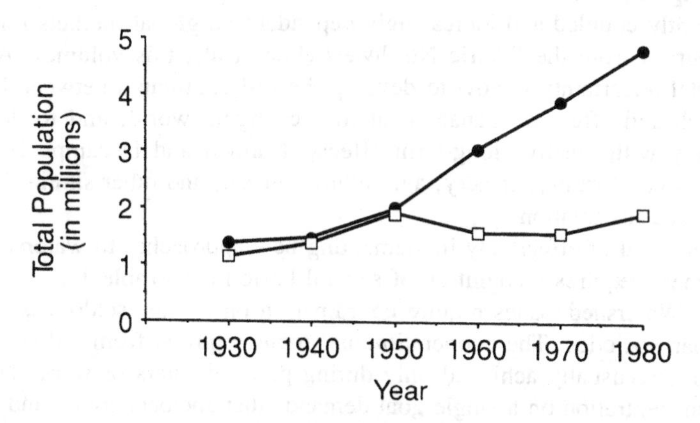

FIGURE 1.2. Trends in the total rural (□) and urban (●) populations for Oregon and Washington, 1930–80. Data from Population Census 1980.

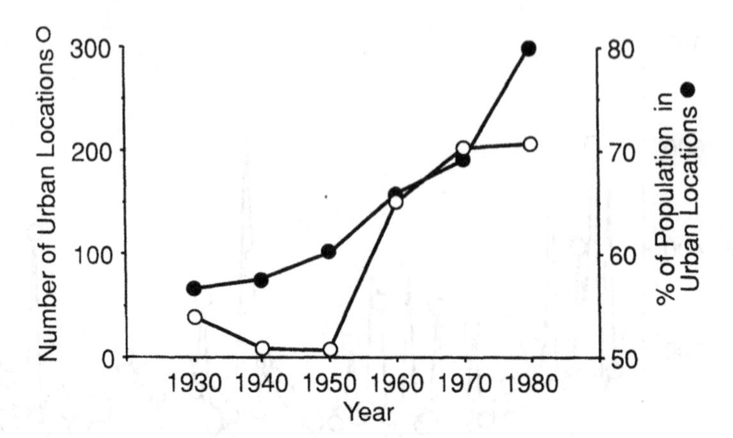

FIGURE 1.3. Number of urban locations (O; population greater than 2,500) and percentage of the total population in urban locations (●) in Oregon and Washington, 1930–80. Data from Population Census 1980.

estry-fishery interactions (Salo and Cundy 1987), the second dealt with forestry-riparian wildlife issues (Raedeke 1988), and the third was concerned with the ecology and management of aquatic-terrestrial ecotones (Naiman and Décamps 1990). The next symposium, planned for 1992, will address social and economic influences in implementing watershed management concepts and plans.

The region's scientists, planners, managers, and citizens have an opportunity to make a difference in our collective future, provided that wise decisions are made in the next few years. These decisions will mean the difference between a long-term vitality for the socioeconomic and environmental systems, as opposed to favoring one over the other, or both of them declining together. The vitality of socioeconomic and environmental systems is tightly coupled and increasingly dependent on global markets for natural resources from the Pacific Northwest (Lee et al., this volume). A fundamental uncertainty is how to develop the full continuum between basic research and effective management in a changing world, and to do it efficiently with positive long-term effects (Naiman and Décamps 1991). No longer can forestry, fishery, agricultural, urban, and other similar issues be resolved in isolation.

The goal of effectively implementing new approaches to watershed management requires recognition of several basic facts (Table 1.1).

1. Watershed issues require coordination on a scale seldom achieved in human societies. The cooperation and focus required from a diverse population are usually achieved only during great disasters or wars. This level of concentration on a single goal demands that socioeconomic and political barriers be crossed efficiently and effectively. Most current management strategies are not at a scale commensurate with issues at the watershed scale.

Table 1.1. Essential conditions for implementing new approaches to watershed management.

1. The scope of the issues demands unparalleled cooperation between industry, governmental agencies, private institutions, and academic organizations.
2. The increasing tendency to resort to technical solutions (e.g., hatcheries, silviculture) must be augmented with increased habitat protection and preservation of fundamental components of long-term watershed vitality.
3. The complexity of information management and the scope of experimental manipulations needed often exceed the capacity of individual institutions.
4. The current tendency to seek conceptual solutions at the expense of data-driven decisions must be reversed.
5. Intra- and interagency inconsistencies in environmental regulations must be corrected.
6. Human activities are a key element of ecosystem vitality and must be integrated with environmental considerations before long-term sustainability of the biosphere can be achieved.

Local control or management by system components (e.g., fish, wildlife, timber production) takes precedence over systemwide needs. An example where such cooperation and focus are required is the Columbia River Basin (28 major dams and 1,750 kilometers of reservoirs) with regard to preserving genetically diverse fish stocks and avoiding looming conflicts over water allocation (Lee 1989).

2. As a society, we tend to address issues of environmental change by using technology (e.g., hatcheries, engineering) while relying on incomplete or inadequate information. Technology cannot resolve issues when it is isolated from a fundamental understanding of the properties of natural, social, and ecological systems. In addition, effective monitoring of ecological and social responses to management decisions is seriously needed to complement technological innovations, and the knowledge we gain from that effort must be incorporated into future decisions. Technology is not the solution, but it can be part of the solution.

3. Effective watershed management requires that information be collected and analyzed on scales that most groups and institutions are reluctant (or unable) to address. Streams in the Pacific Northwest provide a good example. West of the Cascade Mountain crest there are about 2.5 km of streams per square kilometer compared with 0.5/km^2 east of the mountains. Across the states of Oregon and Washington there are nearly 531,000 km (332,000 miles) of streams requiring some level of managerial scrutiny. Efforts of this scope cannot be accomplished by traditional methods such as extensive field surveys. New approaches must be developed. In addition, streams are variable systems whose structural and functional features change over time. Long-term data are nearly nonexistent for these features. It is unfortunate that nearly 90% of the information generated about streams as ecological systems

has been accumulated in the past 15 years, far too short a time to understand adequately the important natural features of these systems.

4. There is an increasingly popular impression that the process of watershed management is becoming more important than the actual discovery and implementation of new knowledge. New groups, often formed from the membership of old ones, are organized with new goals before old goals are realized. Up-to-date information is needed to overcome years of neglect by agencies and industry to fund ecological research commensurate with the questions being asked. As a result, resource information is not generally available to resolve current issues (such as cumulative effects). Will the information be there in the future? Only if long-term commitments are made by institutions for effective interdisciplinary research and education. A better balance must be sought.

5. Finally, two essential points: first, inconsistencies in agency regulations within watersheds must be overcome (such as protection of riparian forests on forested but not agricultural land); and, second, *understanding* and *integrating* human activities into environmental systems must be achieved in the near future if the region and the globe are to attain any form of sustainability (Ruckelshaus 1989, Turner et al. 1990).

Symposium Philosophy and Proceedings

The spirit of the symposium was to leave professional and personal biases, as well as previous conflicts, elsewhere. Issues surrounding natural resources in the Pacific Northwest are increasingly complex and contentious. Yet, resolving these issues requires cooperation among all philosophical groups and the development of innovative approaches. Participants were asked to focus on the future, and on innovative and practical approaches for making life better for everyone in the region. This latter phrase captures in a few words what is hoped to be accomplished by developing new perspectives for watershed management. It is impossible to go back in time and correct all the human transgressions on the landscape; we have only today's landscape and the opportunities this landscape provides for the future.

The proceedings are divided into three broad subject areas.

Global and National Perspectives. These are addressed in the chapters by P.G. Risser on watershed management in a changing global environment, J.F. Franklin on integrating riparian zones into the "new persepctives" for forestry, R.G. Lee on temporal and spatial scales in analyzing biosocial systems, and J.A. Stanford and J.V. Ward on ecosystem management in large watersheds regulated by dams and diversions.

Elements of Integrated Watershed Management. Aspects of this complicated subject are examined in the chapters by R.J. Naiman and his colleagues on fundamental elements of ecologically healthy watersheds, by P.A. Bisson and his colleagues on long-term trends in fish abundance in Pacific

Northwest streams, by J. Dolph and her colleagues on modeling precipitation-runoff relations in the Columbia River Basin under a changing climatic regime, by C.A. Simenstad and his colleagues on the impacts of watershed management on the land-ocean margin using the Columbia River as an example, by F.J. Swanson and his colleagues on concept, model, and data needs for watershed-landscape studies, by J. Pastor and C.A. Johnston on integrating new and existing technologies for a landscape perspective, by K.N. Johnson on new approaches for national forest planning, and by C.D. Oliver and his colleagues on the integration of modern management tools, improved ecological knowledge, and current silvicultural approaches.

Innovative Approaches for Mitigation and Restoration of Watersheds. This topic is addressed in the chapters by R.M. Rice on the science and politics of forestry best management practices, by W.F. Megahan and colleagues on watershed management trends for forested lands, by K.L. Johnson on watershed management trends for rangelands, by W. Elmore on innovative rangeland grazing procedures to improve both economic and environmental concerns, by R.A. Soltero and his colleagues using the Spokane River (Washington) as a case study of adaptive management, by E.H. Marston on the role of western land grant universities in developing a new perspective for watershed management, and by R.G. Lee and his colleagues on developing a socioecologic prespective.

Ultimately, if technical approaches articulated at this symposium are to be successful, there must be an integration of basic science with managerial applications for the resolution of issues. This symposium addresses only the initial steps in developing the full continuum—from concept, to research, to education, to application. The fundamental challenge, as we make the transition from an exploitative society to a sustainable society, is to learn how to integrate socioeconomic needs with environmental vitality. The Pacific Northwest has become the testing site to evaluate if this can be achieved, and the results are being watched closely by regions around the world facing similar challenges.

Acknowledgments. We thank the members of the symposium planning committee for their enthusiasm, guidance, and support: R.L. Beschta (Oregon State University), P.A. Bisson (Weyerhaeuser Company), G. Ice (National Council of the Paper Industry for Air and Stream Improvement), M. Johnson (Environmental Protection Agency), R. Lackey (Environmental Protection Agency), J.A. Stanford (University of Montana), F.J. Swanson (U.S. Forest Service, Pacific Northwest Research Station), and D. Wallace (Department of Ecology, State of Washington).

Financial assistance was graciously provided by the National Council of the Paper Industry for Air and Stream Improvement, Oregon State University, Department of Natural Resources (Washington), Department of Ecology (Washington), Department of Wildlife (Washington), U.S. Forest Service's Pacific Northwest Research Station, U.S. Environmental Protection Agency's Environmental Research Laboratory (Corvallis, Oregon), University of Washington Center for Streamside Studies, and the Washington Forest Protection Association.

All articles were evaluated for scientific merit by anonymous reviewers. Their comments and suggestions were particularly valuable to the authors and the editor. The technical editing was accomplished by Leila Charbonneau and the production coordination by Mary Smith of the University of Washington Institute of Forest Resources. Carla Warth provided staff support during the editorial process. Their assistance is greatly appreciated.

Special appreciation is extended to Beverly Gonyea and Betty Johanna of the University of Washington College of Forest Resources Continuing Education office and to the students and staff of the Center for Streamside Studies. Their efforts in making arrangements and in providing logistic support will be pleasantly remembered by all those in attendance.

References

Benke, A.C. 1990. A perspective on America's vanishing streams. Journal of the North American Benthological Society **9**:77–88.

Brown, L.R., W.V. Chandler, C. Flavin, J. Jacobson, C. Pollock, S. Postel, L. Starke, and E.C. Wolf. 1987. State of the world, 1987. Norton, New York, New York, USA.

Brown, L.R., A. Durning, C. Flavin, H. French, J. Jacobson, N. Lenssen, M. Lowe, S. Postel, M. Renner, J. Ryan, L. Starke, and J. Young. 1991. State of the world, 1991. Norton, New York, New York, USA.

Lee, K.N. 1989. The mighty Columbia: experimenting with sustainability. Environment, July-August: 6–11, 30–33.

Lee, K.N., and J. Lawrence. 1986. Adaptive management: learning from the Columbia River Basin fish and wildlife program. Environmental Law **16**:431–460.

Naiman, R.J., and H. Décamps, editors. 1990. The ecology and management of aquatic-terrestrial ecotones. UNESCO, Paris, and Parthenon Publishing Group, Carnforth, United Kingdom.

Naiman, R.J., and H. Décamps, editors. 1991. Summary: landscape boundaries in the management and restoration of changing environments. Pages 130–137 *in* M.M. Holland, P.G. Risser, and R.J. Naiman, editors. The role of landscape boundaries in the management and restoration of changing environments. Chapman and Hall, New York, New York, USA.

Population Census 1980. Characteristics of population. Volume 1, Chapter A. Number of inhabitants, parts 39 and 49, Oregon and Washington. PC80–1-A39, PC80–1-A49. United States Government Printing Office, Washington, D.C., USA.

Raedeke, K.J., editor. 1988. Streamside management: riparian wildlife and forestry interactions. Contribution 59, Institute of Forest Resources, University of Washington, Seattle, Washington, USA.

Ruckelshaus, W.D. 1989. Toward a sustainable world. Scientific American, September: 166–175.

Salo, E.O., and T.W. Cundy, editors. 1987. Streamside management: forestry and fishery interactions. Contribution 57, Institute of Forest Resources, University of Washington, Seattle, Washington, USA.

Shepard, M.P., C.D. Shepard, and A.W. Argue. 1985. Historic statistics of salmon production around the Pacific Rim. Fisheries and Oceans Canada. Manuscript Report 1819. Vancouver, British Columbia, Canada.

Turner II, B.L., W.C. Clark, R.W. Kates, J.F. Richards, J.T. Mathews, and W.B. Meyer, editors. 1990. The earth as transformed by human action. Cambridge University Press, Cambridge, England.

Walters, C. 1986. Adaptive management of renewable resources. Macmillan, New York, New York, USA.

2

Impacts on Ecosystems of Global Environmental Changes in Pacific Northwest Watersheds

PAUL G. RISSER

Abstract

The ecosystems of the Pacific Northwest are part of the global environment. As such, not only are they affected by global processes but they also, in turn, affect regional and global processes. Changing climate may influence the distribution of species and their growth and also the dynamics of water and nutrients in specific watersheds and across the landscape. Management of natural resources must account for global processes in an organized and coherent manner to maintain their long-term vitality.

Key words. Climate change, trace gases, productivity, management, vegetation, Pacific Northwest.

Introduction

The Pacific Northwest is a geographically diverse region where natural resources are particularly significant. Forestry, fisheries, recreation, and agriculture are important industries economically, and they all depend on sound management of the region's natural resources. However, the best management practices for these resources cannot be developed and implemented by considering just the region itself, because the behavior of many Pacific Northwest processes is driven by processes that operate at the global scale. This connection means that the Pacific Northwest cannot be understood in isolation, and that some lessons learned elsewhere about global processes can be beneficial in addressing the regional issues of the Pacific Northwest.

The purpose of this article is to discuss representative changes in the global atmospheric and climatic processes that affect vegetation, and indicate how these processes may have an impact on the management of the natural resources in the Pacific Northwest. In some cases, describing the probable impacts on Pacific Northwest ecosystems is not a problem because studies done elsewhere can be applied directly to the local conditions. But in other cases, when differences in environmental conditions exist between the Pacific Northwest and the location of the initial research study, direct applications are more speculative and uncertain. This analysis identifies important issues about managing the natural resources in which final decisions require investigation by regional scientists.

Changes in the Global Atmospheric Constituents

The concentrations of the long-lived greenhouse gases—carbon dioxide (CO_2), methane (CH_4), chlorofluorocarbons (CFCs), and nitrous oxide (N_2O)—in the global atmosphere are increasing substantially, largely because of emissions from various human activities. Carbon dioxide, produced from burning fossil fuel and conversion of forest lands, has been responsible for over half of the enhanced greenhouse effect (Rodhe 1990); water vapor, also a greenhouse gas, is expected to increase from evaporation and transpiration in response to global warming and will further heighten the potential for warming of average global temperatures. Without substantial changes in human activities, it is expected that these changes in atmospheric constituents will increase the global mean temperature during the next century at the rate of about 0.3°C per decade (with an uncertainty of 0.2° to 0.5°C per decade). As a consequence, there will be a global mean temperature increase of about 1°C by the year 2025 and 3°C before the end of the next century (Houghton et al. 1990).

Since changes in the climate will not be uniform across continents, there will be locations where the temperature does not increase substantially and some where the soil moisture will be lower or higher than currently. Thus in the Pacific Northwest it is not known whether significant changes in the climate regime will occur. The large-scale models used to predict climate changes do not yet have sufficient resolution to predict local climate changes. Because of inadequate inventories of biomass and because of the coarse resolution of general circulation models (GCMs), the global carbon budget is not well understood (Tans et al. 1990). As in predictions of climate change, our present predictions of carbon fluxes become even less accurate at regional levels and are probably impossible for local situations.

Despite this uncertainty at regional and local scales, it is clear that ecosystems will not only be affected by these climatic conditions, but, equally important, ecosystem processes will also contribute to these global processes (Cates and Keeney 1987, Crutzen and Andreae 1990, Matthews and Fung 1987). Current annual estimates for increases in CO_2, CH_4, and N_2O are 0.5%, 0.9%, and 0.3%, respectively. The relative global contributions of human activities to these increases are 60%, 15%, and 5%, respectively (Earthquest 1990). In other words, natural processes also contribute to the production of greenhouse gases, and in the cases of CH_4 and N_2O, the major contributions are from natural processes. Thus the ecosystems of the Pacific Northwest are not only influenced by these global processes but contribute to them.

The current concentration of atmospheric CO_2 is about 350 ppm, increasing from about 315 ppm in 1958. In addition to the indirect effects that increasing atmospheric CO_2 is expected to have on the regional patterns of temperature and moisture, there will be direct effects on terrestrial ecosystems. Among the most conspicuous are first-order physiological and morphological responses, such as changes in plant photosynthesis and respiration. These responses affect plant growth rates as well as stomatal dynamics. Since stomata control water conductance, ambient CO_2 controls water use efficiency in plants (Strain 1985). As the concentration of CO_2 increases, plant stomata tend to close, resulting in a decrease in the loss of water from the leaf by transpiration and an increase in the water use efficiency of the plants, especially in C_4 plants (Cure 1985). Important C plants in the Pacific Northwest are three-awn (*Aristida longiseta*), shadscale (*Atriplex confertifolia*), saltgrass (*Distichlis stricta*), and dropseed (*Sporobolus cryptandrus*). Thus the vegetation of the Pacific Northwest will be affected by the atmospheric CO_2 concentrations. It is possible that the more humid regions will respond more by changes in plant growth rates, whereas the more arid regions will be influenced by water stress controlled by stomatal conductance.

Initially, plants may respond to high levels of CO_2 because of increased carboxylation of ribulose 1, 5-biphosphate (RuBP) carboxylase. Both CO_2 and O_2 compete for this primary enzyme. With increasing atmospheric CO_2, the CO_2 has a competitive advantage, and in C_3 plants, the rate of photosynthesis may increase by 10 to 15% (Strain 1985). In C_4 plants, however, the primary CO_2 fixing enzyme, phosphoenolpyruvate (PEP) carboxylase, is not inhibited by O_2; photorespiration is negligible; and increasing CO_2 does not appreciably increase the rate of photosynthesis. Thus, depending on the mixture of C_3 and C_4 plants, elevated levels of CO_2 may affect the species composition of the vegetation because of the different responses to CO_2. In the more arid regions of the Pacific Northwest, C_3 grasses might be expected to increase at the relative expense of the warm-season C_4 grasses. These vegetational responses, however, may be quite complex. For example, vegetation responses depend on the successional status of the plant community and the interactions between CO_2, temperature, and moisture (Bazzaz and Carlson 1984, Carter and Peterson 1983, Trolley and Strain 1985, Zangerl and Bazzaz 1984). Currently it is impossible to predict general vegetational responses because only a few intact ecosystems have been studied.

The increase in primary production induced by elevated concentrations of CO_2 (CO_2 fertilization effect) may be transient as the plants become acclimated to higher levels of CO_2 (Besford and Hand 1989). Tissue and Oechel (1987) used experimental enclosures to expose the Arctic tussock tundra, dominated by *Eriophorum vaginatum*, to elevated levels of CO_2. During the first year, the tundra demonstrated increased rates of carbon fixation. But in subsequent years, the plants returned to the original rates of biomass accumulation even in elevated levels of CO_2. It is unknown whether the alpine tundra or other vegetation in the Pacific Northwest would respond in a similar fashion.

A similar study was conducted using open-top chambers in a *Scirpus olneyi* dominated subestuary salt marsh of Chesapeake Bay (Curtis et al. 1989, and pers. comm.). Elevated CO_2 increased carbon assimilation by increased photosynthesis and decreased respiration. In contrast to the Arctic tundra study, in this salt marsh ecosystem there was no reduction of the CO_2 stimulation of carbon assimilation through three years (1987–89) of elevated CO_2 treatment.

The reasons for acclimation or nonacclimation of different vegetation types are not entirely clear; thus it is currently difficult to predict just how regional ecosystems might respond. However, in environments of rising CO_2 there appear to be some consistent patterns: the availability of nutrients may ultimately limit the capacity of plants to assimilate carbon; any increase in photosynthetic carbon accumulation may be compensated by increased respiration in plants in response to higher carbohydrate content and any increases in ambient temperature; and decomposition rates may increase because of increased temperature. Regions consisting of various combinations of nutrient-poor and nutrient-rich habitats might be expected to demonstrate different patterns of carbon assimilation and storage.

There are several other interactions between changing climate and vegetation. Increased ambient CO_2 concentrations cause plant tissues to have a higher ratio of carbon to nitrogen, possibly resulting in reduced decomposition rates. In those ecosystems, such as Pacific Northwest coniferous forests, which contain large amounts of litter on the soil surface, decreasing rates of decomposition might have significant effects not only on nutrient dynamics but on the amount of vegetative reproduction. Also, in vegetation with high C:N ratios, herbivores must consume more to compensate for the lower nitrogen content (Fajer 1989). Especially in forests where leaf herbivory is significant, elevated CO_2 levels might cause significant differences in herbivory rates.

From this discussion it is evident that there are many changes in the global atmosphere that could directly affect the biological processes in the Pacific Northwest, particularly the growth and nutrient dynamics of both humid and semi-arid ecosystems. In addition to the examples presented, there are many other potential interactions that have not been discussed. Figure 2.1 depicts the organizational structure of the scientific program currently envisaged in the International Geosphere-Biosphere Program on terrestrial ecosystems (Global Change 1990). Note that this discussion has focused only on the upper left-hand corner of the diagram. A complete investigation of the interactions between the terrestrial and atmospheric components of the global system would involve focusing on all the linkages. In applying this structure to organizing a research and management program for the Pacific Northwest,

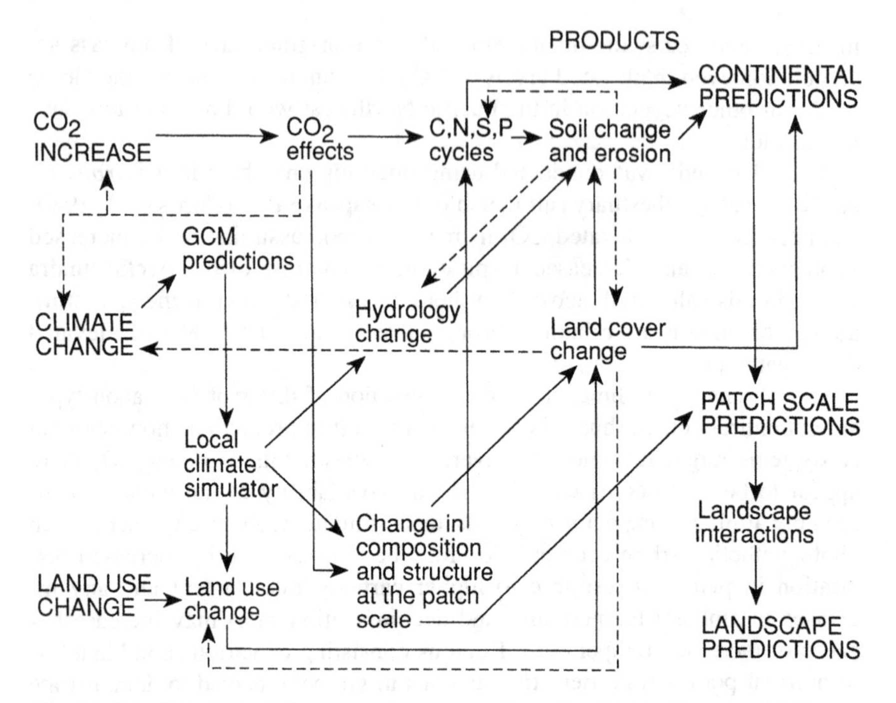

FIGURE 2.1. Structure of the research program from the Global Change and Terrestrial Ecosystems Core Project, International Geosphere-Biosphere Program. (*Note.* From Global Change, IGBP Report No. 12, p. 6.1-4, 1990. Used by permission.)

particular attention must be paid to the interactions between land use and climate change. Discussion here has focused on atmospheric constituents because of their global nature. However, land use activities are also global, in part because these manifestations can occur over large spatial scales and in part because land use practices in aggregate become a global process.

Interactions of global processes occur at different time and space scales (Rosswall et al. 1988). Indeed, a major research challenge is to develop models or process studies that permit the faithful extrapolation of localized measurements to regional or global processes, such as developing techniques that permit the application of results obtained in the H.J. Andrews Experimental Forest (Oregon, USA) to the entire appropriate region in the Pacific Northwest. A number of key research questions at various spatial and temporal scales, and the managerial reasons for their importance, have been described in the two recent reports of the U.S. National Committee for the International Geosphere-Biosphere Program, 1988 and 1990. Especially in the topical areas of water-energy exchanges, nutrient dynamics, and trace gas fluxes, key unknown processes are noted and the requisite experiments are described.

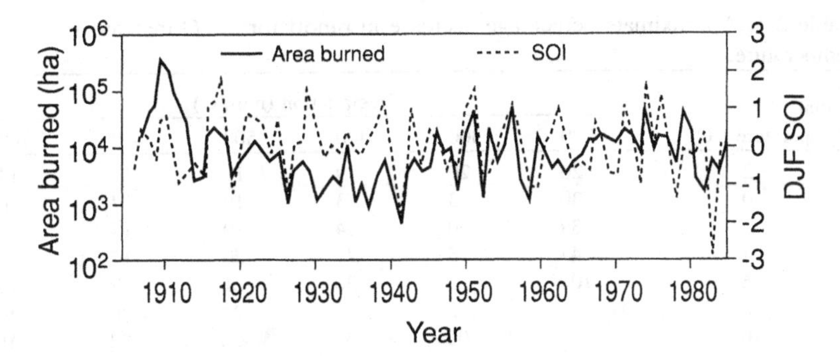

FIGURE 2.2. Time series of the area burned in Arizona and New Mexico (logarithm) and the average December through February (DJF) Southern Oscillation index (SOI), 1905–85 (*Note*. From Swetnam and Betancourt, *Science*, **249**. Copyright 1990. Used by permission.)

Global Climate Processes and Vegetation Management

Changes in the environment at the global scale may have dramatic local or regional effects on the natural resources, particularly where these effects are mediated through other processes (Swanson et al., this volume). For example, various climate-change scenarios from general circulation models propose different patterns of rainfall, growing season temperatures, and soil moisture levels across North America and the globe. These variables not only affect plant growth and reproduction, but they also affect the frequency and intensity of forest fires. For example, in southwestern North America, the annual weather cycle includes a variably wet and cool fall, winter, and early spring, an arid early summer, and a monsoonal rainy period from early July into September. Lightning fires begin in the spring and peak in late June and early July. The winter-spring precipitation determines the amount and moisture content of the fuel in these forests, which are dominated by piñon pine (*Pinus edulis*), ponderosa pine (*P. ponderosa*), and Douglas-fir (*Pseudotsuga menziesii*). When there is high pressure in the southeastern Pacific Ocean (La Niña), the Southern Oscillation index (SOI) is high and there is low winter and spring precipitation in the southwestern United States. During El Niño years, when the SOI is low, there is higher winter and spring rainfall.

Swetnam and Betancourt (1990) used tree-ring analysis and fire scars to examine the relationships between the SOI and fires in the U.S. Southwest over a 300-year period. In 1982–83, the El Niño episode was very severe and there were few forest fires in the Southwest, though there was considerable burning in Indonesia and Australia. As shown in Figure 2.2, until about 1960, there was a very strong correlation between the SOI during December, January, and February (DJF) and the amount of the forested area

Table 2.1. Approximate percentage decrease in runoff for a 2°C increase in temperature.

Initial	Precipitation (mm/yr)					
Temperature (°C)	200	300	400	500	600	700
−2	26	20	19	17	17	14
0	30	23	23	19	17	16
2	39	30	24	19	17	16
4	47	35	25	20	17	16
6	100	35	30	21	17	16
8		53	31	22	20	16
10		100	34	22	22	16
12			47	32	22	19
14			100	38	23	19

Source: Data in Revelle and Waggoner (1989).

in Arizona and New Mexico that was burned. After 1960 the pattern was not so clear, probably because fire suppression permitted accumulations of fuel and perhaps because of the influence of human-caused fires. The frequency and severity of fires is important to the status of vegetation. In New Mexico, for example, ponderosa pine forests are maintained with ground fires that occur every few years. Upon cessation of fires, either because of global weather patterns or human activities, the fuel load increases, leading to crown fires, which in turn permit succession to more shade-tolerant tree species. Although the effects of El Niño are not likely to be identical in the Pacific Northwest, the ingredients for a similar analysis are all present. That is, lightning-driven fires are common and the successional sequences in the vegetation are affected by initial species composition, frequency of burning, and the fuel loads on the forest floor. Thus managerial decisions cannot be based on rote application of current models for predicting the consequences of burning regimes, but must take into account the changing natural patterns of climate conditions.

Global climate processes also affect the hydrologic properties of watersheds, in part by changes in vegetation, as discussed above. In addition, the proportion of water which is either retained in or lost from the watershed is affected by the soil conditions (Schlesinger et al. 1990). For example, in high-latitude tundra environments warmer winters may result in greater snowfall and increased summer runoff. In the arid regions, small changes in the precipitation rates may result in great reductions in the amount of runoff. In the Pacific Northwest, there are significant areas of both alpine tundra and arid lands. Revelle and Waggoner (1989) suggest, for example, that a 2°C rise in temperature and a 10% reduction in precipitation would result in a 76% reduction in runoff in the Rio Grande region and a 40% reduction in runoff in the Upper Colorado area. As Table 2.1 suggests, the impact of a 2°C rise in temperature would have the greatest effect on runoff in regions that receive less than 40 cm of precipitation annually. However, with local validation, this approach could be used to predict the probable impacts of changing rainfall patterns on the runoff conditions in appropriate mesic to dry regions of the Pacific Northwest.

Management of forest and grassland fires and control of runoff are two of the more important issues, not only in the Pacific Northwest but throughout the world. It is clear from this discussion that global climate, in the form of long-term trends in temperature and moisture and in cyclical events such as the El Niño-La Niña sequence, can have major impacts on natural resources and the ways in which these resources must be managed.

Global Process and the Distribution of Vegetation

As the global climate changes, it is expected that so will the geographical distribution of biological resources. Analyses from previous climatic conditions associated with various advances in glaciation indicate that plants migrate across the landscape in response to the prevailing climate. However, because individual plant and animal species have different responses to changing environmental conditions, different species will move at different rates (Davis 1984, 1988). Also, dominant plant species and their influence on carbon and nitrogen cycles may affect the success of other migrating species (Pastor and Post 1988). During the Holocene, some vegetation types were much larger than they are today and some were more restricted than they are today (Davis 1988).

These differential rates of success mean that with changes in the global climate pattern, the vegetation types and associated animals in the Pacific Northwest might be quite different in the future. McNeely (1990) has developed some generalizations about the types of organisms that are likely to benefit from or be threatened by climate change. Among species that might benefit are: plants that respond to CO_2 enrichment and grow more vigorously, insects that thrive on rapidly growing plants, long-lived trees that outlast short cycles of unfavorable climate, species that reproduce quickly with new opportunities (r-strategists), and species whose sex ratios are affected by climate and thereby become adapted to the new conditions.

Those species that are likely to be threatened include: rare or threatened species that are already in jeopardy, migratory species that require appropriate conditions throughout the entire migratory pathway, aquatic species whose life cycle is tied to the timing of ice melts, genetically impoverished species that have a limited ability to adapt, specialized species that depend on only a few species for food, and montane and alpine species that would face competition from species migrating from lower elevations.

These categories act as a preliminary guide to potential impacts of climate change on the Pacific Northwest ecosystems. In addition to these categories of responses, it will be necessary to consider not only climate change but how human utilization of natural resources affects the size, number, and geographic distribution of animals (Brown and Maurer 1989). Moreover, ecotones (boundaries between ecosystem types) may prove to be particularly sensitive to climate change. Consequent changes in community composition as well as habitat size and variability will depend not only on the climate itself but on interactions with topography and the physical environment (Neilson 1991).

Major questions arise as to whether species will be able to migrate with changing climate or whether their adaptive mechanisms will be sufficient to permit them to remain in their current locations. Some model predictions indicate that the expected change in temperature will require species to move much more rapidly than the evidence suggests they did during the Holocene (e.g., approximately 20 km per decade). Also, soil conditions, pollinators, or symbionts in new habitats defined by climate may be inadequate for species that otherwise might be expected to migrate with changing climate. Finally, human activities and land use might act as a barrier to migration, with especially unfortunate implications for species now supported in relatively isolated natural areas (Committee on International Science's Task Force on Global Biodiversity 1989, Peters and Darling 1985).

There have been several attempts to predict the future geographical distribution of species that might be expected with changing climate. Some of these approaches have focused on individual species (Davis 1988) while others have focused on vegetation types. An example of the latter is the study by Emanuel et al. (1985) which used the Holdridge system for classifying vegetation types. These types are related to ambient environmental conditions; moreover, this approach does not allow any dynamic interactions among the vegetation, soil, and climate components of the system. Using a mathematical model, the distribution of these types was predicted within a scenario of enriched atmospheric CO_2 and consequent changes in temperature and moisture. The vegetation types in the Pacific Northwest moved farther north. In general, the results showed an increase in the more arid vegetation types and, in North America, a northward migration of most vegetation categories.

Changing global environmental conditions may affect the distribution of species in the Pacific Northwest as well as throughout the world. The influences of global climate change would be particularly important if migration routes were closed because of inadequate soil conditions or because of land use barriers. More subtle effects than just the migration of individual species would be the gradual loss of productivity or vigor of key species due to slowly changing environmental conditions, or changes in the genetic composition of the populations.

Pacific Northwest Natural Resources: Research and Management

The natural resources of the Pacific Northwest are important to the region, nation, and the world. Although many management decisions can be made on the basis of local conditions, this discussion has demonstrated that these

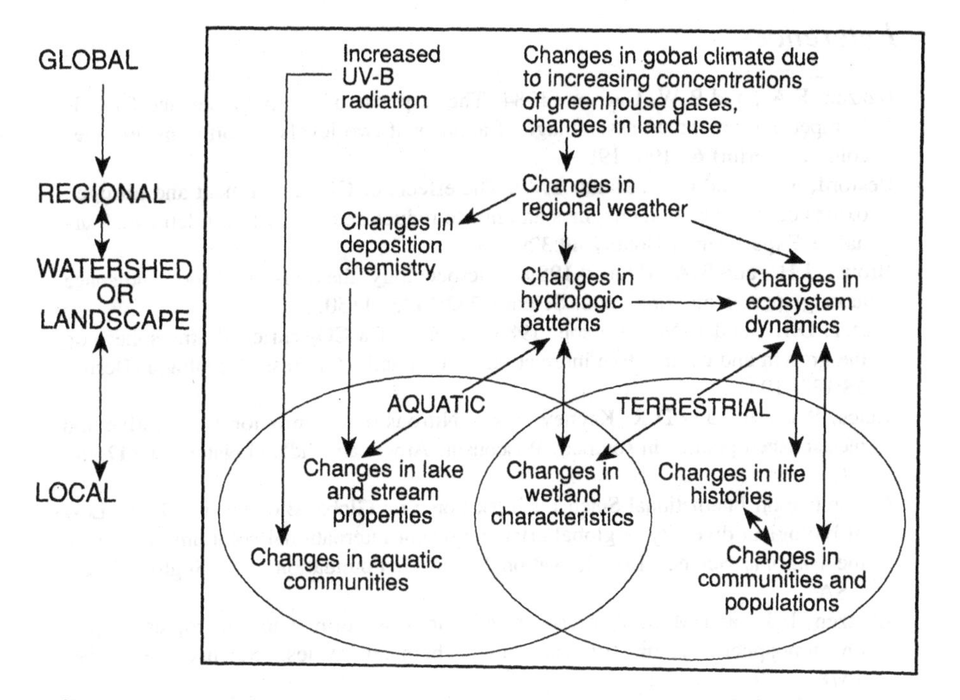

FIGURE 2.3. Structure of the proposed Colorado Rockies Global Change Program, U.S. National Park Service.

resources are part of the biosphere and as such are influenced by and, in turn, influence global processes. Though most of the discussion has dealt with terrestrial systems, there is also strong evidence of ecosystem responses to global change in aquatic systems (Minshall 1988, Schindler et al. 1990).

As the Pacific Northwest enhances its efforts to manage the natural resources of the region, a framework should be established that explicitly recognizes the global nature of many fundamental processes. Figure 2.3 is a general diagram produced within the National Park Service as it begins a Global Change Research Program. This diagram is not presented as a specific model to be emulated in the Pacific Northwest, although there is no inherent reason that it cannot be adopted. Indeed, this or a similar model should be immediately employed to organize a coherent research program directed at answering the inevitable questions about how to manage the natural resources under changing global environmental conditions. The model recognizes the importance of spatial scale and organizes the interactions among several natural resources. A similar systematic approach would be useful in identifying the questions that require additional research and also those topical areas in which there is now adequate knowledge for implementing management strategies that incorporate global processes.

References

Bazzaz, F.A., and R.W. Carlson. 1984. The response of plants to elevated CO_2. I. Competition among an assemblage of annuals at two levels of soil moisture. Oecologia (Berlin) **62**:196–198.

Besford, R.T., and D.W. Hand. 1989. The effects of CO_2 enrichment and nitrogen oxides on some Calvin enzymes and nitrite reductase on glasshouse lettuce. Journal of Experimental Botany **40**:329–336.

Brown, J.H., and B.A. Maurer. 1989. Macroecology: the division of food and space among species on continents. Science **243**:1145–1150.

Carter, D.R., and K.M. Peterson. 1983. Effects of a CO_2-enriched atmosphere on the growth and competitive interaction of a C_3 and a C_4 grass. Oecologia (Berlin) **58**:188–193.

Cates, R.L., Jr., and D.R. Keeney. 1987. Nitrous oxide emission from native and reestablished prairies in southern Wisconsin. American Midland Naturalist **117**:35–42.

Committee on International Science's Task Force on Global Biodiversity. 1989. Loss of biological diversity: a global crisis requiring international solutions. Report to the National Science Board, National Science Foundation, Washington, D.C., USA.

Crutzen, P.J., and M.O. Andreae. 1990. Biomass burning in the tropics: impact on atmospheric chemistry and biogeochemical cycles. Science **250**:1669–1678.

Cure, J.D. 1985. CO_2 doubling responses: a crop survey. Pages 101–116 *in* B.R. Strain and J.D. Cure, editors. Direct effects of increasing CO_2 on vegetation. United States Department of Energy DOE/ER-0238, Washington, D.C., USA.

Curtis, P.S., B.G. Drake, P.W. Leadley, W.J. Arp, and D.F. Whigham. 1989. Growth and senescence in plant communities exposed to elevated CO_2 concentrations on an estuarine marsh. Oecologia (Berlin) **78**:20–26.

Davis, M.B. 1984. Climatic instability, time lags, and community disequilibrium. Pages 269–284 *in* W.C. Clark and R.E. Munn, editors. Community ecology. Harper and Row, New York, New York, USA.

Davis, M.B. 1988. Ecological systems and dynamics. Pages 69–106 *in* Committee on Global Change, editors. Toward an understanding of global change. National Academy Press, Washington, D.C., USA.

Earthquest. 1990. Volume 4, Number 2. Office for Interdisciplinary Studies. International Center for Atmospheric Research, Boulder, Colorado, USA.

Emanuel, W.R., H.H. Shugart, and M.P. Stevenson. 1985. Climatic change and the broad-scale distribution of terrestrial ecosystem complexes. Climatic Change **7**:30–43.

Fajer, E.D. 1989. The effects of enriched CO_2 atmospheres on plant-insect herbivore interactions: growth responses of larvae of the specialist butterfly, *Junonia coenia* (Lepidoptera:Nymphalidae). Oecologia (Berlin) **81**:514–520.

Global Change. 1990. The International Geosphere-Biosphere Programme: a study of Global Change. The initial core projects. Report 12. Stockholm, Sweden.

Houghton, J.T., G.J. Jenkins, and J.J. Ephraums, editors. Climate change: the IPCC scientific assessment. Cambridge University Press, Cambridge, England.

Matthews, E., and I. Fung. 1987. Methane emission from natural wetlands: global distribution, area, and environmental characteristics of sources. Global Biogeochemical Cycles **1**:61–86.

McNeely, J.A. 1990. Climate change and biological diversity: policy implications. Pages 406–429 *in* M.M. Boer and R.S. de Groot, editors. Landscape-ecological impact of climate change. IOS Press, Amsterdam, The Netherlands.

Minshall, G.W. 1988. Stream ecosystem theory: a global perspective. Journal of the North American Benthological Society **7**:263–288.

Neilson, R.P. 1991. Climatic constraints and issues of scale controlling regional biomes. Pages 31–51 *in* M.M. Holland, P.G. Risser, and R.J. Naiman, editors. The role of landscape boundries in the management and restoration of changing environments. Chapman and Hall, New York, New York, USA.

Pastor, J., and W.M. Post. 1988. Response of northern forests to CO_2-induced climate change. Nature **334**:55–58.

Peters, R.L., and J.D.S. Darling. 1985. The greenhouse effect and nature reserves. BioScience **35**:707–717.

Revelle, R.R., and P.E. Waggoner. 1989. Effects of climate change on water supplies in the western United States. Pages 151–160 *in* D.E. Abrahamson, editor. Challenge of global warming. Island Press, Washington, D.C., USA.

Rodhe, H. 1990. A comparison of the contributions of various gases to the greenhouse effect. Science **248**:1217–1219.

Rosswall, T., R.G. Woodmansee, and P.G. Risser. 1988. Scales and global change: spatial and temporal variability in biospheric and geospheric processes. SCOPE 35. John Wiley and Sons, Chichester, England.

Schindler, D.W., K.G. Beaty, E.J. Fee, D.R. Cruikshank, E.R. DeBruyn, D.L. Findlay, G.A. Linsey, J.A. Shearer, M.P. Stainton, and M.A. Turner. 1990. Effects of climatic warming on lakes of the central boreal forest. Science **250**:967–970.

Schlesinger, W.H., J.F. Reynolds, G.L. Cunningham, L.E. Huenneke, W.M. Jarrell, R.A. Virginia, and W.G. Whitford. 1990. Biological feedbacks in global desertification. Science **247**:1043–1048.

Strain, B.R. 1985. Background on the response of vegetation to atmospheric enrichment. Pages 118–154 *in* B.R. Strain, and J.D. Cure, editors. Direct effects of increasing CO_2 on vegetation. United States Department of Energy DOE/ER-0238, Washington, D.C., USA.

Swetnam, T.W., and J.L. Betancourt. 1990. Fire-Southern Oscillation relations in the southwestern United States. Science **249**:1017–1020.

Tans, P.P., I.Y. Fung, and T. Takahashi. 1990. Observational constraints on the global atmospheric CO_2 budget. Science **247**:1431–1438.

Tissue, D.T., and W.C. Oechel. 1987. Responses of *Eriophorum vaginatum* to elevated CO_2 and temperature in the Alaskan tussock tundra. Ecology **68**:401–410.

Trolley, L.C., and B.R. Strain. 1985. Effects of CO_2 enrichment and water stress on gas exchange of *Liquidambar styraciflua* and *Pinus taeda* seedlings grown under different irradiance levels. Oecologia (Berlin) **65**:166–172.

United States National Committee for the International Geosphere-Biosphere Program. 1988. Toward an understanding of global change: initial priorities for U.S. contributions to the International Geosphere-Biosphere Program. National Research Council, National Academy Press, Washington, D.C., USA.

United States National Committee for the International Geosphere-Biosphere Program. 1990. Research strategies for the U.S. Global Change Research Program. National Research Council, National Academy Press, Washington, D.C., USA.

Zangerl, A.R., and F.A. Bazzaz. 1984. The response of plants to elevated CO_2. II. Competitive interactions among annual plants under varying light and nutrients. Oecologia (Berlin) **62**:412–417.

3

Scientific Basis for New Perspectives in Forests and Streams

JERRY F. FRANKLIN

Abstract

New perspectives that involve development of management approaches that integrate ecological and economic values are especially important in riparian zones because of the numerous ecological linkages between terrestrial and aquatic ecosystems. The scientific bases for alternative management regimes for forests and associated streams lie primarily in expanded knowledge of the importance of ecosystem complexity, of biological legacies in reestablishing ecosystems following major disturbances, and of landscape perspectives (larger spatial and temporal scales). Each of these three areas of knowledge is reviewed and then applied in developing some alternative approaches to managing forests and associated riparian zones. Maintenance or re-creation of structurally diverse managed forests is an important principle of New Perspectives concepts at the stand level. Structures such as large trees, snags, and down logs are focal points in management, because they can act as surrogates for organisms and functions that are often difficult to quantify. Important considerations at the landscape level include special attention to riparian habitats (including headwaters), creation of an interconnected system of reserved areas, and selection of appropriate patch sizes for managed areas. Substantial progress is occurring in development and field trials of New Perspective concepts at both stand and landscape levels. A high level of collaboration between scientists, managers, and public interest groups is essential for this process to be successful.

Key words. New Forestry, riparian, biological diversity, biological legacies, landscape ecology, fragmentation, cumulative effects.

Introduction

Today, conflict provides the context for natural resource management. We are faced as a society and as resource managers with ever-increasing and increasingly diverse demands upon a limited resource base. The inevitable conflicts are epitomized by such extraordinarily difficult issues as preserving northern spotted owl (*Strix occidentalis caurina*) habitat versus cutting timber in the old-growth forests, and sustaining anadromous fish runs versus generating power or using water for irrigation from our major river systems.

The traditional approach to resolving conflicts over resource values has been to allocate lands and waters among apparently incompatible uses. Some areas are designated as wilderness, national parks, or national scenic rivers for retention in a natural state, while other lands are committed to the production of commodities such as timber. While both commodity and wild lands are generally recognized as providing a variety of benefits to society, physically separating management activities on these lands by allocation has been the primary approach. This emphasis has impeded efforts by agencies, such as the U.S. Forest Service, to develop and apply a "multiple use" concept.

Allocation is proving to have less and less value as an approach to resolving resource conflicts, due to the decreasing resource base and the increasing demands of a growing and more affluent population (see Lee, this volume). There is also greater recognition that various ecological processes, such as those associated with sustainable productivity and biological diversity, permeate our lands and waters, regardless of their actual or designated uses (Amoros et al. 1987).

The development of alternative approaches to resource management—approaches that better integrate maintenance of ecological values with commodity production—is one response to the dilemma (see Lee et al., this volume). These approaches are known by many labels, including "ecological" or "new" forestry and the Forest Service's New Perspectives Program. They are more often conceptual than prescriptive, pointing to the need for specific management activities to be in concert with the multiplicity of resources, environments, and landowner goals. New knowledge and technology are combined with old tools and experience to provide a broader spectrum of management tools (Franklin and Maser 1988). Most important, these approaches have a sound scientific basis in ecological science, although some of the applications or proposed practices can be viewed as working hypotheses or experiments (Hopwood 1991, Oliver and Hinckley 1987, Walters 1986).

I explore the relevance of New Perspectives to riparian zones in this chapter, beginning with a brief review of the ecological roles played by riparian environments; I take the broader view of the riparian area as the streamside influence zone. Then follows an overview of recent scientific research relevant to the design of ecologically oriented management practices—ecosystem complexity, biological legacies, and landscape ecology. I conclude with a section on key issues in the integration of ecological and commodity values and some specific practices—at the stand, stream reach, and landscape levels—that might be used to address those issues. My purpose is to illustrate the potential contribution of alternative management practices in managing for a mix of resource values in both upland and riparian environments.

Ecological Benefits Provided by Riparian Environments

Riparian portions of our northwestern forest landscapes provide numerous ecological links between the forest and the aquatic ecosystem (Gregory et al. 1991*a*, Agee 1988, Heede 1985, Naiman 1990, Naiman et al. 1988, 1989). Riparian vegetation controls much of the environmental regime of stream ecosystems; this is less true of larger streams and rivers which greatly influence the nature of the riparian vegetation. Quantity and seasonal timing of light levels are most often determined by type and amount of streamside vegetation along small- and medium-size (up to fourth-order) streams. Light levels are critical to a variety of ecological processes as diverse as primary productivity, which is light-limited in heavily shaded streams (Gregory et al. 1991*b*), and feeding by fish (Wilzbach et al. 1986, Cummins 1974). Stream temperature is also strongly influenced by riparian vegetation; shading to maintain stream temperatures below lethal levels for fish was an early justification for preserving forest corridors and remains an important factor in warmer parts of the region (Hunt 1988, Agee 1988).

Riparian zones are the source of extremely important structural components of the aquatic ecosystem. Woody debris is often the dominant element in the physical structure of streams (Bisson et al. 1987); specifically, providing coarse woody debris for the stream channels is a particularly critical role of the riparian forest (Maser et al. 1988, Swanson et al. 1976, 1984; Keller and Swanson 1979, Harmon et al. 1986). The structural complexity resulting from woody debris is important in determining such stream-reach characteristics as ability to retain allochthonous inputs, store sediments, and detain water (Harmon et al. 1986, Sullivan et al. 1987, Bisson et al. 1987, Bilby 1981). Large woody debris can be directly responsible for the creation of stepped stream profiles and a variety of habitats, such as debris jams and sediment accumulations (sand or gravel bars); wood and wood-related materials may account for 50% or more of the habitats in small, densely forested stream reaches (Franklin et al. 1981, Harmon et al. 1986, Gregory et al. 1991*b*, Grant et al. 1990). These materials are important invertebrate resources (Anderson et al. 1978, 1984). Furthermore, large woody debris can strongly influence habitat diversity in large streams and small rivers through its effect on their hydraulic characteristics (Grette 1985, Bilby and Likens 1980).

Riparian vegetation provides important nutritional substrate for aquatic ecosystems (Gregory et al. 1991*b*, Triska et al. 1982). The allochthonous inputs that dominate small streams are the main source of energy and an important source of nutrients for the aquatic ecosystem. Research is making us increasingly aware of the large variety of species and life-forms that are present, as well as the high degree of spatial heterogeneity in natural streamside and riparian vegetation (Oliver and Hinckley 1987, Nilsson et al. 1989, Gregory et al. 1991*b*). One direct consequence of this richness is allochthonous inputs with higher levels of compositional and temporal diversity (Conners and Naiman 1984, Melillo et al. 1984). For example, herbaceous components of riparian vegetation typically senesce earlier in the season, contain higher nutritional content, and are more readily processed by the aquatic community than inputs from deciduous trees and shrubs which, in turn, are of higher quality and are more readily processed than needles and litter from coniferous trees (Gregory et al. 1991*b*, Melillo et al. 1983, Conners and Naiman 1984). Therefore, streamside zones that have a diversity of herbaceous, shrub, and tree communities generate more diverse allochthonous inputs qualitatively and temporally than those dominated by a single vegetation type.

Streamside zones also provide important and specialized habitat for many elements of biological diversity, a function that is disproportionately high for the area they occupy. Many plant and animal species are known to have their primary habitat requirements met within riparian environments (Raedeke 1988, Décamps et al. 1987, Rochelle et al. 1988). The existence of vascular plant species dependent on the special moisture and temperature of the streamside zones is well known; some of these may be equally dependent on the pattern of chronic disturbance associated with floodplain environments. Many species of both vertebrate (Murphy 1979) and invertebrate (Lattin 1990) animals are identified as riparian species. Other animals, including many invertebrates, divide their life cycles between riparian and upland habitats (Merritt and Cummins 1978) and still others, including many species of bats, make essential daily use of both conditions (West 1988, Cross 1988). In addition to direct use of riparian habitats, streamside corridors are hypothesized to be routes for the movement or migration of various animal species (Raedeke 1988), although this use has not been well documented.

Disturbance regimes in stream ecosystems are important in maintenance of both species and processes; furthermore, the roles of chronic events (e.g., annual flooding) and episodic ones (e.g., debris flows and high intensity floods) are different (Gregory et al. 1991*b*, Lamberti et al. 1991). Episodic disturbances are most important in shaping the riparian zone and its vegetation. Substantial import, movement, and export of woody debris and sediments occur during major storm episodes. Shifts in channel morphology and woody debris are more limited with chronic flooding, although annual events do provide for the regular creation of freshly disturbed habitats for plant colonization. In any case, both chronic and episodic disturbances are important elements of riparian zones, and their contrasting roles need to be recognized in New Perspectives management.

One conclusion based on existing ecological research is that the structurally and compositionally diverse streamside zones are well suited to produce the desired mixture of "ecological services" for the associated aquatic ecosystems. It also appears that natural streams or reaches—those free of major human influences—are more likely to have high levels of complexity than those that have been managed (Bryant 1983, Triska et al. 1982). This translates into more diverse, productive, and resilient ecosystems. These conclusions are similar to those for upland areas, which are addressed in the next section.

The Scientific Basis for Alternative Management Regimes

The expanding ecological knowledge of terrestrial and aquatic ecosystems, particularly in the Pacific Northwest, is a major factor both in identifying problems with existing practices and in offering solutions. Therefore, a discussion of the scientific underpinnings of New Perspectives is a good place to begin (Hopwood 1991). Any proposed management practice, however, including most current approaches, must be considered a working hypothesis until its effectiveness is verified.

Only recently have we begun to examine natural forests and associated streams as ecosystems. Although there has been a long history of silvicultural, autecological, and fisheries research, studies of natural forest ecosystems started just a little over two decades ago, with National Science Foundation support of the International Biological Program's (IBP) Coniferous Forest Biome Project in 1969 (Edmonds 1982). This project and subsequent ecosystem research programs, including the Long-Term Ecological Research (LTER) project at H. J. Andrews Experimental Forest (Franklin et al. 1990), have yielded a wealth of information. Other programs, such as the Old-Growth Wildlife Habitat Program (Ruggiero et al. 1991), have greatly expanded our knowledge of biological diversity and its requirements. I will review some of this information under three topical headings: (1) ecosystem complexity, (2) "biological legacies," or aspects of ecosystem regeneration following catastrophic disturbances, and (3) landscape ecology perspectives. While these subject areas are not new, the richness of the scientific information base is, as well as its relevance to forest management issues.

Ecosystem Complexity

Scientific studies have shown that the natural forests and streams are far more complex than we had imagined in terms of their composition, function, and structure (Franklin et al. 1981, Edmonds 1982, and Gregory et al. 1991*b*). In particular, the natural mature and old-growth forests are far more than just young forest stands grown senescent. Such forests have distinctive properties and functions.

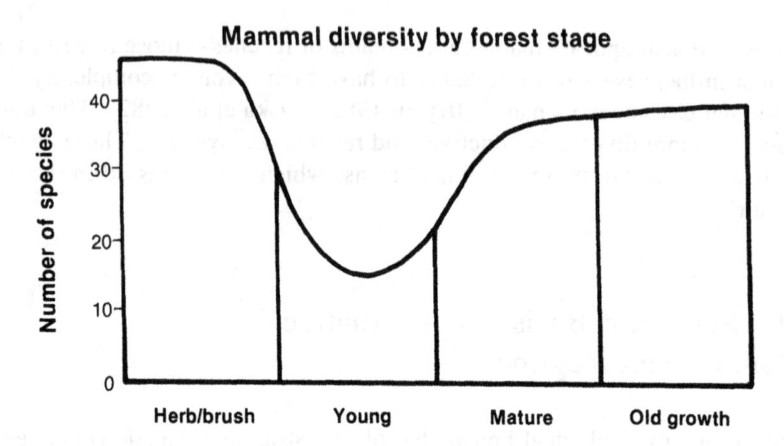

FIGURE 3.1. Number of species of mammals associated with different successional stages in Douglas-fir forests (data based on Harris 1984). Young forests are typically 20 to 80, mature 80 to 200, and old-growth 200 years old.

Compositional Diversity

Natural forests and streams are rich in species, but this varies greatly with successional stage. Mammalian species provide an example of successional changes in species diversity in Douglas-fir (*Pseudotsuga menziesii*) forests (Figure 3.1) (Harris 1984, Brown 1985*a, b*). Species richness is highest early in succession, drops to much lower species numbers following forest canopy closure, and then recovers to intermediate levels of diversity in the mature and old-growth stages. This is a common pattern for many groups of organisms, including birds (Brown 1985*a, b;* Thomas 1979), fishes, and many classes of invertebrates. Vascular plant species behave similarly. High levels of plant species richness occur in the open ecosystem prior to tree canopy closure; this diversity is a mixture of surviving forest species and weedy pioneer species (Schoonmaker and McKee 1988, Halpern and Franklin 1990, Franklin 1990*a*).

Although the early successional (precanopy closure) stage typically has more total species, many of the species found in the mature and old-growth forests have specialized habitat requirements. These species appear to require conditions that are found primarily in later stages of forest succession; consequently, they are found primarily in those forests. The northern spotted owl (*Strix occidentalis caurina*) is a well-known example (Thomas et al. 1990). In his research on old-growth Douglas-fir forests of the Pacific Northwest, Carey (1989) has identified five vertebrate taxa that are dependent on these forests, seven that are closely associated with them, and eight that are possibly associated. Many are also found in low abundance in natural young forests (Ruggiero et al. 1991) for reasons that will be discussed later.

Of equal or even greater importance is the high diversity of so-called lower organisms (such as invertebrates, fungi, and microbes) present in natural forests and streams. This "invisible," or "hidden," diversity involves groups of organisms that are poorly known and—except for pests and pathogens—receive little attention from the public and resource managers. Yet there are far more species of these organisms than the vertebrates (Schowalter 1989, Moldenke 1990). Knowledge of these lesser organisms is limited, but available evidence indicates that their diversity is high in natural forests (at least older forests) and significantly reduced in managed forests. For example, Schowalter (1989) found 61 species of arthropods in the canopy of an old-growth forest and only 16 species in an adjacent young stand (e.g., plantation). Furthermore, most of the old-growth invertebrates were species that prey upon or parasitize other kinds of invertebrates; in terms of total arthropod numbers (not species), the ratio of predators + parasites to herbivores was 1:4 in the old-growth stand and 1:1,000 in the young stand.

Natural forests and streams, including old-growth forests, then, appear to be biologically diverse ecosystems. High levels of species diversity are characteristic of organismal groups that have been studied. Many of these appear to be specialized in their habitat requirements. Many, especially the lower organisms, provide for important ecosystem functions, including maintaining the vitality of the forest and stream ecosystems they inhabit.

Functional Characteristics

Investigations of ecosystem functions such as productivity, nutrient cycling and retention, and regulation of hydrologic cycles reveal a richness of process as well as organisms in natural forests and streams (Franklin et al. 1981, Gregory et al. 1991*b*, Triska and Cromack 1980). Productivity is the most basic ecosystem process. Studies have documented that mature and old-growth forests in the Pacific Northwest are particularly productive biologically (Grier and Logan 1977, Fujimori et al. 1976). The large leaf areas found in these forests (Franklin and Waring 1980) confirm the high levels of gross production, since trees do not retain leaves that lack a net benefit in terms of photosynthesis. The difference between young and old forests appears to be that in older forests much of the productivity is utilized in respiration (estimated, in Grier and Logan 1977, at greater than 90%) rather than for production of additional wood. Furthermore, much tree growth is offset by tree mortality, although most old-growth stands in the Douglas-fir region probably continue to gradually accumulate wood (DeBell and Franklin 1987) and, almost certainly, total organic matter (stored carbon) for at least five to eight centuries (Spies et al. 1988).

Streams associated with old-growth forests typically have less primary productivity than those in open areas because of low levels of sunlight (Gregory et al. 1991*b*, Triska et al. 1982). However, total carbon inputs are typically higher in the old-growth, due to allochthonous inputs.

The nitrogen (N) cycle, particularly as a source of N inputs, provides another example of the richness of processes within natural forests. Thirty years ago it was thought that most inputs of N in coniferous forests were contributed by free-living cyanobacteria. Now at least four locales for N-fixation have been revealed by ecosystem research. An early discovery by the IBP was N-fixation by large foliose lichens, such as *Lobaria oregana*, that occur in canopies of old-growth Douglas-fir (Denison 1979, Carroll 1980). Current estimates place annual N-fixation at 5 to 9 kg/ha in a typical old-growth forest. Other sites that can experience significant N-fixation include rotting wood (e.g., large logs) (Cornaby and Waide 1973, Sharpe and Milbank 1973), the rhizosphere (regions immediately adjacent to tree roots), and leaf litter (Heath et al. 1987). Levels of N-fixation within intact coniferous stands are substantially less than those that can occur within stands of red alder (*Alnus rubra*) or other trees with N-fixing symbionts; annual N-fixation in such stands can exceed 200 kg/ha (Trappe et al. 1967).

Complexity is particularly evident in the numerous linkages revealed by ecosystem research—the richness of food and material webs and flows, or functional relationships. The below-ground portion of the forest provides an excellent example. Although only 20% of the biomass is below ground, 25 to 70% of the photosynthate produced by the plants may be required for below-ground maintenance because of high rates of turnover in fine roots and mycorrhizae (Harris et al. 1980). Such findings underline the importance of green plants as energy sources fueling the soil subsystem, and consequently the reciprocal dependence of trees and soils (Perry et al. 1989)—something typically not appreciated by foresters. The mycorrhizae and mycorrhizal relationships also provide direct linkages among trees and between forest overstory and understory. Through fungi and other organisms, the soil subsystem is in fact a highly interlinked living system.

The forest canopy is a second subsystem illustrating the complex relationships of forest ecosystems. The canopies in old-growth Douglas-fir forests represent immense surface areas; leaf areas of 8 to 14 m^2/m^2 of ground surface are typical (Franklin and Waring 1980), perhaps 50 to 100% more than young stands on comparable sites. Documented foliar surface areas of over 4,000 m^2 (all sides) have been measured on individual old-growth Douglas-fir (Pike et al. 1977, Massman 1982). Large and diverse amounts of habitat are provided by the forest canopies, which are typically continuous from the top of the crown to the ground; hence canopies are capable of providing niches for a very large array of organisms.

Canopies also represent the major interface between the forest and the atmosphere. From one perspective they are giant atmospheric scavengers which condense large amounts of moisture and precipitate dust and other atmospheric particulates, bringing these materials into the ecosystem. In some forest areas condensation from fog and low clouds adds significantly to moisture inputs, and consequently to streamflow. For example, Harr (1982) found that fog drip from old-growth forest canopies produced a net 882 mm increase in precipitation over that experienced in adjacent open areas at a mid-elevation site in the northern Oregon Cascade Range. Of course, in other localities lacking abundant fog and low clouds, forest removal may result in increased streamflow as evapotranspiration is reduced.

Structural Characteristics

Structural complexity and diversity in natural forests provide the key to much of the richness of organisms, habitat, and processes (Franklin et al. 1981). Some of this complexity can be defined in terms of individual structural features, as in many current definitions of old-growth forests (Old Growth Definition Task Group 1986, Franklin and Spies 1984, 1991*a, b*). These individual structures include large old-growth trees, large snags or standing dead trees, and large down logs.

Large individual old-growth trees are not only dominant visual elements of the old-growth forest but also carry out critical processes and provide diverse and essential habitat (Franklin et al. 1981). Old-growth Douglas-fir trees will attain diameters of 1 to 2 m and heights of 50 to 90 m; furthermore, they are highly individualistic, having been shaped over centuries by their genetic heritage, site conditions, competition, and the effects of various kinds of disturbances. They function as primary producers, and the canopies and boles provide habitat for a large number of epiphytic organisms, including nitrogen-fixing lichens, and for a large and diverse community of invertebrates (Franklin et al. 1981, Lattin 1990, Moldenke 1990, Schowalter 1989). The large old trees are also the source of two other key structural components on the forest floor: the large standing dead trees and large logs.

Intermediate-size trees of the shade-tolerant species, such as western hemlock (*Tsuga heterophylla*), western redcedar (*Thuja plicata*), and Pacific silver fir (*Abies amabilis*), are also important structural components of the old-growth forest (Franklin et al. 1981, Spies and Franklin 1991). These species provide a range of tree sizes from seedlings to large individuals which are canopy codominants, producing a many-layered or continuous canopy from the ground to the top of the crown. In fact, light levels at the old-growth forest floor are typically controlled by the density and size of shade-tolerant trees rather than by the dominant Douglas-fir trees.

Large snags and down logs (>10 cm diameter), collectively known as coarse woody debris (CWD), represent two other important structures found in natural forests (Maser and Trappe 1984, Harmon et al. 1986, Maser et al. 1988, Franklin et al. 1981). The importance of snags to wildlife has been recognized by biologists for some time (Brown 1985*a, b;* Thomas 1979), but recognition of the ecological benefits of CWD on the forest floor and in the streams has emerged more recently. These benefits range all the way from geomorphic functions, in influencing erosional processes; to biological diversity, in providing habitat for a broad array of animal and plant organisms; to providing long-term sources of energy and nutrients for these systems.

The change in attitude toward CWD reflects a dramatic recognition that dead trees are as important to ecological functioning in a forest as live trees (Franklin et al. 1987). Moreover, the dead tree structures may perform terrestrial and aquatic functions for many centuries because of their slow rates of decay or disappearance from ecosystems (Harmon et al. 1986). And those functions change throughout the "lifetime" of a dead log.

Overall structural heterogeneity is also an important feature of almost all natural forests and, of course, related riparian habitats; spatial heterogeneity is particularly notable in old-growth forests. So while the individual structural components of the forest—large trees, snags, and down logs—are important, a natural forest cannot be reduced simply to those individual structures: the forest as a whole has both vertical and horizontal critical structural attributes. One important component of stand-level structural diversity is related to canopy density. There are locales within the stand where light levels are higher (canopy gaps) and there is rich development of the understory. There are also areas where a dense overstory, such as of western hemlock and western redcedar, produces heavily shaded locales essentially barren of understory plants. This variability in light conditions, as well as below-ground competition for moisture and nutrients, contributes to the incredible complexity and richness of understories in old-growth forests. These diverse understories can be critical for some organisms; for example, the old-growth forests provide essential winter habitat for the Sitka black-tailed deer (*Odocoileus hemionus sitkensis*) in the coastal Sitka spruce (*Picea sitchensis*) and western hemlock (*Tsuga heterophylla*) forests of southeastern Alaska (Alaback 1984, Schoen 1990, Schoen and Kirchoff 1990). Research by Alaback and others is showing that the development and maintenance of diverse understory plant communities in forest stands is very complex, not simply a matter of manipulating crown density or light levels.

Riparian habitats associated with natural forests are, if anything, even more heterogenous than the upland areas (see, e.g., Triska et al. 1982, Franklin et al. 1981, Gregory et al. 1991*b*, Pringle et al. 1988, Bisson et al. 1982). There are high levels of environmental heterogeneity (e.g., light) associated with the location and activity of the stream channel, and both coniferous and deciduous tree species are typically present. Coarse woody debris is much more aggregated in the riparian habitat than in the adjacent uplands (Harmon et al. 1986). As in the uplands, the structural heterogeneity of the riparian is very important in providing for many microhabitats or niches, which in turn sustain a greater diversity of organisms and functions.

Biological Legacies

The processes by which ecosystems recover from catastrophic disturbances—particularly the living organisms and dead organic materials that are "passed on" from the original, destroyed ecosystem to the new, regenerating ecosystem—provide an important key to understanding how nature perpetuates complexity and richness (Franklin 1990*a*). When Mount St. Helens (Washington, USA) erupted on May 18, 1980, it provided scientists with a unique object lesson and a grand experimental area to study pattern and process in ecosystem recovery. At first they assumed that the area had been essentially sterilized and would recover through primary succession processes.

In fact, most species of organisms survived within the Mount St. Helens landscape (Franklin et al. 1985, 1988). Within two weeks of the eruption, scientists encountered in the devastated zone large numbers of organisms that had survived using a wide variety of strategies. Those living below ground, or which could regenerate from parts protected below ground, or those that were buried in a snowbank or in the mud at the bottom of a lake, or in any of several other environments, were able to survive. Surviving seeds, spores, and full-size organisms were important elements in the early recovery at Mount St. Helens. In addition to the living legacy, there was an immense legacy of dead organic matter, much of which was in those biologically derived structures so important for ecosystem function: snags, logs, and large soil aggregates.

The importance of biological legacies, both living and dead, stimulated several groups of scientists, including the H.J. Andrews Forest Ecosystem Group, to consider ecosystem responses to other catastrophic events. What happens in forests following fire, windstorm, flood, avalanche, or outbreaks of pests and pathogens? While most natural catastrophes kill trees and other organisms, they typically leave behind most of the carbon in the form of snags and down logs. These disturbances also leave behind large legacies of living organisms, including trees, because most disturbances are patchy.

It is apparent, because of these biological legacies, that most natural forest ecosystems do not start "from scratch" following a major disturbance. Young natural forests typically exhibit substantial structural diversity and species richness. Evidence of these legacies is found, for example, in 80- to 145-year-old stands of Douglas-fir regenerated following catastrophic wildfires in the mid-1800s and early 1900s; these natural stands included scattered large old trees, large snags, and abundant down logs as structural components (Spies et al. 1988, Spies and Franklin 1991). Many such naturally regenerated forest stands are actually forests of two or more ages, or mixed structure stands, not the even-aged and -sized stands often described. Long-term patterns in quantity and quality (decay state) of CWD following catastrophic disturbances are also well understood for Douglas-fir forests; young and old stands alike will have high levels while the mature stands (100 to 200 years old) will have minimal levels (Figure 3.2) (Spies and Cline 1988).

The concept of biological legacies is not new, but it has been nearly ignored in the ecological and silvicultural textbooks. Ecological science has emphasized the need for migration and reestablishment of individual species in barren areas; so while scientists knew about biological legacies, their significance was not appreciated. Perhaps one reason for this is the historical

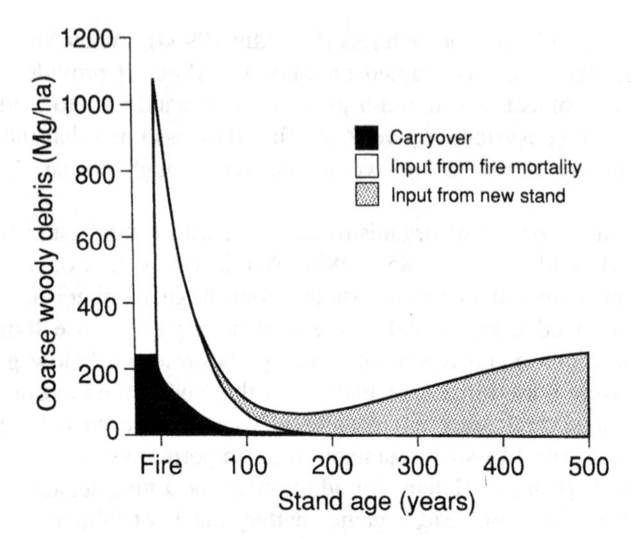

FIGURE 3.2. Levels of coarse woody debris in relation to forest age for Douglas-fir and western hemlock forests The high levels in young forest are the result of the legacy of dead trees from the preceding stand (from Maser et al. 1988).

emphasis on old-field succession in ecological research. Such environments offer minimal levels of legacies compared with other types of secondary forest seres, such as those that follow fire, windstorm, insect epidemic, flood, or avalanche. Another reason may be the forester's traditional orientation to the reforestation of devastated areas (see Oliver et al., this volume). Rapid restocking of sites with trees is one approach where human intercession often does assist nature; this result can also be seen at Mount St. Helens, where tree planting has reestablished forests more quickly than natural reseeding.

Comparisons of biological legacies do make clear some critical differences between natural disturbances and traditional even-aged harvest cutting methods. For example, the effects of clearcutting are not ecologically comparable to the effects of most natural disturbances, including wildfire. Levels of biological legacies are typically high following natural disturbances, leading to rapid redevelopment of compositionally, structurally, and functionally complex ecosystems. Traditional approaches to clearcutting purposely eliminate most of the structural and much of the compositional legacy in the interest of efficient wood production.

Catastrophic disturbances to upland areas typically lead to creation of significant biological legacies in associated riparian and aquatic habitats. At Mount St. Helens, streams received major inputs of large woody debris within areas of blown-down forest (Franklin et al. 1985, 1988). Similar effects in riparian habitats—such as high loadings of woody debris—can be seen following wildfire, windstorms, and insect outbreaks. However, significant time may elapse between the death of standing trees and their delivery to adjacent streams; for example, Engelmann spruce (*Picea engelmannii*) killed in bark beetle (*Dendroctonus rufipennis*) outbreaks in Colorado typically stood for several decades before falling over.

Some disturbances—such as debris torrents or landslides—leave little in the way of biological legacies, and disturbances of this type are more common within riparian areas than in the uplands (Leopold et al. 1963). The materials removed may become a legacy (deposit) somewhere else in the landscape, but a relatively barren site is typically created.

Landscape Ecology Perspectives

Landscape ecology is the third area of science underpinning a new perspective on forest and stream management. The landscape perspective refers to the need to consider larger spatial and temporal scales than have traditionally been considered in forestry. It means thinking beyond the individual stands or patches or reaches of streams to drainages, mosaics of patches, and long-term changes in these mosaics. Many of the critical issues facing forest managers today must be approached at these larger scales, as outlined below in a discussion of the cumulative effects and forest fragmentation issues. Some foresters have thought on larger spatial and longer temporal scales, as is reflected in the forestry logging plans developed in the 1950s (Ruth and Silen 1950, Silen 1955), but the issues addressed had to do with logging rather than ecology. What is needed now is a more comprehensive approach to managing forest and stream resources, using concepts and tools from the emerging discipline of landscape ecology (Forman and Godron 1986; see also Pastor and Johnston, this volume, and Swanson et al., this volume).

Early examples of landscape-level perspectives in natural resources typically dealt with management of large game animals. Managers who worked with grizzly bear (*Ursus horribilis*) and elk (*Cervis elaphus*) quickly recognized the need to consider areas covering thousands of hectares and a mix of habitats. The "greater ecosystem" concepts associated with Yellowstone National Park provide one current example of this perspective (Agee and Johnson 1988).

Unfortunately, in forestry we are learning much about landscape ecology from experiences with dysfunctional landscapes—landscapes that are not working very well—rather than from studying healthy ones. Landscape disfunction includes the phenomena commonly referred to as cumulative effects and forest fragmentation.

Cumulative Effects

Cumulative effects are diverse phenomena resulting from the collective impacts of management activities (Geppert et al. 1984, Peterson et al. 1987, Sonntag et al. 1987). These are negative or undesirable consequences that can involve the additive effects of a repeated single activity, such as forest harvest, or the synergistic or additive effects of two or more activities. Cumulative effects may show threshold behavior: the cumulative impacts are minimal up to some critical point, after which major changes occur—through time at a particular site or through time and across space.

In the forest landscapes of the Pacific Northwest, cumulative effects most often concern impacts of forest harvest and road building on hydrology (e.g., frequency and intensity of flood events) and on fisheries (e.g., sediment production and destruction of spawning habitat). Roads and recent clearcuts tend to be particular problems because of a much higher probability of undesirable hydrologic or geomorphic events in these areas, such as landslides (Sidle et al. 1985) or higher peak flows associated with rain-on-snow flood events (Harr 1986).

Mass soil movements of various types are primarily management-related sources of sediment impacting streams and rivers in many watersheds of the Pacific Northwest (Naiman et al., this volume). Damage to fish habitat, particularly spawning and rearing habitat for anadromous fisheries, is of particular concern (Cederholm and Reid 1987, Bisson and Sedell 1984). Road systems are a primary source of mass soil movements (Sidle et al. 1985, Swanson et al. 1987). Although improved road location and construction methods and reconstruction of older roads reduce the probability of mass soil movements, the potential remains much higher than in unroaded areas (Sidle et al. 1985, Swanson et al. 1987). Similarly, recently clearcut areas in mountainous topography have a significantly higher probability of landslides than unlogged areas, quite apart from roads (Sidle et al. 1985, Swanson et al. 1987, 1989). An important contributing factor is that the root systems of the logged forest, which function as soil binders, decompose before full replacement by the regenerating forest. The period following clearcutting when mass movements are most likely to occur varies across the region (Table 3.1) (Sidle et al. 1985, Swanson et al. 1987). The regional differences shown in Table 3.1 reflect differences in the decay rate of the old root systems and the regeneration rate of the new root system in these contrasting forest environments. Therefore, both the extent and the types of roads and cutover areas at vulnerable ages raise significant issues regarding cumulative effects in the mountain landscapes of the Pacific Northwest.

The frequency and intensity of floods resulting from rain-on-snow events create another cumulative effects issue in the Pacific Northwest (Harr 1986). Much of the Cascade Range and the Olympic Mountains lies within a "transitional" or "warm" snow zone, largely within an elevational belt (350 to 1,100 m in western Oregon; Harr 1986) that develops a snowpack during winter cold periods. A subsequent warm front with warm air and rain can melt much of this snowpack. High streamflows, commonly known as rain-on-snow flood events, sometimes result from the combined runoff from rainfall and the melting snowpack; the snowpack can contribute over a third of the water under some weather conditions (Harr 1986). Many of the major

Table 3.1. Relation of landslides to time since clearcutting (percentage of inventoried slides).

Time since Clearcutting (years)	Location: Oregon Coast Range* (%)	Oregon Cascade Range† (%)	Idaho Batholith‡ (%)
0 to 3	63	46	24
4 to 10	29	42	41
≥11	8	12	35
Total	100	100	100

*Gresswell et al. (1979) for Mapleton, Oregon, USA, area.
†F.J. Swanson, unpublished data for H. J. Andrews Experimental Forest, on file at the Forestry Sciences Laboratory, 3200 Jefferson Way, Corvallis, Oregon, USA.
‡Megahan et al. (1978).

flood events along the northwestern coast of North America, from northern California to southeastern Alaska, are of this type. These meteorological events also trigger major episodes of landslides and channel erosion (Harr 1986).

Clearcutting often increases the size of peak flows during rain-on-snow events relative to flows from areas of mature and old-growth forest (Harr 1986, Berris and Harr 1987, Harr et al. 1989). Old-growth forests typically intercept a large portion of the snow in the canopies. Much of this intercepted snow melts and drips to the ground, infiltrating the soil; additional snow is lost to sublimation and evaporation. The snowpack that does form is protected by the forest canopy from direct exposure to the atmosphere. In contrast, deeper accumulations of snow (and much greater amounts of water) may occur on recently cutover areas; in one study, water equivalents were two to three times greater in a clearcut than in an adjacent old-growth plot (Berris and Harr 1987). Furthermore, the snowpack in the clearcut is fully exposed to turbulent warm air and rain; it is actually latent and sensible heat transfer from the warm, moist air mass to the snow surface during a storm that produces most of the melting (Berris and Harr 1987). Consequently, melting can occur more rapidly in open areas than in forested areas where wind speed and turbulence at the snow surface are low.

The contribution of recent clearcuts to rain-on-snow flood events can be considerable. Measured water outflow was 21% greater from a clearcut than from an old-growth forest plot in the central Oregon Cascade Range during the largest rain-on-snow event studied by Berris and Harr (1987). In the northern Washington Cascade Range, clearcut plots produced 84 and 90% more runoff than paired areas in mature forest (Harr et al. 1989). Clearcutting also decreases the return interval for a water input event of a given size; in other words, it increases the frequency for a flood event of a given magnitude (Harr 1986, Harr et al. 1989). Finally, hydrologic recovery for clearcut areas appears to be longer than the 20 to 25 years originally hypothesized. For example, preliminary results from a study in northern Washington suggest that hydrologic recovery is only about 50% complete in 25-year-old forest plantations (Harr et al. 1989).

Increased frequencies of high flood flows, debris torrents, and dam-break floods can dramatically affect aquatic environments, simplifying some stream reaches (especially those with steeper gradients) and burying low-gradient reaches in debris. An important management implication is recognizing that such events can degrade stream and river ecosystems even where riparian forest corridors are maintained. Hence any overall management strategy must consider not only riparian corridors but other activities, such as road density and construction and maintenance standards, that generate channel disturbances.

Cumulative effects of clearcutting and road systems on hydrologic and sediment regimes are therefore a serious landscape issue in the Pacific Northwest. The cumulative area of clearcuts and the early successional forest during a given period need to be carefully considered if managers wish to avoid high probabilities of specific geomorphic and hydrologic events. Cumulative impacts have, in fact, forced the U.S. Forest Service to limit timber sales in portions of the Mount Baker-Snoqualmie National Forest (United States Forest Service, Pacific Northwest Region 1990).

Forest Fragmentation

Extensive forest fragmentation provides another example of landscape disfunction, hence a stimulus to consider larger spatial and temporal scales. Fragmentation results when a large area of continuous forest becomes a mosaic of high contrast, small forest patches. Such forest fragments often lack the habitat conditions found in larger areas of intact forest and are more vulnerable to disturbances such as windthrow and wildfire (Franklin and Forman 1987, Perry 1988, Hansen et al. 1990, Gardner and O'Neill 1991).

The dispersed patch clearcutting system selected for cutting Douglas-fir forests on federal lands in the 1940s is a major factor in the fragmentation of the northwestern forest landscape (Franklin and Forman 1987). This approach to creating forest pattern is sometimes known as the "checkerboard" system because of its appearance halfway through a cutting cycle in an idealized application (Figure 3.3). The detrimental effects of dispersed patch clearcutting include the fragmentation of the forest matrix early in the cutting cycle (at about the 30% cutover point) and the creation of a large amount of high contrast, forest-cutover edge.

Edge effects are a major problem when a forest matrix is broken into small patches surrounded primarily by cutover areas. Such patches of residual forest have a modified environment throughout, lacking true interior forest conditions. For example, in the Douglas-fir region, the patch size selected for management is typically 10 to 15 ha. Recent studies show that the more extreme environmental conditions of a clearcut extend into the forest, in-

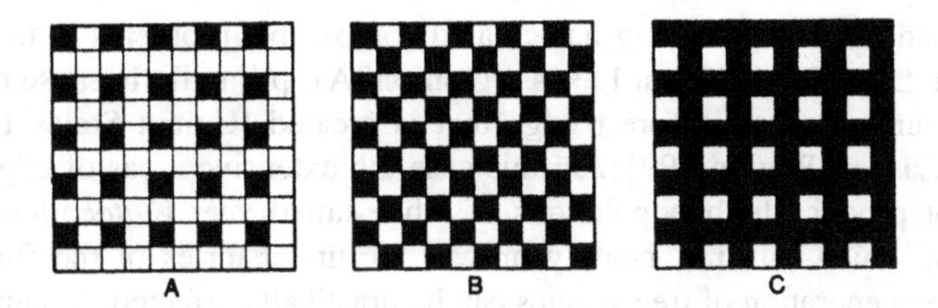

FIGURE 3.3. Dispersed patch clearcutting is sometimes referred to as the "checkerboard" system because the direction for dispersion of cutting areas produces a checkerboardlike pattern at around the 50% cutover level.

fluencing the microclimate of the old-growth forest for 200 m or more in the case of sensitive parameters such as relative humidity and wind. In extreme cases, such as a southerly exposed forest edge, influences may penetrate for more than 300 m (Chen et al. 1990, 1991; Chen 1991). Biotic processes are also influenced by the modified microenvironment; for example, rates of tree mortality increase dramatically (Chen et al. 1990). Forest patches have to attain sizes of 50 ha or more before a significant amount of unmodified interior forest condition is created (Figure 3.4).

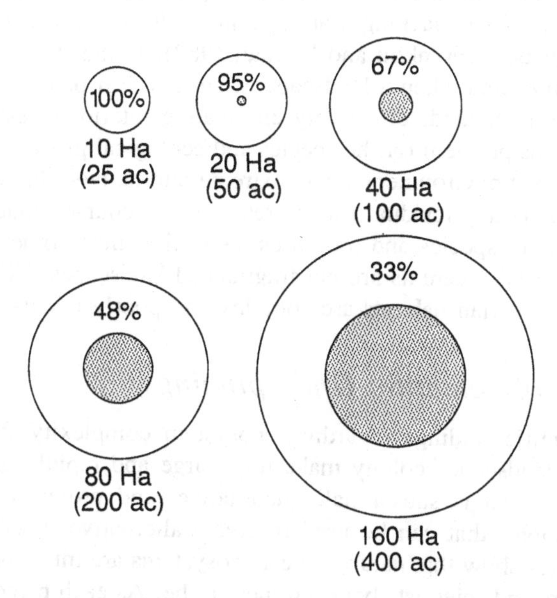

FIGURE 3.4. Edge effects can drastically reduce the amount of interior habitat within a forest patch surrounded by cutovers. Microclimatic measurements in Douglas-fir forests suggest that edge influences often extend 200 m or more into a forested patch. Illustrated here are the percentages of interior habitat associated with various forest patch sizes based on a 240 m area of influence.

Dispersed patch clearcutting has created comparable problems in the forests of the Lake States (Great Lakes region, USA), primarily because of the large amount of clearcut-forest edge that is created (United States Forest Service, Eastern Region 1991). In this case the extensive areas of edge environment produce high populations of white-tailed deer (*Odocoileus virginianus*), which, in turn, heavily browse the understories of the forested patches. Regeneration of tree species can be drastically affected, as can population levels of several other plant species of special interest (United States Forest Service, Eastern Region 1991).

Effects of fragmentation on riparian habitats appear to be more complex and have not, in any case, received as much study as in upland areas. One major question has to do with the relative cumulative effects of dispersed versus concentrated forest cuttings on hydrologic and geomorphic events (Grant 1990). Dispersed cutting is often proposed as a management strategy that minimizes cumulative effects. However, the answer to the question of dispersed versus aggregated cutting appears to be highly dependent on scale—that is, what sizes of watershed and cutover are under consideration (Grant 1990). Negative impacts of fragmentation on aquatic environments relate to the way it tends to chronically disturb essentially all drainages in a landscape in perpetuity and to require maximum densities of active roads.

In summary, fragmentation is a generic problem whereby an area is divided into patches lacking sufficient size to produce suitable habitat for interior-forest species or lacking viability (an ability to persist in the face of windthrow) or both (Franklin and Forman 1987). In the Pacific Northwest, fragmentation occurs when a landscape of small forest patches interspersed with cutovers is created, effectively eliminating interior forest conditions. This is a serious problem for the species and ecological processes dependent on interior forest environments. The extreme alternative—large continuous clearcuts lacking any residual forest areas—is, of course, equally inappropriate for interior species and processes (as well as many other organisms), even though these clearcuts are not fragmented landscapes. Effects of fragmentation on riparian habitats are complex and poorly understood.

Summary of Scientific Underpinnings

Recent scientific findings regarding ecosystem complexity, biological legacies, and landscape ecology make up a large and rapidly growing body of knowledge that is substantial, quantitative, and provides some fundamental principles that can be used to create alternative forestry practices. These findings show that natural forest ecosystems are much more complex than foresters and scientists believed them to be. As each piece and process has been identified and studied, its significant and sometimes essential role in the functioning of the forest has also become apparent. This makes clear that organisms, structures, or processes generally cannot be discarded without ecological consequences. Hence forest simplification—the basis for most current commodity management—needs to be approached with caution and humility.

Nature provides for the rapid recreation of ecologically complex young forests through the mechanism of biological legacies. These carryovers of live and dead organic materials and the processes associated with them help ensure that natural young forests are structurally and compositionally diverse. In effect, nature perpetuates ecosystems rather than simply reforesting with stands of trees. This finding suggests that management of forests should give greater attention to what is retained on the site at the time of harvest. Much larger land areas—landscapes—and time periods must be considered in forest planning. Without these grander perspectives, forest resources will almost certainly experience undesirable cumulative impacts.

What Are Some Alternatives Under New Perspectives?

How well have foresters and other resource managers incorporated this new knowledge into philosophy and practice? Some concepts, such as maintenance of coarse woody debris in forests and streams, have been accepted relatively rapidly, particularly on public forest lands. However, professional managers have been much slower to incorporate other ecological concepts, probably partly because of a general reluctance to accept the idea that traditional practices have significant environmental costs.

Traditional forest practices are based on principles of simplification and homogenization of the forest resource at the tree, stand, and landscape levels in order to achieve economical wood fiber production (Franklin et al. 1986). Approaches based on these principles were consistent with the state of forest science as it existed at midcentury. Consequently, foresters developed a management paradigm for Douglas-fir in the 1940s based on clearcutting, broadcast slash burning (including the disposal of large woody debris), and planting of conifer monocultures. These approaches also reflected a midcentury societal emphasis on production of wood fiber.

Circumstances have changed. As partly outlined in the preceding section, our understanding of forest ecosystems and landscapes has improved dramatically. We are learning that what enhances efficient production and harvesting of wood fiber in the short term does not always enhance other forest values and long-term productivities. Society has become concerned with many other forest values, including the maintenance of biological diversity.

With expanded knowledge and objectives, the goal becomes one of using ecological knowledge to create alternative forest management systems (Perry and Maghembe 1989). That is what New Perspectives (*sensu* this conference), the Forest Service's New Perspectives Program, and New Forestry are all about (Franklin 1989, 1990*b*). Management approaches must be developed which better integrate the maintenance of ecological values with production of commodities (see Lee et al., this volume, and Stanford and

Ward et al., this volume). These approaches must incorporate the best ecological science as well as accurately reflect societal values and objectives. They should also provide feedback mechanisms so that new technical information and societal objectives are continually incorporated.

Two additional philosophical points on New Perspectives should be made before discussing specific practices. First, there is no intent here of throwing out the old tools of forestry, such as traditional clearcutting, slash burning, and tree planting techniques. The idea is to add new knowledge and techniques to the existing approaches, thereby generating more management options. Second, some "new" practices resemble older, traditional forest practices. Logging systems that maintain mixed-structure stands have been utilized by foresters in both Europe and North America for over a century. They have not, however, been widely advocated or applied in western North America, partly because there has been little scientific rationale for using more complicated silvicultural systems.

Moreover, it is important to recognize that the new forestry practices being advocated differ significantly from traditional forestry in their philosophical and scientific roots, regardless of any superficial resemblances. Management of forests as whole ecosystems, rather than as collections of trees, is the basis for New Perspectives. For example, the retention of green trees in a cutting program, discussed below, is far different in objectives, application, and consequences from a partial cutting program that emphasizes the economically valuable trees in a stand. To suggest that foresters have previously tried and discarded New Forestry practices is nonsense, since the ecological knowledge to even conceptualize them is of such recent origin (Hopwood 1991).

Management systems are needed at both the stand and landscape levels that incorporate the new knowledge and the added societal objectives for forest lands and associated waters. Hence the following discussion is divided into stand and landscape approaches. Riparian habitats are relevant at both levels.

Stand-Level Approaches

In forest stands managed for some level of commodity production there is a general New Forestry principle: maintain or create stands that are structurally and compositionally diverse. Within the constraints of objectives and stand conditions, maintain as much diversity as possible rather than simplify the stand. Structural diversity is usually the goal, because structure is normally closely correlated to organisms and processes; that is, the structure provides the necessary conditions for organisms and ecological processes. The general principle of maintaining structural diversity should be kept in mind during the following discussion; this will reduce the tendency to have difficulty with specific practices being inappropriate to a particular stand or region. The exact set of silvicultural practices—the treatments developed to create or maintain structural diversity—will vary with forest type and condition, environment, and specific management objectives (see Oliver et al., this volume).

Young Stand Management

Through management practices, young stands provide many opportunities or pathways for developing elements or attributes that are important in enhancing diversity. An initial step would be aggressive efforts to create stands of mixed composition. Although natural reseeding of multiple species ordinarily reduces the potential for forest monocultures, intensive management tends to force the ecosystem toward a single favored species. Maintaining a mixture of species can greatly enhance ecological values, such as the ability to provide habitat for a broad array of organisms. For example, occasional hardwoods, such as a bigleaf maple (*Acer macrophyllum*), add significant structural and species diversity (e.g., in epiphytes, Coleman et al. 1956) to a conifer-dominated stand. Hardwoods such as alders (*Alnus* spp.) bring an additional benefit of nitrogen-fixation (Trappe et al. 1967).

Richer mixtures of conifers can also be valuable. The cedars, such as *Thuja* and *Chamaecyparis*, help improve site quality in addition to producing valuable wood products. All Cupressaceae are foliar calcium accumulators (Kiilsgaard et al. 1987) and produce high quality litter. In turn, the litter contributes to higher base saturation and rates of nitrogen mineralization, reduced acidity, and production of more biologically active mull humus conditions (Alban 1967, Turner and Franz 1985).

Foresters need to think much more seriously about species mixtures in designing managed forests both for ecologic and economic objectives. Prescriptions for precommercial and commercial thinning increasingly reflect a new appreciation for the maintenance of "minor" species. Continuing attention to species mixtures in both reforestation and stand improvement prescriptions will be important.

Delaying the process of canopy closure can also be of value in some young stands. Intensive forest management has traditionally sought rapid canopy closure (early full occupancy of the site by commercial trees). In fact, some foresters clearly favor instant canopy closure—the year following broadcast slash burning—and consider the open period of early succession wasted from the standpoint of tree productivity! Yet the period prior to tree canopy closure is an important one ecologically. This stage is rich in plant and animal species, as noted earlier, and many of these organisms are valued by humans. For example, many game species use the precanopy closure stage (Brown 1985*a, b*). Nitrogen-fixing vascular plant species tend to be most common at this time. Canopy closure is probably the most dramatic and, in terms of some processes, most traumatic single event in the life of the stand, other than its ultimate destruction by some catastrophe. Many aspects of the forest, including its composition and functioning, change rapidly and significantly at the time of canopy closure.

Canopy closure can be delayed by maintaining wider tree spacings to obtain more of the ecological values associated with precanopy closure stages of forest succession. Reduced tree planting densities and heavy thinning (mostly precommercial) can be used to achieve this objective. Furthermore, spacing studies show that wide spacing can be maintained in young stands with little or no sacrifice in commercial wood volume production (Reukema 1979, Reukema and Smith 1987). Tree pruning can be utilized in order to produce high quality wood in these more open-grown stands. Interestingly, New Zealand foresters use the cost-effective approach of widely spaced young stands and pruning in their management of Monterey pine (*Pinus radiata*) plantations.

All of the above young-stand management concepts have direct application to riparian habitats if riparian zones are committed to wood fiber production. Streamside forests of mixed composition are almost always preferable to forests of a single species from the standpoint of most aquatic ecosystem functions. For example, they produce a more diverse stream environment by virtue of inputs of allochthonous materials that vary in nutrient quality and timing of delivery to the aquatic system. A coniferous component that produces persistent large woody debris, such as Douglas-fir or western redcedar, is clearly desirable. Alternatives to dense, uniform coniferous canopies in riparian forests are also desirable (Salo and Cundy 1987, Raedeke 1988, Naiman and Décamps 1990). Again, these comments apply to riparian areas managed for wood production.

Maintaining Coarse Woody Debris

Practices that provide for a continuous supply of CWD—including large standing dead trees and logs on the forest floor—are important in maintaining stand-level structural diversity. There can be no question about the value of these structures for many species and ecological processes in both terrestrial and aquatic systems; this has been documented in many studies (Harmon et al. 1986, Maser and Trappe 1984, Maser et al. 1988). Practices that would contribute to this objective include retention of large woody debris at the time of harvest cutting and creation of snags and logs from green trees.

Retention of snags and large logs is a particularly effective practice in maintaining CWD when harvesting in young and mature stands of natural-origin and old-growth forests. Natural stands typically have significant amounts of CWD for potential biological legacies. Retention of large, downed wood is greatly enhanced by eliminating or modifying the practice of removing unmerchantable logs, which has been extensively applied on federal forest lands. Specifications for stream cleanup following logging, which called for channel cleaning in the 1960s, have been drastically altered during the last 10 to 15 years to provide for retention of CWD. Current streamside management on forests such as the Willamette National Forest (Oregon, USA) have used CWD production as a major rationale for retaining streamside forests. Provisions for CWD have also been a major consideration in revision of state forest practice regulations in the Pacific Northwest.

Snag retention is a more controversial practice than the maintenance of logs. There are concerns about worker safety, as well as increased difficulty in logging, which result in increased costs (Atkinson in Hopwood 1991). Snags also create potential problems in fire protection because of their tendency to produce firebrands once ignited. Nevertheless, snag retention on cutover areas is increasing, particularly on federal lands. Two approaches used to reduce the hazards while retaining snags are: (1) clustering snags in groups or patches rather than in a dispersed pattern, and (2) creating snags from green trees following cutting.

Many current forest harvest prescriptions for maintenance of CWD on national forest lands involve retaining green trees as snag sources, some existing snags, and unmerchantable logs (Figure 3.5). A key criterion in these prescriptions is to leave snags and trees sufficient in number to maintain a minimal snag level, such as five snags per ha >53 cm (21 inches) in diameter, during the next 80-year forest rotation. Evidence is already emerging that snag retention on cutover areas does provide wildlife benefits (A. Hansen, pers. comm.). Green trees can also be used as sources for CWD in stands that lack either large snags or logs, as is the case in many intensively managed stands. This is a particularly valuable practice for restoring structure to stands and landscapes that have been simplified by past practices.

Maintenance of CWD in streamside management zones is obviously of direct benefit to riparian habitats, given the central role of large wood in the structure and function of stream and river ecosystems (Gregory et al. 1991*b*, Harmon et al. 1986, Maser et al. 1988). The forest adjacent to live channels is particularly critical in small streams (McDade et al. 1990). River systems have greater ability to entrain large wood by bank cutting processes (Lienkaemper and Swanson 1987), but this assumes that the forested terraces support large trees. Hence it is important to maintain or develop riparian forests as source areas of large CWD for both streams and rivers.

Maintenance of appropriate quantities and qualities of CWD in managed stands is, in fact, much more complex than simply periodically providing for a few dead trees. Different tree species provide snags and logs with different characteristics, and therefore different abilities to provide for various habitat and other functions in both aquatic and terrestrial ecosystems. Tree species are not all equal in terms of CWD roles. Further, CWD typically needs to be present in various stages of decay. Numerous questions exist as to the quantities and spatial distribution of CWD required to achieve specific management objectives. Although a need for some level of CWD is desirable, how much is enough? Refining an approach to CWD will remain a challenge for both scientists and managers for many years to come (Maser et al. 1988).

FIGURE 3.5. Retention of structures from the cutover stand at the time of harvest helps maintain organisms and functions associated with later successional stages in forest development. In the Douglas-fir region such structures typically include the large green trees, large snags, and large down logs illustrated here (Willamette National Forest, Oregon, USA).

Green Tree Retention

Retention of living trees on cutover areas is another practice that can create higher levels of structural diversity in managed stands. This approach may be referred to as partial cutting, partial retention, or green tree retention, in an effort to distinguish it from both clearcutting and selection cutting (Franklin 1990*c*). Green tree retention involves reserving a significant percentage (10 to 40%) of the living trees, including some larger or dominant individuals, at the time of harvest for retention through the next rotation. The density, composition, condition, and distribution of the retained trees vary widely depending on management objectives, initial stand conditions, and other constraints. The general objective, however, is to sustain a more structurally diverse stand than could be obtained through even-aged management. Such partial-cutting prescriptions have not been widely used or recognized in forestry. Silvicultural textbooks discuss related concepts such as "shelterwood with reserves." Partial-cut stands contrast markedly with traditional even-aged stands, which involve only a single age class (Figure 3.6).

Many objectives can be achieved by retaining green trees on harvested areas. They can be used as sources of snags and logs, especially where safety concerns do not allow for retention of snags, and they can also provide desired wildlife habitat.

Green trees can function as refugia and inocula for some of that "hidden," or "invisible," diversity described earlier. For example, many species of the rich invertebrate fauna found in old-growth forests do not disperse well; many are soil dwelling, not canopy or stem species (Lattin 1990). Such organisms typically do not easily recolonize areas once habitat has been altered by clearcutting. Refugia for these kinds of organisms can be provided by leaving host trees which then become an inoculum or seed source in the new stand. The same concept is applicable to mycorrhizae-forming fungal species. At least some of these fungi can disappear in a short time if all potential host species are eliminated (Perry et al. 1987). When some of their hosts are left behind, the fungal communities conserved inoculate the young stands. This concept is a counterpoint to a common complaint about partial cutting: that green trees cannot be retained because they are sources of pests and pathogens. Most of the invertebrates and fungi in forest stands are, in fact, essential components that should be retained; retention of green trees is one tool with which to achieve that objective.

Green tree retention also alters the microenvironment of the cutover area. That is what shelterwood cutting is all about; the overstory moderates microclimate, encouraging tree regeneration where the environment on a clearcut is severe due to heat or frost. Obviously, what works for trees works for other organisms as well; other components of forest ecosystems will certainly survive better on partial cuttings than on clearcuts. Perhaps as important is that many organisms move more readily or safely through a patch or a landscape with green trees than they would through a clean-cut envi-

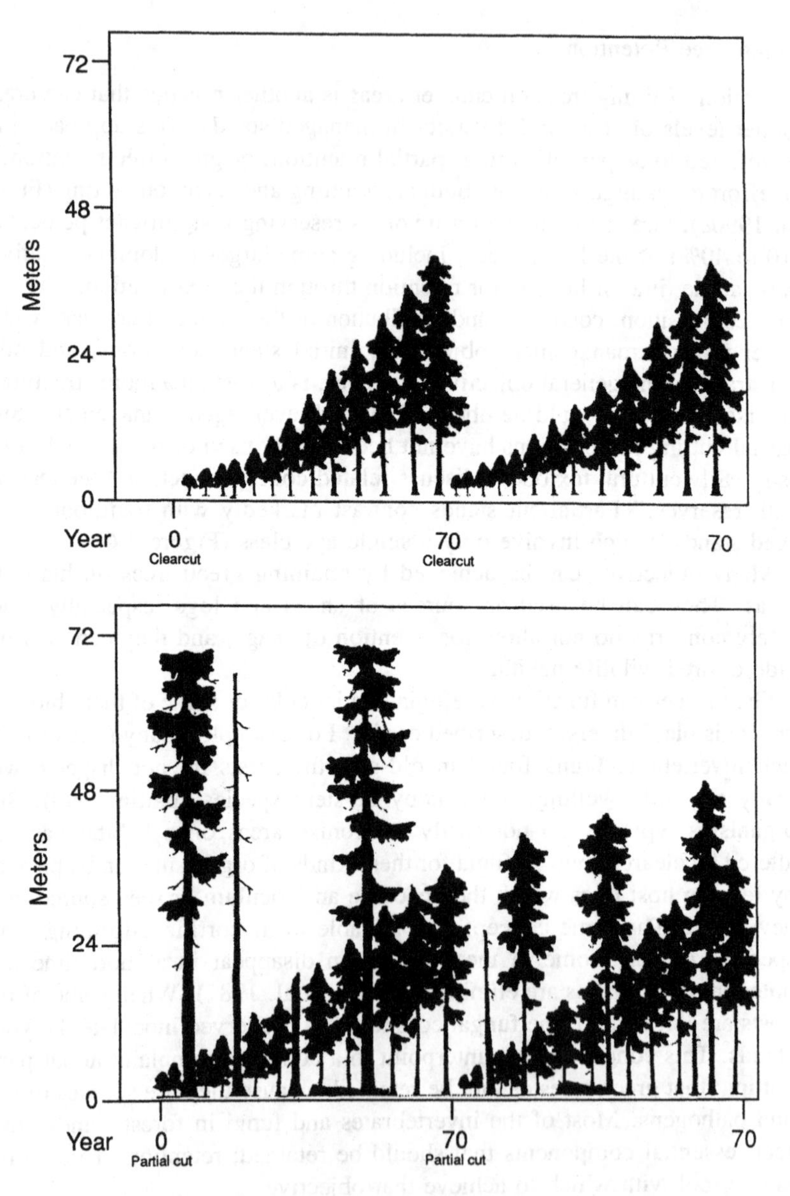

FIGURE 3.6. Idealized contrast in structure between a system based on clearcutting and even-aged management (top) and partial cutting and multi-aged management (bottom). The level of green tree retention under the second scenario can be quite small (10 to 20% of the stand) and still produce a structurally diverse stand.

ronment because of the ameliorated environment, protective cover, or both. Perhaps the connectivity in the landscape matrix can be improved by green tree retention reducing the isolating effects of cutover areas.

Retention of green trees can be used as a strategy to grow large, high quality logs during the next rotation (Figure 3.7). For example, mature (80- to 250-year-old) Douglas-fir can show substantial growth (Williamson 1973). Dominant trees in these age classes that are left behind have significant growth potential. Hence green trees could provide both economic and ecologic benefits in scenarios involving simultaneous management of a stand with a low density of large trees (e.g., western redcedar) on a long rotation and multiple short rotations of a species (e.g., western hemlock) managed for wood fiber production.

Green tree retention may have limited application within riparian zones, since harvest cutting is generally not proposed for these areas. However, if harvest cutting is proposed in a riparian zone, green tree retention is a relevant practice. Tree species that produce larger and more persistent wood, such as Douglas-fir and western redcedar, should be favored where generation of CWD is a concern. Hardwoods produce CWD that persists for a much shorter period (Harmon et al. 1986). Blowdown will typically be a more serious threat in the riparian zone than in the adjacent uplands, due to moister soils and shallow tree rooting; however, a major function of riparian forests is to provide CWD for the stream and floodplain (Maser et al. 1988).

The strategy of green tree retention on upland areas may offer direct benefits to stream ecosystems, although evidence on the following examples is not conclusive. Retention could reduce the potential for landslides through the maintenance of root strength. This concept underlies prescriptions for forest retention on stream headwalls in portions of the Oregon Coast Range (Sidle et al. 1985). Partial cutting might also be utilized to reduce the impact of forest harvest on the intensity of rain-on-snow flood events. This would presumably require sufficient trees for significant snow interception and alteration of the thermal balance on cutover areas.

Partial retention has immediate relevance to the northern spotted owl in the Pacific Northwest. Most natural disturbances, whether wind or fire, leave behind a significant component of green trees as well as a legacy of snags and logs. Even though true old-growth conditions do not develop until after 200 years, many natural stands begin to provide habitat suitable for the northern spotted owl and other late successional species at 70 or 80 years. For example, a large forest area on the Olympic Peninsula (Washington, USA) was blown down in a windstorm in 1921. Today these forests are dominated by 70-year-old western hemlock trees, but many such stands contain large residual old-growth trees, snags, and logs—and provide suitable habitat for the northern spotted owl. There are other examples of predominantly young but structurally diverse forests in the Pacific Northwest that provide suitable habitat for owls.

FIGURE 3.7. The number, species composition, and condition of trees left under partial cutting prescriptions vary widely according to management objectives, stand conditions, and environment. In this cutting, 25 to 30 ha of dominant mature Douglas-fir trees of superior vigor have been retained so as to provide for production of high quality wood during the next rotation while also fulfilling ecological functions (H. J. Andrews Experimental Forest, Willamette National Forest, Oregon, USA).

Partial cutting systems could be used to create similar mixed-structure stands suitable as habitat for owl and other species characteristic of late successional forests (see Oliver et al., this volume). For example, partial cutting using a rotation of 120 years might provide suitable habitat for such organisms for 40 to 50 years. Such an approach has much merit as an alternative to using rotations of 250 to 300 years, which would typically be necessary to recreate old-growthlike forest structures following a traditional "clean" clearcutting operation.

Prescriptions for green retention cutting will vary widely, depending on many factors, including objectives and stand conditions. For example, to provide for a minimal number of snags and logs it may be necessary to retain as few as 10 to 18 green trees 50 cm in diameter per hectare. Creation of mixed-structure forests suitable for late successional species may require retention of 20 to 40 green trees per hectare, depending on the forest age.

The appropriate spatial distribution of retained green trees (e.g., concentrating them in patches or dispersing them throughout the cutover area) is another important issue currently under study. In part, the answer depends on objectives and constraints. For example, green trees retained in patches may function more effectively as refugia for invertebrates and may minimize impacts on logging and other forestry operations (Figure 3.8). On the other hand, well-distributed snags and down logs are desirable from the standpoint of maintaining habitat for some species of wildlife and in contributing to maintenance of soil productivity.

Partial retention may have extensive application to areas adjacent to riparian management zones. Obviously, partial cutting prescriptions could be utilized to maintain a greater structural and compositional legacy when cutting is planned within the riparian zone. Partial retention in such areas can be used to buffer the riparian leave area, improving (1) the ability to provide interior environmental conditions and (2) the viability of the area, such as its ability to resist windthrow. Partial cuts would also provide a more gradual transition between intensively managed upland areas and protected riparian habitats. Aggregated green tree retention can also be used in upland areas to protect critical source areas of water and sediment.

Partial cutting may have applications for enhancing riparian zones even where harvest cutting is excluded. It could be used to release conifers from hardwood competition, for example, where larger conifer trees are desired as structural elements of a riparian stand. It could also be used to increase levels of CWD in and outside of the channel. The value and specifics of such silvicultural practices in the riparian zone are clearly important for New Perspectives research and development programs (Rainville et al. 1985).

Landscape-Level Issues in New Perspectives

The basic principle of New Perspectives at the landscape level is consideration of larger spatial and longer temporal scales while planning management activities. Recent research and experience make clear the need for

FIGURE 3.8. Green trees and standing dead trees retained on cutover areas can be dispersed or clumped in distribution. There are ecological benefits from both approaches (suggesting a mixed pattern might be in order), but aggregating the structures is often advantageous from the standpoint of logging costs and hazards and overall management efficiency (cutover on Plum Creek Timber Company lands, Cougar Ramp Cutting, Oregon, USA, with 15% of green trees retained).

broadened perspectives in assessing potential ecological consequences of various management alternatives. Managers must systematically think beyond the 10 ha patch, the kilometer stream reach, and planning horizons of 5 to 10 years. This is necessary not only to achieve long-term resource management objectives but to avoid undesirable future conditions, such as those associated with cumulative effects and habitat fragmentation. In considering landscape-level issues it is typically inappropriate to consider the upland or terrestrial and the streamside or aquatic components separately; they are far too interrelated at this geographic scale. But each does have its own distinct structural and functional properties, so a different management paradigm is appropriate for these two major components.

It is important to note in introducing this section that scientists know much less about landscape-level processes than they do about stand- or reach-level processes. Therefore, the following material generally addresses issues rather than offering prescriptions.

Management of Streamside Zones

Managing streamside management zones in their geomorphic and vegetative context is an important, widely recognized landscape-level concern (see Naiman et al., this volume). It involves several issues, including the desirability of matching the width and shape of riparian corridors with landform features rather than using a simple, standard prescription (Swanson et al. 1988, Beschta and Platts 1986, Potts and Anderson 1990). It also means treating different conditions individually. Many different stream types exist in a region such as the Pacific Northwest (Naiman et al. 1991). Most managers, scientists, and regulators would agree that each riparian area needs to be considered individually, and not blanketed under some common prescription.

Explicitly recognizing the longitudinal nature of our riverine ecosystems is essential in placing streamside management in its proper geomorphic context (Leopold et al. 1963). This linkage requires managers to pay close attention to sources and sinks of materials, such as sediments and large woody debris, and the circumstances under which they are generated, transported, and deposited, particularly when they are critical to downstream conditions. For example, areas of high fish productivity in coastal Oregon stream systems are often maintained by periodic debris flows from tributary drainages; these deposits provide the desired stepped profile and sorted gravels in the main stream (Benda 1985; Bisson et al., this volume). In southeastern Alaska large woody debris, an essential component of productive fish rearing habitat in estuaries, comes from mature and old-growth coniferous forest growing on islands and in floodplains along upstream channels and rivers; if all the floodplain forests are logged, the source of this structural material is eliminated, depriving the estuarine rearing habitats of their essential structure (Maser et al. 1988).

Systems of Preserved Areas and Habitats

Designing and implementing a network of reserved or protected areas within a matrix of timber-oriented or commodity lands is another major landscape-level concern. Some protected lands are needed to maintain the full array of ecological values, even within landscapes devoted primarily to timber production. Currently, numerous areas on public lands—riparian corridors, unstable soil areas, habitat areas for specific organisms, and Research Natural Areas—are preserved to achieve specific ecological objectives. Similarly, on state and private lands there are riparian management zones (RMZs) and upland management areas (UMAs) (Armour 1989). Often lacking is a systematic approach to designing these areas, especially with respect to their geographic distribution and connectivity (Naiman et al., this volume). For example, streams could have significant RMZs but if headwater areas lack sufficient protection, debris torrents and related phenomena can damage aquatic values. Basin-scale planning of silvicultural activities is needed to incorporate these kinds of components into a larger, functioning landscape (see Oliver et al., Pastor and Johnston, and Swanson et al., this volume).

Riparian corridors, including RMZs, are critical to this concept (Budd et al. 1987). They contribute directly to the maintenance of ecological processes and biological diversity in aquatic ecosystems and in the landscape as a whole (Naiman and Décamps 1990; Naiman et al., this volume). Furthermore, streamside corridors are geographically the dominant connecting elements in most networks for forested reserve areas, regardless of debates over their effectiveness as movement corridors for upland species.

Developing a comprehensive strategy for preserving biological diversity in our forest and range landscapes is, in fact, a much broader and more difficult challenge, particularly because of the large and diverse array of species requiring special consideration (see, e.g., Cissel 1990). Many of the vertebrate species are reasonably well known in terms of their habitat requirements, which often include late successional forest for some elements of their life history (Ruggiero et al. 1991). Furthermore, the size and spacing of suitable habitat enclaves required for the various plant and animal species, large and small, are highly variable. Some species, for example the northern spotted owl, have large home ranges and superior dispersal ability; therefore, large and widely spaced habitat preserves are appropriate, such as the habitat conservation areas (HCAs) of the Interagency Scientific Committee (Thomas et al. 1990). Other vertebrates, such as the tailed frog (*Ascaphus truei*), appear to have very limited dispersal ability and require smaller, highly localized habitat enclaves in order to maintain well-distributed populations. Invertebrate species exhibit a similar wide range in dispersal abilities; a significant number characteristic of old-growth Douglas-fir forests do have limited mobility (Lattin 1990, Moldenke 1990).

A comprehensive strategy for maintaining biological diversity will, therefore, require an array of habitat preserves of widely varying size and spacing. Large, widely spaced preserves, such as wilderness and other roadless areas, national parks, and HCAs represent the broadest approach. Reserved areas, such as riparian corridors, UMAs, and Research Natural Areas, provide a more closely spaced net of small- to medium-size reserves. Individual structures like large living trees, snags, and down logs, retained on the commodity producing lands, provide the finest net of habitat enclaves, or "preserves," in this multiscale strategy for maintaining biological diversity.

Regardless of the approach, biologists are beginning to recognize that the maintenance of biological diversity will not be achieved solely, or even primarily, using large preserves (J. Brown, University of New Mexico, pers. comm.). Preserved lands are too limited in area and are typically located in the less productive and less diverse parts of our landscape. For example, in the Pacific Northwest, low elevation forests have the highest vertebrate diversity (Harris 1984) and, almost certainly, invertebrate diversity as well. But most preserved forests are at higher elevations. Therefore, any strategy for maintaining biological diversity or ecosystem sustainability has to include modifications of commodity producing lands and stream management that incorporate these concerns.

Patch Size and Geographic Pattern of Management Activities

Other critical landscape-level issues are: (1) patch size, whether for managed or preserved areas; and (2) temporal pattern, whether to disperse or concentrate management activities over periods of a decade or more. These two issues are considered together because they are so interrelated.

Public forest lands within the Douglas-fir region traditionally have been managed using patch clearcuts of 10 to 15 ha in a dispersed landscape pattern. This approach produces a fragmented forest landscape with large amounts of clearcut-forest edge and little habitat suitable for interior forest species (Franklin and Forman 1987). Patch sizes are not adequate to provide for either interior forest conditions or the habitat requirements of some of the old-growth-related vertebrates, such as the pine marten (*Martes americana*) and the northern spotted owl.

The fragmented landscape is also predisposed to some kinds of catastrophic disturbance. For example, in the Bull Run watershed of the northern Oregon Cascade Range, wind damage associated with clearcut boundaries and road rights-of-way resulted in a blowdown of hundreds of hectares of residual forest areas (Franklin and Forman 1987). Dispersed patch clearcutting also requires the creation and maintenance of an extensive, permanent road system with continuing impacts on stream ecosystems, including sustained higher levels of sediments and peak flow alteration by roads serving as extensions of the drainage network (Harr 1982).

The net effect of dispersed patch clearcutting is chronic disturbance of all of the landscape in perpetuity. This human-imposed pattern contrasts with natural patterns of disturbance in the Douglas-fir region, which are usually large but infrequent, allowing extensive disturbance-free periods for ecosystem recovery (Swanson et al. 1982).

Landscape-level approaches to forest cutting patterns which provide alternatives to the dispersed patch clearcutting as currently practiced are now being designed and tested. One approach is to enlarge the patch size for both reserved and cutover areas to reduce edge effects and maintain interior forest conditions. For example, cutting could be concentrated in portions of the landscape—such as subdrainages of a few hundred to a few thousand hectares—over a period of 10 to 20 years. One consequence would be larger cutover areas with higher levels of within-stand structural diversity maintained within the cutover area by retaining green trees and CWD. Alternative approaches to transportation and logging systems could be considered as a part of this strategy. For example, lower-standard roads could be constructed and then eliminated following the period of forest cutting and regeneration. An approach of this type, which concentrates management impacts in time and space, would produce somewhat more intense episodic disturbances of the aquatic environments, and allow long recovery periods free from chronic perturbations.

Of course, there are problems inherent in taking such an approach to management. Creating large continuous clearcut areas lacking compositional and structural diversity, and devoid of preserved areas, is undesirable—particularly taking into account the cumulative effects of concentrated cutting patterns on aquatic ecosystems. Initial research indicates that the spatial scale is of critical importance in assessing the impact of concentrated cutting on cumulative effects (Grant 1990). While a 100 ha cutover may have little impact at the scale of a fourth-order drainage, it will clearly be a major "cumulative" effect for an impacted second-order drainage. It is also clear that recognizing primary source areas for sediments and avoiding them whenever possible is probably more critical than total area subject to cutting. The trade-offs between various approaches to dispersed and concentrated harvest cutting are complex, poorly understood, and currently far from resolved (Grant 1990), making this a critical area of landscape research (see Swanson et al., this volume).

Alternatives to continued dispersed clearcutting are also being considered and sometimes applied on National Forest landscapes that already have a high degree of fragmentation. These approaches, sometimes referred to as minimum fragmentation alternatives, strive to maintain large residual forest patches by placing new cuttings adjacent to existing cutover areas (Hemstrom 1990*a*, *b*). Such alternatives maintain options while long-term landscape designs are worked out.

Larger patch size is also a major consideration in new approaches to management of forest lands in the Lake States. Dispersed patch clearcutting in this region has produced large amounts of edge habitat, high populations of white-tailed deer, and heavy browsing of forest understories in the residual forest areas. Larger patch sizes (for both reserve and harvest areas) and reduced fragmentation are also major considerations in management alternatives being considered in current environmental impact analyses for areas of the Chequamegon and Nicolet National Forests in Wisconsin, USA. In one of the first of these to reach the decision stage, the chosen alternative for the Sunken Camp Management Area on the Chequamegon is one that emphasizes "providing large units of habitat for plant and animal species" (United States Forest Service, Eastern Region 1991).

Decisions on issues such as patch size and the geographic distribution of management activities will vary dramatically with the characteristics of the landscape and organisms of interest. The previous discussion has concerned landscapes in the Douglas-fir region and Lake States; approaches appropriate to these areas are not necessarily applicable to other forested regions. For example, forest landscapes in many areas—large portions of the Klamath Mountains of southwestern Oregon and northwestern California, of the Sierra Nevada Mountains in California, and of the southern Rocky Mountains in Colorado—are often fine-scale mosaics of relatively small (0.5 to 10 ha) and low contrast patches, largely because of their history of frequent, low severity fire. Other landscapes, such as those in eastern Washington, often involve medium-scale mosaics of forested and nonforested patches. Although patches often have high contrast, forest patch edges may be sealed by canopy development, producing interior forest environments in much smaller patches than might otherwise be expected. Still other landscapes, such as those associated with lodgepole pine (*Pinus contorta*) forests and large episodic fire events, resemble Douglas-fir landscapes in patch size and contrast.

Managers can look to the natural landscape and its disturbance regime as one guide to the appropriate design of managed landscapes for a particular area. This is not a proposal to mindlessly adopt nature's design as our model. For example, we would certainly not care to emulate the scale of many of the natural catastrophes that have occurred in the Douglas-fir region, where hundreds of thousands of hectares are "treated" in a single event. However, a study of the natural landscapes does recognize that organisms present in a region are adapted to that landscape and its disturbance regime, and almost certainly will reveal elements that need to be incorporated into the managed landscape.

Overview: Alternative Management Approaches Under New Perspectives

In this article I have outlined some scientific bases for new management approaches to forested landscapes and general principles for New Forestry at the stand and landscape levels, and I have offered some examples of specific approaches that are currently being tested. Clearly, there are several important issues that require careful examination by scientists and managers before major policy changes are adopted. One merit of many proposed New Forestry practices is that, unlike traditional practices of the past few decades, most actually maintain future management options.

At the stand level we are progressing rapidly in developing and implementing alternative management approaches. Extensive research is needed on the effectiveness of various practices in achieving ecological objectives as well as accurate appraisals of their economic impacts. The relationship of attributes of interest to different management practices—that is, how biological diversity responds to various densities of retained trees—is a developing research area. Indeed, a generic question will be the level in structural retention where a harvested area continues to behave like a forest rather than a gap. The answer to this question will vary, of course, with the resource, organism, or function of interest—making the generic question even more interesting!

Some have suggested that many New Perspectives stand-level practices have not been scientifically tested, and should therefore not be widely utilized (Atkinson in Hopwood 1991). These criticisms have some validity, since we have only recently begun to test approaches such as green tree retention. However, there is considerable evidence that supports innovation. First, many examples of structurally diverse stands, created by accidents of nature and man, provide strong direction on how suitable stands for late successional species can be created. Among these is the example of northern spotted owls living in dominantly young, mixed-structure stands. Second, despite current protestations to the contrary, foresters have never done adequate scientific testing of concepts before adopting them as basic professional tenets. In the 1940s, autecological studies of Douglas-fir provided some scientific basis for the development of dispersed patch clearcutting. However, it took nearly two decades to perfect the original concept—that is, to prove it scientifically with regard to regeneration practice. It was another two decades before landscape-level consequences were analyzed and the system was found to have some basic flaws. As another example, there are few sites, if any, where foresters have proved that repeated intensive forest cropping is possible without ultimate reductions in site productivity.

In actuality, the clear failure of traditional intensive forest practices to maintain many forest organisms and values provides the strongest evidence supporting the need for alternative or new silvicultural practices. While we may not know precisely the effectiveness of some of the new techniques, we do know that traditional approaches are not working well for many values, and are no longer socially acceptable (Lee, this volume). If a large segment of our forest fauna is dependent on CWD and CWD is being eliminated from intensively managed forests and streams, then it does not take a detailed scientific study to know that we are not providing for that faunal component. Similarly, it seems obvious that a silvicultural system that maintains at least some level of CWD is going to have a much higher probability of providing for at least some elements of the dependent fauna.

Landscape-level issues are often more complex and less amenable to prescriptive solutions. Our understanding of landscapes and the tools available for their study and design is much more limited than it is for stands and stream reaches. A great deal of research is needed in this area. Equally important is the need for extensive dialogue and collaboration between managers, scientists, and public interest groups on landscape issues. Topics in need of serious attention include: the potential contribution of alternative landscape patterns, larger patch sizes, or longer rotations to cumulative effects; the relative impacts of episodic and chronic disturbances on physical processes, ecosystem health, and biological diversity; and alternatives to the creation of extensive permanent road systems in steep mountainous topography.

Resource managers, scientists, and public interest groups alike must reexamine their basic assumptions about forests, fisheries, biological diversity, and related issues. At the landscape level there are currently no prescriptions, only issues. Furthermore, the unique nature of each landscape makes it likely that management decisions will necessarily be highly individualistic.

At least as a starting point, we can look to the natural landscape for some idea of what patch sizes, edge types, and disturbance regimes might be appropriate.

There are currently many important questions regarding the New Perspectives landscape and the form that it should take. One set of questions involves cumulative effects and the trade-offs between dispersed and spatially concentrated management. A related and extremely interesting question concerns the relative impacts of episodic and chronic disturbances on aquatic ecosystems and organisms, including patterns of recovery and such physical processes as sediment production.

Conclusions

Scientific studies of terrestrial and aquatic ecosystems and landscapes during the last two decades have drastically altered our views on their composition, structure, and function. The natural ecosystems have proved to be far more intricate than we had imagined. Particularly relevant and important elements of this expanded knowledge concern riparian zones: the special importance of riparian habitats as essential habitats for many organisms and processes and the high level of integration or linkage between riparian and upland ecosystems.

The expanded ecological knowledge, in combination with increased societal concerns for ecosystem sustainability, biological diversity, and other ecological values, propels natural resource managers toward making major changes in philosophy and practice. These changes are well encompassed in the New Perspectives approach.

The New Perspectives approach calls for new levels of collaboration between managers, scientists, and public interest groups as we attempt to redesign management methods based on necessarily incomplete information. In many cases, practices will have to precede definitive testing. But, as with current New Forestry practices, they can have a strong base in current ecological science. It is critical that managers and scientists address these questions together, and work to develop a systematic approach to feeding new knowledge back into management practices (Oliver and Hinckley 1987). Only in this way can further traumatic upheaval be avoided.

The future, certainly the near future, will not be a time of rote prescription in resource management. The present is a time for reexamination, innovation, testing, and the building of new working relationships. And, viewing the riparian environment, we can see that the New Perspectives approach does have rubber boots. With additional efforts in research, definition of objectives, and development of innovative approaches to riparian management, it can have a pair of chest waders!

Acknowledgments. The author gratefully acknowledges intellectual and data contributions from his colleagues and students associated with the H.J. Andrews Experimental Forest and the University of Washington. Reviews of the manuscript by Robert J. Naiman, Robert Bilby, Fred Swanson, Charlotte Pyle, Dean Berg, and Art McKee were especially helpful. This paper is a contribution from the Olympic Natural Resources Center.

References

Agee, J.K. 1988. Successional dynamics in forest riparian zones. Pages 31–43 *in* K.J. Raedeke, editor. Streamside management: riparian wildlife and forestry interactions. Contribution 59, Institute of Forest Resources, University of Washington, Seattle, Washington, USA.

Agee, J.K., and D.R. Johnson, editors. 1988. Ecosystem management for parks and wilderness. University of Washington Press, Seattle, Washington, USA.

Alaback, P.B. 1984. A comparison of old-growth forest structure in the western hemlock-Sitka spruce forests of southeast Alaska. Pages 219–225 *in* W.R. Meehan, T.R. Merrell, Jr., and T.A. Hanley, editors. Fish and wildlife relationships in old-growth forests. American Institute of Fishery Research Biologists, Juneau, Alaska, USA.

Alban, D.H. 1967. The influence of western hemlock and western redcedar on soil properties. Dissertation. Washington State University, Pullman, Washington, USA.

Amoros, C., J. Rostan, G. Pautou, and J. Bravard. 1987. The reversible process concept applied to the environmental management of large river systems. Environmental Management **11**:607–617.

Anderson, N.H., J.R. Sedell, L.M. Roberts, and F.J. Triska. 1978. The role of aquatic invertebrates in processing of wood debris in coniferous forest streams. American Midland Naturalist **100**:64–82.

Anderson, N.J., R.J. Steedman, and T. Dudley. 1984. Patterns of exploitation by stream invertebrates of wood debris (xylophagy). Internationale Vereinigung für theoretische und angewandte Limnologie, Verhandlungen **22**:1847–1852.

Armour, C. 1989. Characterization of RMZ's and UMA's with respect to wildlife habitat. 1988 field report. Washington Department of Wildlife, Olympia, Washington, USA.

Benda, L.E. 1985. Behavior and effects of debris flows on streams of the Oregon coast range. Pages 153–162 *in* A symposium on the delineation of landslide, flashflood, and debris flow hazards in Utah. Report uwr4g-8503. Utah Water Research Lab, Utah State University, Logan, Utah, USA.

Berris, S.N., and R.D. Harr. 1987. Comparative snow accumulation and melt during rainfall in forested and clear-cut plots in the western Cascades of Oregon. Water Resources Research **23**:135–142.

Beschta, R.L., and W.S. Platts. 1986. Morphological features of small streams: significance and function. Water Resources Bulletin **22**:369–379.

Bilby, R.E. 1981. Role of organic debris dams in regulating the export of dissolved and particulate matter from a forested watershed. Ecology **62**:1234–1243.

Bilby, R.E., and G.E. Likens. 1980. Importance of organic debris dams in the structure and function of stream ecosystems. Ecology **61**:1107–1113.

Bisson, P.A., R.E. Bilby, M.D. Bryant, C.A. Dolloff, G.B. Grette, R.A. House, M.L. Murphy, KV. Koski, and J.R. Sedell. 1987. Large woody debris in forested streams in the Pacific Northwest: past, present, and future. Pages 143–190 *in* E.O. Salo and T.W. Cundy, editors. Streamside management: forestry and fishery interactions. Contribution 57, Institute of Forest Resources, University of Washington, Seattle, USA.

Bisson, P.A., J.L. Nielsen, R.A. Palmason, and L.E. Grove. 1982. A system of naming habitat types in small streams, with examples of habitat utilization by salmonids during low streamflow. Pages 62–73 *in* N.B. Armantrout, editor. Aquisition and utilization of aquatic habitat inventory information. Western Division, American Fisheries Society, Portland, Oregon. The Hague Publishing, Billings, Montana, USA.

Bisson, P.A., and J.R. Sedell. 1984. Salmonid populations in streams in clearcut vs. old-growth forests of western Washington. Pages 121–129 *in* W.R. Meehan, T.R. Merrell, Jr., and T.A. Hanley, editors. Fish and wildlife relationships in old-growth forests. American Institute of Fishery Research Biologists, Juneau, Alaska, USA.

Brown, E.R., technical editor. 1985*a*. Management of wildlife and fish habitats in forests of western Oregon and Washington. Part 1. Chapter narratives. United States Forest Service, Pacific Northwest Region, and United States Bureau of Land Management.

Brown, E.R., technical editor. 1985*b*. Management of wildlife and fish habitats in forests of western Oregon and Washington. Part 2. Appendices. United States Forest Service, Pacific Northwest Region, and United States Bureau of Land Management.

Bryant, M.D. 1983. The role and management of woody debris in west coast salmonid nursery streams. North American Journal of Fisheries Management **3**:322–330.

Budd, W.W., P.L. Cohen, P.R. Saunders, and F.R. Steiner. 1987. Stream corridor management in the Pacific Northwest: management strategies. Environmental Management **11**:592–605.

Carey, A.B. 1989. Wildlife associated with old-growth forests in the Pacific Northwest. Natural Areas Journal **9**:151–162.

Carroll, G.C. 1980. Forest canopies: complex and independent subsystems. Pages 87–107 *in* R.H. Waring, editor. Forests: fresh perspectives from ecosystem analysis. Proceedings of the 40th Annual Biology Colloquium. Oregon State University Press, Corvallis, Oregon, USA.

Cederholm, C.J., and L.M. Reid. 1987. Impact of forest management on coho salmon (*Oncorhynchus kisutch*) populations of the Clearwater River, Washington. Pages 373–398 *in* E.O. Salo and T.W. Cundy, editors. Streamside management: forestry and fishery interactions. Contribution 57, Institute of Forest Resources, University of Washington, Seattle, Washington, USA.

Chen, J. 1991. Edge effects: microclimatic pattern and biological responses in old-growth Douglas-fir forests. Dissertation. College of Forest Resources, University of Washington, Seattle, Washington, USA.

Chen, J., J.F. Franklin, and T.A. Spies. 1990. Microclimatic pattern and basic biological responses at the clearcut edges of old-growth Douglas-fir stands. Northwest Environmental Journal **6**:424–425.

Chen, J., J.F. Franklin, and T.A. Spies. 1992. Vegetation responses to edge environments in old-growth Douglas-fir forests. Ecological Applications, *in press*.

Cissel, J. 1990. An approach for evaluating stand significance and designing forest landscapes. Coastal Oregon Productivity Enhancement (COPE) Report **3**(4):8–11.

Coleman, B.B., W.C. Muenscher, and D.R. Charles. 1956. A distributional study of the epiphytic plants of the Olympic Peninsula, Washington. American Midland Naturalist **56**:54–87.

Conners, E.M., and R.J. Naiman. 1984. Particulate allochthonous inputs: relationships with stream size in an undisturbed watershed. Canadian Journal of Fisheries and Aquatic Sciences **41**:1473–1484.

Cornaby, B.W., and J.B. Waide. 1973. Nitrogen fixation in decaying chestnut logs. Plant and Soil **39**:445–448.

Cross, S.P. 1988. Riparian systems and small mammals and bats. Pages 93–112 *in* K.J. Raedeke, editor. Streamside management: riparian wildlife and forestry interactions. Contribution 59, Institute of Forest Resources, University of Washington, Seattle, Washington, USA.

Cummins, K.W. 1974. Structure and function of stream ecosystems. BioScience **24**:631–641.

Debell, D.S., and J.F. Franklin. 1987. Old-growth Douglas-fir and western hemlock: a 36-year record of growth and mortality. Western Journal of Applied Forestry **2**(4):111–114.

Décamps, H., J. Joachim, and J. Lauga. 1987. The importance for birds of the riparian woodlands within the alluvial corridor of the River Garonne, Southwest France. Regulated Rivers **1**:301–316.

Décamps, H., and R.J. Naiman. 1989. L'écologie des fleuves. La Recherche **20**:310–319.

Denison, W.C. 1979. *Lobaria oregana,* a nitrogen-fixing lichen in old-growth Douglas-fir forests. Pages 266–275 *in* J.C. Gordon, C.T. Wheeler, and D.A. Perry, editors. Symbiotic nitrogen fixation in the management of temperate forests. Forest Research Laboratory, Oregon State University, Corvallis, Oregon, USA.

Edmonds, R.L., editor. 1982. Analysis of coniferous forest ecosystems in the western United States. Hutchinson Ross, Stroudsburg, Pennsylvania, USA.

Forman, R.T.T., and M. Godron. 1986. Landscape ecology. John Wiley and Sons, New York, New York, USA.

Franklin, J.F. 1989. Toward a new forestry. American Forests, November–December: 37–44.

Franklin, J.F. 1990*a*. Biological legacies: a critical management concept from Mount St. Helens. Pages 216–219 *in* Transactions of the Fifty-fifth North American Wildlife and Natural Resources Conference. Wildlife Management Institute, Washington, D.C., USA.

Franklin, J.F. 1990*b*. Old-growth forests and the new forestry. Pages 1–19 *in* A.F. Pearson and D.A. Challenger, editors. Forests—wild and managed: differences and consequences. Faculty of Forestry, University of British Columbia, Vancouver, Canada.

Franklin, J.F. 1990*c*. Thoughts on applications of silvicultural systems under new forestry. Forest Watch **19**(7):8–11.

Franklin, J.F., C.S. Bledsoe, and J.T. Callahan. 1990. Contributions of the Long-Term Ecological Research Program. BioScience **40**:509–523.

Franklin, J.F., K. Cromack, Jr., W. Denison, A. McKee, C. Maser, J.R. Sedell, F.J. Swanson, and G. Juday. 1981. Ecological characteristics of old-growth Douglas-fir forests. United States Forest Service General Technical Report PNW-118, Pacific Northwest Forest and Range Experiment Station, Portland, Oregon, USA.

Franklin, J.F., and R.T.T. Forman. 1987. Creating landscape patterns by forest cutting: ecological consequences and principles. Landscape Ecology **1**:5–18.

Franklin, J.F., P.M. Frenzen, and F.J. Swanson. 1988. Re-creation of ecosystems at Mount St. Helens: contrasts in artificial and natural approaches. Pages 1–37 *in* J. Cairns, Jr., editor. Rehabilitating damaged ecosystems. Volume 2. CRC Press, Boca Raton, Florida, USA.

Franklin, J.F., J.A. MacMahon, F.J. Swanson, and J.R. Sedell. 1985. Ecosystem responses to the eruption of Mount St. Helens. National Geographic Research **2**:198–216.

Franklin, J.F., and C. Maser. 1988. Looking ahead: some options for public lands. Pages 113–122 *in* C. Maser, R.F. Tarrant, J.M. Trappe, and J.F. Franklin, editors. From the forest to the sea: a story of fallen trees. United States Forest Service General Technical Report PNW-GTR-229. Pacific Northwest Forest and Range Experiment Station, Portland, Oregon, USA.

Franklin, J.F., H.H. Shugart, and M.E. Harmon. 1987. Tree death as an ecological process. BioScience **37**:550–556.

Franklin, J.F., and T.A. Spies. 1984. Characteristics of old-growth Douglas-fir forests. Pages 10–16 *in* New forests for a changing world. Proceedings of the 1983 SAF Convention, Portland, Oregon, October 16–20. Society of American Foresters, Washington, D.C., USA.

Franklin, J.F., and T.A. Spies. 1991*a*. Composition, function, and structure of old-growth Douglas-fir forests. *In* L.F. Ruggiero, K.B. Aubry, A.B. Carey, and M.H. Huff, technical coordinators. Wildlife and vegetation of unmanaged Douglas-fir forests. United States Forest Service General Technical Report PNW-GTR, *in press*.

Franklin, J.F., and T.A. Spies. 1991*b*. Ecological definitions of old-growth Douglas-fir forests. *In* L.F. Ruggiero, K.B. Aubry, A.B. Carey, and M.H. Huff, technical coordinators. Wildlife and vegetation of unmanaged Douglas-fir forests. United States Forest Service General Technical Report PNW-GTR, *in press*.

Franklin, J.F., T.A. Spies, D. Perry, M. Harmon, and A. McKee. 1986. Modifying Douglas-fir management regimes for nontimber objectives. Pages 373–379 *in* C.D. Oliver, D.P. Hanley, and J.A. Johnson, editors. Douglas-fir: stand management for the future. Contribution 55, College of Forest Resources, University of Washington, Seattle, Washington, USA.

Franklin, J.F., and R.H. Waring. 1980. Distinctive features of the northwestern coniferous forest: development, structure, and function. Pages 59–86 *in* R.H. Waring, editor. Forests: fresh perspectives from ecosystem analysis. Proceedings of the 40th Annual Biology Colloquium. Oregon State University Press, Corvallis, Oregon, USA.

Fujimori, T., S. Kawanabe, H. Saito, et al. 1976. Biomass and primary production in forests of three major vegetation zones of the northwestern United States. Journal of the Japanese Forestry Society **58**(10):360–373.

Gardner, R.H., and R.V. O'Neill. 1991. Pattern, process, and predictability: the use of neutral models for landscape analysis. Pages 287–307 *in* M.G. Turner and

R.H. Gardner, editors. Quantitative methods in landscape ecology. Springer-Verlag, New York, New York, USA.

Geppert, R.R., C.W. Lorenz, and A.G. Larson. 1984. Cumulative effects of forest practices on the environment. Washington State Department of Natural Resources, Olympia, Washington, USA.

Grant, G.E. 1990. Hydrologic, geomorphic and aquatic habitat implications of old and new forestry. Pages 35–53 *in* A.F. Pearson and D.A. Challenger, editors. Forests—wild and managed: differences and consequences. Faculty of Forestry, University of British Columbia, Vancouver, Canada.

Grant, G.E., F.J. Swanson, and M.G. Wolman. 1990. Pattern and origin of stepped-bed morphology in high-gradient streams, western Cascades, Oregon. Geological Society of America Bulletin **102**:340–352.

Gregory, S.V., G.A. Lamberti, and K.M. Moore. 1991*a*. Influence of valley landforms on stream ecosystems. Proceedings of California Riparian Systems: Protection, Management, and Restoration for the 1990's. University of California Press, Davis, California, USA.

Gregory, S.V., F.J. Swanson, W.A. McKee, and K.W. Cummins. 1991*b*. An ecosystem perspective of riparian zones. BioScience **41**:540–551.

Gresswell, S., D. Heller, and D.N. Swanston. 1979. Mass movement response to forest management in the central Oregon coast ranges. United States Forest Service Resource Bulletin PNW-84. Pacific Northwest Forest and Range Experiment Station, Portland, Oregon, USA.

Grette, G.B. 1985. The role of large organic debris in juvenile salmonid rearing habitat in small streams. M.S. Thesis. University of Washington, Seattle, Washington, USA.

Grier, C.C., and R.S. Logan. 1977. Old-growth *Pseudotsuga menziesii* communities of a western Oregon watershed: biomass distribution and production budgets. Ecological Monographs **47**:373–400.

Halpern, C.B., and J.F. Franklin. 1990. Physiognomic development of *Pseudotsuga* forests in relation to initial structure and disturbance intensity. Journal of Vegetation Science **1**:475–482.

Hansen, A., J. Peterson, and E. Horwath. 1990. Do wildlife species respond to stand and edge type in managed forests of the Oregon Coast Range? Coastal Oregon Productivity Enhancement Program (COPE) Report **3**(2):3–4.

Harmon, M.E., W.K. Ferrell, and J.F. Franklin. 1990. Effects on carbon storage of conversion of old-growth forests to young forests. Science **247**:699–702.

Harmon, M.E., J.F. Franklin, F.J. Swanson, P. Sollins, S.V. Gregory, J.D. Lattin, N.H. Anderson, S.P. Cline, N.G. Aumen, J.R. Sedell, G.W. Lienkaemper, K. Cromack, Jr., and K.W. Cummins. 1986. Ecology of coarse woody debris in temperate ecosystems. Advances in Ecological Research **15**:133–302.

Harr, R.D. 1982. Fog drip in the Bull Run municipal watershed, Oregon. Water Resources Bulletin **18**:785–789.

Harr, R.D. 1986. Effects of clearcutting on rain-on-snow runoff in western Oregon: a new look at old studies. Water Resources Bulletin **22**:1095–1100.

Harr, R.D., B.A. Coffin, and T.W. Cundy. 1989. Effects of timber harvest on rain-on-snow runoff in the transient snow zone of the Washington Cascades. Interim final report submitted to Timber, Fish and Wildlife (TFW) sediment, hydrology and mass wasting steering committee for Project 18 (rain-on-snow). United States

Forest Service, Pacific Northwest Forest and Range Experiment Station, Portland, Oregon, USA.

Harris, L.D. 1984. The fragmented forest: island biogeography theory and the preservation of biotic diversity. University of Chicago Press, Chicago, Illinois, USA.

Harris, W.F., D. Santantonio, and D. McGinty. 1980. The dynamic belowground ecosystem. Pages 119–129 *in* R.H. Waring, editor. Forests: fresh perspectives from ecosystem analysis. Proceedings of the 40th Annual Biology Colloquium. Oregon State University Press, Corvallis, Oregon, USA.

Heath, B., P. Sollins, D.A. Perry, and K. Cromack, Jr. 1987. Asymbiotic nitrogen fixation in litter from Pacific Northwest forests. Canadian Journal of Forest Research **18**:68–74.

Heede, B.H. 1985. Interactions between streamside vegetation and stream dynamics. Pages 54–58 *in* Proceedings of Riparian Ecosystems and Their Management: Reconciling Conflicting Uses. Tucson, Arizona, USA. United States Forest Service General Technical Report RM-120, Rocky Mountain Forest and Range Experiment Station, Fort Collins, Colorado, USA.

Hemstrom, M.A. 1990*a*. Alternative timber harvest patterns for landscape diversity. Coastal Oregon Productivity Enhancement (COPE) Report **3**(1):8–11.

Hemstrom, M.A. 1990*b*. New forestry—how will it look on the landscape? Pages 27–43 *in* A.F. Pearson and D.A. Challenger, editors. Forests—wild and managed: differences and consequences. Faculty of Forestry, University of British Columbia, Vancouver, Canada.

Hopwood, D. 1991. Principles and practices of new forestry. Land Management Report 71. Ministry of Forests, Victoria, British Columbia, Canada.

Hunt, R.L. 1988. Management of riparian zones and stream channels to benefit fisheries. United States Forest Service General Technical Report NC-122, North Central Forest Experiment Station, St. Paul, Minnesota, USA.

Keller, E.A., and F.J. Swanson. 1979. Effects of large organic material on channel form and fluvial processes. Earth Surface Processes **4**:361–380.

Kiilsgaard, C.W., S.E. Greene, and S.G. Stafford. 1987. Nutrient concentrations in litterfall from some western conifers with special reference to calcium. Plant and Soil **102**:223–227.

Lamberti, G.A., S.V. Gregory, L.R. Ashkenas, R.C. Wildman, and K.M.S. Moore. 1991. Stream ecosystem recovery following a catastrophic debris flow. Canadian Journal of Fisheries and Aquatic Sciences **48**:196–207.

Lattin, J.D. 1990. Arthropod diversity in Northwest old-growth forests. Wings **15**(2):7–10.

Leopold, L.B., M.G. Wolman, and J.P. Miller. 1963. Fluvial processes in geomorphology. W.H. Freeman, San Francisco, California, USA.

Lienkaemper, G.W., and F.J. Swanson. 1987. Dynamics of large woody debris in streams in old-growth Douglas-fir forests. Canadian Journal of Forest Research **17**:150–156.

Maser, C., and J.M. Trappe. 1984. The fallen tree: a source of diversity. Pages 335–339 *in* New forests for a changing world. Proceedings of the 1983 SAF Convention, Portland, Oregon, October 16–20. Society of American Foresters, Washington, D.C., USA.

Maser, C., R.F. Tarrant, J.M. Trappe, and J.F. Franklin, editors. 1988. From the forest to the sea: a story of fallen trees. United States Forest Service General

Technical Report PNW-GTE-229, Pacific Northwest Forest and Range Experiment Station, Portland, Oregon, USA.

Massman, W.J. 1982. Foliage distribution in old-growth coniferous tree canopies. Canadian Journal of Forest Research **12**:10–17.

McDade, M.H., F.J. Swanson, W.A. McKee, and J.F. Franklin. 1990. Source distances for coarse woody debris entering small streams in western Oregon and Washington. Canadian Journal of Forest Research **20**:326–329.

Megahan, W.F., N.F. Day, and T.M. Bliss. 1978. Landslide occurrence in the western and central northern Rocky Mountain physiographic province in Idaho. Pages 116–139 *in* C.T. Youngberg, editor. Forest soils and land use. Proceedings of the Fifth North American Forest Soils Conference, August 1978.

Melillo, J.M., R.J. Naiman, J.D. Aber, and K.N. Eshleman. 1983. The influence of substrate quality and stream size on wood decomposition dynamics. Oecologia (Berlin) **58**:281–285.

Melillo, J.M., R.J. Naiman, J.D. Aber, and A.E. Linkins. 1984. Factors controlling mass loss and nitrogen dynamics of plant litter decaying in northern streams. Bulletin of Marine Science **35**:341–346.

Merritt, R.W., and K.W. Cummins, editors. 1978. An introduction to the aquatic insects of North America. Kendall-Hunt, Dubuque, Iowa, USA.

Moldenke, A. 1990. One hundred twenty thousand little legs. Wings **15**(2):11–14.

Murphy, M.L. 1979. Predator assemblages in old-growth and logged sections of small cascade streams. Master's Thesis. Oregon State University, Corvallis, Oregon, USA.

Naiman, R.J. 1990. Forest ecology: influence of forests on streams. Pages 151–153 *in* 1991 McGraw-Hill Yearbook of Science and Technology. McGraw-Hill, New York, New York, USA.

Naiman, R.J., and H. Décamps, editors. 1990. The ecology and management of aquatic-terrestrial ecotones. UNESCO, Paris, and Parthenon Publishing Group, Carnforth, United Kingdom.

Naiman, R.J., H. Décamps, and F. Fournier, editors. 1989. Role of land/inland water ecotones in landscape management and restoration, proposals for collaborative research. UNESCO, Vendome, France.

Naiman, R.J., H. Décamps, J. Pastor, and C.A. Johnston. 1988. The potential importance of boundaries to fluvial ecosystems. Journal of the North American Benthological Society **7**:289–306.

Naiman, R.J., D.G. Lonzarich, T.J. Beechie, and S.C. Ralph. 1991. General principles of classification and the assessment of conservation potential in rivers. Pages 93–123 *in* P.J. Boon, P. Calow, and G.E. Petts, editors. River conservation and management. John Wiley and Sons, Chichester, England.

Nilsson, C., G. Grelsson, M. Johansson, and U. Sperens. 1989. Patterns of species richness along riverbanks. Ecology **70**:77–84.

Old Growth Definition Task Group. 1986. Interim definitions for old-growth Douglas-fir and the mixed-conifer forests in the Pacific Northwest and California. United States Forest Service Research Note PNW-447, Pacific Northwest Research Station, Portland, Oregon, USA.

Oliver, C.D., and T.M. Hinckley. 1987. Species, stand structures, and silvicultural manipulation patterns for the streamside zone. Pages 259–276 *in* E.O. Salo and T.W. Cundy, editors. Streamside management: forestry and fishery interactions.

Contribution 57, Institute of Forest Resources, University of Washington, Seattle, Washington, USA.

Perry, D.A. 1988. Landscape pattern and forest pests. Northwest Environmental Journal **4**:213–228.

Perry, D.A., M.P. Amaranthus, J.G. Borchers, S.L. Borchers, and R.E. Brainerd. 1989. Bootstraping in ecosystems. BioScience **39**:230–237.

Perry, D.A., and J. Maghembe. 1989. Ecosystem concepts and current trends in forest management: time for reappraisal. Forest Ecology and Management, **26**:123–140.

Perry, D.A., R. Molina, and M.P. Amaranthus. 1987. Mycorrhizae, mycorrhizospheres, and reforestaton: current knowledge and research needs. Canadian Journal of Forest Research **17**:929–940.

Peterson, E.B., Y.H. Chan, N.M. Peterson, G.A. Constable, R.B. Caton, C.S. Davis, R.R. Wallace, and G.A. Yarranton. 1987. Cumulative effects assessment in Canada: an agenda for action and research. Canadian Environmental Assessment Research Council, Ottawa, Ontario, Canada.

Pike, L.H., R.A. Rydell, and W.C. Denison. 1977. A 400-year-old Douglas-fir tree and its epiphytes: biomass, surface area, and their distributions. Canadian Journal of Forest Research **7**:680–699.

Potts, D.F., and B.K.M. Anderson. 1990. Organic debris and the management of small stream channels. Western Journal of Applied Forestry **5**(1):25–28.

Pringle, C.M., R.J. Naiman, G. Bretschko, J.R. Karr, M.W. Oswood, J.R. Webster, R.L. Welcomme, and M.J. Winterbourn. 1988. Patch dynamics in lotic systems: the stream as a mosaic. Journal of the North American Benthological Society **7**:503–524.

Raedeke, K.J. 1988. Ecology of large mammals in riparian systems of the Pacific Northwest forests. Pages 113–132 *in* K.J. Raedeke, editor. Streamside management: riparian wildlife and forestry interactions. Contribution 59, Institute of Forest Resources, University of Washington, Seattle, Washington, USA.

Raedeke, K.J., editor. 1988. Streamside management: riparian wildlife and forestry interactions. Contribution 59, Institute of Forest Resources, University of Washington, Seattle, Washington, USA.

Rainville, R.P., S.C. Rainville, and E.L. Lider. 1985. Riparian silvicultural strategies for fish habitat emphasis. *In* Proceedings of the 1985 Society of American Foresters National Convention, Fort Collins, Colorado, USA.

Reukema, D.L. 1979. Fifty-year development of Douglas-fir stands planted at various spacings. United States Forest Service Research Paper PNW-254.

Reukema, D.L., and H.G. Smith. 1987. Development over 25 years of Douglas-fir, western hemlock, and western redcedar planted at various spacings on a very good site in British Columbia. United States Forest Service Research Paper PNW-RE-381, Pacific Northwest Forest and Range Experiment Station, Portland, Oregon, USA.

Rochelle, J.A., K.J. Raedeke, and L.L. Hicks. 1988. Introduction: management opportunities for wildlife in riparian areas. Pages 135–138 *in* K.J. Raedeke, editor. Streamside management: riparian wildlife and forestry interactions. Contribution 59, Institute of Forest Resources, University of Washington, Seattle, Washington, USA.

Ruggiero, L.F., K.B. Aubry, A.B. Carey, and M.H. Huff, technical coordinators. 1991. Wildlife and vegetation of unmanaged Douglas-fir forests. United States Forest Service General Technical Report PNW-GTR, *in press*.

Ruth, R.H., and R.R. Silen. 1950. Suggestions for getting more forestry in the logging plan. United States Forest Service Research Note 72, Pacific Northwest Forest and Range Experiment Station, Portland, Oregon, USA.

Salo, E.O., and T.W. Cundy, editors. 1987. Streamside management: forestry and fishery interactions. Contribution 57, Institute of Forest Resources, University of Washington, Seattle, Washington, USA.

Schoen, J.W. 1990. Forest management and bear conservation. Presentation, Fifth International Congress of Ecology International Symposium on Wildlife Conservation, Yokohama, Japan.

Schoen, J.W., and M.D. Kirchoff. 1990. Seasonal habitat use by Sitka black-tailed deer on Admiralty Island, Alaska. Journal of Wildlife Management **54**:371–378.

Schoonmaker, P., and A. McKee. 1988. Species composition and diversity during secondary succession of coniferous forests in the western Cascade Mountains of Oregon. Forest Science **34**:960–979.

Schowalter, T.D. 1989. Canopy arthropod community structure and herbivory in old-growth and regenerating forests in western Oregon. Canadian Journal of Forest Research **19**:318–322.

Schowalter, T.D., S.G. Stafford, and R.L. Slagle. 1988. Arboreal arthropod community structure in an early successional coniferous forest ecosystem in western Oregon. Great Basin Naturalist **48**:327–333.

Sharpe, R.F., and J.W. Milbank. 1973. Nitrogen fixation in deteriorating wood. Experimentia **29**:895–896.

Sidle, R.C., A.J. Pierce, and C.L. O'Loughlin. 1985. Hillslope stability and land use. Water Resources Monograph 11. American Geophysical Union, Washington, D.C., USA.

Silen, R.R. 1955. More efficient road patterns for a Douglas-fir drainage. The Timberman **56**(6):82–88.

Sonntag, N.C., R.R. Everitt, L.P. Rattie, D.L. Colnett, C.P. Wolf, J.C. Truett, A.H.J. Dorcey, and C.S. Holling. 1987. Cumulative effects assessment: a context for further development. Canadian Environmental Assessment Research Council.

Spies, T.A., and S.P. Cline. 1988. Coarse woody debris in forests and plantations of coastal Oregon. Pages 5–24 *in* C. Maser, R.F. Tarrant, J.M. Trappe, and J.F. Franklin, editors. From the forest to the sea: a story of fallen trees. United States Forest Service General Technical Report PNW-GTR-229.

Spies, T.A., and J.F. Franklin. 1991. The structure of natural young, mature and old-growth Douglas-fir forests in Washington and Oregon. *In* L.F. Ruggiero, K.B. Aubry, A.B. Carey, and M.H. Huff, technical coordinators. Wildlife and vegetation of unmanaged Douglas-fir forests. United States Forest Service General Technical Report PNW-GTR, *in press*.

Spies, T.A., J.F. Franklin, and T.B. Thomas. 1988. Coarse woody debris in Douglas-fir forests of western Oregon and Washington. Ecology **69**:1689–1702.

Sullivan, K., T.E. Lisle, C.A. Dolloff, G.E. Grant, and L.M. Reid. 1987. Stream channels: the link between forests and fishes. Pages 39–97 *in* E.O. Salo and T.W. Cundy, editors. Streamside management: forestry and fishery interactions. Contribution 57, Institute of Forest Resources, University of Washington, Seattle, Washington, USA.

Swanson, F.J., L.E. Benda, S.H. Duncan, G.E. Grant, W.F. Megahan, L.M. Reid, and R.R. Ziemer. 1987. Mass failures and other processes of sediment production in Pacific Northwest forest landscapes. Pages 9–38 *in* E.O. Salo and T.W. Cundy, editors. Streamside management: forestry and fishery interactions. Contribution 57, Institute of Forest Resources, University of Washington, Seattle, Washington, USA.

Swanson, F.J., M.D. Bryant, G.W. Lienkaemper, and J.R. Sedell. 1984. Organic debris in small streams, Prince of Wales Island, southeast Alaska. United States Forest Service General Technical Report PNW-166, Pacific Northwest Forest and Range Experiment Station, Portland, Oregon, USA.

Swanson, F.J., J.L. Clayton, W.F. Megahan, and G. Bush. 1989. Erosional processes and long-term site productivity. Pages 67–81 *in* D.A. Perry, R. Meurisse, B. Thomas, et al., editors. Maintaining the long-term productivity of Pacific Northwest forest ecosystems. Timber Press, Portland, Oregon, USA.

Swanson, F.J., S.V. Gregory, J.R. Sedell, and A.G. Campbell. 1982. Land-water interactions: the riparian zone. Pages 267–291 *in* R.L. Edmonds, editor. Analysis of coniferous forest ecosystems in the western United States. Hutchinson Ross, Stroudsburg, Pennsylvania, USA.

Swanson, F.J., T.K. Kratz, N. Caine, and R.G. Woodmansee. 1988. Landform effects on ecosystem patterns and processes. BioScience **38**:92–98.

Swanson, F.J., G.W. Lienkaemper, and J.R. Sedell. 1976. History, physical effects, and management implications of large organic debris in western Oregon streams. United States Forest Service General Technical Report PNW-56, Pacific Northwest Forest and Range and Experiment Station, Portland, Oregon, USA.

Thomas, J.W., editor. 1979. Wildlife habitats in managed forests: the Blue Mountains of Oregon and Washington. United States Department of Agriculture Agricultural Handbook 553.

Thomas, J.W., E.D. Forsman, J.B. Lint, E.C. Meslow, B.R. Noon, and J. Verner. 1990. A conservation strategy for the northern spotted owl: report of the Interagency Scientific Committee to address the conservation of the northern spotted owl. United States Forest Service, Bureau of Land Management, Fish and Wildlife Service, and National Park Service, Portland, Oregon, USA.

Trappe, J.M., J.F. Franklin, R.F. Tarrant, and G.M. Hansen. 1967. Biology of alder. Proceedings of a symposium held at the Northwest Scientific Association, Pullman, Washington, April 14–15, 1967. United States Forest Service, Pacific Northwest Forest and Range Experiment Station, Portland, Oregon, USA.

Triska, F.J., and K. Cromack, Jr. 1980. The role of wood debris in forests and streams. Pages 171–190 *in* R.H. Waring, editor. Forests: fresh perspectives from ecosystem analysis. Proceedings of the 40th Annual Biology Colloquium. Oregon State University Press, Corvallis, Oregon, USA.

Triska, F.J., J.R. Sedell, and S.V. Gregory. 1982. Coniferous forest streams. Pages 292–332 *in* R.L. Edmonds, editor. Analysis of coniferous forest ecosystems in the western United States. Hutchinson Ross, Stroudsburg, Pennsylvania, USA.

Turner, D.P., and E.H. Franz. 1985. The influence of western hemlock and western redcedar on microbial numbers, nitrogen mineralization, and nitrification. Plant and Soil **88**:259–267.

United States Forest Service, Eastern Region. 1991. Final environmental impact statement for the sunken camp area, management area 351.

United States Forest Service, Pacific Northwest Region. 1990. Land and resource management plan, Mount Baker-Snoqualmie National Forest. Seattle, Washington, USA.

Walters, C. 1986. Adaptive management of renewable resources. Macmillan, New York, New York, USA.

West, S.D. 1988. Introduction: riparian systems and wildlife. Pages 59–60 *in* K.J. Raedeke, editor. Streamside management: riparian wildlife and forestry interactions. Contribution 59, Institute of Forest Resources, University of Washington, Seattle, Washington, USA.

Williamson, R.L. 1973. Results of shelterwood harvesting of Douglas-fir in the Cascades of western Oregon. United States Forest Service Research Paper PNW-161, Pacific Northwest Forest and Range Experiment Station, Portland, Oregon, USA.

Wilzbach, M.A., K.W. Cummins, and J.D. Hall. 1986. Influence of habitat manipulations on interactions between cutthroat trout and invertebrate drift. Ecology **67**:898–911.

4

Ecologically Effective Social Organization as a Requirement for Sustaining Watershed Ecosystems

ROBERT G. LEE

Symbiotic relationships mean creative partnerships. The earth is to be seen neither as an ecosystem to be preserved unchanged nor as a quarry to be exploited for selfish and short-range economic reasons, but as a garden to be cultivated for the development of its own potentialities of the human adventure.

René Dubos (1976)

Abstract

The social sciences can make significant contributions to solving watershed management problems. Sustainable watershed management requires knowledge about ecologically effective forms of social organization. Including humans as a component of the ecosystem permits scientists and policy makers to consider how resource management activities affect biophysical processes regulating ecosystems. A major reason for the failure of human societies to develop sustainable resource management activities has been the limitations on their ability to acquire and process ecological information. Difficulty in maintaining adequate information on the state of ecological systems originates in the inability of people to develop an effective cognitive map of their environment. Institutional structure has a major influence on cognitive learning of environments, and institutional arrangements determine the scale of human social organization and the incentives for people to learn and adopt ecologically sustainable practices. Institutionalization of sustainable resource and ecosystem management practices will require better information about the appropriate scale and form of social organization. Small, flexible institutional units may be best suited for the adaptive learning necessary to achieve sustainable resource management.

Key words. Sustainability, resource management, institutions, environmental learning, watershed management, social adaptability.

Introduction

Conceptual separation of humans and natural ecosystems is reflected in the thinking of most natural resource management professions, including forestry, wildlife management, fisheries, range management, and watershed management (Burch 1971). Such thinking can deny the reality of the human element in local, regional, and global ecosystems (Bonnicksen and Lee 1982, Klausner 1971, Vayda 1977). As complex organisms with highly developed cultural abilities to modify their environment, humans directly or indirectly affect almost all terrestrial and aquatic ecosystems (Bennett 1976). Consequently, information for managing watershed ecosystems is incomplete without consideration of human institutions and activities.

Sociologists have studied the relationships between human societies and the land base or ecosystems on which they depend for over 60 years (Field and Burch 1990). These studies are distinguished by (1) a holistic perspective that sees people and their environments as interacting systems, (2) flexible approaches that permit either the environment or human society to be treated as the independent variable in analyzing of society-environment relations, and (3) accumulation of a substantial body of knowledge about how the future welfare of a society is influenced by its uses (or misuses) of land and water (Firey 1990). A more comprehensive approach to stimulating rapid accumulation of knowledge has been promoted in recent years, including the development of a Natural Resources Research Group in the Rural Sociological Society, a biennial Symposium on Society and Resources Management, a new journal (*Society and Natural Resources*), and a series of edited volumes (Miller et al. 1987, Lee et al. 1990).

As a result, social science expertise has been successfully applied to several contemporary resource management problems, including social impact assessment (Burch and DeLuca 1984, Finsterbusch and Wolf 1980, Wenner 1987), public involvement (Carroll 1988, Wondolluck 1988), stability of resource-dependent communities (Lee et al. 1990, Machlis and Force 1988), residential settlement in nonurban environments (Blahna 1990, Fortmann and Starrs 1990, Bradley 1984), and recreational carrying capacity (Burch 1984, Moore and Brickler 1987, Stankey et al. 1985). Other problems also require the application of substantial social science expertise. This article extends accumulated sociological and anthropological knowledge to the problem of sustainable natural resource management. The concept of sustainability originated with attempts to manage biologically renewable resources such as fisheries and forests. According to this purely physical concept, "sustainability means using no more than the annual increase in the resource without reducing the physical stock ... using the interest earned from a savings account but leaving the principal invested to continue to generate interest in the future" (Dixon and Fallon 1989:74). A biologically determined harvest rate called the *maximum sustained yield* was assumed to continue indefinitely with the adoption of appropriate harvest and regeneration practices.

But physical sustainability has proved to be far too simple a concept to guide policy development and implementation. Left unanswered are questions of social and individual welfare involving choices about who will benefit and when (Norgaard 1988). The distribution of benefits within and between generations calls for difficult policy choices, especially when population growth will reduce future per capita resource consumption and there is uncertainty whether technological advances can increase efficiency in resource supply (Dixon and Fallon 1989).

The problems of intergenerational equity are central to the Brundtland Commission report, *Our Common Future*, since it defines sustainable development as that which "meets the needs of the present without compromising the ability of future generations to meet their own needs" (World Commission on Environment and Development 1987:8). Sustainable development implies human activities that address the "limitations imposed by the present state of technology and social organization on environmental resources and by the ability of the biosphere to absorb the effects of human activities." The Commission report clearly states that substantial legal, institutional, and economic changes are necessary to achieve sustainability. Humans, including their industrial activities, are considered to be integral parts of the biosphere.

Hence sustainability is fundamentally a problem of human social organization and technology, not simply management of the physical environment and its biological processes. Technology and social organization can limit what is done to adapt to the constraints imposed by ecological processes. But technology, together with the social and industrial activities it supports, also provides some of our greatest opportunities for harmonizing human activities with larger ecological processes.

This article examines ecologically effective social organization as a requirement for ensuring the sustainability of watershed ecosystems. It begins with a review of some reasons people fail to develop sustainable ecological activities because of structural limitations in their ability to acquire and process information. This is followed by a detailed discussion of how humans have institutionalized ecological processes—how institutional arrangements can help people overcome problems in processing information. Institutional arrangements that may be more appropriate for ensuring sustainability are explored. Examples are interwoven with this discussion to illustrate unsuccessful and successful approaches to institutionalizing the sustainability of watershed ecosystems.

Sources of Failure to Institutionalize Sustainability

History is replete with the failures of societies to perpetuate ecological processes supporting human populations (Thomas 1956). Decline of classical civilizations in the Middle East and North Africa and contemporary deforestation in the tropics are only two examples (Perlin 1989). Although changes in climate may have played a role in deforestation and in failures of agriculture, social and economic factors have often been more important influences because of their impact on land use decisions. We see this in the short-term political and economic expediency of contemporary land use decisions. There is no reason to believe that social and economic factors were not equally important historically. What are some of these factors and how did they affect sustainability?

Two primary requirements for sustainability are the use of ecological information in decision making and, given such information, successful control over human activities. We will begin with a discussion of how inadequate information flow can limit sustainability and then will turn to the institutional regulation of human activities that affect ecological processes.

Information Flow Pathologies

McGovern et al. (1988) discuss "information flow pathologies" that have limited the successful adaptation of people to ecological conditions. They use the term *information flow* to express the assumption that humans react "not to the real world in real time, but to a cognized environment filtered through expectations and a world view which may or may not value close tracking of local environmental indicators" (p. 245). They note that even the most technologically advanced modern societies have difficulty maintaining adequate information on the state of ecological systems, including problems of maintaining current, accurate, and properly scaled (localized) information. Chandler (1990) pioneered new opportunities for studying how modern land managers can learn about the ecological systems they manipulate when he extended McGovern's work to the study of traditionally derived agrosilvicultural systems in China.

Seven factors may result in maladaptive information flow. McGovern et al. (1988:245) listed the first six, Lee (1991) suggested the seventh.

1. *False Analogy*. The managers' cognitive model of ecosystem characteristics (potential productivity, resilience, and stress signals) is derived from another ecosystem, whose surface similarities mask critical threshold differences from the managers' ecosystem.

2. *Insufficient Detail*. The managers' cognitive model is overgeneralized, and does not adequately allow for the range of spatial variability in an ecosystem whose patchiness is better measured in resilience than in initial abundance.

3. *Short Observational Series*. The managers lack a sufficiently long memory of events to track or predict variability in key environmental factors over a multigenerational period, and are subject to chronic inability to separate short-term and long-term processes.

4. *Managerial Detachment*. The managers are socially and spatially distant from agricultural producers who carry out managerial decisions at the

lowest level and are normally in closest contact with local-scale environmental feedbacks.

5. *Reactions Out of Phase*. Partly as a result of the last two factors, the managers' attempts to avert unfavorable impacts are too little and too late, or they apply the wrong remedy.

6. *S.E.P. (Someone Else's Problem)*. Managers at many levels may perceive a potential environmental problem but feel no obligation to take action, since their own particular short-term interests are not immediately threatened.

7. *Ideological Beliefs*. Managers conform to ideological beliefs shared by generalized publics and overlook particular ecological details and management practices. Unquestioned moral commitments to the principles of capitalism, socialism, environmentalism, and other ideologies can divert the managers' attention from the problems of attending to particular ecological conditions (Lee 1991, Schiff 1966).

The first three causes for maladaptive information flows are most likely to be encountered when people first colonize a region, but diminish as they "learn" the new ecosystem (McGovern et al. 1988). The fourth, fifth, and sixth factors appear most often in highly differentiated societies with complex public or private institutional arrangements for managing ecosystems. The seventh can be found in societies at all stages of development, but is frequently revealed in the behavior of large public land management bureaucracies in contemporary societies (Schiff 1966).

The fourth, fifth, and sixth pathologies can be reduced by altering the institutional arrangements to make decision makers more responsive to localized ecological conditions. The effects of the seventh can be diminished by increasing the authority and responsibility (including real accountability) of localized ecosystem managers and improving the integration of scientific learning with decision making.

Institutionalization of Behavior in Relation to Ecological Processes

The sociological concept of *institutionalization* can make a significant contribution to understanding how the processing of ecological information is affected by human organization. Institutionalization involves the development of persistent patterns of human behavior expressed as formalized rules, laws, or customs or as informal rituals and patterns of social interaction or interaction with the nonhuman environment (Berger and Luckmann 1966). Just as repeated patterns of human social interaction are institutionalized, human manipulation of ecological processes reflects regularized patterns of human behavior that are similarly institutionalized. Two examples will suffice.

Shifting cultivation has persisted relatively unchanged for thousands of years. Even-aged management of forests involving clearcut harvesting became the dominant form of industrial wood production on both public and private lands in North America during the 1950s. When shifting cultivation is practiced in its more traditional forms, it involves making relatively small clearings in a forest so that food crops can be grown until soil fertility is exhausted, at which point the plot is largely abandoned and the forest recovered through natural succession. Over many generations land managers learned to adapt cropping cycles and practices to particular ecological conditions. Although the general pattern of rotation cropping was institutionalized, particular practices were not so highly prescribed that trial and error could not be used to adapt this management regime to localized conditions.

Industrial wood production under an even-aged management regime involves clearing large areas of forest and controlling species composition so that biomass accumulation occurs in species of trees with the highest commercial value. Like intensive agriculture, it short-circuits natural processes of succession, simplifies ecological structure, and channels the flow of energy and cycling of materials along pathways that are most productive of commercially valuable products (Kimmins 1987). Managers learned very quickly how to minimize production costs in order to maintain or increase profit margins.

Institutionalization of industrial wood production constrained learning to economic considerations and may have prevented an adequate flow of information on how best to adapt to ecological conditions (longer-term ecological adaptability was "someone else's problem").

Wherever ecological processes are appropriated and patterned by human society, we can refer to them as *institutionalized ecological processes*. This term refers to the ways in which humans regulate structural components of ecosystems or alter the flow of energy or cycling of materials.

The concept of institutionalization also helps us understand how regulated ecological processes are stabilized and persist relatively unchanged for long periods. Walter Firey, a distinguished sociologist who studied natural resources, dedicated his career to investigating the conditions under which the human use of ecological systems could be sustained. Firey (1963:150) referred to basic issues that underlie sustainability when he stated "there are many kinds of activities which, by their very nature, require some kind of orientation on the part of human agents to a remote future." He struggled with the same problems we find so troubling when he said:

> Thus the cultivation of certain perennial tree crops, such as the olive, cocoa, and pecan, presupposes many years of care before the cultivator will reap any marketable crop at all. Sustained yield management of forests in several European countries has involved reproduction cycles of more than a century. Amortization of capital investments in some mining and plantation enterprises often transcends the span of a single generation. Maintenance of soil fertility in peasant cultures, such as those of Europe and China, has imposed costs upon generations who have never realized any compensation for their trouble.

Firey sought to explain how societies motivated people to work for objectives that would not be realized during their lifetimes, and posed two questions (pp. 150–151): "is this sacrificial effort by one generation for the welfare of another generation the function of explicit future-referring values? Or is it rather an epiphenomenal manifestation of certain structural properties of the social orders in question?"

Firey understood American culture and its unquestionable commitment to the ideal of natural resources conservation—a value that by the 1930s had taken on the force of a "moral imperative." Yet he observed great disparity between the idealistic commitment to conservation and the actual behavior of people who managed natural resources. Studies of soil and water management showed that the future-referring values were insufficient to motivate farmers to practice conservation. Farmers did not necessarily implement the idealistic values of the culture they shared.

Firey (1960, 1963) concluded that conservation behavior requires at least two conditions, in addition to being biologically possible: (1) individuals must internalize values, and (2) these values must be articulated socially in ways that motivate conformity; that is, they must be both expedient (gainful for the individual or group) and psychologically satisfying (maintain self-esteem and group identification). In short, they must be institutionalized.

Firey noted that values that do not become institutionalized in the form of ongoing social relationships can have only an ideological status. He observed that conservation values in contemporary American culture are largely ideological. "Sustainability" seems to have acquired a similar ideological status as a political slogan popularized during the 1980s (Dixon and Fallon 1989).

There has been very little progress in understanding that values must be institutionalized before they can affect behavior. Yet this principle was well understood before Firey's time, and was stated by Erich Zimmerman (1951:376), the noted resource economist: "all perennial culture, but particularly the planting of trees, rests on the stability of social institutions. No one would be foolish enough to spend a decade or more . . . to build up an olive grove which can bear fruit for a century unless he feels reasonably sure of a reward for himself and his descendants."

The record of international development efforts in forestry has convincingly documented the importance of stable institutional conditions for attaining sustainability. People in developing countries have not been willing to plant and tend long-maturing crops such as trees when the chances of realizing gains are diminished by unstable land tenure arrangements, inadequate control over fire and grazing, and an inability to enforce property rights in land or trees (Fortmann 1988). They have also abandoned centuries-old silvicultural systems when the ground rules of tenure, rights, and control were disrupted by unpredictable change (Chandler 1990).

In summary, sustainability is only possible if humans behave in ways that do not eliminate essential ecological options for future generations. Existing behavior may be institutionalized in nonsustainable patterns, such as road building on steep, unstable slopes. Or it may be rechanneled in ways that will provide incentives for people to learn how to adapt their behavior to ecological processes. But, most important, successful institutionalization of behavior results in people following routines and believing that these routines are morally right. There is far less need for coercion and formal social control when people voluntarily, or habitually, adhere to patterns of behavior (Berger and Neuhaus 1977).

Alternative Institutional Arrangements for Managing Watershed Ecosystems

Divisibility of Institutionalized Ecological Processes

Many of the problems in watershed management are fundamentally sociological in nature, since they involve issues of scale in the divisibility of institutionalized ecological processes. The contribution Freeman and Lowdermilk (1981) made to understanding the divisibility of irrigation technology provides a useful framework for understanding how ecological processes are institutionalized at different scales. An institutional pattern of behavior is divisible if it can be organized in both small and large spatial units. It is perfectly divisible if regulation of the ecological function is insensitive to scale. For example, the growth of individual trees or the building of residences is highly divisible because landowners with small plots as well as large tracts can plant and care for trees or build houses. By contrast, an indivisible ecological function is one in which there exists some spatial threshold below which it is not possible to regulate the ecological process. Regulation of atmospheric carbon is highly indivisible because it involves global cycling driven by atmospheric processes.

Divisibility is not always easily determined. Conventional patterns of social, economic, and political behavior affect the degree of divisibility in the regulation of ecological processes. Divisibility is relatively high in societies that have retained an autonomous ecological role for families and small communities (Padoch 1986), as contrasted with centralized command economies where state regulation has replaced localized decision-making authority (Chandler 1990). Divisibility may be low when centralization is essential for mobilizing the capital or social organizational requirements for resource development and utilization (Freeman and Lowdermilk 1981). A major irrigation project involving dams, aqueducts, and terracing of agricultural plots is a relatively indivisible agricultural system (Smith 1978). The functional necessity for larger scale in irrigation projects can be contrasted with the conventional structure of large-scale corporate silviculture in many industrialized and industrializing countries. Since corporations are conventional instruments for mobilizing capital, it is often assumed that large-scale corporate ownership is essential for capital-intensive silviculture. However, instruments for mobilizing capital are generally insensitive to scale—allowing small owners to be equally successful in making capital-intensive investments when political and economic conditions are suitable for small-scale investments.

The spatial organization of ecological processes can provide inflexible thresholds of divisibility. The discovery of ecological processes such as habitat requirements for animal species operating at a landscape scale demonstrates the importance of intermediate degrees of divisibility (Lee et al., this volume). Regional and global ecological processes are even less divisible. The management of an entire watershed is an indivisible process, even though the management of situated objects such as trees may be divisible (Franklin, this volume; Naiman et al., this volume). Many of the problems of watershed management, especially issues involving cumulative effects, originate in the difficulties of integrating divisible processes across an entire watershed landscape. The fact that this is fundamentally a problem of integrating institutional processes at different thresholds of divisibility has received little attention. But, as shall be discussed below, attempts to use large landownership units to facilitate integration at the watershed scale have not always met with success on either large public or private ownerships.

Private and Public Goods

Another important dimension of institutional arrangements for regulating ecological processes is the distinction between private and public (or collective) goods (Freeman and Lowdermilk 1981). A good is *private* if its benefits can be captured by the owners and denied to all other members of the community. A private good is one for which the investor as the owner has the incentive to invest because those who do not invest cannot derive benefits (there are no "free riders"). Timber production or home ownership are examples.

A good is considered to be *public* (or collective) if benefits cannot be denied to people who do not invest in producing it. For example, scenery and clean air are public goods because there are no convenient ways of excluding benefits to people who do not help bear the costs of creating scenic vistas or protecting air quality. Fish habitat enhancement and river system planning are examples of public goods at a watershed scale. A rational, calculating individual would choose not to share in costs of fishery enhancement or river basin planning if the benefits from these investments could be captured by others who do not pay for production costs. The easiest way of limiting access to a resource is to adopt institutional arrangements that make it possible to restrict benefits to those who contribute.

However, sustainability requires that future beneficiaries be considered when making management decisions. The most reliable way of eliciting commitments from those who are yet to be born is to ensure that institutions are stable (Firey 1963). Institutional stability can be ensured by honoring inherited institutions such as private property, rights to protection, and basic human rights. Rational investors will continue to make commitments that

Table 4.1. Alternative institutional arrangements in watershed management.

Type of Good	Divisibility of Space: High (small scale)	Moderate (medium scale)	Low (large scale)
	1	2	3
Private	Trees	Individual tree farm	Corporate tree farm
	Homesites	Subdivision	Private utility
	4	5	6
Public	Silt dam	Fisheries habitat enhancement	River basin planning
	Fishing and hunting access	Community resource management	National forest management

yield future benefits as long as they can be assured that essential institutions are stable (Zimmerman 1951).

Divisibility and the nature of goods can be combined in a table showing how they define institutional arrangements. Table 4.1 displays alternative institutional arrangements for regulating ecological processes in managed watersheds. Cell 1 in the table combines high spatial divisibility with private goods. Regulation of ecological processes occurs at a relatively small scale with all the advantages of a market for private goods. Economic incentives can efficiently channel human behavior because (1) users who do not pay can be excluded, (2) necessary credit and technical assistance are accessible, (3) more powerful members of a community cannot monopolize most of the available resources, and (4) sufficient numbers of owners are involved to provide the discipline of competition (Freeman and Lowdermilk 1981, Savas 1977).

Market systems have the advantage of regulating ecological processes in ways that can reduce most of the information flow pathologies. Competition forces accountability and self-monitoring, reducing the chances that a false analogy or ideological commitment will be perpetuated for long. The scale of operations is small and can encourage attention to detail. Longer-term ownership commitments (especially intergenerational institutions that guarantee or require inheritance of real property) can encourage longer observational periods. When ownership and management are combined, there can be fewer problems of managerial detachment and the assignment of responsibility for problems to someone else. Longer observational periods increase the chances that reactions to problems will not be out of phase.

However, land uses such as uncontrolled residential development or speculation in forest or agricultural land involving frequent turnover in ownership can result in reactions that are out of phase, short observation periods, greater detachment, and a tendency to leave the next owner with unsolved problems.

The main disadvantages of competitive market systems are the difficulty in disciplining actions that go beyond ownership boundaries and the emphasis on present benefits (which cause negative externalities) (Savas 1977). Small spatial scale and emphasis on the present can result in a failure to appropriately regulate ecological processes that are larger in scope or more extended in time than the concerns of a landowner. These are problems of institutional structure rather than information processing. One example of landowner behavior having adverse impacts beyond ownership boundaries is the disruption of the ecological functioning of riparian systems caused by erosion and transport of organic and inorganic materials; another is the cumulative impact of small-scale vegetation management on wildlife species that have landscape-scale habitat requirements (Lee et al., this volume). Loss of genetic diversity is an example of a problem originating in emphasis on present benefits.

Attempts to correct for these disadvantages have involved increasing the scale of ownership and ecological regulation, and also changing property rights to incorporate greater responsibility for respecting the rights of others, including the general public and future generations.

In cell 2 (see table), ecological regulation by larger ownership units can provide a scale of operations large enough to absorb some of the costs of meeting public responsibilities (environmental costs can be internalized). Individual tree farm owners generally hold enough land to dedicate some of it to watershed protection, windbreaks, wildlife habitat, and other ecological functions that benefit others in addition to themselves. Residential subdivisions are generally developed at a scale that allows internalization of costs associated with water supply systems, sewage treatment plants, careful road design and maintenance, and dedication of land to open space.

Yet these relatively small ownership units are still responsive to the competitive market conditions that can minimize most of the information processing pathologies. They are large enough to display environmental variation, thus affording opportunities for trial-and-error learning. But they are not so large that one person cannot get to know the land base in detail and monitor its responses to management practices over long periods. False analogies are more readily corrected when monitoring of land management practices does not produce the expected feedback. Cumulative learning can occur when managers have a long tenure. Reactions to ecological events can be phased more appropriately when owners are also managers and can be flexible in their responses to unpredictable or periodic events.

However, undesirable effects that extend beyond ownership boundaries and effects of emphasis on present benefits are still likely to be seen as someone else's problem. Other than institutional arrangements for securing public goods (to be discussed below), there are no effective mechanisms for limiting these effects.

There is a widely shared belief that when it comes to ecological processes, large-scale regulation is better regulation. The incentive to see the undesirable effects discussed above as someone else's problem can be reduced and sufficient capital can often be mobilized to invest in creating future as well as present benefits. For example, large industrial forest ownerships (see cell 3 in the table) have been justified on the basis of the impersonal, long-term commitments of corporate organizations and the availability of sufficient capital to ensure future benefits and protect the environment. Balanced against these potential advantages are several problems of information flow that accompany larger organizations.

Larger organizations generally require formalized decision processes that take the form of abstract rules. Such standard operating procedures increase the chance that information flow will be distorted by false analogies and inattention to detail (Schiff 1966). The separation of ownership from management increases managerial detachment. Personnel management in large organizations generally involves frequent transfers, thereby limiting the number of years a manager can dedicate to learning a particular land area. Lack of long-term observation can increase the chances of reacting out of phase.

Centralized organizational decisions may also be affected by short-term goals or ideological beliefs that result in a failure to learn how land responds best to treatments. Examples of short-range goals in private forest management include the rapid harvesting of timber to stave off the purchase of a publicly held corporation with high assets but moderate or low profits, or liquidation of assets to pay off short-term loans or bonds used to purchase lands with utilizable resources. Ideological commitments to the principle of private property can blind decision makers to necessary responsibilities of private ownership, including monitoring the effects of management practices.

Institutional arrangements for managing ecological processes providing public goods also range in scale (see cell 4 in the table). Where ecological processes are relatively divisible, subunits of communities provide regulation. Joint family enterprises or cooperatives are examples. The appropriate scale of social organization is determined by the necessity of ensuring that collectivities of beneficiaries will share in paying the costs of management. The cooperation of farmers in building and maintaining a check dam to keep debris from obstructing an irrigation system is an example of a relatively divisible process.

The fact that private owners cooperate to solve the problem of users benefiting without paying is an important feature of many small-scale collective enterprises. Such cooperation becomes increasingly difficult with the increasing size of ecological processes and increasing number of private owners (Savas 1977). Institutionalization of ecological processes, especially as it involves the internalization of conservation ethics, seems to work best at the level of relatively small social organizations where disciplining of behavior is regulated by personal interactions, personal identity, and pride in maintaining a reputation as a sound and respected local citizen (Korten and Klauss 1984). This is clearly illustrated by Smith's (1978) work on community regulation of irrigation systems in Japan.

Like divisible private ownership, such small-scale collective regulation can be effective in limiting most of the information flow pathologies. Cooperation can increase the efficiency with which people learn how to regulate ecological processes, since it can facilitate more effective exchange of information among managers and provide opportunities for accumulating a collective memory that can extend observations and better distinguish short- and long-term processes (Chandler 1990). There can still be serious problems of attributing problems to someone else, but collective responsibility can help overcome the human tendency to focus on short-term interests.

Cooperation among a small number of owners increases the chances that management activities will be institutionalized and ensures investors that they can depend on their kin, friends, or neighbors to make contributions in the future. Moreover, exercise of social controls by a local community can eliminate most of the detrimental effects generated by those who refuse to cooperate (Chandler 1990).

The tendency for people to make present commitments to future-referring values is perhaps most developed for collective enterprises at the scale of communities (Firey 1960, Smith 1978, Berger and Neuhaus 1977). Ecological processes of intermediate divisibility (cell 5 in the table) are large enough to encompass one or more watershed ecosystems. Most of the information processing pathologies can be limited where the participants in a process of collective governance are individual or family landowners. Cooperation can capture all the information processing advantages of a decentralized market system while also limiting the tendency to treat undesirable effects and future beneficiaries as someone else's problem.

The social structure of Japanese mountain villages illustrates how watersheds can be effectively regulated when individual ownership is coupled with community resource management. Land is owned by individual families, but decisions on land use are made by the community through local mechanisms of democratic governance. Decisions as to what crops are grown where and when are informed by shared ecological knowledge of the effects of slope, aspect, soil productivity, stability, and moisture, as well as other factors. Communities with prosperous economies and stable populations readily make long-term investments in growing sugi (*Cryptomeria japonica*), tea, fruit trees, or oak trees (as substrate for growing shitaki mushrooms) because they are assured of long-term institutional stability. Literacy, advanced technical training, and an educational infrastructure provide the capacity for accelerated learning of ecological processes. This has enabled some tree and rice farmers to achieve economic and community stability by diversifying resource production to include mushrooms, mountain vegetables, fruit trees, tourism, and miscellaneous value-added wood products. Communities dependent on the production of wood from Japanese national forests exhibited far less resiliency, since their options for land use were limited to the collective national values of wood production, watershed protection, and forest preservation (Lee, unpublished data).

Large-scale collective ownership or regulation of ecological processes (see cell 6 in the table) has long been assumed to be the best institutional arrangement for managing large river basins or watershed ecosystems (Selznick 1949). The United States, Canada, and many other nations have retained substantial areas of undeveloped land in public ownership with the objective of providing public goods in both the present and the future. Retention of these lands in collective ownership has succeeded in providing significant options for contemporary resource use and allocation. Whether they will continue to be as successful in providing public goods is uncertain.

Regardless of these benefits, the large public organizations used to manage extensive ecosystems suffer from the same problems of information flow as large private organizations. When highly bureaucratized, such as the U.S. Forest Service (Kaufman 1960), large organizations can suffer from information flow problems as great, or even greater, than large private organizations.

Reliance on abstract decision-rules, handbooks, and frequent transfer of personnel may develop a manager's mind to the point where it embodies the organization and ensures conformity (Kaufman 1960). But such an "organizational mind" increases the chances that a manager will use false analogies, rely on overgeneralized models, have short observation periods, experience managerial detachment, exhibit out-of-phase reactions, and, often quite appropriately, make attributions of cause for management problems to someone else's decisions.

Large organizations also suffer from a tendency to rely on a sense of mission defined by ideological beliefs (Twight 1983). For almost 70 years, public land management organizations were ideologically committed to eliminating all fires from wildland ecosystems in the United States (Schiff 1962). A more generalized ideological commitment to economic and ecological stability also led public land management agencies to ignore the importance of disturbances and spatial variation in ecological systems (Schiff 1966). A post-World War II commitment to timber production on national forest lands appears to have been motivated by a similar ideological belief in the primacy of wood production (Clary 1986). The Forest Service is now struggling to chart a future that will deemphasize wood production and embrace the primacy of "ecological values" (Franklin 1989; Franklin, this volume). Time will tell whether the Forest Service will simply trade one ideology for another by adopting a commitment to promoting "ecological values," with the consequent limitations on its ability to develop an accurate cognitive map that includes humans as a component of ecosystems.

Conclusions

Since humans are an integral part of ecological systems, watershed management cannot achieve ecosystem sustainability without addressing the problems of human organization. As has been shown by sociological studies for a wide variety of resource management problems, knowledge about ecologically effective forms of human organization is as important as knowledge of biology or hydrology. Yet the science of human organization in ecological systems is far less developed.

This article has coupled two social science approaches that are essential for advancing our knowledge about watershed ecosystem management. Cognitive anthropology can help us understand how people learn to manage complex ecological processes because it provides us with the means for studying how people develop accurate cognitive maps of their environments. An institutional approach to sociology can enable us to understand how and why people will conserve options for future generations when individual rationality would lead them to get as much as they could in the present—and thus to understand why it is not possible to maintain ecological legacies without also maintaining cultural and institutional legacies.

This synthesis of institutional and cognitive analysis yields promising opportunities for future research. To begin with, the generalizations summarized in Table 4.1 need to be reformulated as hypotheses and challenged by empirical research. The most interesting hypotheses involve the possibility that (1) small-scale institutions for regulating ecological processes may have a better capacity than large-scale organizations to overcome information flow pathologies and (2) a hierarchical system of regulation involving local communities as the primary collective governance units may be the most efficient and effective means for institutionalizing sustainable ecological processes, because an ecological identity and conscience are more likely to be products of community life than of regional or national collectivities (Korten and Klauss 1984, Berger and Neuhaus 1977).

The promise of sustainable development embodied in the report of the World Commission on Environment and Development (1987) cannot be realized unless individual initiative is harnessed to serve the purposes of flexible and adaptive management of ecological processes. We have much to learn about the institutional conditions that are best suited to rapid, adaptive environmental learning. The literature summarized in this chapter suggests that large government as well as large private land management organizations may be poorly suited for adaptive environmental learning. Both appear to have been sources of social, economic, and ecological instability. Further study must determine whether smaller and more flexible institutional units are better suited for the rapid, adaptive learning that will be necessary to achieve sustainability, or whether there are ways of restructuring large organizations to serve this purpose.

What we do know with certainty is that sustainable watershed management begins by building ecologically effective human organizations. This fact alone must stand as a centerpiece of a new perspective on watershed management.

Acknowledgments. The author thanks Dr. Paul Chandler, Dr. Robert J. Naiman, and two unidentified reviewers for their constructive comments on this article. Dr. Jerry Franklin provided valuable assistance with the development of ideas about the scale of ecological regulation.

References

Bennett, J.W. 1976. The ecological transition: cultural anthropology and human adaptation. Pergamon Press, New York, New York, USA.

Berger, P.L., and T. Luckmann. 1966. The social construction of reality. Doubleday, New York, New York, USA.

Berger, P.L., and R.J. Neuhaus. 1977. To empower people: the role of mediating structures in public policy. American Enterprise Institute, Washington, D.C., USA.

Blahna, D.J. 1990. Social bases for resource conflicts in areas of reverse migration. Pages 159–178 *in* R.G. Lee, D.R. Field, and W.R. Burch, Jr., editors. Community and forestry: continuities in the sociology of natural resources. Westview Press, Boulder, Colorado, USA.

Bonnicksen, T.M., and R.G. Lee. 1982. Biosocial systems analysis: an approach for assessing the consequences of resource policies. Journal of Environmental Management **15**:47–61.

Bradley, G.A., editor. 1984. Land use and forest resources in a changing environment. University of Washington Press, Seattle, Washington, USA.

Burch, W.R., Jr. 1971. Daydreams and nightmares: a sociological essay on the American environment. Harper and Row, New York, New York, USA.

Burch, W.R., Jr. 1984. Much ado about nothing—some reflections on the wider and wilder implications of social carrying capacity. Leisure Sciences **6**:487–496.

Burch, W.R., Jr., and D.R. DeLuca. 1984. Measuring the social impact of natural resource policies. University of New Mexico Press, Albuquerque, New Mexico, USA.

Carroll, M.S. 1988. A tale of two rivers: comparing NPS-local interactions in two areas. Society and Natural Resources **1**:317–333.

Chandler, P. 1990. Ecological knowledge in a traditional agroforest management system among peasants in China. Dissertation. University of Washington, Seattle, Washington, USA.

Clary, D. 1986. Timber and the Forest Service. University of Kansas Press, Lawrence, Kansas, USA.

Dixon, J.A., and L.A. Fallon. 1989. The concept of sustainability: origins, extensions and usefulness for policy. Society and Natural Resources **2**:73–84.

Dubos, René. 1976. Symbiosis between earth and humankind. Science **193**:459–462.

Field, D.R., and W.R. Burch, Jr. 1990. Rural sociology and the environment. Social Ecology Press, Madison, Wisconsin, USA.

Finsterbusch, K., and C.P. Wolf, editors. 1980. Methodology of social impact assessment. Hutchinson Ross, Stroudsburg, Pennsylvania, USA.

Firey, W. 1960. Man, mind and land: a theory of resource use. Free Press, Glencoe, Illinois, USA.

Firey, W. 1963. Conditions for the realization of values remote in time. Pages 147–159 *in* E. A. Tiryakian, editor. Sociological theory, values, and sociocultural change: essays in honor of Pitirim A. Sorokin. Free Press, Glencoe, Illinois, USA.

Firey, W. 1990. Some contributions of sociology to the study of natural resources. Pages 15–26 *in* R.G. Lee, D.R. Field, and W.R. Burch, Jr., editors. Community and forestry: continuities in the sociology of natural resources. Westview Press, Boulder, Colorado, USA.

Fortmann, L. 1988. Whose trees? Property dimensions of forestry. Westview Press, Boulder, Colorado, USA.

Fortmann, L., and P. Starrs. 1990. Power plants and resource rights. Pages 179–194 *in* R.G. Lee, D.R. Field, and W.R. Burch, Jr., editors. Community and forestry: continuities in the sociology of natural resources. Westview Press, Boulder, Colorado, USA.

Franklin, J. 1989. Toward a new forestry. American Forests, November-December: 37–44.

Freeman, D.M., and M.K. Lowdermilk. 1981. Sociological analysis of irrigation water management: a perspective and approach to assist decision making. Pages 153–173 *in* C.S. Russell and N.K. Nicholson, editors. Public choice and rural development. Resources for the Future Research Paper R-21. Johns Hopkins University Press, Baltimore, Maryland, USA.

Kaufman, H. 1960. The forest ranger: a study in administrative behavior. Johns Hopkins University Press, Baltimore, Maryland, USA.

Kimmins, H. 1987. Forest ecology. Macmillan, New York, New York, USA.

Klausner, S. 1971. On man and his environment. Jossey-Bass, San Francisco, California, USA.

Korten, D.C., and R. Klauss. 1984. People-centered development: contributions toward theory and planning frameworks. Kumarian Press, West Hartford, Connecticut, USA.

Lee, R.G. 1991. Scholarship versus technical legitimation: avoiding politicization of forest science. *In* R.A. Leary, editor. Proceedings of Philosophy and Methods of Forest Research, XIX Congress of the International Union of Forestry Research Organizations, Montreal, Canada, August 7, 1990.

Lee, R.G., D.R. Field, and W.R. Burch, editors. 1990. Community and forestry: continuities in the sociology of natural resources. Westview Press, Boulder, Colorado, USA.

Machlis, G.E., and J.E. Force. 1988. Community stability and timber-dependent communities: future research. Rural Sociology **53**:220–234.

McGovern, T.H., Bigelow, T. Amorosi, and D. Russell. 1988. Northern Islands, human error, and environmental degradation: a view of social and ecological change in the medieval North Atlantic. Human Ecology **16**:225–270.

Miller, M.L., R.P. Gale, and P.J. Brown, editors. 1987. Social science in natural resource management systems. Westview Press, Boulder, Colorado, USA.

Moore, S.D., and S.K. Brickler. 1987. A planning approach to social carrying capacity research for Aravaipa Canyon Wilderness, Arizona. Pages 167–180 *in* M.L. Miller, R.P. Gale, and P.J. Brown, editors. Social science in natural resource management systems. Westview Press, Boulder, Colorado, USA.

Norgaard, R. 1988. Sustainable development: a co-evolutionary view. Futures, December: 606–620.

Padoch, C. 1986: Agricultural site selection among permanent field farmers: an example from East Kalimantan, Indonesia. Journal of Ethnobiology **6**:279–288.

Perlin, J. 1989. A forest journey: the role of wood in the development of civilization. Norton, New York, New York, USA.

Savas, E.S., editor. 1977. Alternatives for delivering public services: toward improved performance. Westview Press, Boulder, Colorado, USA.

Schiff, A.L. 1962. Fire and water: scientific heresy in the U.S. Forest Service. Harvard University Press, Cambridge, Massachusetts, USA.
Schiff, A.L. 1966. Innovation and administrative decision making: the conservation of land resources. Administrative Science Quarterly **11**:1–30.
Selznick, P. 1949. TVA and the grass roots: a study in sociology of the formal organization. University of California Press, Berkeley, California, USA.
Smith, R.J. 1978. Kurusu: the price of progress in a Japanese village, 1951–1975. Stanford University Press, Stanford, California, USA.
Stankey, G.H., D.N. Cole, R.C. Lucas, M.E. Peterson, and S.F. Frissell. 1985. The Limits of Acceptable Change (LAC) system for wilderness planning. United States Forest Service General Technical Report INT-176, Intermountain Forest and Range Experiment Station, Ogden, Utah, USA.
Thomas, W.L., Jr., editor. 1956. Man's role in changing the face of the earth. Volume 1. University of Chicago Press, Chicago, Illinois, USA.
Twight, B.W. 1983. Organizational values and political power: the Forest Service versus the Olympic National Park. Pennsylvania State University, State College, Pennsylvania, USA.
Vayda, A.P. 1977. An ecological approach in cultural anthropology. Pages 3–8 *in* W.R. Burch, Jr., editor. Readings in ecology, energy and human society: contemporary perspectives. Harper and Row, New York, New York, USA.
Wenner, L.N. 1987. The practice and promise of social science in the U.S. Forest Service. Pages 63–83 *in* M.L. Miller, R.P. Gale, and P.J. Brown, editors. Social science in natural resource management systems. Westview Press, Boulder, Colorado, USA.
Wondolluck, J.M. 1988. Public lands and conflict resolution: managing national forest dispute resolution. Plenum, New York, New York, USA.
World Commission on Environment and Development. 1987. Our common future. Oxford University Press, Oxford, England.
Zimmerman, E. W. 1951. World resources and industries. Revised edition. Harper and Row, New York, New York, USA.

5

Management of Aquatic Resources in Large Catchments: Recognizing Interactions Between Ecosystem Connectivity and Environmental Disturbance

J.A. STANFORD AND J.V. WARD

Abstract

Management within catchment basins must be approached with an empirically based understanding of the natural connectivity and variability of structural and functional properties of riverine ecosystems. Rivers are four-dimensional environments involving processes that connect upstream-downstream, channel-hyporheic (groundwater), and channel-floodplain (riparian) zones or patches, and these differ temporally. Natural and human disturbances, including biotic feedback (such as predation, parasitism, and other food web dynamics), interact to determine the most probable biophysical state of the catchment ecosystem. Human disturbances can be quantitatively determined by deviations from an observed biophysical state (baseline), but usually this requires long-term ecological data sets. A case history of the Flathead River-Lake system in Montana (USA) and British Columbia (Canada) is summarized to illustrate how disturbances interact at the catchment level of organization. Owing to the natural complexities of catchment ecosystems and the cumulative effects of human disturbances, the rationale and logistics of obtaining long-term data often seem intractable and excessively expensive. The naive alternative is to derive and implement simplistic procedures that are agency specific and often result in management actions that interfere with each other. We argue that integrated management at the catchment level is needed and propose some simple principles, beginning with broader based collegiate training for prospective managers.

Key words. Ecosystem, river, catchment, drainage basin, management, disturbance, natural resources, watershed, Flathead River, Montana.

Introduction

Professor Noel Hynes first synthesized the concept of ecological connectivity in the context of river systems in his Baldi Lecture at the 19th Congress of the International Society for Pure and Applied Limnology (Hynes 1975). He eloquently described how rivers are a manifestation of the biogeochemical nature of the valleys they drain, and he proposed that understanding the inherent connectivity between terrestrial and lotic biotopes would lead to important predictions about the future structure and function of river ecosystems.

In the nearly two decades since that seminal lecture, several paradigms (reviewed by Cummins et al. 1984) emerged from scores of studies that examined spatial and temporal aspects of geomorphic, hydrologic, thermal, and riparian influences on biotic attributes (e.g., diversity, zonation, food web associations, bioproduction) of rivers. The river continuum concept (Vannote et al. 1980, Minshall et al. 1985) provided a template for examining how biotic attributes of rivers change within the longitudinal gradient from headwaters to ocean confluence. The serial discontinuity concept (Ward and Stanford 1983*a*) provided a construct for the propensity of rivers to predictably reset biophysical attributes in relation to distance downstream from on-channel impoundments. Comparison of organic matter budgets in streams in different biomes provided the basis for the riparian control concept and demonstrated the extreme importance of allochthonous debris (wood and leaves) in lotic systems (Cummins et al. 1984, Harmon et al. 1986, Webster and Benfield 1986, Ward et al. 1990, Gregory et al. 1991). The nutrient spiraling concept (Webster and Patten 1979, Newbold et al. 1983) led to an understanding of how plant growth nutrients are transformed from dissolved to particulate states during translocation from upstream to downstream reaches. Lastly, the ecotone concept (Naiman and Décamps 1990, Holland et al. 1991) has fostered greater understanding of the extreme importance and potential predictive power related to transformations and fluxes of materials that occur within boundaries between functionally interconnected patches that form the riverine landscape. In many ways the ecotone concept integrates the other paradigms by emphasizing the functional connectivity inherent in all ecosystems.

Studies articulating these paradigms and other syntheses of stream ecology (Lock and Williams 1981, Barnes and Minshall 1983, Dodge 1989, Stanford and Covich 1988, Yount and Niemi 1990), plus a great number of other research projects, have largely verified Hynes's proposition that the streams are tightly coupled with catchment characteristics. Drainage basins or catchments (i.e., the river valley in Hynes's context) may indeed be characterized as ecosystems composed of a mosaic of terrestrial "patches" (Pickett and White 1985) that are connected (drained) by a network of streams. Of course, the lotic environment itself is a smaller scale patchwork or mosaic of habitats in which materials and energy are transferred (connected) through dynamic, biodiverse food webs. In most catchments, on-channel lakes and floodplain aquifers dramatically increase the complexity of the ecosystem, in contrast to the contemporary view of rivers as dynamic channels bounded by a riparian corridor (Sedell et al. 1989).

In this chapter we discuss the catchment in ecosystem terms (Lotspeich 1980, Naiman and Sedell 1981), stressing the ecological coupling that characterizes aquatic components of catchments, and discuss natural and human disturbances that influence biophysical connectivity. We describe how management actions can work at cross purposes when the interactions of natural and human disturbances are not considered from a catchment ecosystem viewpoint, and we discuss the difficulties of assessing cumulative effects of human perturbations. We use the Flathead River (British Columbia, Montana) as an example of a large river ecosystem influenced or partly uncoupled by a myriad of anthropogenic effects and competing management bureaucracies and interests. Finally, we propose an alternative general approach to natural resource management—an approach that begins with revised college curricula for training resource managers as conservators of ecological connectivity in river ecosystems.

Habitat Dimensions, Ecological Connectivity, and Natural Disturbance within River Ecosystems

In the United States, the term *watershed* is often misused in the context of river basin research and management. By proper definition, the watershed is the ridgeline or elevation contour that delimits drainage basins or catchments. The catchment is bounded by the watershed, and since water flows downstream from the watershed through the catchment, thereby integrating influences of natural and human disturbances within the catchment, we use the watershed as the natural ecosystem boundary.

Obviously, in these terms an ecosystem may be very small, such as a first-order catchment (sensu Strahler 1957), or it may be very large, encompassing entire river systems (e.g., the 671,000 km^2 catchment of the Columbia River, USA). Choice of ecosystem dimension (i.e., catchment size) is logically determined by the question being examined or the resource being managed.

The time frame encompassing the research question or management problem is of course also important. In geologic time, as a result of orogeny and erosion, watersheds were bisected and catchments reorganized, clearly having enormous zoogeographic consequences (Stanford and Ward 1986). In a much shorter time frame, engineering projects artificially connected catchments via transwatershed diversions of rivers in many areas (Stanford and Ward 1979, Davies and Walker 1986), allowing differently adapted organisms to commingle (Guiver 1976) or greatly accelerating immigration of

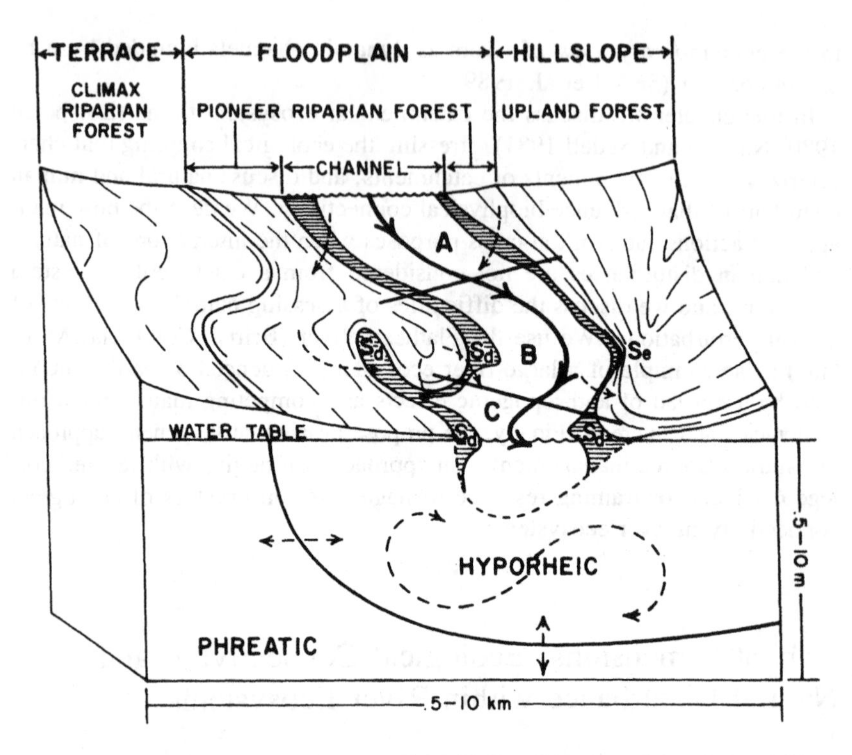

FIGURE 5.1. Major landscape features of the Kalispell Valley of the Flathead River, Montana, USA, showing the three primary spatial dimensions (lateral, longitudinal or altitudinal, and vertical) which are dynamically molded through time (the fourth dimension) by fluvial processes. Biota may reside in all three spatial dimensions: riparos (streamside or riparian), benthos (channel), hyporheos (river-influenced groundwater), and phreatos (true groundwater). The hatched area is the varial zone, or the area of the channel that is periodically dewatered as a consequence of the average amplitude of the discharge regime. Major channel features include a run (A), riffle (B), and pool (C); Sd refers to sites of sediment deposition and Se refers to a major site of bank erosion. The heavy solid line is the thalweg, and broken lines conceptualize circulation of water between benthic, hyporheic, and phreatic habitats.

nonnative biota introduced by other means (Stanford and Ward 1986, Mooney and Drake 1986).

Given that catchments may be referred to as ecosystems and that the ecosystem is dynamic in time and space as well as in its relation to environmental problem solving, it is fundamentally important to recognize the major structural features and dimensions of river ecosystems (Figure 5.1). Ecologists have appreciated for many years the importance of microhabitats encompassed by the run-riffle-pool sequence as influencing the distribution and abundance of biota within the river channel. Zonation of biota within the longitudinal continuum has long been recognized as a fundamental feature of the lotic environment (Hynes 1970), although explanations of specific distribution patterns often remain contentious (Alstad 1982, 1986; Thorp et al. 1986). Within the last decade, the connection between riparian zones, including the surficial floodplain dynamics, and ecological structure and function has been clearly demonstrated (see reviews in Décamps and Naiman 1989, Dodge 1989, Gregory et al. 1991). The importance of microbial transformation and transport of solutes in groundwaters has been shown in relation to plant growth nutrients for channel biotopes in streams (Stanford and Ward 1988, Ford and Naiman 1989, Dahm et al. 1991, Stream Solute Workshop 1990, Grimm et al. 1991, Valett et al. 1991); and penetration of groundwaters (i.e., the hyporheic zone, Figure 5.1) by amphibiotic stream biota has been documented (Schwoerbel 1967, Stanford and Gaufin 1974, Williams and Hynes 1974, Bretschko 1981, Danielopol 1984, Pugsley and Hynes 1986, Stanford and Ward 1988). But the presence of large-scale hyporheic zones, and the critical importance of groundwater – surface water interchange as a major landscape feature of catchments, have only recently been demonstrated (Stanford and Ward 1988, Danielopol 1989, Gibert et al. 1990).

River floodplains are often, if not always, penetrated by interstitial, subterranean flow (Figure 5.2). Water penetrates (downwells) at the upstream end of the floodplain, flows through unconfined aquifers at rates determined by the porosity of the substrata and the slope of the floodplain, and eventually upwells to the surface some distance downslope. Location of aquifer discharge is often related to bedrock outcrops or encroaching canyon walls (knickpoints in Figure 5.2). Effluent groundwaters may enter the channel directly or emerge as floodplain springbrooks that exhibit seasonally dynamic hydrology controlled by flow entering the floodplain from the river and from tributaries. These springbrooks usually occur in abandoned meander channels blocked at the upstream end by natural deposition of alluvium and woody debris. They have been referred to as wall-base channels in locations where they erupt from the substratum of old channels originally constrained by contact with the terrace or canyon walls (Peterson and Reid 1984). However, variations on this general theme may occur, depending on floodplain geomorphology and catchment hydrology (Amoros et al. 1982). Since spates frequently may overflow these springbrooks (in the Flathead River, Montana, these systems are flooded on about a ten-year return frequency; J. Stanford et al., unpublished), woody debris often accumulates, providing structurally complex lotic habitat. Moreover, relative to the main river channel, these springbrooks are characterized by fairly stable flows, moderated temperature regimes, high water clarity, and elevated concentrations of plant growth nutrients, particularly nitrate and soluble reactive phosphorus. As a result, standing crops of attached algae and zoobenthos can exceed biomass in the channel by several orders of magnitude. Juveniles of native cutthroat trout (*Oncorhynchus clarkii*) are abundant (J. Stanford et al., unpublished).

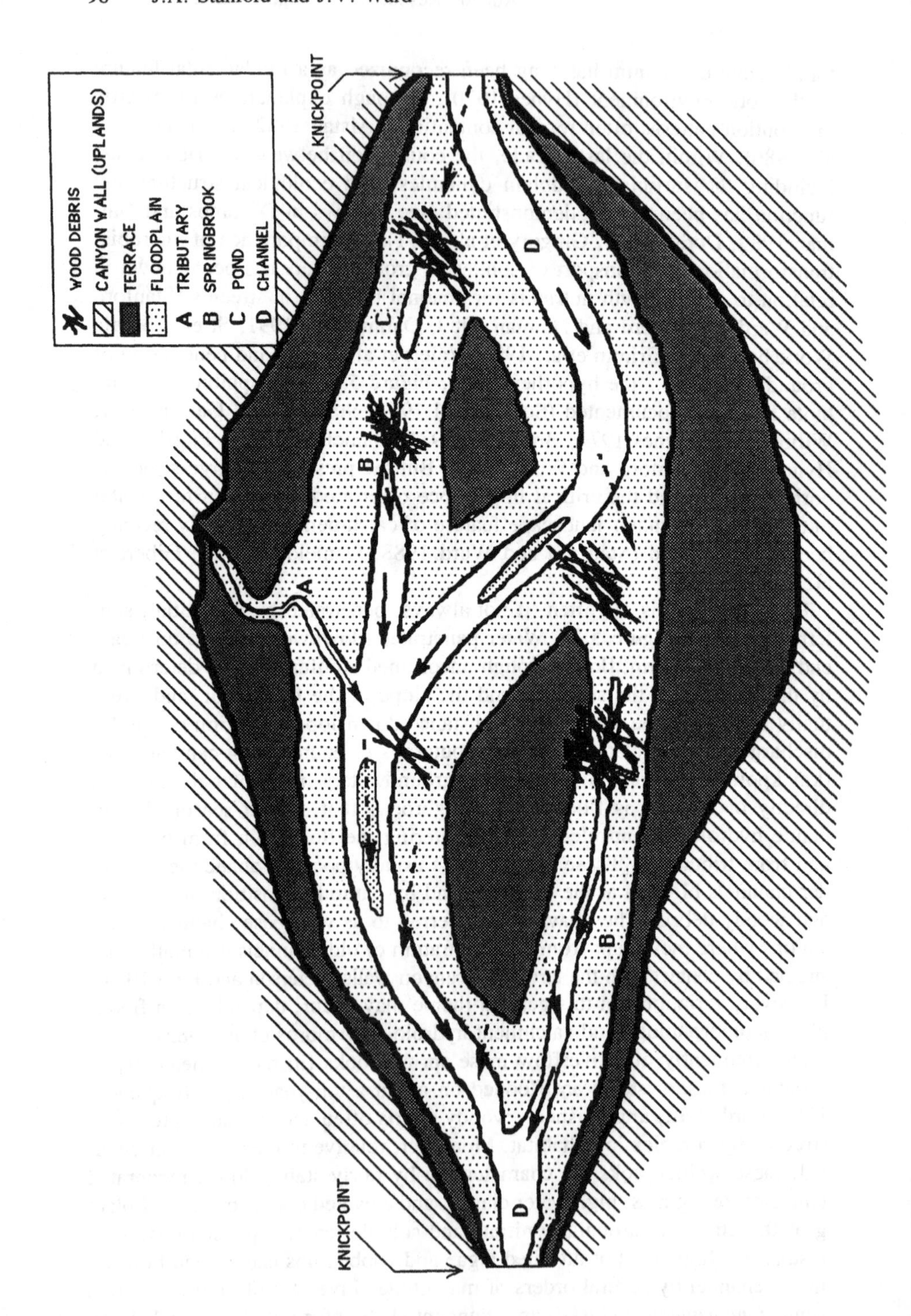
WOOD DEBRIS
CANYON WALL (UPLANDS)
TERRACE
FLOODPLAIN
A TRIBUTARY
B SPRINGBROOK
C POND
D CHANNEL
KNICKPOINT
KNICKPOINT
A
B
B
C
D
D

Therefore, it appears that these springbrooks are "hot spots" of bioproduction, although this relation has yet to be thoroughly documented.

Wall-base streams are known to be critically important as spawning and rearing habitats for salmonids in Pacific Northwest streams (Peterson and Reid 1984); and recent analyses suggest that aggraded floodplains and upwelling groundwaters historically were key production areas for anadromous salmonids (*Oncorhynchus* spp.) and resident bull charr (*Salvelinus confluentus*) in the Columbia River system (James Sedell, U.S. Forest Service, pers. comm.). In the Flathead River, Montana, native bull charr adults migrate upstream from Flathead Lake to spawn in specific habitats of fourth-order tributaries (Figure 5.3; see also Fraley and Shepard 1989). Juveniles remain in riverine habitats for two or three years before migrating downstream to Flathead Lake, where they mature. This phenology is termed *adfluvial*. Primary bull charr spawning sites are the groundwater upwelling zones of aggraded floodplain segments, which usually occur downstream from major altitudinal transitions (knickpoints) in the river continuum. Bull charr select only fourth-order streams that are not regulated by on-channel lakes, apparently in response to temperature criteria (J. Stanford, unpublished).

These observations emphasize that the riverine ecosystems are truly four dimensional, with longitudinal (upstream-downstream), lateral (floodplain-uplands), and vertical (hyporheic-phreatic) dimensions (Figure 5.1); since these spatial dimensions are transient or dynamic over time as a consequence of relativity, temporality is the fourth dimension (Ward 1989). Within a given stream reach, distribution and abundance of organisms form a multivariate function of the structural and functional attributes of channel (fluvial), riparian (floodplain, shoreline), and hyporheic (groundwater) habitats as they interact within time and space with the geomorphology and hydrology of the catchment. Clearly, catchments may be characterized as patch-dynamic systems (Pringle et al. 1988, Townsend 1989), and ecological connectivity of patches is a fundamental feature.

Many riverine organisms may traverse all three spatial dimensions in the process of completing life cycles (high connectivity), whereas others may be relatively stationary (low connectivity). For example, in the Flathead River, Montana, a gravel-bottom system with expansive intermontane floodplains characterized by substantial interstitial flow (Figure 5.2), certain specialized stoneflies (Insecta: Plecoptera) reside within floodplain groundwaters during the entire larval stage. Indeed, hundreds of these crepuscular stoneflies have been collected in single samples taken from groundwater monitoring wells 2–3 km from the river channel, demonstrating the enormous volume of the

FIGURE 5.2. Simplified plan view of an intermontane floodplain of a gravel-bed river on the Middle Fork of the Flathead River, Montana, USA. The floodplain is formed on the aggraded slope between bedrock constrictions (knickpoints) of the river channel. Riparian forests are well developed (mature) on the terrace, intergrade into upland forests, and are in various successional stages on the floodplain.

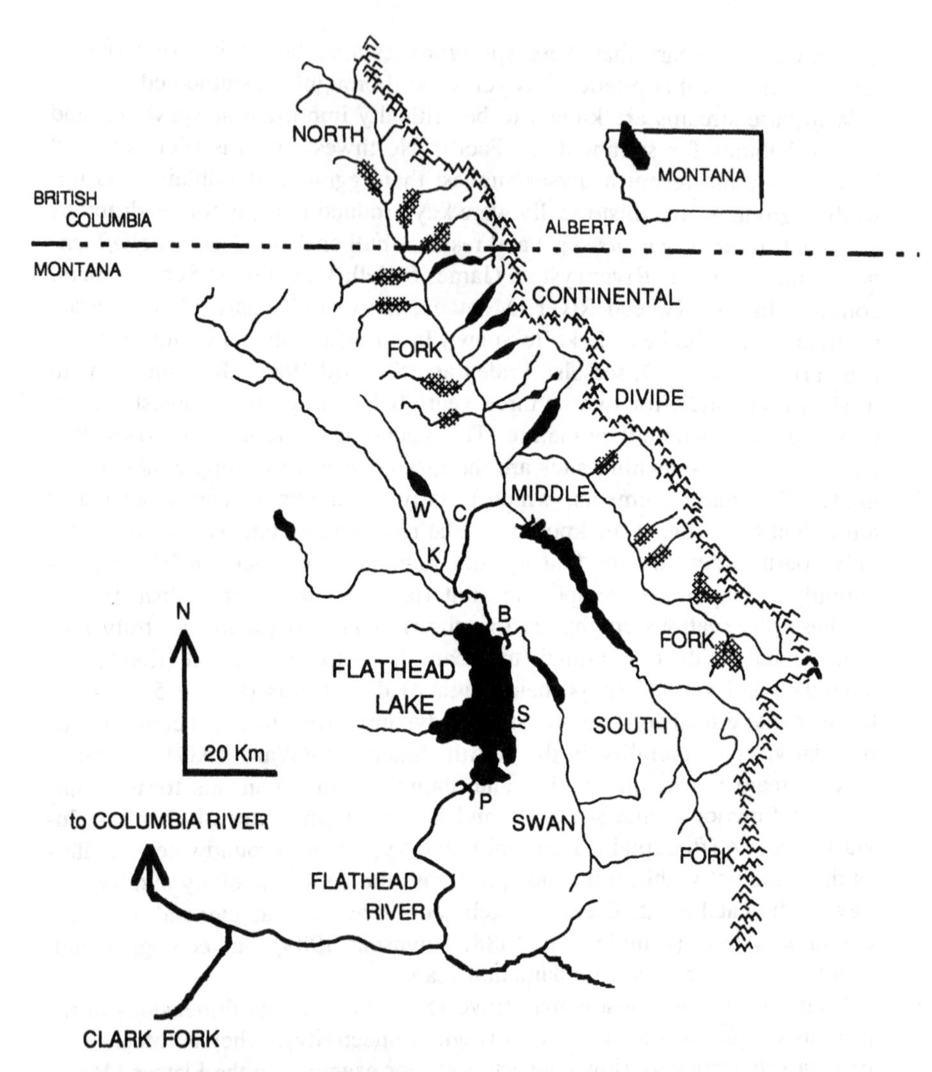

FIGURE 5.3. The Flathead River catchment basin in Montana, USA, and British Columbia, Canada. Primary spawning sites for adfluvial bull charr (*Salvelinus confluentus*) from Flathead Lake are shown on tributary creeks of the North and Middle Forks by cross hatching. Towns include Kalispell (K), Whitefish (W), Columbia Falls (C), Bigfork (B), and Polson (P). The Flathead Lake Biological Station (S) is on the east shore of the lake. The hydroelectric dam on the Swan River near Bigfork is a small run-of-the-river facility, whereas the other two dams in the system are much larger. See text for further explanation.

hyporheic zone in this river. They are the top consumers in a speciose (80+ species) groundwater food web. Yet these stoneflies emerge as winged adults from the river channel and fly into the riparian vegetation to mate and produce eggs. The eggs are deposited in the river channel, followed by larval immigration into the hyporheic zone (Stanford and Ward 1988). Many other riverine insects, which commonly characterize the rhithron (cold, swift-flowing, gravel-cobble substratum) habitat of western USA rivers, also depend on riparian vegetation during the flight period, but the larval stage is completed within the channel. Most noninsect zoobenthos and periphyton (attached algae) are essentially obligate channel inhabitants, although they, like most fish species and insect larvae, are often distinctly segregated by temperature, flow, substratum, or behavioral criteria within the altitudinal gradient of the stream continuum (e.g., bull charr distribution in Figure 5.3; see also Resh and Rosenberg 1984, Matthews and Heins 1987).

Biodiversity and bioproduction in rivers are related to a plethora of factors that interact bioenergetically (Figure 5.4) to determine reproductive success of individuals (e.g., the P and C compartments of Figure 5.4) attempting to coexist. Phenologies (life histories) are highly evolved and sensitive to environmental change. Consequently, disturbance events (e.g., floods, droughts, fires, disease epidemics, invasions by exotic species) reduce reproductive success and, hence, bioproduction; thus connectivity of lotic food webs is naturally decreased (Figure 5.5). Our main point is that for a particular species to survive, either as a resident of the catchment or as an immigrant, enough individuals must realize a net energy gain to meet phenological requirements which permit conservation of the gene pool (i.e., net positive contribution to riverine bioproduction). Bioproduction at the ecosystem level of organization is controlled by the same plethora of environmental factors; although in the case of riverine fishes, especially anadromous species, harvest by humans often is more pervasive than other environmental disturbances.

The degree of structural (Figures 5.1 and 5.2) and functional (e.g., flux of organic and inorganic materials and energy between consumer groups, Figure 5.4) connectivity determines the most probable biophysical state of the ecosystem at any given time. For many scientists this implies that tightly coupled ecosystems are highly evolved, undisturbed, and essentially in equilibrium. However, circumspection of equilibrium concepts is waning in contemporary ecology (Murdoch 1991), owing to the realization that natural and anthropogenic disturbances occur too frequently in most catchments to allow equilibrium models at any level of organization to be realistic (Resh et al. 1988; Naiman et al., this volume). Disturbance events alter structural and functional connectivity (Figure 5.5); the instantaneous biophysical status of ecosystems is usually more analogous to a quasi-equilibrium (sensu Schumm and Lichty 1956; see also Huston 1979).

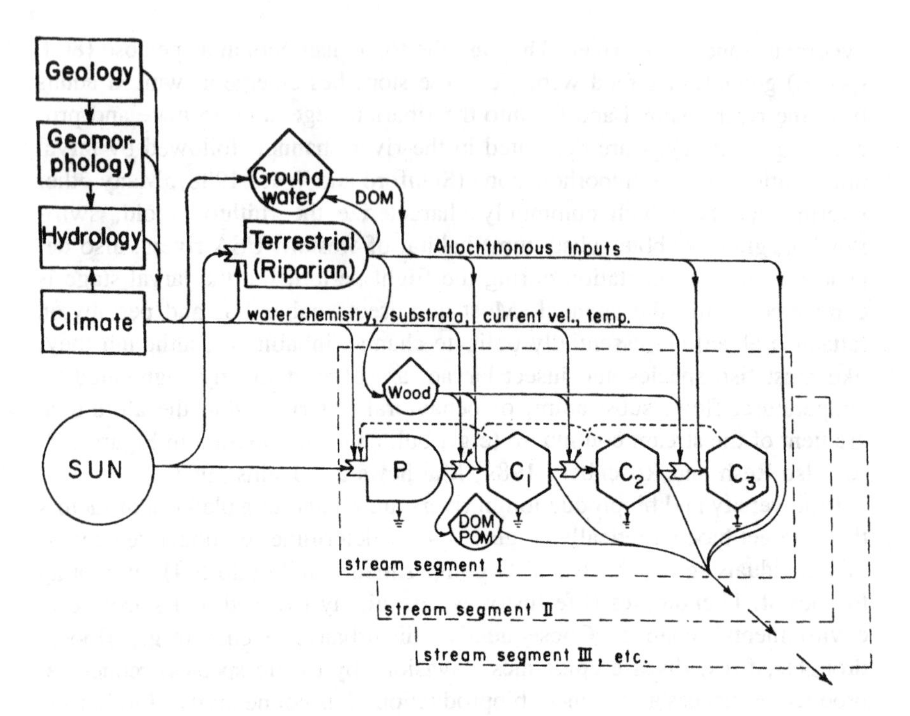

FIGURE 5.4. Energetics of successive segments of a stream ecosystem (from Benke et al. 1988). Solar energy provides energy for primary production in both the terrestrial ecosystem and the stream. Climate, geology, geomorphology, and hydrology have interdependencies, and all directly affect both the terrestrial and stream systems. The terrestrial system, with indirect input through the groundwater, provides allochthonous resources for the stream consumers, including important substrata (wood) and food (leaf litter, DOM, organisms). P = primary producer module. C_1, C_2, C_3 = consumer modules. Symbols after Odum (1983). Solid arrows are energy flows or energy regulators. Dashed lines are biotic feedback regulators.

Human Disturbances and Loss of Ecological Connectivity

How much disturbance can occur in a catchment before ecosystem resilience (i.e., the ability to recover from disturbance, Odum et al. 1979) is exceeded and ecosystem structure and function are permanently altered (Yount and Niemi 1990)? How much is attributable to natural interannual variation? That such questions were articulated years ago but remain largely unanswered is, of course, problematic for researchers and especially for managers attempting cumulative impact assessments at the catchment level.

We have argued (Ward and Stanford 1983*b*) that natural interannual variation in catchments is encompassed by Connell's (1978) intermediate disturbance hypothesis. Connell suggested that biodiversity is maximized by

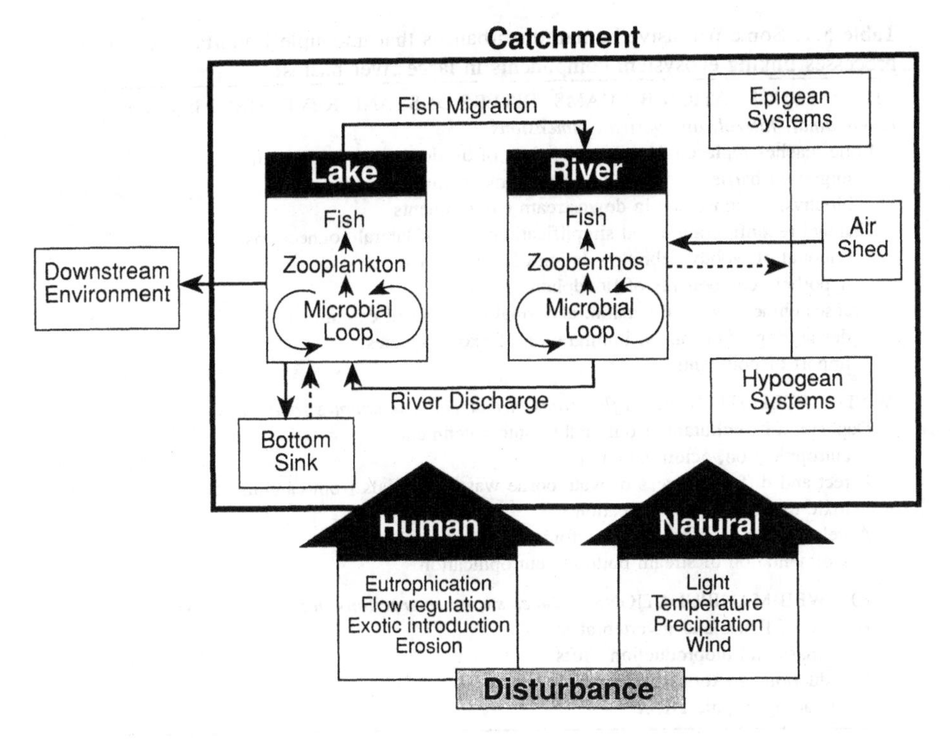

FIGURE 5.5. Ecological connectivity of the Flathead River-Lake ecosystem.

ecosystems that are "adapted" to disturbance events of intermediate intensity and duration. Intermediate might be loosely quantified in catchment terms as less than a 100 year flood event and more dynamic than constant flow, for example, from a spring or a storage reservoir. In other words, it is intuitive that a most probable state of quasi-equilibrium can be maintained by natural, intermediate disturbances until the occurrence of a major disturbance event on the scale of a volcanic eruption or hurricane. Events of that magnitude can completely restructure ecosystems. However, recovery is more rapid than once thought (e.g., recovery of streams following the 1980 eruption of Mount St. Helens in the Cascade Range, USA, is occurring decades sooner than expected).

In many ways the idea of natural disturbance controls on stream ecosystem structure and function—however intuitive—remains hypothetical for lack of long-term data to test inferences. Indeed, the National Science Foundation decided nearly a decade ago to support long-term ecological research (LTER) at a variety of sites in different biomes so that accurate data describing interannual variation and ecosystem responses to environmental change could be evaluated quantitatively. The objective was to initiate work on hypotheses requiring data sets of five years or more and, at the same time, set up a

Table 5.1. Some pervasive human disturbances that uncouple important ecological processes linking ecosystem components in large river basins.

STREAM REGULATION BY DAMS, DIVERSIONS, AND REVETMENTS: *uncouples longitudinal, lateral, and vertical dimensions*
Lotic reaches replaced by reservoirs: loss of up-downstream continuity, migration barrier, flood and nutrient sink, stimulates biophysical constancy in downstream environments
Channel reconfiguration and simplification: loss of lateral connections, removal of woody debris, isolation of riparian and hyporheic components of floodplains
Transcatchment water diversion: abnormal coupling of catchments, dewatering of channels, immigration of exotic species, import of pollutants
WATER POLLUTION: *alters flux rates of materials, uncouples food webs*
Deposition of pollutants from airshed into catchment: eutrophication, acidification
Direct and diffuse sources of waterborne waste materials from catchment: toxic responses, eutrophication
Accelerated erosion related to deforestation and roading: sedimentation of stream bottoms, eutrophication
FOOD WEB MANIPULATIONS: *induces strong interactions that alter food webs*
Harvest of fishes and invertebrates: biomass and bioproduction shifts
Introduction of exotic species: cascading trophic effects

network of sites where basic biophysical data would be systematically gathered for decades (Likens 1989, Franklin et al. 1990). These studies have already greatly contributed to understanding ecosystem connectivity, although data are not yet of sufficient scope to resolve many of the landscape- and patch- specific hypotheses proposed in the LTER program (Swanson and Sparks 1990). Moreover, these data are very site specific and tied to falsification of hypotheses that are clearly of great scientific importance but may be rather narrow in scope from the point of view of many managers.

Even though the scientific community has a long way to go before ecosystem response to natural environmental changes is fully understood, the human disturbances of catchments are often more extreme than natural events in frequency, intensity, and duration. In case after case, ecosystems in the catchment sense presented herein have been essentially uncoupled by the cumulative impacts and interactions of human disturbances (Table 5.1) (see also Ward and Stanford 1989). Perhaps the most pervasive disturbance is encompassed by the combined effects of channelization, revetment, and harvest of riparian timber within major river corridors. It has often been written that we may never know the true nature of channel-floodplain connectivity of large (> eighth order) rivers in the temperate latitudes because cultural development of the industrial nations was so dependent on these rivers as commercial waterways and because the attendant effects were so ecologically devastating (cf. Regier et al. 1989). In most, if not all cases, precious little information about the connectivity of these large rivers was recorded before major human disturbances took place. However, several carefully researched case histories provide insightful syntheses of the interactive effects of human and natural disturbances on the ecology of river systems (reviewed in Davies and Walker 1986).

Rather than attempt to summarize the important inferences of these and many other studies chronicling human disturbance in catchments, we present below a single case history of a large catchment that retains numerous pristine attributes but is threatened by a variety of interactive effects. In this case an ecosystem-level understanding might be very productive in fostering a new management ethic. The goal is to sustain the natural ecological connectivity of the system. We use this example to set the stage for articulation of some new approaches to that goal that may be useful elsewhere.

A Case History of Interactive Effects on Ecosystem Connectivity

Background

The Flathead River-Lake ecosystem in northwestern Montana provides a good example of a tightly coupled system where natural and human disturbances are clearly interactive. Understanding of this catchment is based on ecological studies by scientists at the Flathead Lake Biological Station (a field station of the University of Montana), where biophysical data have been routinely collected since 1896, and a wide variety of management-oriented research has been conducted by tribal, state, and federal agencies (reviewed by Stanford and Hauer 1992). Salient points are summarized here.

This 22,000 km^2 catchment is dominated by runoff from the myriad tributaries that feed the sixth-order Flathead River (mean annual discharge = 340 m^3/s), which flows through 496 km^2 Flathead Lake (Figure 5.3). Water quality in this river-lake system is extremely good; solute concentrations and bioproduction are uniformly low (oligotrophic); waters are usually highly transparent (Secchi disc readings in Flathead Lake routinely exceed 15 m autumn and winter); and native fisheries are healthy. Fewer than 80,000 people reside in the entire catchment, and no major industrial or agricultural sources of pollution currently exist. The Flathead River dominates the inflow of solutes and particulate materials that influence water quality, structure food webs, and drive bioproduction in the lake. For example, the river provides 65% of the annual load of bioavailable phosphorus reaching the lake. Six of the ten native fishes in the lake are adfluvial; that is, they reside in the lake but migrate upstream into tributaries to spawn (cf. bull charr in Figure 5.3). Hence the fishes constitute an upstream feedback loop and enhance the ecological connectivity of the ecosystem (Figure 5.5).

Ecological connectivity of the Flathead system is of course maintained in a quasi-equilibrium status by natural disturbance events (Figure 5.5). For example, the catchment is naturally disturbed by floods. Catchment hydrology is annually dominated by spring snowmelt, and in that sense the hydrograph is very predictable. But the magnitude of the spring spate is highly unpredictable, based on a 90 year period of record. Climatic events alternately juxtapose either continental (cold, dry) or Pacific maritime (warm, wet) air masses over the catchment, determining precipitation patterns. Infrequently and under the extreme moisture conditions in the Pacific front, the two air masses collide directly over the catchment, resulting in intense precipitation. Intermediate levels (10–20 x mean annual flow) of flooding occur on about a 10 year return pattern and almost always during spring; but high magnitude (20–50 x mean annual flow) floods have occurred 17 times during the historical record. The timing and duration of high magnitude floods and other extreme climatic events (Figure 5.5) are stochastic. Another example concerns the occurrence of wildfires caused by lightning strikes during dry periods. Mosaics of successional stages in forest stands characterize the uplands of the catchment as a result of these randomly distributed burns over many decades. Thus natural disturbance is a fundamental feature of this ecosystem and, coupled with the zoogeographic history of the area, is responsible for the generally high biodiversity of plants and animals by preventing dominance by a few species.

However, four generalized classes of human perturbations clearly have affected the natural attributes of this catchment: (1) stream regulation, (2) eutrophication, (3) food web manipulation, and (4) erosion (Figure 5.5). While localized effects may vary and the magnitude of the impacts has not been so severe as to completely compromise ecosystem connectivity, anthropogenic disturbances have degraded natural structure and function.

Stream Regulation

Two large hydroelectric and flood control dams partly regulate flows in the mainstem river and volume in Flathead Lake (Figure 5.3). The spring flood pulse of the Flathead River is predominantly stored behind these dams and discharged during the baseflow period. Owing to the presence of a natural bedrock sill at the outlet, the backshore of Flathead Lake historically was inundated up to about 882.5 m above sea level (masl) during the spring spate; however, the lake returned to base level (878.8 masl) by mid-July. Kerr Dam was built downstream from the sill in 1937 and extends the full pool (881.8 masl) period into late October to facilitate hydropower production. Hungry Horse Reservoir was first filled in 1953 and stores runoff from the entire South Fork subcatchment. Hydropower operations currently cause both flow and temperature problems in the river segments downstream from the dams. The varial zone of the river channel (Figure 5.1) is alternately inundated and dewatered by fluctuating flows related to power production below the dams. As a consequence, the varial zone is quite large and is essentially devoid of aquatic biota. Sluicing of the substratum by clear water flows has removed the smaller particles, leaving larger rocks and cobblestones firmly implanted on the river bottom (a phenomenon of regulated rivers known as armoring; Simons 1979). Capture of flood flows has partly or totally eliminated the natural fluvial disturbances on the floodplains of regulated river segments, thereby allowing senescence or other alteration of riparian plant communities. Since Hungry Horse Dam discharges from the bottom of the reservoir, nutrient concentrations are elevated relative to the free-flowing river segments, and algal mats coat the armored substratum in the minimum flow channel below the varial zone. Stream regulation has reduced the biodiversity in the dam tailwaters by about 80%. Spawning, juvenile recruitment, and growth of resident and adfluvial fishes have also been seriously compromised by extension of the varial zone in both regulated river segments; and cold (5–8°C) summer temperatures in the effluent water from Hungry Horse Reservoir compound the problem in the mainstem river upstream from Flathead Lake (Stanford and Hauer 1992).

Eutrophication

Plant growth in most of the lakes and streams of the Flathead catchment is limited by a general lack of labile nutrients. Most of the waters appear to be phosphorus limited or co-limited by paucity of both nitrogen and phosphorus (Dodds and Priscu 1989). Many alpine and subalpine lakes contain no measurable soluble reactive phosphorus and <20 mg/L nitrate, owing to the lack of these minerals in the Precambrian argillites that dominate the bedrock of much of the catchment. Therefore, bioproduction in these waters is very low (Stanford and Ellis 1988, Stanford and Prescott 1988).

Consequently, abnormally accelerated algal production associated with anthropogenic nutrient enrichment (i.e., eutrophication) is a primary concern, particularly as it relates to degradation of the high quality water in Flathead Lake. Of the total bioavailable phosphorus load entering Flathead Lake annually, 17% is derived from sewage treatment plants in the catchment and 30% is atmospheric deposition. Smoke from homes heated with wood burning stoves and from slash burning may be the primary source of labile phosphorus measured in bulk precipitation samples. In 1983 a lakewide bloom of the noxious blue-green alga *Anabaena flos-aqua* occurred for the first time in Flathead Lake since records began in 1902. The bloom was not severe and it has not recurred, but it did suggest that conditions were near a threshold beyond which major changes in the autotrophic community of the lake might be expected. Recent nutrient bioassays and analyses of long-term mass balance data have supported this inference (Stanford et al. 1983, 1990).

Food Web Manipulation

Since the turn of the century, 17 fish and 2 crustacean species have been purposely introduced into the Flathead catchment, primarily by fishery managers. Most fishes and both crustaceans established viable populations and gradually immigrated widely within the catchment. Today only a very few lakes in Glacier National Park have entirely native food webs, because of their remote localities and the presence of cascades, falls, or other migration barriers that prevented invasion by nonnative species from waters downstream.

These introductions had major impacts on native populations and dramatically restructured food webs in the lakes and streams throughout the catchment. Often effects cascaded through the food webs in ways that were unanticipated and sometimes involved both terrestrial and aquatic species.

For example, the kokanee salmon (*Oncorhynchus nerka*) fishery has undergone extreme fluctuations since the species was introduced into Flathead Lake in 1916. The population expanded rapidly and gradually replaced the native cutthroat trout as the dominant planktivore. Adfluvial kokanee from Flathead Lake spawned primarily in the outlet of McDonald Lake in Glacier National Park (Figure 5.3), where they attracted large numbers of migratory bald eagles (*Haliaeetus leucocephalus*). When the kokanee spawners were abundant (>150,000), so were eagles (>700).

In 1981 the nonnative crustacean *Mysis relicta* immigrated to Flathead Lake from intentional plants made in lakes upstream. Within six years, numbers exceeded 125/m^2. *Mysis* feed on zooplankton near the lake surface at night and rest on the lake bottom during the day. They have reduced zooplankton biomass in the lake by almost an order of magnitude. Kokanee are also dependent on zooplankton, but they prefer to stay near the lake surface, perhaps to avoid predation by piscivorous lake trout (*Salvelinus namaycush*, a nonnative species) and native bull charr. Thus *Mysis* created a trophic restriction for kokanee, and the fishery collapsed in 1987–88. Since 1989, only incidental kokanee spawners have been observed in McDonald Creek and the bald eagles have dispersed elsewhere (Spencer et al. 1991).

Erosion

Soil and other mineral substrata are naturally eroded by fluvial processes within the Flathead catchment, as in all river basins. Owing to the porous nature of the bedrock substrata and extensive tills of glacial origin, very little overland or sheet flow occurs except during extreme precipitation events or during periods of intensely accelerated snowmelt. Streams originate primarily as springbrooks fed by waters that percolate into substrata from precipitation at higher altitudes. Springbrooks coalesce to form the drainage network of the catchment. Therefore, most of the sediment loads carried by the streams and rivers are derived from erosion of stream channels and banks.

The rate of erosion is determined by channel morphology, slope, relative erosiveness of streambank substrata, and the intensity and duration of spates. Most of the sediment load in the system is derived from Tertiary shales deposited as valley fill and Quaternary tills and alluvium. These soils contain nitrogen and phosphorus either within the organic debris or associated with the clay lattice of the mineral particles. Therefore, as much as 60% of the annual riverine nutrient load of the Flathead River may be associated with sediment particles that are transported for short periods, most years during spring runoff, when the rivers and streams of the catchment are flooding. Only about 10% of the nutrients associated with particles can be assimilated by the biota (i.e., only about 10% of the particulate phosphorus is labile or bioavailable; Ellis and Stanford 1986, 1988), and much of the load is deposited either on the river floodplain or into the lakes as a short-term pulse event. In spite of the low nutrient bioavailability, the fertilization effect of the particulates eroded and transported by fluvial processes is significant owing to (1) the oligotrophic nature of the water bodies and (2) the dominance of the hydrograph and nutrient mass balance of both rivers and lakes in the catchment by spring runoff.

Clearly, erosion is a natural process that both shapes the catchment landscape and to some extent fertilizes patches within the landscape. Natural (e.g., lightning-caused fire, insect epidemics, beaver [*Castor canadensis*] and other large herbivore influences) and human (e.g., road building, clearcutting) deforestation increases the seasonal and annual variation in water yield, particularly during spring snowmelt (Hauer and Blum 1991), thereby accelerating erosion of streambanks and increasing sediment loads. Erosion of road surfaces and berms or stream crossings is of particular concern, because unstable roads are known to be major sources of fine particles in some streams in the Flathead catchment, as elsewhere (see Megahan et al., this volume). Accelerated erosion, locally associated with logging and road building, has increased the volume of fine particles within the channel of disturbed streams, clogging interstices and reducing interflow and aeration of the substratum. Speciosity and biomass of zoobenthos may be reduced by 80% in highly sedimented areas compared with adjacent cobble substratum (Spies 1986), and survival of bull charr eggs and juveniles decreases markedly when fines (particles <6.35 mm) exceed 40% of the substratum volume (Weaver and Fraley 1991). Moreover, recent work has shown a clear correlation between sedimentation rates in on-channel lakes and road building activities in the McDonald and Whitefish subcatchments (Spencer 1991). Inflowing riverine sediments apparently fertilize the water column of Flathead Lake in the spring, based on the observation that phytoplankton productivity is highest in years of high runoff and high sediment loading from the catchment (J. Stanford and B. Ellis, unpublished); however, the sediment load has not been apportioned in terms of natural versus human disturbances.

Interactions Between Natural and Human Disturbances: Management Considerations

Many different management jurisdictions exist within the Flathead River Basin. Seventy-two percent of the basin is federally administered, involving the Flathead National Forest (U.S. Department of Agriculture), Glacier National Park (National Park Service), national wildlife refuges (U.S. Fish and Wildlife Service), and the Flathead Indian Reservation trust lands (U.S. Bureau of Indian Affairs). Large areas of state and tribal (Confederated Salish and Kootenai Tribes of the Flathead Indian Reservation) lands exist, with the remainder of the basin primarily in privately held tracts. Hungry Horse Dam is a federal project operated by the U.S. Bureau of Reclamation. Kerr Dam is located within the Flathead Indian Reservation and operated by a private corporation, Montana Power Company, Inc., on the basis of a rental agreement with the Tribes as mandated by the Federal Energy Regulatory Commission. Many other federal, state, and local agencies have statutory authority to manage specific resources in the catchment. Since the headwaters of the North Fork are in British Columbia (Figure 5.3), many provincial and Canadian federal agencies are involved. For example, the authority of the International Joint Commission (organized under the U.S.-Canada Boundary Waters Treaty of 1909) was invoked during 1986–88 to quantify and reference the potential impacts of a large open-pit coal mine (International Joint Commission 1988) proposed by a Canadian subsidiary of an American corporation in Canada (Figure 5.3). This maze of management jurisdictions and associated interactions between natural and human disturbances complicates resource management within the ecosystem.

The threat of deteriorating water quality in Flathead Lake from urban sewage, the proposed Canadian coal mine, and burgeoning road building and clearcutting on federal and private forest lands stimulated management actions designed to implement conservation of natural conditions in the tributaries and to reduce nutrient loading in the lake by about 20% (i.e., to near natural conditions). Actions included a ban on the sale of phosphorus-containing detergents (mandated by the Montana state legislature), construction of new sewage treatment plants to allow phosphorus and nitrogen removal from all urban effluents in the catchment, and voluntary imposition of best management practices (BMPs) to reduce nonpoint sources of nutrients, especially those associated with accelerated erosion in the catchment (mandated by the State Water Quality Bureau, which has statutory authority to enforce water quality laws).

During 1983–90, annual nutrient loads into the lake decreased (least squares regression, $P<0.1$, J. Stanford and B. Ellis, unpublished); and, as noted above, *Anabaena* blooms did not recur. This, of course, suggested that initial management actions were successful. However, construction of a new sewage treatment plant for Kalispell, Montana, which has been the largest point source of bioavailable nutrients in the past, is not yet complete. Moreover, very little, if any, of the reduced nutrient load can currently be related to voluntary BMPs, because their utility in improving water quality has not been quantified empirically in the Flathead Basin. The apparent reduction in nutrient loading and lack of recurring *Anabaena* blooms may be due to at least three other interactive linkages.

First, catchment precipitation has been below average since 1983. Natural loading rates of water and nutrients have been generally lower on an annual basis than occurred earlier in the period of record. However, average concentrations in the river did not change significantly.

Second, operations at Hungry Horse Dam changed from primarily midwinter to summer and fall discharges, in response to economic considerations for hydropower production as related to demands for higher summer flows in the lower Columbia River to more effectively flush smolts of anadromous salmon out to sea (discussed below). Owing to thermal stratification in the reservoir and the hypolimnial (bottom) release mode of the dam, the high volume water masses from Hungry Horse Reservoir are cold (4–7°C) and dense relative to ambient temperatures (unregulated, natural) in the river below the dam and within the epilimnion (surface) of Flathead Lake, which is also thermally stratified in the summer (22°C surface, 4° bottom). Thus summer discharges from Hungry Horse Dam essentially dilute the pollutants entering from the urban and argricultural areas in the Kalispell Valley. Moreover, these cold waters and the nutrient load immediately sink to the lake bottom (underflow) upon entry into Flathead Lake. Since the lake is maintained at full pool during the summer for ease of access by boaters, Kerr Dam must discharge water volumes equal to the inflowing volumes. But Kerr Dam releases water from the surface layers of the lake, owing to its location below the natural outlet sill. The net effect on the limnology of the lake appears to be (1) a significant reduction in the heat budget, (2) cooler surface temperatures during the plant growing season, (3) stripping of plankton and nutrients from the surface by the Kerr withdrawal current, and (4) deposition of a large portion of the summer and fall nutrient load far below the upper portion of the water column that is penetrated by sunlight. Therefore, conditions favorable for sustained algal biomass, especially forms like *Anabaena*, in the epilimnion of the lake may have been compromised by hydropower operations.

Third, food web shifts caused by the collapse of the kokanee fishery may have influenced grazing rates on the algae. Owing to intense predation by *Mysis*, zooplankton biomass decreased almost an order of magnitude in the peak *Mysis* years, 1987–88, compared with pre-*Mysis* measures. During 1988–90, *Mysis* numbers decreased from 125/m^2 in 1987 to 30/m^2 in 1989 (Spencer et al. 1991) and 35/m^2 in 1990 (Spencer 1991), owing to predation by bottom-oriented fishes (whitefishes [*Coregonus* spp.], lake trout, and bull charr). Phytoplankton primary production was the highest on record in 1988 and decreased during 1989–91 in concert with declining *Mysis* numbers. At the same time, cladoceran zooplankton have recovered during periods of thermal stratification. Apparently, large numbers of *Mysis* do not penetrate the thermocline and enter the epilimnion during summer. This thermal refugia from *Mysis* and the lack of kokanee or other surface-dwelling planktivores apparently allowed *Daphnia thorata* to increase, and the inference is that grazing on phytoplankton has also increased (Stanford et al. 1990). These interpretations are based on preliminary analyses of long-term trends in the various data bases for Flathead Lake. Our main point here is simply to reinforce by example the idea that food web dynamics in lakes can be strongly interactive in response to both bottom-up (nutrient supply) and top-down (*Mysis* introduction) effects (see also Carpenter 1988).

Interactions between dam operations, natural circulation patterns, and shoreline erosion in Flathead Lake are also noteworthy. It is exceedingly difficult to move large water masses through Flathead Lake while also maintaining it at full pool elevation. Often the lake exceeds the full pool owing to lack of coordination between the dams coupled with the complexities of wind and temperature-driven internal circulation events and patterns. Flathead Lake is an extremely large, deep lake and therefore its hydrodynamics are profoundly influenced by Coriolis and density currents and circulation patterns in addition to volume regulation by the dams. The lake has a 30 km wind fetch on the long axis, and storms and shoreline erosion rates exceed 2 m per year (lineal cross section) at the north end of the lake where the shoreline is dominated by deltaic sand substratum. Surface and internal seiches are common after storms and may influence the pattern of sediment transport from eroding shorelines. As a consequence of these natural (wind) and human (lake level regulation) disturbances, the 970 ha depositional delta of the Flathead River has entirely eroded into the lake within the last 50 years; littoral and riparian communities of the lakeshore have also been vastly altered, if not partly uncoupled from processes in the main (pelagic) part of the lake (Bauman 1988, Hauer et al. 1988, Lorang et al. 1992).

The negative effects of both Kerr and Hungry Horse operations have been carefully documented (see review by Stanford and Hauer 1992), and a mitigation plan for hydropower impacts (e.g., fluctuating flows and lake levels, temperature changes, migration barriers, habitat and production losses, accelerated lakeshore erosion) on fish and wildlife resources has been proposed to regulatory authorities. In this case, two different regulatory authorities exist. The Federal Energy Regulatory Commission is currently considering a plan related to Kerr Dam, since it is operated by private concerns. Owing to Hungry Horse Dam's operation by a federal agency, mitigation of impacts falls under the mandate of the Northwest Power Planning Act of 1984 for the entire Columbia River Basin, which involves the Northwest Power Planning Council (planning) and the Bonneville Power Administration (research and implementation). The mitigation plans were jointly developed by the state (Montana Department of Fish, Wildlife and Parks), the tribes, and the entities that operate the dams, with input from university scientists and other agency biologists. Proposed actions include: retrofitting Hungry Horse Dam to allow selective withdrawal to facilitate more natural temperature regime downstream; construction of re-regulation dams or operational changes to moderate flow fluctuations from Hungry Horse and Kerr dams; reducing the full-pool level of Flathead Lake to reduce shoreline erosion; revetment of some shorelines to curtail erosion and enhance wetland development for waterfowl; habitat restoration in damaged fish and wildlife production areas; and hatchery supplementation of fishes as replacement for losses associated with hydropower operations at both dams (Fraley et al. 1989, Fraley et al. 1991, Jourdonnais et al. 1990, Stanford and Hauer 1992).

Differences of opinion remain as to whether the various mitigation actions are appropriate or whether they will work as proposed, primarily because the statutory authorities of the two processes are independent and mandate solution of impacts on fish and wildlife without in-depth consideration of other ecosystem interactions, such as influences on timing and magnitude of nutrient loads and connectivity between riverine processes and food web dynamics (Stanford and Hauer 1992). However, the pervasive effects of stream and lake regulation were thoroughly documented and an interagency consultation and public information transfer was effective. This was fostered by forums coordinated by a public information and oversight group called the Flathead Basin Commission. This commission was legislated by the state to bring together agency heads and informed citizens in a manner that stimulated interagency cooperation to fund research, effectively monitor ecosystem indicators (e.g., catchmentwide water quality and population dynamics of important indicator organisms, like the bull charr), and facilitate interactive discussion of results and proposed management actions in a nonstatutory fashion.

The natural ecological connectivity of the Flathead catchment remains largely intact. It is a high priority area for conservation and effective resource management, since large areas are designated as national parks, wildlife refuges, wilderness areas, and tribal lands. Environmental problems exist but they have been quantified, articulated, and periodically reassessed in the process of understanding how this large catchment is influenced by natural and human disturbances. More information is needed, but the presence of a legislated commission to coordinate monitoring of ecosystem conditions by the many different management agencies has proved to be an effective and empirically based forum for considering and implementing alternative actions to protect and enhance ecological connectivity in this large catchment.

Interference Management and the Illusion of Technique

The Flathead experience illustrates the travail of contemporary resource management. Interactive and cumulative effects become seemingly intractable in large and ecologically complex catchments. Managers often want simplistic methodology that will explicitly satisfy an increasingly circumspect public. Unfortunately, in the absence of practical and conceptual understanding of ecosystem structure and function, management actions often produce results significantly different from what was expected. Usually this happens because management questions are not posed in an ecosystem (whole-catchment) context and actions evolve as interferences with the natural ecosystem connectivity. The introduction of *Mysis* as a forage stimulus for sport fishes in a very tightly coupled system interfered with the quasi-equilibrium of the Flathead Lake food web and produced a trophic cascade that ultimately displaced a critically important population of bald eagles.

On a larger scale, influences far downstream may have unanticipated effects on the operations of the two large dams in the catchment. In particular, we are concerned that efforts to increase the runs of Pacific salmon and steelhead downstream in the middle and lower reaches of the Columbia River may interfere with mitigation efforts in the Flathead Basin and other headwater reaches that, because of natural barriers, never contained anadromous fishes. The plight of the anadromous salmonid fishery involves overharvest, continually increasing dominance of runs by cultured stocks (apparently at the expense of naturally reproducing runs, owing to genetic introgression and increased harvest), predation of wild and cultured smolts by resident fishes, highly variable oceanic survival, and passage problems created by the nine mainstem dams (Ebel et al. 1989). Prominent in this discussion is the fact that early summer flood crest of the Columbia River has been eliminated by storage of the spring spate in four large reservoirs (Hungry Horse, Dworshak, Libby, and Mica) in the headwaters. Historically, the flood pulse of the river not only flushed smolts along on their outmigration, it also stimulated bioproduction in the estuarine food web which sustained the fisheries (Simenstad et al., this volume). Recovery plans for the fisheries call for a water budget for the river that mandates "fish flows" that will very likely interact with the economics of hydropower production and the need for flood control in a manner that will introduce a large measure of uncertainty in operations of the headwater dams. Unless the needs of resident fishes directly influenced by these dams have equal priority with downstream objectives, mitigation of resident fish and wildlife in the headwater segments may be compromised by actions for anadromous fishes.

A related problem is the tendency of today's managers to use a standardized methodology that often relies on little or no empirical data, or data that have little or no predictive power at the ecosystem level. Because of the natural complexities of river ecosystems, the intractability of cumulative effects in large catchments, and the cost of long-term data acquisition, managers too often tend to seek simple answers to complex problems. Often this involves nothing more than a formalization and synthesis of "best professional judgment" with no ecological rationale that is empirically based.

For example, one approach in current vogue is to assemble groups of professional hydrologists, biologists, engineers, silviculturists, and foresters to assess or "audit" forest practices (BMPs) as they relate to observed, but not empirically quantified, impacts on water quality. Specific sites are visited, and each person simply provides his or her qualitative judgment as to whether the logging activity has had any impact on the streams draining the area. Again, audit values are apportioned among BMPs on an areal basis and summed up to allow inferences about levels of disturbance to be drawn (Ehinger and Potts 1990).

In the Rocky Mountains and Pacific Northwest, including the Flathead, another popular approach for assessing the impacts of forestry on water and sediment yield is to assemble a series of impact or "risk" values and recovery rates for various land disturbance activities (e.g., roads, skid trails, site preparation, logging method). These values are then apportioned on an areal basis for the catchment and summed to provide a measure of cumulative effects (Klock 1984, United States Forest Service 1988, Cobourn 1989). This approach can be greatly improved when formalized as a true risk analysis (Cairns and Orvos 1990) or Markovian simulation, in which the impact values are based on catchment-specific experiments and the results are expressed in terms of specific forest dynamics such as the mass transfer of water, sediment, or nutrients (Pastor and Johnston, this volume).

Unfortunately, subjective methods or model results are often never verified in terms of actual impact measured *in situ* (e.g., increase in fine sediments, decrease in fish production), and inferences and recommendations can be misleading to those seriously interested in minimizing negative instream effects associated with anthropogenic land disturbances. Clearly, these methods will identify pervasive effects, such as severe sedimentation resulting from roads collapsed into streams or skid crossings that are not bridged. But it is virtually impossible to detect chronic effects (e.g., accelerated water yield and bank erosion, slow reduction in woody debris accumulation, changes in water chemistry and bioproduction, fish habitat alteration) via nonempirical audits. The value of the judgment is lost in formalization of the approach unless the audit result can be verified by temporal and spatial ecological measures obtained within appropriate experimental designs.

Too often standardized techniques or mathematical models are used to evaluate impacts when they have little or no predictive power in terms of ecosystem connectivity. This amounts to an "illusion of technique" (R. Behnke, Colorado State University, unpublished).

A prime example of the illusion of technique is the very popular incremental method (IFIM) that is recommended by the U.S. Fish and Wildlife Service to determine minimum flows to protect fisheries from the effects of stream regulation. The method is based on field surveys that determine the area of the varial zone that is inundated at different instream volumes (i.e., wetted usable area, WUA), along with other physical habitat components (e.g., velocities). These data are then used to drive a sophisticated simulation model involving target species and different flow scenarios to determine minimum flows required to sustain fisheries (Nestler et al. 1989). The model does nothing more than predict physical habitat availability for various life stages of specific fishes, and in some cases it does not appear to do that very well, among other problems (Mathur et al. 1985, Orth and Maughan 1982, Scott and Shirvell 1987, Shirvell 1989, Gan and McMahon 1990). However, the IFIM clearly is a refined and standardized technique and its use has in some instances prevented chronic dewatering of rivers. Our point is that this and other models are not responsive to processes that ultimately determine variability of bioproduction and other important aspects of ecosystem connectivity (Figures 5.4 and 5.5). In spite of warnings to the contrary by the authors of IFIM (and other standardized approaches), the illusion for naive users in this case is that WUA is deterministic, when in fact complex interactions of abiotic and biotic components of a river are naturally stochastic. This is precisely why the ecosystem exists in a quasi-equilibrium state. Naive managers and administrators easily confuse quantification, objectivity, and sophistication with biological reality, and such illusions should not be fostered (R. Behnke, Colorado State University, unpublished; Lee, this volume).

A more rationale approach is to recognize and appreciate the complexities of river catchments and utilize standardized tools and models in the limited sense for which they were designed. It is not likely that any model or other deterministic construct will ever accurately predict ecosystem structure and function at the catchment scale. But model building is one very effective way to plan and articulate the need for collection of long-term ecological data that will ultimately explain observed variability caused by natural and human disturbances. In almost all assessments of cumulative impacts at the catchment level, long-term empirical data describing ecosystem structure and function are required as baselines to firmly quantify environmental change.

Integrated Management

In this age of desktop computer power and electronic communication, it is paradoxical that interference management should occur. However, as communication power has burgeoned, so have agency bureaucracies. For example, the Bureau of Reclamation has run out of dam sites and is now attempting to add supervision of fish and wildlife resources of western rivers to its official mandate (our observation). Indeed, we think that many state and federal agencies are purposely fostering an insular approach to resource management. Each wants to do ecological research, develop and follow standardized management criteria and procedures for ecological resources, and, most important, minimize influence of other agencies. Local and regional fragmentation of management authority is guaranteed to result in interference management, which in turn fragments catchment ecosystems.

The structure and function of catchment ecosystems and the cumulative effects of human disturbances are in fact intractable without an integrated analysis based on long-term data (Magnuson 1990). No single agency can effectively deal with the plethora of management/research problems on a large catchment scale. Yet the bureaucracies and their individual mandates are firmly entrenched, as are the public groups that are increasingly sensitized by the negative effects of interference management and the illusion of technique.

What should be done? If human disturbances are to be managed for the purpose of maintaining natural ecological connectivity at the catchment scale, management agencies must cooperate to minimize interferences. Cooperation is needed for collection of long-term data that will allow BMPs and other management actions to be quantified and adjusted before they interfere with each other. That level of cooperation requires effective information transfer, continual education, and independent coordination.

State-of-the-art ecology almost always originates from research at the university level or in agency research centers closely allied with universities. Although university-based research is also often very insular, we note a recent trend toward interdisciplinary work at the ecosystem level. The long-term research initiatives of the National Science Foundation described above have greatly fostered this trend. It may therefore be expected that university research will provide guidance for a new integrated management ethic.

However, we note three fundamental problems. First, creative research is currently compromised by dwindling funding at the national level and particularly at the state level. Part of the problem is rooted in the growing tendency of agencies to attempt their own basic and applied research in opposition to cooperatives with universities. Second, we perceive a growing gulf between agencies and universities because it is often university scientists who point out flaws and interferences in agency management actions (see also Marston, this volume). Third, universities are not currently producing management specialists in the natural resource arena who are astutely attuned to ecosystem connectivity. Graduates are trained primarily to do basic research, and in most cases that training is highly specialized. We should not be surprised that agencies are becoming insular in their approach to management. Moreover, we should not be surprised that agencies tend to attempt ecological manipulations (e.g., introductions of exotic species, hatchery supplementation of wild populations) rather than focusing management on public education and regulation of human disturbances.

Conducting research and managing resources should be distinguished as separate but complimentary activities. The successful manager must understand ecosystem connectivity and must be able to translate research findings into holistic resource management. It is also the manager's job to involve the public in the decision-making process by communicating how proposed actions relate to the whole and will thereby serve to reconnect severed interactive pathways.

Because those making high-level management decisions must (1) comprehend ecosystem connectivity at the catchment level, (2) be familiar with the relevant primary literature, (3) determine when additional problem-oriented research is needed, and (4) translate all of the above into appropriate managerial decisions while effectively communicating with the public, their proper training is indeed a formidable task. University curricula in natural resource management need to be revamped to foster an understanding of such matters as economic and environmental sustainability, cultural needs and influences, demography and political change, and conservation ethics (Marston, this volume) in addition to traditional biology and ecology. Moreover, high level management jobs (e.g., forest supervisors, park superintendents) require more rigorous training. Doctoral programs typically train either researchers or managers. We argue that to properly protect and manage our valued natural resources requires a solid grounding in research plus managerial expertise. We believe that contemporary management problems at the catchment scale are so complex that nothing less than a Ph.D. degree accompanied by a postdoctoral internship program will suffice to train conservators of ecological connectivity in river ecosystems.

This cannot be done by the universities alone. Agencies must return to the university environment for basic research and cut down wasteful duplication of space, equipment, and effort. University scientists must accommodate managers by doing innovative applied research and by providing educational forums that articulate management problems and potential solutions to students and agency personnel. Some of the cooperatives between a few universities and regional research units in the National Park Service, U.S. Forest Service, and U.S. Geological Survey have been somewhat successful in this regard. However, we envision formal cooperatives at the level of local Forest Service districts and state fish and game regional offices and involving many, if not all, research universities.

We emphasize that education and effective management of natural resource issues also must formally involve the public. Many management interferences and failures could have been avoided simply by the quality control afforded by an *a priori* public forum. A template for success in this regard is a state legislated catchment commission composed of all pertinent agency heads (e.g., forest supervisor, park superintendent, local land use planner, fish and game agency, tribal resource administrator, county commissioners) and at least an equal number of informed citizens who equitably represent the various publics (e.g., industry, agriculture, urban development, conservation). University scientists should be used as advisers or sources of basic information in analyzing and guiding the process. One fairly successful example is the Flathead Basin Commission described above.

In summary, we propose several important principles of integrated management at the catchment level.

1. *The major objective should be to conserve and enhance ecological connectivity.* Processes and disturbances within the catchment are interconnected biophysically in time and space.

2. *The key management questions should define the catchment scale.* However, for very large catchments (e.g., the Columbia River Basin) no

good formulas for success currently exist. Coordination and representation can become quickly fragmented or politicized because there are too many participants at the same table. We suggest that, if possible, the focus should be areas more the size of the Flathead catchment, as described above. The inference is that if ecosystem connectivity can be conserved in all subcatchments of very large drainage basins, the ecological integrity of the entire system should remain stable. Or, at least, an approach to problem solving in very large catchments should be forthcoming from an integration of subcatchment data and knowledge.

3. *A research and monitoring agenda should be established that will provide long-term data bases that may be used to separate variability due to natural and human disturbances (e.g., precipitation, discharge, nutrient loading, primary productivity, population trends of indicator organisms such as the bull charr in the Flathead case history).* University scientists should be utilized independently and in cooperatives with agency research and management personnel to plan monitoring programs and collect and interpret data. If planned properly, monitoring programs can be both an ongoing evaluation of BMPs and an assessment of environmental change at the catchment level. The latter may be expected to provide insights into the effects of regional or global influences on the catchment.

4. *Management actions should be examined from an ecosystem point of view.* A formal evaluation is needed of the risks that management actions portend and alternatives should be developed that can be activated if monitoring or research data suggest that interferences are manifested.

5. *A mechanism (we recommend a commission) should be provided that brings managers, researchers, and public groups into a forum for open debate.* The objective is education and information transfer before management actions are implemented.

Conclusion: Reconnecting Catchment Ecosystems

Ecology as a science has evolved into an understanding of landscapes as interconnected patches that vary in scale from a single rock in a stream to whole catchments (Gillis 1990; Naiman et al., this volume). Research is focused on processes, time frames, and disturbances that control the transfer of materials and energy through catchment landscapes. Management in this context refers to actions that limit interference of human disturbances to the extent that catchment ecosystems are sustained in a natural quasi-equilibrium.

In many catchments, human disturbance has eliminated or severely compromised natural connectivity. Catchment management in the future may logically involve reconnecting patches into landscapes. One example might be reestablishing floodplain springbrooks as functional patches (e.g., as important rearing areas for salmonids). This may involve removing revetments and allowing flood-pulse events to reconnect the channel and the floodplain

(Figure 5.2). Integrated forests, agricultural lands, and urban management can provide many other avenues to allow damaged catchment ecosystems to recover.

Threats to catchments usually manifest measurably in aquatic habitats as problems related to stream regulation, eutrophication and other forms of water pollution, food web changes, and accelerated sedimentation. These phenomena can be used as benchmarks that integrate the environmental health of the catchment if the data are gathered systematically over long periods. Analysis of trends in such data can reveal how leaky or unconnected the system may be and provide clear insights where management actions can be effective in reconnecting the system. This effort can best be accommodated by insightful, integrated management.

References

Alstad, D.N. 1982. Current speed and filtration rate link caddisfly phylogeny and distributional patterns on a stream gradient. Science **216**:533–534.

Alstad, D.N. 1986. Dietary overlap and net-spinning caddisfly distributions. Oikos **47**:251–252.

Amoros, C., M. Richardot-Coulet, and G. Patou. 1982. Les "Ensembles Fonctionelles": des entités écologiques qui traduisent l'évolution de l'hydrosystème en integrant la géomorphologie et l'anthropisation (exemple du Haut-Rhone français). Revue Géographie de Lyon **57**:49–62.

Barnes, J.R., and G.W. Minshall, editors. 1983. Stream ecology: Application and testing of general ecological theory. Plenum, New York, New York, USA.

Bauman, C.H. 1988. Effects of nutrient enrichment and lake level fluctuation on the shoreline periphyton of Flathead Lake, Montana. M.A. Thesis. University of Montana, Missoula, Montana, USA.

Benke, A.C., C.A.S. Hall, C.P. Hawkins, R.H. Lowe-McConnell, J.A. Stanford, K. Suberkropp, and J.V. Ward. 1988. Bioenergetic considerations in the analysis of stream ecosystems. Journal of the North American Benthological Society **7**:480–502.

Bretschko, G. 1981. Vertical distribution of zoobenthos in an alpine brook of the Ritrodat-Lunz study area. Internationale Vereinigung für theoretische und angewandte Limnologie, Verhandlungen **21**:873–876.

Cairns, J., Jr., and D.R. Orvos. 1990. Developing an ecological risk assessment strategy for the Chesapeake Bay. Pages 83–98 *in* M. Haire and E.C. Krome, editors. Perspectives on the Chesapeake Bay: advances in estuarine studies. Chesapeake Bay Consortium, Gloucester Point, Virginia, USA.

Carpenter, S.R., editor. 1988. Complex interactions in lake communities. Springer-Verlag, New York, New York, USA.

Coboum, J. 1989. Is cumulative watershed effects analysis coming of age? Journal of Soil and Water Conservation **44**:267–270.

Connell, J.H. 1978. Diversity in tropical rain forests and coral reefs. Science **199**:1302–1310.

Cummins, K.W., G.W. Minshall, J.R. Sedell, C.E. Cushing, and R.C. Petersen. 1984. Stream ecosystem theory. Internationale Vereinigung für theoretische und angewandte Limnologie, Verhandlungen **22**:1818–1827.

Dahm, C.N., D.L. Carr, and R.L. Coleman. 1991. Anaerobic carbon cycling in stream ecosystems. Internationale Vereinigung für theoretische und angewandte Limnologie, Verhandlungen, *in press*.

Danielopol, D.L. 1984. Ecological investigations on the alluvial sediments of the Danube in the Vienna area—a phreatobiological project. Internationale Vereinigung für theoretische und angewandte Limnologie, Verhandlungen **22**:1755–1761.

Danielopol, D.L. 1989. Groundwater fauna associated with riverine aquifers. Journal of the North American Benthological Society **8**:18–35.

Davies, B.R., and K.F. Walker, editors. 1986. The ecology of river systems. Dr. W. Junk, Dordrecht, The Netherlands.

Décamps, H., and R.J. Naiman. 1989. L'écologie des fleuves. La Recherche **20**:310–319.

Dodds, W.K., and J.C. Priscu. 1989. Ammonium, nitrate, phosphate, and inorganic carbon uptakes in an oligotrophic lake: seasonal variations among light response variables. Journal of Phycology **25**:699–705.

Dodge, D.P., editor. 1989. Proceedings of the International Large River Symposium (LARS). Special Publication of the Canadian Journal of Fisheries and Aquatic Sciences **106**:1–629.

Ebel, W.J., C.D. Becker, J.W. Mullan, and H.L. Raymond. 1989. The Columbia River: toward a holistic understanding. Pages 205–219 *in* D. P. Dodge, editor. Proceedings of the International Large River Symposium (LARS). Special Publication of the Canadian Journal of Fisheries and Aquatic Sciences **106**:1–629.

Ehinger, W., and D. Potts. 1990. On-site assessment of best management practices as an indicator of cumulative watershed effects in the Flathead Basin. Flathead Basin Water Quality and Fisheries Cooperative, University of Montana School of Forestry, Missoula, Montana, USA.

Ellis, B.K., and J.A. Stanford. 1986. Bio-availability of phosphorus fractions in Flathead Lake and its tributary waters. Final Report for the U.S. Environmental Protection Agency, Open File Report 091–86, Flathead Lake Biological Station, University of Montana, Polson, Montana, USA.

Ellis, B.K., and J.A. Stanford. 1988. Phosphorus bioavailability of fluvial sediments determined by algal assays. Hydrobiologia **160**:9–18.

Ford, T.E., and R.J. Naiman. 1989. Groundwater-surface water relationships in boreal forest watersheds: dissolved organic carbon and inorganic nutrient dynamics. Canadian Journal of Fisheries and Aquatic Sciences **46**:41–49.

Fraley, J.J., B. Marotz, J. Decker-Hess, W. Beattie, and R. Zubik. 1989. Mitigation, compensation, and future protection for fish populations affected by hydropower development in the upper Columbia system, Montana, USA. Regulated Rivers: Research and Management **3**:3–18.

Fraley, J.J., B. Marotz, and J. DosSantos. 1991. Fisheries mitigation plan for losses attributable to the construction and operation of Hungry Horse Dam. Montana Department of Fish, Wildlife and Parks, Kalispell, Montana, and Confederated Salish and Kootenai Tribes, Pablo, Montana, USA.

Fraley, J.J., and B.B. Shepard. 1989. Life history, ecology and population status of migratory bull trout (*Salvelinus confluentus*) in the Flathead Lake and River system, Montana. Northwest Science **63**:133–143.

Franklin, J.F., C.S. Bledsoe, and J.T. Callahan. 1990. Contributions of the long-term ecological research program. BioScience **40**:509–523.

Gan, K., and T. McMahon. 1990. Variability of results from the use of PHABSIM in estimating habitat area. Regulated Rivers: Research and Management **5**:233–239.

Gibert, J., M. Dole-Olivier, P. Marmonier, and P. Vervier. 1990. Surface water-groundwater ecotones. Pages199–225 *in* R.J. Naiman and H. Décamps, editors. The ecology and management of aquatic-terrestrial ecotones. UNESCO, Paris, and Parthenon Publishing Group, Carnforth, United Kingdom.

Gillis, A.M. 1990. The new forestry. BioScience **40**:558–562.

Gregory, S.V., F.J. Swanson, and W.A. McKee. 1991. An ecosystem perspective of riparian zones. BioScience, *in press*.

Grimm, N.B., S.G. Fisher, H.M. Valett, and B.H. Stanley. 1991. Contribution of the hyporheic zone to the stability of an arid lands stream. Internationale Vereinigung für theoretische und angewandte Limnologie, Verhandlungen, *in press*.

Guiver, K. 1976. Implications of large-scale water transfers in the UK: the Ely Ouse to Essex transfer scheme. Chemistry and Industry **4**:132–135.

Harmon, M.E., J.F. Franklin, F.J. Swanson, P. Sollins, S.V. Gregory, J.D. Lattin, N.H. Anderson, S.P. Cline, N.G. Aumen, J.R. Sedell, G.W. Lienkaemper, K. Cromack, Jr., and K.W. Cummins. 1986. Ecology of coarse woody debris in temperate ecosystems. Advances in Ecological Research **15**:133–302.

Hauer, F.R., and C. Blum. 1991. The effect of timber management on stream water quality. Flathead Basin Forest Practices, Water Quality and Fisheries Study. Open File Report 121–91, Flathead Lake Biological Station, University of Montana, Polson, Montana, USA.

Hauer, F.R., M.S. Lorang, J.H. Jourdonnais, J.A. Stanford, and A.E. Schuyler. 1988. The effects of water regulation on the shoreline ecology of Flathead Lake, Montana. Open File Report 100–88, Flathead Lake Biological Station, University of Montana, Polson, Montana, USA.

Holland, M.M., P.G. Risser, and R.J. Naiman, editors. 1991. The role of landscape boundaries in the management and restoration of changing environments. Chapman and Hall, New York, New York, USA.

Huston, M. 1979. A general hypothesis of species diversity. American Naturalist **113**:81–101.

Hynes, H.B.N. 1970. The ecology of running waters. University of Toronto Press, Toronto, Ontario, Canada.

Hynes, H.B.N. 1975. The stream and its valley. Internationale Vereinigung für theoretische und angewandte Limnologie, Verhandlungen **19**:1–15.

International Joint Commission. 1988. Impacts of a proposed coal mine in the Flathead River Basin. International Joint Commission, Washington, D.C., USA.

Jourdonnais, J.H., J.A. Stanford, F.R. Hauer, and C.A.S. Hall. 1990. Assessing options for stream regulation using hydrologic simulations and cumulative impact analysis: Flathead River Basin, USA. Regulated Rivers: Research and Management **5**:279–293.

Klock, G.A. 1984. Modeling the cumulative effects of forest practices on downstream aquatic ecosystems. Journal of Soil and Water Conservation **40**:237–241.

Likens, G.E., editor. 1989. Long-term studies in ecology: approaches and alternatives. Springer-Verlag, New York, New York, USA.

Lock, M.A., and D.D. Williams, editors. 1981. Perspectives in running water ecology. Plenum, New York, New York, USA.

Lorang, M.S., J.A. Stanford, F.R. Hauer, and J.H. Jourdonnais. 1992. Dissipative and reflective beaches in a large lake and the physical effects of lake level regulation. Journal of Ocean and Shoreline Management, *in press*.

Lotspeich, F.B. 1980. Watersheds as the basic ecosystem: this conceptual framework provides a basis of a natural classification system. North American Journal of Fisheries Management **2**:138–149.

Magnuson, J.J. 1990. Long-term ecological research and the invisible present. BioScience **40**:495–501.

Matthews, W.J., and D.C. Heins, editors. 1987. Community and evolutionary ecology of North American stream fishes. University of Oklahoma Press, Norman, Oklahoma, USA.

Mathur, D.L., W.H. Bason, E.J. Purdy, Jr., and C.A.A. Silver. 1985. Critique of the instream flow incremental methodology. Canadian Journal of Fisheries and Aquatic Sciences **42**:825–831.

Minshall, G.W., K.W. Cummins, R.C. Petersen, C.E. Cushing, D.A. Burns, J.R. Sedell, and R.L. Vannote. 1985. Developments in stream ecosystem theory. Canadian Journal of Fisheries and Aquatic Sciences **42**:1045–1055.

Mooney, H.A., and J.A. Drake. 1986. Ecology of biological invasions of North America and Hawaii. Springer-Verlag, Berlin, Germany.

Murdoch, W.W. 1991. Equilibrium and non-equilibrium paradigms (meeting review). Bulletin of the Ecological Society of America **72**:49–51.

Naiman, R.J., and H. Décamps, editors. 1990. The ecology and management of aquatic-terrestrial ecotones. UNESCO, Paris, and Parthenon Publishing Group, Carnforth, United Kingdom.

Naiman, R.J., and J.R. Sedell. 1981. Stream ecosystem research in a watershed perspective. Internationale Vereinigung für theoretische und angewandte Limnologie, Verhandlungen **21**:804–811.

Nestler, J.M, R.T. Milhous, and J.B. Layzer. 1989. Instream habitat modeling techniques. Pages 295–315 *in* J.A. Gore and G.E. Petts, editors. Alternatives in regulated river management. CRC Press, Boca Raton, Florida, USA.

Newbold, J.D., J.W. Elwood, R.V. O'Neill, and A.L. Sheldon. 1983. Phosphorus dynamics in a woodland stream ecosystem: a study of nutrient spiralling. Ecology **64**:1249–1265.

Odum, E.P., J.T. Finn, and E.H. Franz. 1979. Perturbation theory and the subsidy-stress gradient. BioScience **29**:249–352.

Odum, H.T. 1983. Systems ecology. John Wiley and Sons, New York, New York, USA.

Orth, D.J., and O.E. Maughan. 1982. Evaluation of the incremental methodology for recommending instream flows for fishes. Transactions of the American Fisheries Society **111**:413–445.

Peterson, N.P., and L.M. Reid. 1984. Wall-base channels: their evolution, distribution, and use by juvenile coho salmon in the Clearwater River, Washington. Pages 215–226 *in* J.M. Walton and D.B. Houston, editors. Proceedings of the Olympic Wild Fish Conference. Fisheries Technology Program, Peninsula College, Port Angeles, Washington, USA.

Pickett, S.T.A., and P.S. White, editors. 1985. The ecology of natural disturbance and patch dynamics. Academic Press, Orlando, Florida, USA.

Pringle, C.M., R.J. Naiman, G. Bretschko, J.R. Karr, M.W. Oswood, J.R. Webster, R.L. Welcomme, and M.J. Winterbourn. 1988. Patch dynamics in lotic systems: the stream as a mosaic. Journal of the North American Benthological Society **7**:503–524.

Pugsley, C.W., and H.B.N. Hynes. 1986. Three-dimensional distribution of winter stonefly nymphs, *Allocapnia pygmaea*, within the substrate of a southern Ontario river. Canadian Journal of Fisheries and Aquatic Sciences **43**:1812–1817.

Regier, H.A., R.L. Welcomme, R.J. Steedman, and H.F. Henderson. 1989. Rehabilitation of degraded river ecosystems. Pages 86–97 *in* D.P. Dodge, editor. Proceedings of the International Large River Symposium (LARS). Special Publication of the Canadian Journal of Fisheries and Aquatic Sciences **106**:1–629.

Resh, V.H., A.V. Brown, A.P. Covich, M.E. Gurtz, H.W. Li, G.W. Minshall, S.R. Reice, A.L. Sheldon, J.B. Wallace, and R.C. Wissmar. 1988. The role of disturbance in stream ecology. Journal of the North American Benthological Society **7**:433–455.

Resh, V.H., and D.M. Rosenberg, editors. 1984. The ecology of aquatic insects. Praeger Publishers, New York, New York, USA.

Schumm, S.A., and R.W. Lichty. 1956. Time, space and causality in geomorphology. American Journal of Science **263**:110–119.

Schwoerbel, J. 1967. Das hyporheische Interstitial als Grenzbiotop zwischen oberirdischem und subterranem Ökosystem und seine Bedeutung für die Primär-Evolution von Kleinsthöhlenbewohnern. Archiv für Hydrobiologie, Supplement **33**:1–62.

Scott, D., and C.S. Shirvell. 1987. A critique of the instream flow incremental methodology and observations on flow determination in New Zealand. Pages 27–43 *in* J.F. Craig and J.B. Kemper, editors. Regulated streams: advances in ecology. Plenum, New York, New York, USA.

Sedell, J.R., J.E. Richey, and F.J. Swanson. 1989. The river continuum concept: a basis for the expected ecosystem behavior of very large rivers? Pages 49–55 *in* D.P. Dodge, editor. Proceedings of the International Large River Symposium (LARS). Special Publication of the Canadian Journal of Fisheries and Aquatic Sciences **106**:1–629.

Shirvell, C.S. 1989. Ability of PHABSIM to predict chinook salmon spawning habitat. Regulated Rivers: Research and Mangement **3**:277–289.

Simons, D.B. 1979. Effects of stream regulation on channel morphology. Pages 95–111 *in* J.V. Ward and J.A. Stanford, editors. The ecology of regulated streams. Plenum, New York, New York, USA.

Spencer, C.N. 1991. Historical water quality changes evaluated through analysis of lake sediments. Flathead Basin Forest Practices/Water Quality and Fisheries Cooperative. Open File Report 123–91, Flathead Lake Biological Station, University of Montana, Polson, Montana, USA.

Spencer, C.N., B.R. McClelland, and J.A. Stanford. 1991. Shrimp stocking, salmon collapse and eagle displacement: cascading interactions in the food web of a large aquatic ecosystem. BioScience **41**:14–21.

Spies, M. 1986. Benthic matter and aquatic invertebrates, North Fork Flathead River, Montana, USA. Open File Report. Flathead Lake Biological Station, University of Montana, Polson, Montana, USA.

Stanford, J.A., and F.R. Hauer. 1992. Mitigating the impacts of stream and lake regulation in the Flathead River catchment, Montana, USA: an ecosystem perspective. Aquatic Conservation: Marine and Freshwater Ecosystems **2**, *in press*.

Stanford, J.A., and A.P. Covich, editors. 1988. Community structure and function in temperate and tropical streams. Journal of the North American Benthological Society **7**:261–529.

Stanford, J.A., and B.K. Ellis. 1988. Water quality: status and trends. Pages 11–32 *in* Our clean water: Flathead's resource of the future. Conference Proceedings. Flathead Basin Commission, Governor's Office, Helena, Montana, USA.

Stanford, J.A., B.K. Ellis, and C.N. Spencer. 1990. Monitoring water quality in Flathead Lake, Montana. Open File Report 115–90, Flathead Lake Biological Station, University of Montana, Polson, Montana, USA.

Stanford, J.A., and A.R. Gaufin. 1974. Hyporheic communities of two Montana rivers. Science **185**:700–702.

Stanford, J.A., and G.W. Prescott. 1988. Limnological features of a remote alpine lake in Montana, including a new species of Cladophora (Chlorophyta). Journal of the North American Benthological Society **7**:140–151.

Stanford, J.A., T.J. Stuart, and B.K. Ellis. 1983. Limnology of Flathead Lake. Flathead River Basin Environmental Impact Study, U.S. Environmental Protection Agency, Helena, Montana, USA.

Stanford, J.A., and J.V. Ward. 1979. Stream regulation in North America. Pages 215–236 *in* J.V. Ward and J.A. Stanford, editors. The ecology of regulated streams. Plenum, New York, New York, USA.

Stanford, J.A., and J.V. Ward. 1986. The Colorado River system. Pages 353–374 *in* B. Davies and K. Walker, editors. Ecology of river systems. Dr. W. Junk Publishers, Dordrecht, The Netherlands.

Stanford, J.A., and J.V. Ward. 1988. The hyporheic habitat of river ecosystems. Nature **335**:64–66.

Strahler, A.N. 1957. Quantitative analysis of watershed geomorphology. Transactions of the American Geophysical Union **38**:913–920.

Stream Solute Workshop. 1990. Concepts and methods for assessing solute dynamics in stream ecosystems. Journal of the North American Benthological Society **9**:95–119.

Swanson, F.J., and R.E. Sparks. 1990. Long-term ecological research and the invisible place. BioScience **40**:502–508.

Thorp, J.H., J.B. Wallace, and T.J. Georgian. 1986. Untangling the web of caddisfly evolution and distribution. Oikos **47**:253–256.

Townsend, C.R. 1989. The patch dynamics concept of stream community ecology. Journal of the North American Benthological Society **8**:36–50.

United States Forest Service. 1988. Region 5 soil and water conservation handbook (FSH 2509.22). Cumulative offsite watershed effects analysis. United States Department of Agriculture, San Francisco, California, USA.

Valett, H.M., S.G. Fisher, and E.H. Stanley. 1992. Physical and chemical characteristics of the hyporheic zone of a sonoran desert stream. Journal of the North American Benthological Society, *in press*.

Vannote, R.L., G.W. Minshall, K.W. Cummins, J.R. Sedell, and C.E. Cushing. 1980. The river continuum concept. Canadian Journal of Fisheries and Aquatic Sciences **37**:130–137.

Ward, G.M., A.K. Ward, C.N. Dahm, and N.G. Aumen. 1990. Origin and formation of organic and inorganic particles in aquatic systems. Pages 27–56 *in* R.S. Wotton, editor. The biology of particles in aquatic systems. CRC Press, Boca Raton, Florida, USA.

Ward, J.V. 1989. The four-dimensional nature of lotic ecosystems. Journal of the North American Benthological Society **8**:2–8.

Ward, J.V., and J.A. Stanford. 1983*a*. The serial discontinuity concept of lotic ecosystems. Pages 29–42 *in* T.D. Fontaine III and S.M. Bartell, editors. Dynamics of lotic ecosystems. Ann Arbor Science, Ann Arbor, Michigan, USA.

Ward, J.V., and J.A. Stanford. 1983*b*. The intermediate disturbance hypothesis: an explanation for biotic diversity patterns in lotic ecosystems. Pages 347–356 *in* T.D. Fontaine III and S.M. Bartell, editors. Dynamics of lotic ecosystems. Ann Arbor Science, Ann Arbor, Michigan, USA.

Ward, J.V., and J.A. Stanford. 1989. Riverine ecosystems: the influence of man on catchment dynamics and fish ecology. Pages 56–64 *in* D.P. Dodge, editor. Proceedings of the International Large River Symposium (LARS). Special Publication of the Canadian Journal of Fisheries and Aquatic Sciences **106**:1–629.

Weaver, T.M., and J. Fraley. 1991. Fisheries habitat and fish populations. Flathead Basin Forest Practices/Water Quality and Fisheries Cooperative. Draft Report. Montana Department of Fish, Wildlife and Parks, Kalispell, Montana, USA.

Webster, J.R., and E.F. Benfield. 1986. Vascular plant breakdown in freshwater ecosystems. Annual Review of Ecology and Systematics **17**:567–594.

Webster, J.R., and B.C. Patten. 1979. Effects of watershed perturbation on stream potassium and calcium dynamics. Ecological Monographs **49**:51–72.

Williams, D.D., and H.B.N. Hynes. 1974. The occurrence of benthos deep in the substratum of a stream. Freshwater Biology **4**:233–256.

Yount, J.D., and G.J. Niemi, editors. 1990. Recovery of lotic communities and ecosystems following disturbance: theory and application. Environmental Management **14**:515–762.

Part 2
Elements of Integrated Watershed Management

6

Fundamental Elements of Ecologically Healthy Watersheds in the Pacific Northwest Coastal Ecoregion

ROBERT J. NAIMAN, TIMOTHY J. BEECHIE, LEE E. BENDA,
DEAN R. BERG, PETER A. BISSON, LEE H. MACDONALD,
MATTHEW D. O'CONNOR, PATRICIA L. OLSON,
AND E. ASHLEY STEEL

Abstract

Characteristics of streams and rivers reflect variations in local geomorphology, climatic gradients, spatial and temporal scales of natural disturbances, and the dynamic features of the riparian forest. This results in a variety of stream types which, when coupled with the many human uses of the Pacific Northwest coastal ecoregion, presents a difficult challenge in identifying and evaluating fundamental, system-level components of ecologically healthy watersheds. Over 20 types of streams are found in western Oregon, Washington, and British Columbia and in southeastern Alaska, a region where extractive forest, agricultural, fishing, and mining industries and a rapidly increasing urban population are severely altering the landscape. Yet stream characteristics remain the best indicators of watershed vitality, provided the fundamental characteristics of healthy streams are accurately known. The premise of this article is that the delivery and routing of water, sediment, and woody debris to streams are the key processes regulating the vitality of watersheds and their drainage networks in the Pacific Northwest coastal ecoregion. Five fundamental components of stream corridors are examined: basin geomorphology, hydrologic patterns, water quality, riparian forest characteristics, and habitat characteristics. Ecologically healthy watersheds require the preservation of lateral, longitudinal, and vertical connections between system components as well as the natural spatial and temporal variability of those components. The timing and mode of interdependencies between fundamental components are as important as the magnitude of individual components themselves.

Key words. Stream, river, watershed, Pacific Northwest, riparian, sediment, woody debris.

Introduction

Watersheds in the Pacific Northwest coastal ecoregion play a vital role in shaping and supporting diverse cultures and professions. Specific characteristics of streams and rivers reflect an integration of countless physical and biological processes, providing a long-term memory of environmental conditions. Streams and adjacent riparian forests provide clean water, habitat for fish and wildlife, recreational opportunities, raw materials, and spiritual values for many American and Canadian cultures (Salo and Cundy 1987, Raedeke 1988, Naiman and Décamps 1990, Gregory et al. 1991). In combination, these features have enormous value but represent substantial managerial difficulties, especially when competing interests vie for limited resources. Difficult issues are magnified when natural ecological processes and their variability in space and time are poorly understood (Décamps and Naiman 1989).

For over a century, human alteration of the Pacific Northwest has proceeded without a basic understanding of watershed dynamics, especially for the stream corridors, or of the long-term consequences of land use alterations. Comprehensive studies of stream corridors and watersheds are rare, and most have been done since 1975 (Likens 1989, Strayer et al. 1986). Investigations of running waters as ecological systems are rudimentary, with major conceptual advancements only in the last decade (Vannote et al. 1980, Newbold et al. 1982, Hynes 1985, Naiman et al. 1988). Longitudinal, lateral, and vertical connections in controlling the ecological vitality of streams and rivers are recent concepts requiring additional investigation. Yet, in combination, these and subsequent studies point to running waters as ecological systems demonstrating considerable variability in space and time and requiring a high degree of connectivity between system components for the maintenance of long-term environmental health (see Stanford and Ward, this volume).

Connectivity is manifested in such features as access to required habitat during the life history of fish and wildlife (Salo and Cundy 1987, Raedeke 1988), stream temperature regimes that synchronize the migration and emergence of aquatic organisms (Sweeney and Vannote 1978, Quinn and Tallman 1987), riparian forests that regulate nutrient and material exchanges between forests and streams (Swanson et al. 1982, Peterjohn and Correll 1984), and hydraulic regimes that are within an accepted range for evolutionary adaptation of specific organisms (Statzner et al. 1988).

Variability is reflected in the wide variety of stream types in the coastal ecoregion (Naiman et al. 1991), resulting from differences in geology, climate, and stream size (Figure 6.1). Important geologic processes include

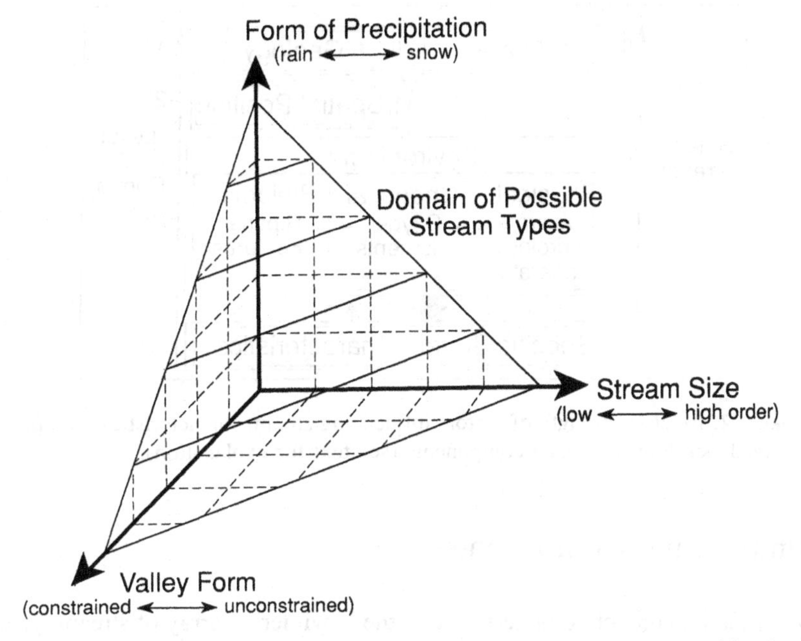

FIGURE 6.1. Principal factors influencing stream characteristics in the Pacific Northwest coastal ecoregion.

volcanism, continental and alpine glaciation, and erosional processes, including mass wasting. Climatic processes include the dominant form of precipitation (e.g., rain, snow, or transition), the seasonal timing, and the runoff patterns. Stream size is a function of the spatial position of the channel in the drainage network and upstream geomorphic and climatic processes. Thus a broad array of stream types may be found within a relatively small area, with each having different biological characteristics and presenting different managerial challenges.

The phrase *ecologically healthy* refers to functions affecting biodiversity, productivity, biogeochemical cycles, and evolutionary processes that are adapted to the climatic and geologic conditions in the region (Karr et al. 1986, Karr 1991). Collectively, these functions can be a measure of system vitality. Some tangible measurements of ecological healthy watersheds include water yield and quality, community composition, forest structure, smolt production, wildlife use, and genetic diversity. Our working hypothesis is that delivery and routing of water, sediment, and woody debris to the stream channel are the key processes determining the ecological health of watersheds in the Pacific Northwest coastal ecoregion. This article will present a broad overview of several fundamental components of ecologically healthy watersheds in this ecoregion, and present a conceptual model of principal interdependencies between those components.

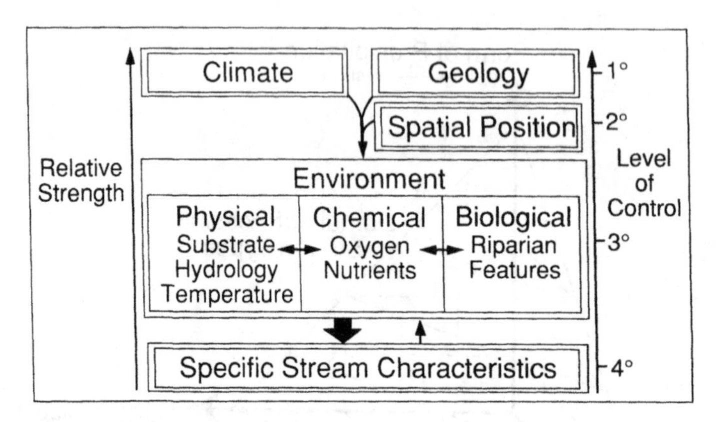

FIGURE 6.2. Relative strength of factors influencing stream characteristics, and principal feedback loops between components (see text for explanation).

Fundamental Components

An efficient approach to understanding the bewildering array of stream types is through a hierarchy of controlling factors (Frissell et al. 1986), with each assigned a ranking based on perceived ability to influence other components (Figure 6.2). However, components with a lower hierarchical ranking may influence higher ranked components through feedback loops (e.g., feeding strategies or disturbance; Starfield and Bleloch 1986, DeAngelis et al. 1986, Naiman 1988).

Based on this analysis, we chose five components that reflect different hierarchical levels of control on stream characteristics, that are integrative of species-specific processes (e.g., population dynamics), and that are essential for maintaining the long-term environmental vitality of drainage networks (Table 6.1). All processes influence either water, sediment, and woody debris delivery to the channel or the routing of those materials through the drainage network, or are responsive to material delivery and routing processes. These five fundamental components are basin geomorphology, hydrologic patterns, water quality, riparian forest characteristics, and habitat characteristics.

The geographic reference for our observations is limited largely to high precipitation basins west of the Cascade Mountains in northern California, Oregon, and Washington but may have application to the coastal regions of British Columbia and southeastern Alaska (Figure 6.3). The majority of rivers draining into the Pacific Ocean along the northwestern coast of North America originate in steep, mountainous terrains of the Northwestern Cordillera (McKee 1972). Along the coast of Washington and Oregon, the Northwestern Cordillera is represented by the Coast Range and the Cascade Mountains to the east. These watersheds have diverse geomorphic processes

Table 6.1. Six fundamental components of ecologically healthy watersheds in the Pacific Northwest Pacific Northwest coastal ecoregion.

Component	Approximate Hierarchical Level	Factors Considered	Sphere of Influence
1. Basin geomorphology	1°-2°	A. Physiographic and geologic setting B. Significant geomorphic processes C. Natural disturbance regimes	Effects all factors except climate
2. Hydrologic patterns	1°-2°	A. Discharge pattern flood characteristics and water storage B. Bedload and sediment routing C. Subsurface dynamics	Channel geomorphology and other physical characteristics, some aspects of chemical regime, riparian forest, and in-channel community dynamics
3. Water quality	3°-4°	A. Biogeochemical processes B. Fundamental parameters	Feedbacks to terrestrial vegetation and direct effects on chemical and biotic characteristics
4. Riparian forest characteristics	2°-3°	A. Light and temperature B. Allochthonous inputs C. Woody debris source	Most aspects of the physical, chemical, and biotic characteristics
5. Habitat characteristics	3°	A. Fish habitat preferences B. Fish community dynamics C. Spatial and temporal dynamics D. Woody debris accumulations E. Wildlife communities F. Trophic pathways	Influence in other biotic communities in stream and strong feedbacks to physical, chemical, and terrestrial dynamics

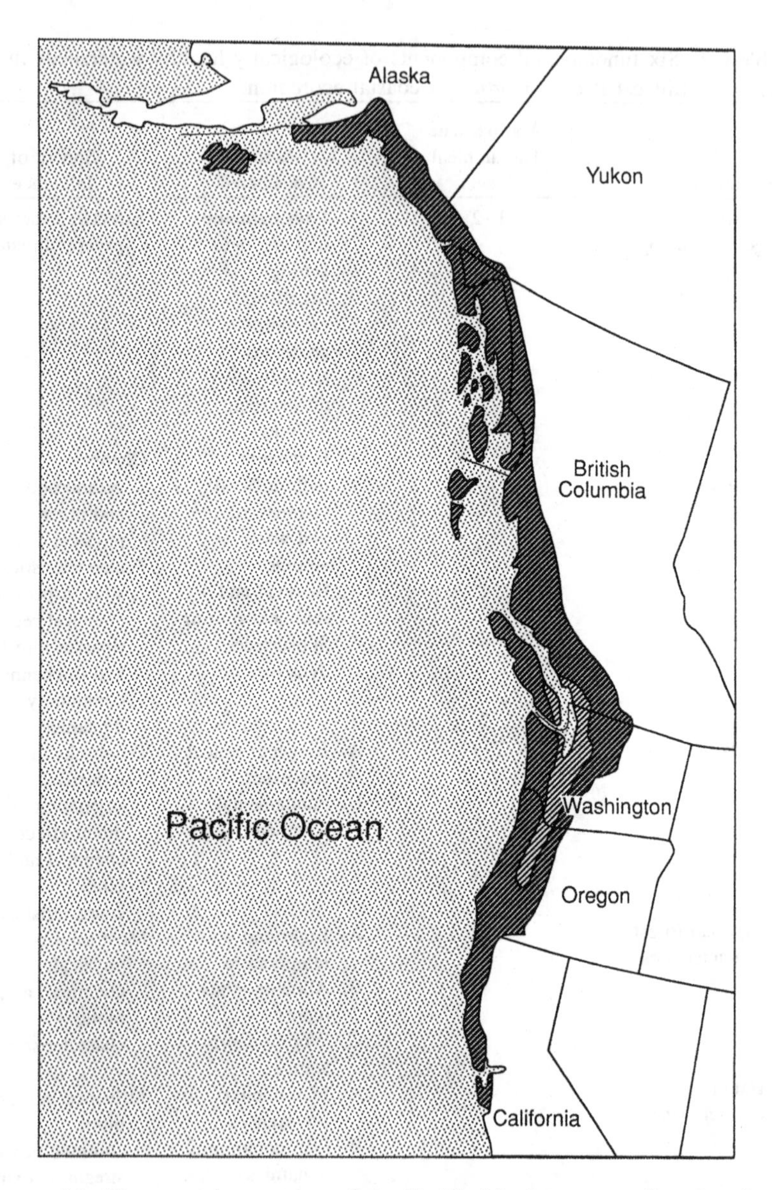

FIGURE 6.3. The coastal ecoregion of the Pacific Northwest as referred to in this article.

that vary along a longitudinal gradient extending from mountain ridges to the sea. Subduction of oceanic crust beneath continental lithosphere along the Pacific Coast results in steep, high elevation mountains. Heavy precipitation due to orographic lifting of moist air from the Pacific Ocean, together with the steep terrain, results in erosion dominated by mass wasting. The geology of the Pacific Northwest mountains contains a diversity of lithologies, including marine sedimentary sequences, metasedimentary rocks, various old and more recent volcanics, and granitic plutons (McKee 1972). Suspect terrain, or rafted continental crust, also contributes to the geologic complexity.

The most recent continental glaciation ended approximately 14,000 years B.P. and significantly altered the landscape of northwestern Washington (north of Centralia), British Columbia, and parts of southeastern Alaska (Crandell 1965). The retreating ice sheets deposited thick layers of till, lake clay, and outwash sand in valley bottoms and the Puget Lowland. Typically, major rivers and many smaller tributaries within these regions continue to incise through glacial sediments.

Basin Geomorphology

Natural Disturbances and Stream Size

Basin geomorphology in ecologically healthy watersheds is affected by the type, frequency, and intensity of natural disturbances. Characteristics of these disturbances reflect the spatial position of channel segments in the drainage network. Processes of material delivery and routing are highly interactive with hydrologic patterns and riparian vegetation.

Geomorphic Processes and Forms in Low-Order Channels.

In the Pacific Northwest, low-order (e.g., first- and second-order) stream segments represent >70% of the cumulative channel length in typical mountain watersheds (Benda et al. 1992). Hence low-order channels are the primary conduits for water, sediment, and vegetative material routed from hillslopes to higher-order rivers. First- and second-order basins are naturally prone to catastrophic erosion because steep slopes adjacent to steep channels favor landslides and debris flows.

In the few studies conducted in the coastal ecoregion, first- and second-order channels in steep bedrock of mountain basins do not transport significant quantities of sediment by water flow; hence these channels have limited amounts of stored alluvium (Figure 6.4, top). Swanson et al. (1982) estimated that fluvial transport accounted for approximately 20% of the total sediment yield from a first-order basin in the central Oregon Cascades. In the Oregon Coast Range, Benda and Dunne (1987), using a sediment budget approach that included specifying sediment routing by debris flow, estimated that fluvial processes accounted for 10 to 20% of the total sediment yield.

Low-order channels are filled primarily with colluvium, characterized by coarse, unsorted sediments, including silts and clays derived from landslides and debris flows (Benda and Dunne 1987). Low-order channels also contain substantial amounts of boulders (>27 cm diameter) and woody debris (>10 cm diameter) because streamflow is not competent to transport the largest

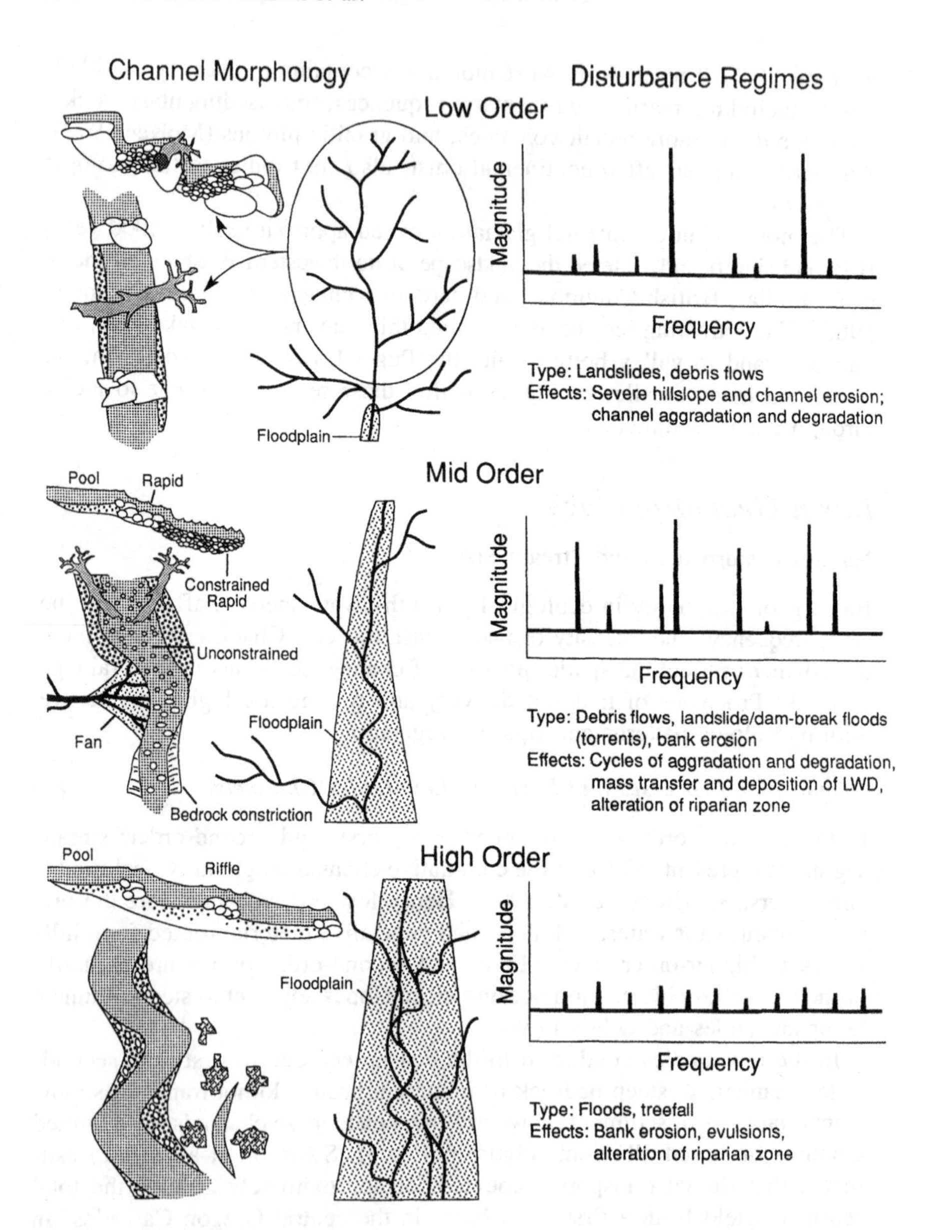

FIGURE 6.4. Generalized channel morphology and disturbance regimes associated with low-, mid-, and high-order streams in the Pacific Northwest coastal ecoregion.

materials. First- and second-order channels have steep gradients (typically >8 degrees), but have lower gradients over shorter (<10 m) lengths due to accumulations of woody debris organized into debris jams which create a stepped longitudinal profile (Swanson and Lienkaemper 1978). Woody debris jams store sediment and contribute to discontinuous sediment transport during storms (Heede 1972, Mosley 1981). Bedforms in steep first- and second-order channels are generally limited to plunge pools in coarse-textured substrate, migrating gravel sheets originating from streamside landslides, and local accumulations of finer sediment upstream from woody debris and boulder obstructions. Woody debris jams can account for the majority of energy loss of flowing water in steep, low-order channels (Beschta and Platts 1986), implying that less energy is available for sediment transport.

Landslides occur on steep slopes during large rainstorms or rain-on-snow events (Swanston and Swanson 1976). Furthermore, wildfires kill vegetation, and loss of root strength may result in landslide and debris flow activity during large rainstorms (see below). Certain landslides at the heads of low-order valleys are transformed into debris flows, eroding accumulated sediment and organic debris from first- and second-order channels (Benda and Dunne 1987). Debris flows are one of the most common forms of mass wasting in mountain watersheds of the coastal ecoregion, and are the principal process transporting sediment and woody debris in first- and second-order channels (Swanson et al. 1985). Recurrence intervals of debris flows have been estimated to be once in 500 years for first-order channels in the central Oregon Cascades (Swanson et al. 1982), and once in 750–1,500 years for first- and second-order channels in the Oregon Coast Range (Benda and Dunne 1987).

Geomorphic Processes and Forms in Mid-Order Channels.

Third- through fifth-order channels, referred to as mid-order channels, are characterized by moderate to steep gradients (1 to 6 degrees), substrates ranging from boulders to gravels (or sands in some streams), and abundant large organic debris in jams and single pieces (Figure 6.4, middle). True alluvial channel systems in the coastal ecoregion begin at third-order channels (e.g., where the majority of sediment and organic debris is transported by water).

Mass wasting often dominates the amount and type of deposition along narrow valley floors of mid-order channels. Debris flow deposits in confined valleys create fans at mouths of first- and second-order basins, and levees and terraces along valley floors (Benda 1990). Streams are often forced toward the opposite valley wall by alluvial or debris flow fans. For example, 65% of stream meanders in a fifth-order basin in the Oregon Coast Range with a sinuosity of >1.1 were maintained by debris flow fans (Benda 1990). Landslides and debris flows also deposit large volumes of sediment and vegetative material (1,000 – 10,000 m^3) directly into mid-order channels. Local channel aggradation and logjams persist at the site of deposition for years to decades (Benda 1988, Perkins 1989). In the central Oregon Coast Range, debris flows control the distribution of boulders in alluvial channels (Benda 1990).

Earthflows often impinge on third- through fifth-order valley floors, causing local constrictions and streamside landslides (Swanson et al. 1985). Deposition of entire trees and large volumes of sediment cause local channel aggradation at the site of earthflows and immediately downstream. Streams within narrow valley floors may impinge directly on hillslopes, causing erosion in the form of rockfall in nonglaciated basins, or landslides or slumps in glacial deposits, which further affects channel morphology.

An important catastrophic process in mid-order channels is the landslide/dam-break flood. Deposits of landslides and debris flows often temporarily dam streams along narrow valley floors (Figure 6.4). A rapid failure of the dam and subsequent release of the ponded water may cause an extreme flood, destroying riparian vegetation and removing in-channel organic debris. Accelerated sediment transport by streamflow follows in the wake of the dam-break flood.

Channel form is influenced by mass wasting deposits (see above) but also reflects normal alluvial processes. Bedforms include alternate pool and riffle morphology at meanders, and migrating gravel sheets originating from local inputs of sediment. In addition, a kind of bedform called scour lobe occurs as flow convergence erodes channel beds, forming pools, and flow divergence downstream deposits bars or gravel lobes (Lisle 1987). Scour around bedrock bends may also produce scour lobes. Channels dominated by large substrate may develop a boulder-cascade or step-pool morphology. This bedform is characterized by organized aggregations of boulders which create a sequence of cascades and boulder pools (Whittaker and Jaeggi 1982, Grant et al. 1990). Pools tend to be deep in relation to channel widths; width-depth ratios are 1:10.

Third- through fifth-order channels transport the majority of sediment delivered to them rather than storing it for long periods, because of relatively steep gradients, well-armored channel banks, and narrow valley floors. Channels with these characteristics tend to contain flood flows. Hence terraces, overflow channels, and oxbow lakes tend to be limited in these areas. Channels are typically single thread, with the exception of diversion of streams around woody debris jams and mass wasting deposits.

Mass wasting influences the temporal distribution of sediments in alluvial channels, either because it deposits sediment directly into alluvial channels or because the deposits provide a source for accelerated sediment transport further downstream (Benda and Dunne 1987). The stochastic nature of sediment supply to alluvial channels by debris flows and landslides may promote significant temporal variation of bedforms. For example, in the Oregon Coast Range, episodic debris flows are responsible for channel aggradation which results in a gravel-bed morphology, while the absence of debris flows results in a mixed bedrock- and boulder-bed morphology at the scale of stream reaches to entire tributaries (Benda 1990). In addition, spatial variability in channel form in mid-order streams can be significant due to frequent deposits of sediment and organic debris from landslides, debris flows, and earthflows.

Geomorphic Processes and Forms in High-Order Channels.

Large rivers of sixth order and higher integrate the diversity of erosional processes in time and space (Figure 6.4, bottom). Hence sediment supply is more steady in time, and as a result the channel form (that which is dependent on the sediment supply rate, such as pools) is more uniform in space. In addition, extensive alluvial terraces and floodplains isolate the river from direct contact with hillslopes and low-order tributary basins, and therefore limit the direct influences of mass wasting.

Large rivers sort sediment by size or selectively transport it along the longitudinal gradient from third- through sixth- and higher-order channels. As gradient decreases and channels widen, transport of large sediment decreases. The coarsest sediment is found in upper watersheds, often adjacent to mass wasting deposits; the finest sediment, such as sand and small gravel, is in the lower reaches (Brierley and Hickin 1985). Particle comminution or abrasion further limits sediment size downstream. General exceptions to this decline in substrate size downvalley include local accumulations of larger particles at tributary confluences, aggradation of finer sediment behind debris jams, and landslide and debris flow deposits.

Large discharge and easily erodible banks in large rivers favor the development of meandering floodplain channels, creating alternating pool and riffle morphology (Dunne and Leopold 1978). Other common bedforms in large rivers include mid-channel bars formed by organic or inorganic obstructions, and transverse bars formed by flow separation due to changing channel geometry (Figure 6.4, bottom). The scale and dimension of the pool-riffle morphology depend on bank height and composition, size and type of riparian vegetation, size of bedload, discharge regime, and so forth. Rivers tend to be slightly deeper and significantly wider than steep stream channels upvalley; width to depth ratios range from $\sim$10 to $>$20.

Lateral migration of rivers occurs continuously and does not depend on extreme events, though migration may occur more rapidly during large floods. Evulsions are common and multiple thread channels are formed, often during flood events, because of weak bank deposits. This leads to the development of floodplains kilometers wide containing numerous active and semi-active channels (Sedell and Froggatt 1984). Meander cutoffs create oxbow lakes, the size and number depending on meander history of the river, width of the floodplain, and groundwater characteristics of the alluvial plain. Wetlands become numerous within and along cutoff meanders and oxbow lakes. If the hydrologic regime is characterized by a high flood frequency, lateral migration may also occur more frequently. Hence vegetation patterns along these types of rivers will contain young successional stages and perhaps a greater heterogeneity of stand ages.

Small tributaries, originating from steep areas or as springs, cross alluvial terraces and floodplains with larger rivers. These small channels are similar, though smaller in scale, to larger alluvial channels, and they have been referred to as wall-base channels (Peterson 1982*a*, *b*). Wall-base channels meander through easily erodible alluvial sediments and their bedforms are dominated by alternate pools and riffles, scour lobes, and plunge pools.

Erosion and Sedimentation in Ecologically Healthy Watersheds

While erosion and sedimentation are often viewed negatively from a biological point of view, they are essential to the ecological functioning of aquatic and terrestrial communities because they provide the sources and the surfaces necessary for habitat. In mountain regions in particular, erosion and sedimentation are often violent (e.g., landslides, debris flows, landslide/dam-break floods, and snow avalanches) and produce mortality among terrestrial and aquatic organisms. Geomorphic surfaces in streams or on land in the wake of these powerful processes often evolve into productive and biologically attractive sites because of the revitalization of geochemical cycles, introduction of buried and unburied organic debris, and opening of forest canopies, thus increasing sunlight. In ecologically healthy watersheds of the Pacific Northwest, these violent geomorphic processes vary considerably in extent and frequency. Disturbance regime is the term generally used in discussing the results of these processes.

A disturbance is any significant fluctuation in the supply or routing of water, sediment, or woody debris which causes a measurable change in channel morphology and leads to a change in a biological community (Pickett and White 1985). The disturbance regime, therefore, is the type, frequency, magnitude and spatial distribution of changes in biological communities. Changes in the supply or the routing of sediment and organic debris are usually the result of local mass wasting or large floods. Some specific effects of mass wasting deposits on valley floors and channels have been covered in the previous sections on low- and mid-order channels. Floods not associated with mass wasting have less of an influence on low- to mid-order rivers (Grant 1986).

Although fluvial transport of sediment is often limited by supply of sediment, large floods can generate considerable bank erosion. Moreover, during floods, sediment transport resulting from bank erosion in low-order channels reaches its maximum. Hence the potential for bedload waves propagating from low- to mid-order streams exists during floods. As these waves pass downstream, they may locally cause accelerated bank erosion, thereby creating a positive feedback mechanism or a complex fluvial response. Given the potential for significant sediment transport by floods, watershed influences on the magnitude and timing of annual (and less frequent) floods are contributing determinants of natural stream characteristics. The influence of flood regimes (associated largely with climate and elevation) and watershed characteristics (such as geomorphology) is discussed in the following section.

Erosion in mountain terrains of the coastal ecoregion is dominated by mass wasting. Therefore, triggering mechanisms such as rainstorms and wildfires were responsible for episodic transfers of sediment and organic debris from hillslopes and low-order channels to third- and higher-order streams and rivers (Swanson et al. 1982, Benda 1990). Past tense is used because wildfires are typically suppressed, while timber harvest activities have created a new pattern of disturbance across landscapes. Little is known about frequencies, magnitudes, and spatial distributions of historic wildfires and rainstorms, and therefore characteristics of natural erosional patterns (or disturbance regimes) in large watersheds are not well understood.

Wildfires, of varying intensities and in various locations, are controlled, to some extent, by climate and local topography (Teensma 1987). In Mount Rainier National Park, Washington, Hemstrom and Franklin (1982) estimated average recurrence intervals for stand-resetting wildfires to be ~450 years. In coastal forests of the Olympic Peninsula and Oregon Coast Range, a recurrence interval of a large stand was about two centuries (Agee 1990). In contrast, recurrence intervals of rainstorms that initiate landslides and debris flows are a few decades or less (Pierson 1977). A wildfire followed by large storms has the potential of inducing a spate of erosion that may result in sedimentation, and therefore a large channel disturbance over stream reaches to entire drainage networks (Benda 1990).

Erosion- and sedimentation-related disturbances in channels exhibit spatial variability because storms and fires act on different areas at different times. The scale of heterogeneity may range from individual small basins (and channels) to a major portion of a large watershed (numerous channels).

Several aspects of disturbance regimes are important to the functioning of biological communities in mountain watersheds. Unfortunately, knowledge of natural disturbance regimes is limited because of the length of time required for the processes to operate (100 to 1,000 years) and therefore to be observed by humans, and because recent use has altered the disturbance regimes in ways not fully understood. Nevertheless, we are able to conceptualize some attributes of natural disturbance regimes.

Disturbances related to erosion and sedimentation in low- to mid-order basins are thought to be characterized by high magnitude and low frequency because of the large number of mass wasting events and the potential for wildfires to range over these basins (Swanson et al. 1982, Benda 1990) (Figure 6.4). Within larger watersheds (mid-order, Figure 6.4) asynchronous combinations of fires and rainstorms act to limit the magnitude of the sediment and organic debris movement downstream. Though events may not be coupled in time within a larger watershed, they occur (collectively) more frequently because of the greater area and therefore greater probability of occurrence. This creates a disturbance regime, as far as sediment flux is

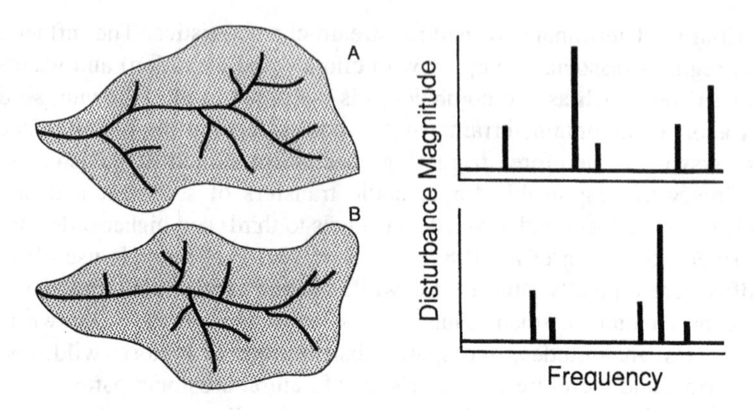

FIGURE 6.5. In ecologically healthy watersheds there is a natural spatial and temporal heterogeneity of disturbances between subbasins.

concerned, characterized by lower magnitude and increasing frequency (Figure 6.4, middle). In high-order rivers, the integration of many occurrences of erosion and the isolation of larger systems from direct influences of mass wasting conspire to further reduce the magnitude of sediment-related disturbances while increasing their frequency (Figure 6.4, bottom).

Spatial heterogeneity of erosion and sedimentation due to asynchronous fires and rainstorms may lead to adjacent basins having decoupled patterns of disturbance (Figure 6.5). In this example, the degree of disturbance can be thought of as time since the last event and the magnitude of that event. This potential habitat variability can be viewed as a habitat diversity. At the scale of a larger watershed, stochastic occurrences of mass wasting lead to basins having a unique variety of disturbance regimes. This results in a spatial and temporal distribution of subbasins in states of alternating low, intermediate, and high disturbance.

Therefore, the type, intensity, and frequency of erosional events and their spatial distribution across landscapes are important considerations to understanding the relationships between geomorphic process, form, and ecological functioning of watersheds. The temporal and spatial scales at which these processes occur, however, complicate their study. Our minimal knowledge of natural disturbance regimes limits our understanding of the functioning of ecologically healthy watersheds over long periods and large spatial scales, thus precluding accurate environmental assessments of the long-term effects of land use in watersheds in the coastal ecoregion.

Hydrologic Patterns

Hydrologic patterns of ecologically healthy watersheds in the coastal ecoregion are strongly related to the timing and quantity of flow, characteristics of seasonal water storage and source area, and the dynamics of surface-subsurface (e.g., hyporheic) exchanges.

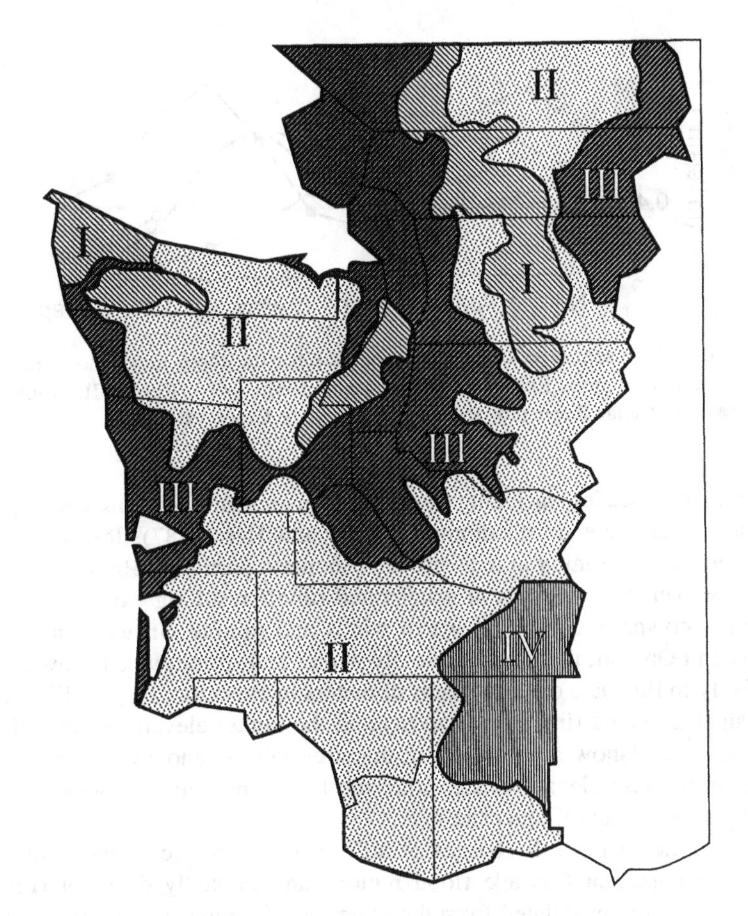

FIGURE 6.6. Flood regions of western Washington, USA.

Timing and Quantity of Flow

Distinct flood regimes in the coastal ecoregion of Washington have been identified in two analyses of regional stream gauging data (Bodhaine and Thomas 1964, Cummans et al. 1975). A map of these flood regions provides perspective on the spatial variability of runoff intensity (Figure 6.6).

Comparison of the Stillaguamish River (Region I), Nooksack River and Cedar River (Region II), and Pilchuck Creek (Region III) provides an example of basic differences in amount and form of precipitation, seasonal timing, and water storage within a region (Figure 6.7). Region I comprises the western Cascades and the northwestern portion of the Olympic Peninsula. The Cascade Mountain portion of Region I is located in the atmospheric convergence zone east of the Strait of Juan de Fuca where air masses moving north through the Puget Lowland collide with air moving east through the Strait, causing heavy precipitation. Region II covers most of the rest of

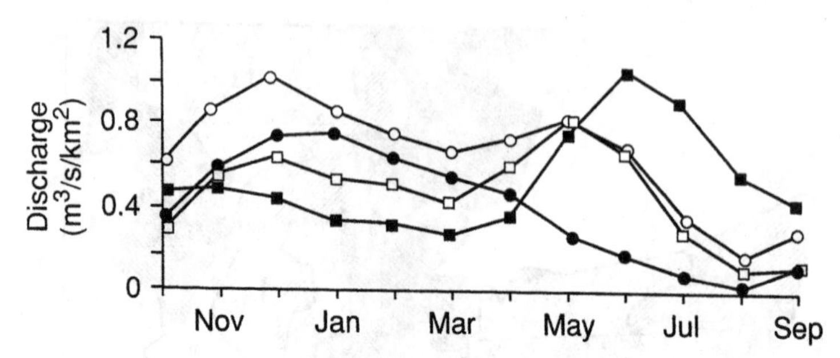

FIGURE 6.7. Representative annual flow regimes for the three major flood regions of western Washington. Region I: Stillaguamish River (-O-); Region II: Nooksack River (■) and Cedar River (□); Region III: Pilchuck Creek (●).

the western Cascades, including intermediate mountain elevations where rain-on-snow events are most common (Berris and Harr 1987). Rain-on-snow events are also common in the Cascade Mountain portion of Region I. Rain-on-snow events are periods of accelerated runoff resulting from melting of a warm snowpack caused by warm winds and rainfall. In western Washington and Oregon, these occur in early winter and spring, when snow levels are likely to fluctuate over hundreds of meters in elevation. Region III covers the Puget Lowland (in a rain-shadow) and the higher elevations west of the Cascade crest (snow dominated); in this area rain-on-snow events are least common because elevations are either too low to produce frequent snow or too high to produce frequent rain.

The predicted peak discharges of a 25-year recurrence interval flood in the three dominant Cascade flood regions are markedly different (Figure 6.8). Discharges calculated from the regression formulae of Cummans et al. (1975) (standardized to units of discharge per unit watershed area with standard errors of 32, 39, and 50% in Regions I, II, and III, respectively) show that Region I produces the highest discharge, Region III the lowest, and Region II an intermediate amount. In addition, smaller drainage areas produce more runoff per unit area than larger ones, regardless of region. This reflects the decreasing likelihood of a storm covering the entire watershed as watershed area increases, and the increase in floodplain storage (e.g., lakes and wetlands) as watershed size and stream order increase.

The predicted peak discharges of 2-year, 25-year, and 100-year recurrence interval floods in Region I show a similar pattern of discharge per unit watershed area (Figure 6.9). As before, the discharge was standardized and calculated (after Cummans et al. 1975) for hypothetical watersheds of three different sizes. Standard errors of the estimates are 25, 32, and 40% for the 2-year, 25-year, and 100-year recurrence interval floods, respectively. Predicted flood discharge decreased with increasing drainage area. The 25-year flood was about twice the magnitude of the 2-year flood; the 100-year flood was about triple the magnitude of the 2-year flood (Figure 6.9).

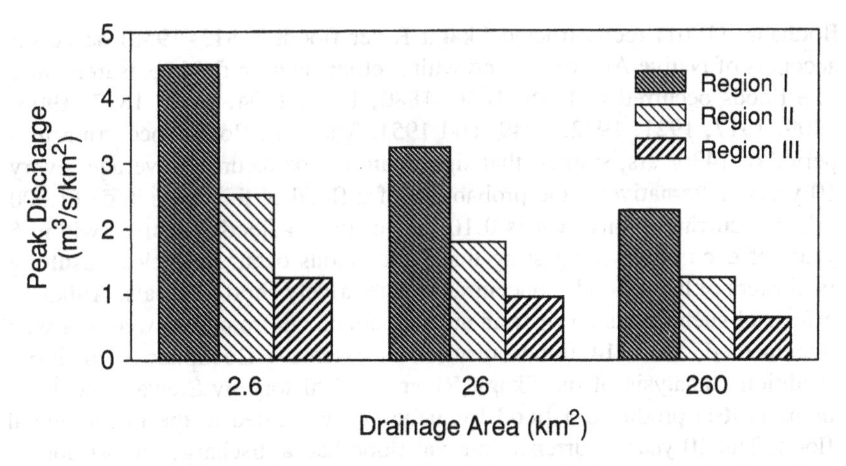

FIGURE 6.8. Estimated 25-year recurrence interval peak discharges for the three dominant flood regions in western Washington, by drainage area.

The significance of floods to high-order channels and associated riparian forests is as a disturbance creating heterogeneous habitat and as a recharge source for alluvial aquifers. Identifying a characteristic flood threshold at which significant disturbance occurs for a given stream, location, or region may be desirable to assess ecological effects. Once identified, such thresholds could define the spatial and temporal limits of disturbance in floodplain and riparian environments. This concept is best explored in the context of the following case study.

The potential for defining an ecologically significant flood recurrence interval is suggested by a study of the Skagit River, Washington. Stewart and

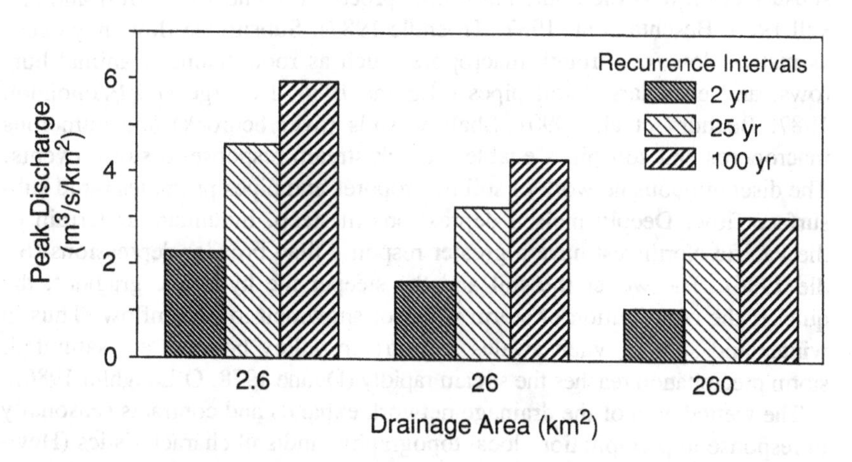

FIGURE 6.9. Predicted peak discharge of 2-year, 25-year, and 100-year recurrence interval floods in Region I (western Washington, USA) by drainage area.

Bodhaine (1961) reconstructed Skagit River floods (1815–1958) based on accounts of Native Americans and white settlers and on field measurements. The floods occurred in 1815, 1856, 1880, 1882, 1894, 1896, 1897, 1906, 1909, 1917, 1921, 1932, 1949, and 1951. These 14 floods, occurring in a period of 143 years, suggest that significant floods recur, on average, every 10 years. Alternatively, the probability of a flood of this magnitude (4,000 m^3/s) occurring in any year is 0.10. Six of the 14 floods occurred within 5 years of each other, suggesting that large floods occur in cycles, resulting in clusters of large floods concentrated in relatively short intervals. Although effects of such floods on riparian and floodplain environments were not well documented, these 14 floods resulted in extensive floodplain inundation. Additional analysis of the Skagit River flood history by Stewart and Bodhaine (1961) produced a flood frequency curve scaled to the mean annual flood. The 10-year recurrence interval flood had a discharge almost double the mean annual flood—suggesting, perhaps, a threshold. Further hydrological and ecological analyses would be necessary to determine whether the 10-year flood recurrence interval can be matched with patterns of riparian vegetation succession, thus providing evidence of an ecologically significant flood threshold.

Runoff Processes

Runoff processes influence quantity, quality, and timing of surface and subsurface flow. Water routing influences riparian vegetation, nutrient inputs, and stream productivity. Processes acting on runoff before the flow reaches the channel are numerous (Dunne 1978). In the heavily forested Pacific coastal ecoregion, overland flow is usually not an important process because infiltration capacities greatly exceed precipitation intensity (Harr 1976). Lateral subsurface flow is the dominant runoff process (Dunne 1978, Burt and Arkell 1986, Beschta et al. 1987, Troendle 1987). Subsurface flow may occur as matrix flow, or through macropores such as root channels, animal burrows, and even larger soil pipes (Higgins 1984, Roberge and Plamondon 1987, Beaudry et al. 1990). Shallow soils (over bedrock) and numerous macropores and soil pipes enable a quick stream response to storm events. The discontinuous network of soil macropores and pipes promotes rapid subsurface flow. Deeply incised depressions in steep, mountainous terrain of the Pacific Northwest have a greater response than broader depressions (Sidle 1986). The wetter the soil and the steeper the hydraulic gradient, the quicker the contribution of storm water or snowmelt to streamflow. Thus in winter and spring, when swales and stream valley bottoms are saturated, storm precipitation reaches the stream rapidly (Dunne 1978, O'Loughlin 1986).

The wetted area of the drainage network expands and contracts seasonally in response to precipitation, local topography, and soil characteristics (Hewlett and Nutter 1970, Dunne 1978, Burt and Arkell 1986). Investigations in humid mountainous regions throughout the world substantiate the applica-

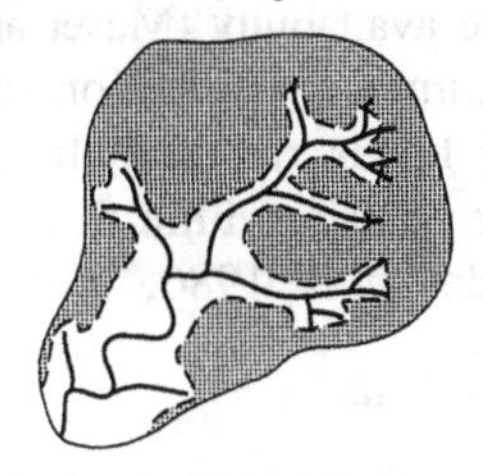

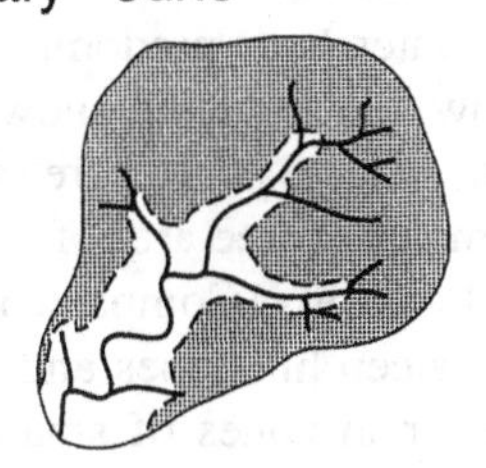

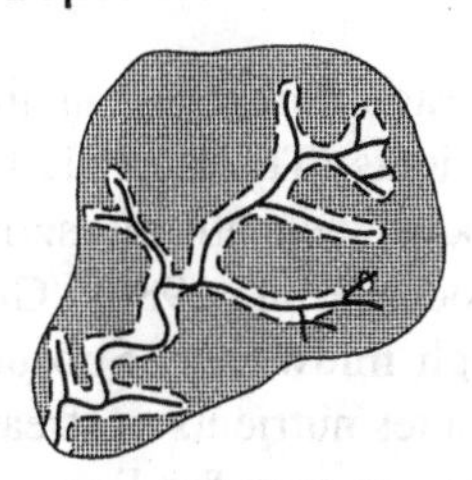

FIGURE 6.10. Variable source area concept for water availability and storage for three hydrologic seasons : October – January, February – June, and July – September.

bility of this variable source area concept (Swistock et al. 1989, Wolock et al. 1989, Sunada and Hong 1988, Fritsch et al. 1987, Rhodes et al. 1986, O'Loughlin 1986, Satterlund 1985, Troendle 1985, Bruijnzeel 1983) (Figure 6.10). These studies conducted outside the Pacific Northwest show enough similarities to extend the concept to the Pacific Northwest coastal ecoregion.

Generally, variable source area has been applied only to runoff in small watersheds. However, the concept should be extended to larger watersheds or to explain other processes, such as nutrient cycling and characteristics of riparian forests that are important to the stream ecosystem (Rhodes et al. 1986, Oliver and Hinckley 1987, Burt 1989). The dynamics of the variable source area are important where runoff timing and duration influence ecological processes. For example, during summer drought, source areas contract as drainage from hillslopes decreases (O'Loughlin 1986). As soil temperature rises, biological activity increases in the shallow soil mantle until it is eventually reduced by low moisture availability (Maser and Trappe 1984, Waring and Schlesinger 1985). In autumn, precipitation increases and the variable source area expands. Water levels rise and the groundwater recharge phase begins. Increased water contact within the soil enhances the capture of carbon and nutrients (Rhodes et al. 1986, Wolock et al. 1989).

Runoff Processes in Low-Order Watersheds.

Variability in channel pattern and flow regimes creates a habitat mosaic along the longitudinal and lateral axes of streams (Minshall 1988, Pringle et al. 1988). At high elevations, low-order watersheds predominate. Topography, soil attributes, and initial source of flow (e.g., glacial, snow, seeps, hillslope runoff) influence the flow regime and variable source area for the low-order channels. Low-order watersheds are major source areas for downstream surface water and for recharge of alluvial aquifers (Compana and Boone 1986). However, water storage is limited by steep hillslopes and shallow soils adjacent to the channel. Exceptions occur in zones of small floodplains and wet meadows where local subsurface water may contribute to streamflow, in alpine areas during snowmelt, and in permanent snowfields or glacier-fed streams.

Subsurface areas may act as a storage compartment in winter and as a source for organic and inorganic nutrients (Triska et al. 1989*a*). Deep percolation and groundwater recharge occur beneath the snowpack (Compana and Boone 1986, Munter 1986, Rhodes et al. 1986). Groundwater levels and soil moisture remain high through midwinter. Subsurface flow is sufficient to sustain baseflow and contributes nutrients to streams during winter.

At high elevations, the variable source area for lower-order streams expands substantially during snowmelt as soil layers become saturated (Figures 6.4 and 6.10). When soil is fully saturated, overland flow is generated by continued snowmelt. Overland flow is important for providing additional nutrients to the stream. For example, nitrate-nitrogen concentrations are higher in overland flow than in groundwater flow, with the highest concentrations occurring during peak discharge (Rhodes et al. 1986). This nitrogen pulse is important for providing nutrients to the nitrogen-limited streams of the coastal ecoregion.

In midsummer, water levels recede and the variable source area contracts as soil water is withdrawn and transpired by vegetation (Harr 1976, Walters et al. 1980). The subsurface hydrologic gradient toward the stream increases as stream water levels subside. The subsurface storage compartment drains relatively quickly toward the stream (zone of convergence) transporting nutrients and carbon from the riparian forest (Peterjohn and Correll 1984, Triska et al. 1989*a*, *b*). During drier periods, the variable source area may shrink to the channel, with the channel then acting as the primary storage compartment.

Runoff Processes in Mid-Order Watersheds

Alluvial stream processes dominate in third- through fifth-order watersheds and channels. Variable source area and storage expand as the stream valley widens. However, surface storage in oxbow lakes and overflow channels remains limited (Benda 1990). Water movement from the hillslopes and the level of soil saturation at the hill bottom are still important runoff components (O'Loughlin 1986). One result is formation of wetlands at the base of hillsides. Wetlands are important storage areas for flow, expanding the variable source area through the season (Holland et al. 1990, Carter 1986, Wald and Schaefer 1986). In the upper reaches, riparian wetlands are restricted due to narrowness of the valley. However, wetlands form landward of fluvial floodplain features such as scroll bars, and on alluvial fans deposited by tributaries at the base of hillslopes (L. Mertes, University of California, Santa Barbara, pers. comm.). Beaver (*Castor canadenis*) also contribute to the formation of storage features (Naiman et al. 1988). During dry seasons, some wetlands become part of the variable source area; however, others are perched and do not act as a water or nutrient source to the stream, but as a sink (Beaudry et al. 1990).

Local groundwater aquifers are recharged by ample winter precipitation and snowmelt and are an important component of streamflow and the variable source area throughout the year (Walker 1960, Hewlett and Nutter 1970, Freeze and Cherry 1979, Burt and Arkell 1986). Overbank flooding provides additional recharge to groundwater and wetlands. Flooding duration is shorter than in higher-order channels. For example, the two-year flood duration in a mid-order stream (at 500–1,000 m elevation) is approximately 1–2 days. The two-year flood duration for higher-order streams is approximately 3–4 days (estimated from the USGS – Water Resource Data Reports, 1950–86). Aggradation due to mass wasting or episodic debris flows can increase the flood stage and duration for mid-order streams.

Runoff Processes in High-Order Watersheds

Higher-order (>sixth-order) watersheds and channels occur at lower elevations where the dominant form of precipitation is rain. As the stream valley broadens, the variable source area expands and remains more spatially and temporally stable (Figures 6.4 and 6.10). Tributaries still exhibit seasonal expansion and contraction, but not to the degree seen in lower-order streams at higher elevations. Flow hydrographs (Figure 6.7) indicate that winter storm and snowmelt are the dominate flow generators. Tributaries confined to lower elevations show no influence from rain-on-snow events; higher-order streams are still influenced by rain-on-snow events.

At low and moderate flow conditions, the main river channel meanders across the floodplain (Leopold et al. 1964). Over time, cutoff channels, oxbow lakes, meander scrolls, and other backwater and high water channels develop on the broad valley floor. Along some reaches, logjams redirect flow into secondary channels, or the river forms new channels (Walker 1960, Lienkaemper and Swanson 1987). Increases in depression storage serve to lengthen the time of ponding, and to dampen flood peaks, while prolonging the duration of inundation. Nonperched systems supply water to subsurface systems following flood recession (Mitsch and Gosselink 1986, Wald and Schaefer 1986). When floodplains become inundated by overbank and breach flow (Hughes 1980), the riparian vegetation decreases water velocities due to increased resistance, causing an increase in flood stage (Dunne and Leopold 1978, Ponce and Lindquist 1990).

Precipitation, side valley runoff, and groundwater seeps contribute to water inputs and erosion of the floodplain (Dunne 1978, Higgins 1984). Major floods occur during periods of maximum soil moisture and highest subsurface and wetland water levels, thus there is little room for temporary storage. Some floodwaters are placed in long-term storage in groundwater and deeper wetlands. Flooding is not restricted just to snowmelt or storm events. Lesser magnitude floods are due to accumulated sediment, beaver dams, fallen trees, and debris dams (Sedell et al. 1988). Thus, in ecologically healthy watersheds in the Pacific Northwest coastal ecoregion, the valley bottoms are wet or flooded most of the year.

Tributaries, originating from seeps, springs, and wetlands at the base of hillslopes, flow across river terraces. Most are low gradient with stable flow regimes from strong groundwater influences. Other tributaries, classified as valley-wall tributaries (Cupp 1989*a*, *b*; Swanson and Lienkaemper 1982), start near the top of the valley slopes. These have steep gradients and variable flows. Unlike the headwater tributaries originating in the alpine/glacier area, the basins of these streams are heavily forested and contribute less suspended sediment. The flow regime of these tributaries is dominated by late winter and early spring precipitation, whereas the flow regime of the higher-order main channel is regulated by winter storms, snowmelt, and groundwater.

Lowland rivers become a network of islands, wood debris dams, sloughs, oxbow lakes, and beaver ponds (Sedell et al. 1988). The channel pattern becomes anastomosed (multiple channels) due to significant reduction in gradient, sediment inputs from upstream causing channel aggradation and backwater effects (Smith 1973). The backwater effect creates additional overbank flooding, deposition of fines, and buildup of river levees, backswamps, and other floodplain features. The alluvial deposits along the higher-order channels of the western Cascades result in some of the highest specific yields of groundwater in the United States (Martin 1982, Lum 1984, Davis 1988).

Hyporheic Processes and Subsurface Habitat

Streams in the Pacific Northwest coastal ecoregion have three interactive aquatic habitats: surface or in-channel habitat, floodplain habitat, and subsurface or hyporheic habitat. The subsurface habitat, or hyporheic zone, is the interstitial habitat beneath the streambed that is the interface between surface water and the adjoining groundwater. Traditionally, the hyporheic zone has been considered a relatively thin area extending only tens of centimeters vertically and laterally beneath and alongside the stream (Pennak and Ward 1986). However, recent investigations on gravel-bed rivers show that these habitats can extend throughout the alluvial gravels of floodplains. Stanford and Ward (1988) found the average hyporheic habitat to be 3 km wide and 10 m deep in an alluvial floodplain on the Flathead River, Montana. Thus, in areas of extensive alluvial gravel floodplains, the hyporheic zone contributes substantially to total habitat area.

Vertical and lateral dimensions of subsurface water movements are controlled by geologic structure and layering of aquifers in the continental glacial till deposits of Washington, British Columbia, and Alaska. Toth (1963) suggests that there are three distinct systems; local systems, intermediate systems, and regional systems. Under this framework, local systems develop only where there is pronounced typographic relief. Increasing typographic relief is hypothesized to increase the depths and the intensities of the local flow systems.

Occurrence of fractured bedrock, as found throughout the Cascades, provides an avenue for upwelling of regional groundwater systems (Freeze and Cherry 1979). In lower-gradient alluvial valleys, the local water system may stagnate, allowing regional or intermediate systems to dominate or mix with local systems. Boundaries between systems are located at the highest and lowest elevations of local hills and depressions (Toth 1963). Thus the stream channel may serve as a zone of convergence between different groundwater systems. Chemical characteristics of the groundwater systems are different and are reflected in the variability of chemical parameters found within the stream or along the banks.

Hyporheic areas are important regulators of nutrient inputs to streams. The hyporheic zone, as a retention or storage compartment, provides a medium for biotic processing (Hynes 1983, Bencala 1984, Grimm and Fisher 1984, Dahm et al. 1987, Stanford and Ward 1988, Triska et al. 1989*a*, *b*). Nutrient and organic fluxes within the hyporheic zone are hypothesized to be a function of the direction and type of groundwater or surface water influence. Close to the channel, groundwater and stream water mix (Triska et al. 1989*b*, Vervier and Naiman 1992). Triska et al. (1989*b*) found that within 3.5 meters of the channel, at least 80% of the subsurface water was stream water.

The rate of exchange between the subsurface and stream ecosystems varies with the dominant hydraulic process (discharge or recharge). Dominance of groundwater or surface water depends on the season or magnitude and duration of storms (Compana and Boone 1986, Gilbert et al. 1990, Vervier and Naiman 1992). As surface water rises in the channel, groundwater recharge dominates (Freeze and Cherry 1979, Beaudry et al. 1990).

Hyporheic Processes in Low-Order Channels

Spatial connectivity of hyporheic zones in the steeper bedrock-controlled channels may be discontinuous due to constrained topography (Figure 6.4). Hyporheic zones are limited to small floodplains, meadows, and stretches of stream where coarse sediment is deposited over bedrock. System continuity is further interrupted by mass wasting and debris dam breaks which gouge channels.

Local hyporheic systems are fed by subsurface flow from hillsides. Subsurface flow through porous soils can be significant from forested slopes (Sloan and Moore 1984) where root channels, decayed root holes, worm holes, piping, and animal burrows are common (Sklash and Farvolden 1979, Higgins 1984, Roberge and Plamondon 1987, Beaudry et al. 1990).

In zones of bedrock fracture, mixing between local and regional groundwater systems may occur (S. Burgess, Civil Engineering, University of Washington, pers. comm.). However, in the subsurface ecosystem, local groundwater dominates. Dominance affects several parameters, such as quality of organic matter, water chemistry, and faunal distribution (Triska et al. 1989*a, b*; Gilbert et al. 1990). Relatively high subsurface velocities on hillsides mean less soil contact time and mineralization than in downstream areas (Wolock et al. 1989).

Hyporheic Processes in Mid-Order Channels

As valleys broaden to a wider alluvial floodplain with less topograhic constraints, the spatial connectivity of the hyporheic zone becomes more continuous (Figure 6.4). Local groundwater systems still dominate the subsurface ecosystem. Depth and intensity of the local system should be greatest in this zone (Toth 1963). As the floodplain widens and topographic relief decreases, stagnation of the local system promotes mixing of intermediate and regional systems, especially during drier periods when hydraulic gradients toward the stream are strongest (Freeze and Cherry 1979).

As surface discharge declines, as during the summer drought period, the system is dominated by groundwater discharge. Exceptions occur during storms (Vervier and Naiman 1992). During surface low-water periods the hyporheic zone acts as a source of water, nutrients, and energy to the stream (Wallis et al. 1981, Bencala 1984, Naiman et al. 1987, Ford and Naiman 1989, Triska et al. 1989*a, b*).

Hyporehic Processes in High-Order Channels

In higher-order streams, with wide floodplains and unconstrained valleys, the spatial extent of hyporheic habitat is greater than upstream (Figure 6.4). However, discontinuities in spatial connectivity increase as the influence of local groundwater system decreases. For example, small topographic variations created by fluvial structures such as scroll bars, cutoff channels, oxbow lakes, and wetlands create discontinuities. As a result of local system dynamics and varying topography on the floodplain, alternating recharge and discharge areas are found across the valley (Toth 1963, Winter 1987).

Soil and substrate permeability in the floodplain and channel is variable due to erosional and depositional fluvial processes and deposition of till during the continental glacial period. However, duration of overbank flows and ponding is longer in higher-order watersheds, enhancing opportunities to transmit organic matter and nutrients from surface water to the hyporheic zone. Subsurface storage and retention of nutrients increases as the spatial extent of channel features (e.g., side bars and channel bars) increases (Bencala 1984).

Functions of Hyporheic Zones

Hyporheic zones act as sensitive indicators of ecological health, since processes there substantially influence energy and nutrient resources in riparian forests and aquatic surface systems (Wallis et al. 1981, Hynes 1983, Peterjohn and Correll 1984, Lowrance et al. 1984, Grimm and Fisher 1984, Stanford and Ward 1988, Ford and Naiman 1989, Triska et al. 1989*a, b*; Gibert et al. 1990). Hyporheic zones can act as a sink, storage, or source depending on spatial location and season.

Retention of nitrogen, phosphorus, and organic carbon within the subsurface zone occurs during the recharge phase. Low hydraulic gradients and slower velocities enhance biotic activity (Winter 1987, Triska et al. 1989*a, b*). However, anaerobic processes such as denitrification may dominate during this phase due to saturated soil conditions (Hixson et al. 1990). As the recharge phase shifts to a discharge phase, aerobic processes such as nitrification become dominant. The type and intensity of biochemical processes will influence biodiversity and the spatial distribution of animals using hyporheic habitat.

Numerous and often contradictory hypotheses exist concerning the flux of carbon, nitrogen, and phosphorus from subsurface to surface waters. For example, Fisher and Likens (1973) reported that groundwater diluted organic carbon in stream systems in New Hampshire. Hynes (1983) also hypothesized that hyporheic zones serve as a sink for organic matter in Ontario. Yet groundwater has been found to be a significant source of carbon and nutrients to streams (Hynes 1983, Grimm and Fisher 1984, Dahm et al. 1987, Naiman et al. 1987, Ford and Naiman 1989). Rutherford and Hynes (1987) suggest that the hyporheic zone is too heterogeneous to make a source or sink conclusion. The results, equivocal to date, point to the need for better information about an inherently complicated system that is of fundamental importance for watershed functions (Pinay et al. 1990).

Water Quality

Selection of Fundamental Water Quality Elements

Water quality is a fundamental component of watershed health because it effectively integrates the full range of geomorphic, hydrologic, and biologic processes (Hem 1985). Alterations to any one of these processes will affect one or more water quality parameters (Frere et al. 1982, Peterjohn and Correll 1984). Hence changes in water quality indicate a change in some aspect of the terrestrial, riparian, or in-channel ecosystem. Conversely, water quality affects the aquatic, riparian, and hyporheic ecosystems (Hynes 1966, 1970; Stanford and Ward 1988, MacDonald et al. 1991). These interactions are extremely complex, and recognition of their importance does not simplify the problem of associating an observed change in water quality with a particular cause. In ecologically healthy streams there is considerable spatial and temporal variability in water quality parameters due to the large number of controlling factors and the uneven distribution of these factors in space and time (Feller and Kimmins 1979, Bencala et al. 1984, Keller et al. 1986).

We focus on just five of many water quality elements related to ecologically healthy systems: (1) nitrogen (particularly nitrate-nitrogen), (2) phosphorus (principally phosphates), (3) turbidity, (4) temperature, and (5) intragravel dissolved oxygen. Other important elements related to the ecological health of watersheds, such as buffering capacity (pH and alkalinity), organic nutrients (forms of dissolved organic carbon), and potential toxicants (wastes, insecticides, herbicides), are not considered here. The five elements chosen were selected after discussions with water quality experts and a review of the results of a recent project sponsored by the Environmental Protection Agency to develop guidelines for monitoring the effects of forestry activities on streams in the Pacific Northwest and Alaska (MacDonald et al. 1991). Consideration was limited to the physical and chemical constituents of water even though channel, riparian, and other biological characteristics are equally important for maintaining an ecologically healthy system (MacDonald et al. 1991). The intent is that, taken together, these five elements provide an indication of the basic health of lotic systems in the Pacific Northwest coastal ecoregion.

Role and Expected Values

Nitrogen and phosphorus are typically limiting nutrients in coastal ecoregion streams (Hem 1985). The mass flux of N and P is a function of critical processes such as the efficiency of terrestrial nutrient cycles, flow and transformations of organic material, and erosion of particulate matter (Sollins et al. 1980, Harr and Fredriksen 1988, Martin and Harr 1989). Use of both nitrogen and phosphorus is complementary because phosphorus tends to be sorbed and transported in particulate form, while nitrogen usually is dissolved and transported by subsurface and groundwater flow (Mohaupt 1986). In the absence of other limiting factors such as light, increased concentrations of plant-available nitrogen and phosphorus stimulate primary production (Gregory et al. 1987).

The range of conditions found from southeastern Alaska to northern California, and from the coast to the permanent snow zone, make it difficult to specify expected values. For example, mean annual nitrate-nitrogen concentrations in undisturbed headwater streams range from less than 0.01 mg/L (Harr and Fredriksen 1988, Martin and Harr 1989) to 1.2 mg/L (Brown et al. 1973). Because atmospheric inputs usually are larger than the loss of nitrogen by leaching, there is a small net input of 0.1–2.6 kg N ha^{-1} yr^{-1} in undisturbed forested watersheds west of the Cascades (Feller and Kimmins 1979, Martin and Harr 1989).

Mean annual phosphorus concentrations in small forest streams typically are less than 0.06 mg/L (Brown et al. 1973, Feller and Kimmins 1979, Harr and Fredriksen 1988, Martin and Harr 1989). Annual phosphorus budgets for four forested coastal watersheds range from a net gain of 0.1 kg ha^{-1} yr^{-1} to a net loss of 0.3 kg ha^{-1} yr^{-1} (Feller and Kimmins 1979).

Turbidity is a measure of light scattering by a water sample. In most cases, suspended silt and clay particles are the primary cause of high turbidities, although colored organic compounds, finely divided organic matter, and microorganisms such as plankton also contribute (APHA 1989). Turbidity is useful as an easily measured indicator of suspended sediment concentrations (Kunkle and Comer 1971, Aumen et al. 1989), and hence a first approximation of erosion rates. Suspended sediment has wide-ranging effects on salmonids, invertebrates, and other aquatic organisms (Everest et al. 1987, Chapman and McLeod 1987).

Expected values for turbidity are difficult to specify because turbidity is discharge dependent and extremely variable throughout the region. Hence turbidity standards usually are expressed in terms of an allowable increase over background (Harvey 1989). The absolute values necessary to protect designated uses, such as sight feeding by salmonids, are <25 NTU (EPA 1986); greater values generally are encountered only during major floods (Aumen et al. 1989).

Water temperature greatly affects rates of chemical and biological processes. Although absolute stream temperatures are largely a function of the subsoil environment and climatic conditions (Beschta et al. 1987), stream temperature is a relatively sensitive indicator of riparian conditions (Brown and Krygier 1970, Harr and Fredriksen 1988). As noted previously, stream channel morphology also affects the temperature regime. Temperature is largely a function of discharge and incoming solar radiation, and is relatively predictable for specific locations.

Intragravel concentration of dissolved oxygen (DO) is critical for salmonid reproduction, invertebrates, and other aquatic life. Furthermore, the concentration of intragravel DO integrates numerous other factors, including temperature, bed material particle size, and the deposition of fine sediment and particulate organic matter (MacDonald et al. 1991). In undisturbed alluvial streams the concentration of intragravel dissolved oxygen should approach saturated values; values substantially less than saturation suggest blockage of interstitial water flow (Chapman and McLeod 1987) or high

Table 6.2. Relative importance of factors controlling the observed values of selected water quality elements in coastal ecoregion streams.

	Water Quality Element				
Controlling Factor	Nitrogen	Phosphorus	Turbidity	Temper-ature	Intragravel Dissolved Oxygen
Climatic and atmospheric inputs	High	Low	Moderate	High	Low
Geology and soils	Moderate	High	High	Moderate	High
Stream order	Moderate	Moderate	Moderate	High	Moderate
Constrained or unconstrained channels	High	High	High	Moderate	Moderate
Vegetation	High	Moderate	Moderate	High	Low

oxygen demand from the breakdown of organic materials (Ringler and Hall 1975, Plamondon et al. 1982).

Little data are available for intragravel DO values for ecologically healthy watersheds. Suggested one- and seven-day minimum values for intragravel DO are 5.0 and 6.5 mg O_2/L, respectively (EPA 1986). Idaho is considering an intragravel DO standard of 85% of the saturated value (Harvey 1989), but adoption of this standard has been slowed by uncertainty over intragravel DO values in undisturbed streams, and by the high spatial variability of intragravel DO in a stream segment or even within a salmonid redd (Chapman and McLeod 1987).

Taken together, these five fundamental elements are one indication of the suitability of streams for cold-water fishes and provide an integrated view of watershed health. Other parameters could be supplemented, but these five represent a best *initial* indication of watershed health over the range of environmental conditions found in the Pacific Northwest coastal ecoregion.

Controlling Factors

Expected values of these five fundamental elements are a function of multiple controlling factors, and each element has a unique response to the set of controlling factors. Table 6.2 qualitatively summarizes the relative importance of five factors—climatic and atmospheric inputs, geology and soils, stream order, valley type (constrained versus unconstrained channels), and vegetation—on the selected water quality elements.

The first controlling factor—climatic and atmospheric inputs—strongly affects nearly all water quality elements. In the Bull Run watershed near Portland, Oregon, for example, precipitation accounts for approximately 60% of the dissolved ionic load in surface runoff (Aumen et al. 1989). Atmospheric nitrogen inputs generally exceed nitrogen losses (Feller and Kimmins 1979). Solar radiation is a dominant variable in predicting stream temperatures (Beschta et al. 1987). Turbidity is highly responsive to the size and spacing of storm events (Brown 1983). Climatic and atmospheric inputs also help define basic processes such as the volume and timing of runoff, weathering rates, and the likelihood of mass failures. Thus virtually all water quality parameters are affected by climatic and atmospheric inputs (Risser, this volume), but nitrogen and water temperature generally are more responsive than phosphorus or intragravel DO.

Geology and soils are important factors in determining the amount and type of erosion, hence the levels of turbidity, phosphorus, and indirectly intragravel DO (Everest et al. 1987, Chapman and McLeod 1987). Nitrogen also is relatively sensitive to geology and soils because losses occur primarily in dissolved form, and this is a function of soil and groundwater processes (Feller and Kimmins 1979, Sollins et al. 1980).

Relatively few studies have related values of these fundamental elements to stream order in undisturbed watersheds. In general, increasing stream order reduces temporal variability, but absolute effects are uncertain. In interior Alaska stream order has no effect on phosphorus or turbidity, while nitrogen shows a slight decrease downstream (Hilgert and Slaughter 1988). Changes in stream temperature are more predictable, with the observed temperature generally increasing downstream. Intragravel DO should decline with increasing stream order because the larger volume of water reduces the reaeration rate, mean water temperatures are higher, and the finer bed material associated with higher-order streams reduces subsurface permeability.

Differences in water quality between geomorphically constrained and unconstrained channels will result from differences in subsurface flow paths, sideslope gradients and resultant erosion and transport rates, and width of riparian and hyporheic zones. Turbidity is most likely to be sensitive to valley form, with lesser or indirect effects on other water quality elements.

Vegetation is the final controlling factor. Healthy watersheds in the Pacific coastal ecoregion generally have a dense forest cover, and this helps keep water temperatures and sediment loads in the range suitable for salmonids. In addition, both density and vegetation type affect nitrogen fixation and uptake (Sollins et al. 1980).

Since all five fundamental elements respond directly or indirectly to the same basic driving forces of runoff, erosion, and weathering, there are significant interactions among these elements. The use of several water quality elements is necessary to assess the specific condition (health) of the drainage network, and to fully evaluate the effects of natural and anthropogenic changes.

Riparian Forest Characteristics

The natural characteristics and ecological health of streams and rivers are intimately linked to the surrounding landscape by the biotic and physiochemical properties of the riparian zone. The riparian zone extends from the edge of the average high water mark of the wetted channel toward the uplands

(Figure 6.11). This zone includes terrestrial areas where vegetation and microclimate are influenced by perennial or intermittent water associated with high water tables, and by the ability of soils to hold water. Beyond this is the riparian "zone of influence," a transition area between the riparian zone and the upland forest where vegetation still influences the stream under some conditions (Gregory et al. 1991).

In ecologically healthy watersheds, riparian forest characteristics are strongly controlled by climate (e.g., hydrologic regime), channel geomorphology, and the spatial position of the channel in the drainage network. Historically, riparian forests formed a continuous ribbon of vegetation along stream channels. Upon closer examination, this vegetative ribbon was a mosaic of different stand ages and species from the headwaters to the sea (Maser et al. 1988). Riparian forests once covered large areas, especially in the alluvial lowlands (Sedell et al. 1988).

The width of the riparian zone, and the extent of the forest's influence on the stream, are strongly related to stream size and valley morphology. Small streams possess relatively little riparian vegetation; they are more influenced by vegetation in the upland forest (zone of influence) (Figure 6.11). Mid-order streams and rivers (third to fifth order) typically have a distinct band of riparian vegetation, whose width is defined by long-term (>50 yr) channel dynamics and the annual discharge regime. Large rivers are characterized by well-developed, complex floodplains with long periods of seasonal flooding, oxbow lakes in old river channels, a diverse forest community, and moist soils. Ecologically healthy watersheds require the influence of riparian forests on streams, especially in relation to controlling the light and temperature regimes, providing nourishment for the stream biota, and being a source of large woody debris (Table 6.3).

Light and Temperature

The amount and quality of light reaching streams are determined by forest vegetation height, forest canopy density, stream channel width, and channel orientation in relation to the sun's path in the sky. Light is important for streams because of its influence on water temperature, on primary production by aquatic plants, and on the behavior of organisms.

Seasonal and daily water temperatures are strongly influenced by the amount of solar radiation reaching the stream surface through the forest canopy (Beschta et al. 1987). The temperature of water entering a small stream typically reflects that of the forest's subsoil environment, but changes as water flows downstream. Water temperature is an important factor for environmental vitality because of its controlling influence on the metabolism, development, and activity of stream organisms.

Small forested streams typically receive 1 to 3% of total available solar radiation (Naiman and Sedell 1980, Naiman 1983, 1990). Small streams have relatively cool but stable daily temperatures, low rates of primary pro-

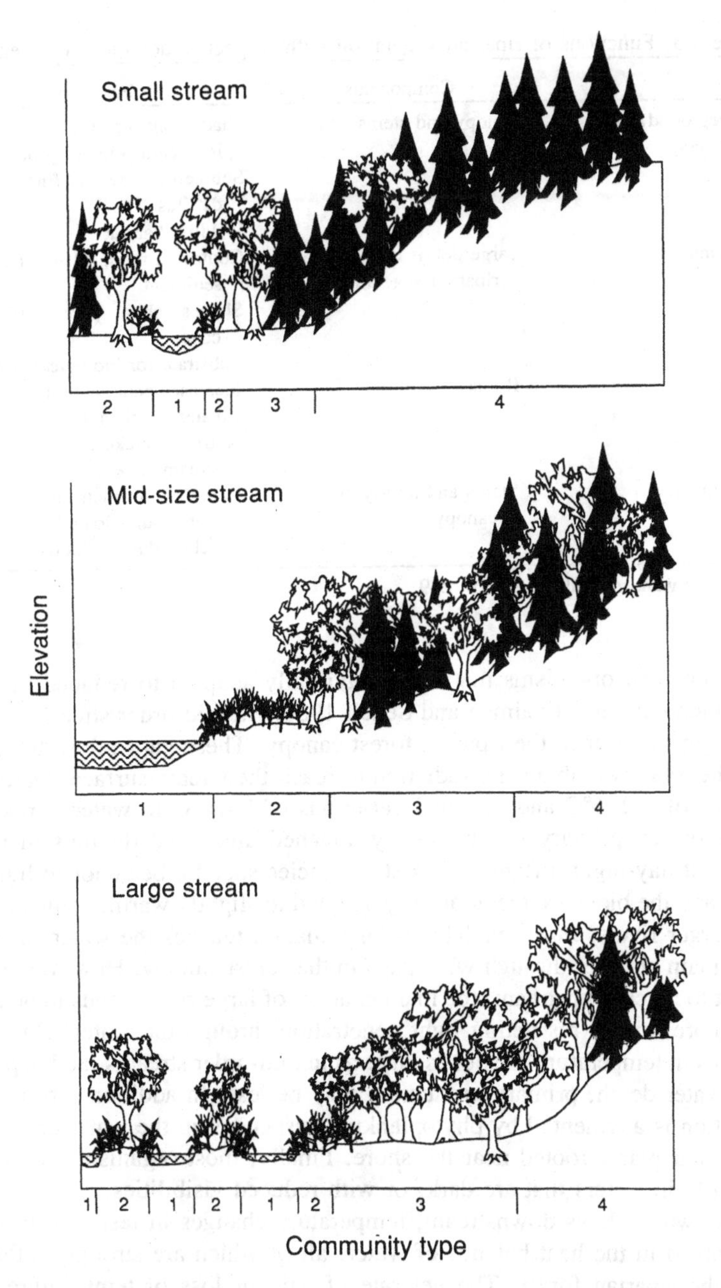

FIGURE 6.11. The natural characteristics of the riparian zone change with stream size. In low- and mid-order streams the links between the riparian forest and the stream are strong. In large rivers the links are not as strong in the main channel but they do remain strong in the secondary channels. Key: (1) active channel, (2) riparian zone, (3) zone of influence, (4) uplands. Note that (2) and (3) make up the complete riparian zone of influence.

Table 6.3. Functions of riparian vegetation with respect to aquatic ecosystems.

Sites	Components	Functions
Aboveground/above channel	Canopy and stems	Shade controls temperature and in-stream primary production Source of large and fine plant detritus Wildlife habitat
In channel	Large debris derived from riparian vegetation	Controls routing of water and sediment Shapes habitat: pools, riffles, cover Substrate for biological activity
Streambanks	Roots	Increases bank stability Creates overhanging banks, cover Nutrient uptake from ground and stream water
Floodplain	Stems and low-lying canopy	Retards movement of sediment, water, and floated organic debris during floods

Source: Swanson et al. (1982), p. 269.

duction, and organisms that are behaviorally adapted to reduced light and cool temperatures (Naiman and Sedell 1980). In mid-order streams and rivers, gaps appear in the riparian forest canopy. These gaps allow 10 to 25% of the total available solar radiation to reach the stream surface. Daily variations of ~2–6°C and seasonal variations of 5–20°C in water temperature may occur, primary production by attached algae and diatoms increases, distinct day-night differences exist in species-specific behavior to light levels, and the biota are metabolically adapted to slightly warmer temperatures. In larger rivers, most available solar radiation reaches the water surface of the main channel through wide gaps in the forest canopy. However, in contrast to reaches upstream, the main channel of large rivers tends to be deeper or more turbid, restricting light penetration through the water. Daily variations in temperature are not as large as in mid-order streams, and depending on water depth, primary production may be less. In addition, primary production is augmented by phytoplankton suspended in the water column and vascular plants rooted near the shore. Finally, most organisms are adapted for life in waters that are dark, or with reduced visibilities.

As water flows downstream, temperature changes in response to factors involved in the heat balance of water, all of which are strongly influenced by the riparian forest. The net rate of gain or loss of temperature is the algebraic sum of net solar radiation, evaporation, convection, conduction, and advection. Net radiation is generally dominated by the amount of direct-beam solar radiation reaching a stream's surface. Heat gain or loss from evaporation and convection depends on the vapor pressure and the temperature gradient at the water surface and the air immediately above the surface, respectively. Wind speed at the air-water interface is also an important controlling variable. Conduction of heat between water in the stream and the channel substrate depends on the type of material making up the substrate. Bedrock channels are more efficient than gravel-bed channels at conducting heat. Heat exchange by advection occurs when tributaries or groundwater of different temperatures mix with the main streamflow, thereby either increasing or decreasing the main stream temperature.

The influence of shading by the riparian forest on the heat balance of a small stream can be enormous because net solar radiation, evaporation, convection, conduction, and advection remain relatively small over a 24-hour period, even in midsummer, relative to groundwater temperature. Should the forest canopy be opened by a disturbance, the net heat exchange can be significantly altered. For example, during winter, streams without riparian canopies may experience lower temperatures because the lack of cover enhances energy losses by evaporation, convection, or long-wave radiation. Long-wave radiation losses are greatest when clear skies prevail, particularly at night, resulting in the formation of surface and anchor ice. During summer, the lack of a forest canopy cover results in large (3–10°C) diel variations in temperature as the amount of direct solar radiation increases (Beschta et al. 1987).

Instantaneous temperatures and cumulative temperatures (e.g., degree-days per unit of time) have a significant influence on biotic characteristics. Instantaneous temperature significantly affects water viscosity, and therefore the amount of energy required to swim. It influences an organism's metabolism, dictating the amount of food required for daily activities and reproductive products. Further, species preferences for temperature influence the ability of an organism to successfully compete for resources, and thereby influencing community composition and abundance. Fish and invertebrates have specific requirements for the number of degree-days needed for egg development and for the timing of reproduction and emergence, thereby reducing competition for food and reproduction sites by subtle differences in phenology or life history strategies (Sweeney and Vannote 1978).

Sources of Nourishment

Annually, riparian forests add large amounts of leaves, cones, wood, and dissolved nutrients to low- and mid-order streams (Gregory et al. 1991). These organic inputs originate as particles falling directly from the forest into the stream channel (or moving downslope along the forest floor by wind and water driven erosion) and as dissolved materials in subsurface water flowing from the hyporheic zone.

The riparian forest is an important regulator of stream productivity through the amounts and qualities of material directly contributed to the stream. Small streams directly receive 300–600 g C/m^2 annually from the forest, with the rate per unit area decreasing as channel width (and the gap in the forest canopy) increases (Conners and Naiman 1984). In deciduous riparian forests >80% of these inputs may be leaves that are delivered over a six-to-eight week autumn period. In coniferous riparian forests ~40–50% of the material may be cones or wood. The chemical quality of the material (i.e., nitrogen and lignin content) strongly influences the rate of decay and subsequent trophic pathways (Melillo et al. 1983, 1984). The complete decay process takes about one year for most high quality materials such as leaves and herbaceous plants and may take several years or decades for low quality materials such as cones and wood (Gregory et al. 1991).

Subsurface water moving from the uplands to the stream also carries large quantities of dissolved organic matter and nutrients essential for stream function. The riparian forests chemically alter these materials as the subsurface water flows pass their root systems. Riparian forests take up nutrients for growth, promote denitrification by subtle changes in the position of oxic-anoxic zones, and modify the chemical composition and availability of carbon and phosphorus (Pinay et al. 1990). Exact mechanisms regulating these processes are not well understood (Triska et al. 1989*a*, *b*). Yet the presence of riparian forests significantly regulates the amount of nitrogen and phosphorus reaching streams from upland areas (Karr and Schlosser 1978, Schlosser and Karr 1981*a*, *b*; Peterjohn and Correll 1984).

Large Woody Debris

Large woody debris (LWD) is the principal factor determining the characteristics of aquatic habitats in low- and mid-order forested streams. The amount of LWD in streams can be substantial, ranging from >40 kg/m^2 in small streams to 1–5 kg/m^2 in large rivers (Harmon et al. 1986). The importance of LWD relates to its ability to control the routing of sediment and water, to shape the formation and distribution of pools, riffles, and cover, and to act as a substrate for biological activity (Swanson et al. 1982; Table 6.3).

Wood boles (>10 cm diameter) enter streams of all sizes from the riparian forest. However, the spatial distribution of LWD varies systematically from small streams to large rivers, reflecting, in part, the balance between stream size and wood size (Bilby 1981). Wood in small streams is large relative to channel dimensions and to peak stream flow. Thus LWD cannot be easily floated and redistributed, and consequently is randomly distributed and often located where it initially fell. But these small channels are often in the steepest part of the drainage network, and are most prone to catastrophic flushing by extreme landslide/dam-break floods (Benda 1990). Mid-order streams are large enough to redistribute LWD but narrow enough that LWD accumulations across the entire channel are common. LWD tends to be concentrated in distinct accumulations spaced several channel widths apart along the stream. In large rivers, LWD is commonly collected in scattered, distinct accumulations at high water and particularly on the upstream ends of islands and at bends in the river (Lienkaemper and Swanson 1987, Potts and Anderson 1990). Natural anchors such as root wads, large limbs, or lodging of LWD between other obstructions improve debris retention and are important considerations for environmental health of the system.

LWD in streams influences channel morphology as well as sediment and water routing. In small streams LWD creates a stair-stepped gradient where the streambed becomes a series of long, low gradient sections separated by relatively short, steep falls or cascades (Grant et al. 1990). Therefore, much of the streambed may have a gradient less than the overall gradient of the valley bottom, because much of the decrease in altitude, and in potential energy, takes place in the short, steep reaches. This pattern of energy dissipation in short stream reaches results in less erosion to bed and banks, more sediment storage in the channel, slower downstream movement of organic detritus, and greater habitat diversity than in straight, even-gradient channels (Bisson et al. 1987).

Comparison of volumes of stored sediment and annual sediment export suggests that small forested streams with natural amounts of LWD annually export often <10% of sediment stored in the channel system (Swanson et al. 1982). LWD makes up ~40% of the obstructions that trap sediment in forested streams (Bilby and Ward 1989). Unfilled storage capacity serves to buffer potential sedimentation impacts on downstream areas when pulses of sediment from the uplands enter stream channels. Scattered LWD in channels reduces the rate of sediment movement downstream, routing sediment through the stream ecosystem slowly, except in cases of catastrophic flushing events or when the storage capacity is filled.

By redirecting water flow, LWD has both positive and negative effects on bank stability, on the lateral geomorphic mobility of channels, and on the stability of aquatic habitats (Keller and Swanson 1979). LWD-related bank stability problems in steep-sided, bedrock-controlled streams result from undercutting of the soil mantle on hillslopes by debris torrents. Undercut slopes are subject to progressive failure by surface erosion and small-scale (<1,000 m^3) mass erosion over a period of years. Both bank instability and lateral channel migration may be facilitated by LWD accumulations in channels with abundant alluvium and minimal bedrock influence. Changes in channel conditions and position often occur as a stream bypasses a LWD accumulation and cuts a new channel. Where channels pass through massive debris accumulations, streamflow may become subsurface much of the year. In areas of active earthflows from the forest, lateral stream cutting may undermine banks, encouraging further hillslope failure and accelerated sediment supply to the channel. On balance, however, LWD generally stabilizes small streams by dissipating energy and by protecting streambanks.

As a result of these mechanisms, LWD helps regulate the distribution and temporal stability of fast-water erosional areas and slow-water depositional sites. LWD and riparian vegetation provide cover and nourishment for all stream organisms, serving as habitat or substrate for substantial biological activity by microorganisms, invertebrates, and other aquatic organisms (Gregory et al. 1991, Naiman 1990). Filter feeding invertebrates, algae, and diatoms attach in large numbers to LWD, significantly influencing nutrient cycling and, consequently, downstream water quality. LWD accumulations on the streambanks provide habitat for small mammals and birds that feed on stream biota (Doyle 1985). An alteration to the supply rate or the size of LWD from the riparian forest has consequences that may take tens to hundreds of years for natural processes to correct.

Habitat Characteristics

In ecologically healthy watersheds, interactions between channel geomorphology, hydrologic pattern, spatial position of the channel, and riparian forest characteristics produce habitat for terrestrial and aquatic organisms. Fundamental habitat features influencing animal population dynamics, productivity, biodiversity, and evolutionary processes are related to riparian forest dynamics, spatial and temporal variability of the habitat, and maintenance of migratory connectivity. In our discussion we offer six examples: salmonid habitat preferences, fish community habitat requirements, watershed-scale patterns in habitat, influence of LWD on habitat development, the potential role of woody debris piles to act as nodes of ecological organization in fluvial corridors, and habitat alterations by wildlife communities. We then discuss an example of food web dynamics shifting in response to changing habitat. In combination, these examples have broad implications for ecosystem health.

Salmonid Habitat Preferences

Each salmonid species indigenous to the coastal ecoregion employs a different life history strategy (Everest 1987), and consequently utilizes a different suite of habitats (Bisson et al. 1982). These differing life history strategies allow salmonids to fully utilize available habitats in a single watershed by segregating habitats spatially and temporally. It is well known that healthy salmonid populations respond to characteristics and location of spawning areas, postemergent rearing areas, summer (low water) rearing areas, winter rearing areas, and estuarine conditions at smolting. For example, rearing habitats in small streams are used differently by juvenile steelhead (*Oncorhynchus mykiss*) and coho salmon (*O. kisutch*). During the summer rearing period, coho tend to occupy pools, whereas steelhead prefer riffles and glides (Bisson et al. 1988), but during winter coho move to wall-base channels (Peterson and Reid 1984) and steelhead occupy terrace tributaries (Scarlett and Cederholm 1984). Furthermore, spawning areas and timing of spawning may nearly overlap among species such as pink salmon (*O. gorbuscha*), chum salmon (*O. keta*), and coho salmon. However, because pink salmon and chum salmon migrate to sea soon after emergence, there is no competition among juveniles of these species while in streams.

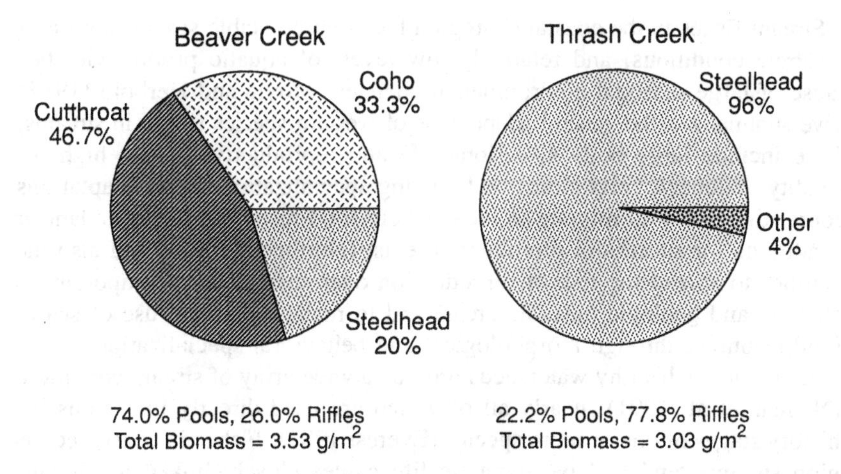

FIGURE 6.12. Juvenile fish habitat preferences in two western Washington streams with different proportions of pool and riffle areas (P. A. Bisson, unpublished data).

In stream reaches (defined as a section of channel >10 m in length), habitat preferences influence which portions of streams are dominated by juveniles of different species during summer (Figure 6.12). However, these influences are also affected by the availability of suitable spawning gravels for each species, and by other habitat factors, such as cover. Though the habitat requirements of juvenile coho salmon, steelhead trout, and cutthroat trout (*O. clarki*) have been extensively studied, species occupying the larger rivers, floodplains, and estuaries during the juvenile life history stages (e.g., chinook [*O. tshawytscha*], chum, and pink salmon) are not well known. However, it is clear that healthy salmonid populations utilize habitats throughout the drainage network during different stages of their life cycles, suggesting that connectivity between habitats is of fundamental importance.

Fish Community Habitat Requirements

Compared with forested ecosystems in eastern North America, fish species diversity in the coastal ecoregion is low. This results from zoogeographic barriers (McPhail 1967, Reimers and Bond 1967), the geologically short time interval since the last glacial period, and an unpredictable and often severe maritime climate (Moyle and Herbold 1987). What the coastal ecoregion streams lack in species richness, however, is partly offset by a remarkable differentiation among locally adapted stocks. For example, Nehlsen et al. (1991) have identified a large number of distinctive stocks of salmonids within California, Oregon, Idaho, and Washington, 214 of which are considered to be currently at risk of extinction. Fish assemblages in coastal streams are therefore characterized by relatively few, but highly adapted and often genetically unique, native species.

Stream fishes in the coastal ecoregion face unpredictable streamflows and substrate conditions, and relatively low levels of aquatic productivity because of large seasonal environmental changes. Moyle and Herbold (1987) have summarized the general adaptations of western species to life in streams; these include large body size, long life and reproductive spans, high fecundity, extensive migrations, and feeding specialization. Such adaptations confer the ability to migrate to areas where food resources are abundant or conditions for successful reproduction exist (Northcote 1978), and also the abilities to spread the risk of reproduction over several years (important in climatic and geologic unstable areas) and to make maximum use of scarce food resources through morphological and behavioral specialization.

Ecologically healthy watersheds provide a wide array of stream conditions (Naiman et al. 1991), nearly all of which are used directly in various life history stages of one or more species (Everest 1987). Fishes in coastal ecoregion streams tend to have dynamic life cycles closely linked to climate, valley form, and the input and transport of sediment and woody debris within the drainage network. Reproduction is associated with specific hydrologic and climatic conditions. Spawning migrations usually occur when streams possess intermediate flows, typically in autumn or spring. Eggs are laid in well-sorted, clean gravels that have a relatively low probability of scouring. Often these substrates are located in unconstrained channels that exhibit some braiding. Fish species that spawn on aquatic vegetation or on mud substrates (Umbridae, Gasterosteidae, some Cyprinidae) locate suitable areas in beaver ponds and in backwaters and sloughs on the floodplain; *connections* between the river mainstem and floodplain channels must be maintained for successful reproduction of these species to occur. Most fishes in the coastal ecoregion avoid spawning in steep, constrained river valleys unless their lack of mobility prevents migration to reaches with a shallower gradient. In such cases, tributary mouths or other geomorphic features characterized by coarse sediment deposits and low gradient may provide easily accessible but localized spawning sites.

Egg and alevin survival within stream gravel depends on the redd remaining relatively free of fine sediment and maintaining good permeability to ensure that developing embryos have an adequate oxygen supply (Chapman 1988). Deposition of fine sediment during incubation reduces survival to emergence through anoxia or physical entrapment. In general, emergence is timed to avoid large freshets that carry fry downstream away from rearing areas. Increases in the magnitude and frequency of storm flows, particularly in spring, can have a significant impact on survival of juvenile salmonids (Hartman et al. 1987). Other taxa such as Cottidae, Cyprinidae, and Catostomidae, which usually spawn in spring and early summer, may be similarly affected by changes in discharge patterns.

Newly emerged fry are weak swimmers and take up foraging stations along the edge of the stream channel behind prominent flow obstructions, including boulders and LWD (Bisson et al. 1987, Moore and Gregory 1988*a*). The complexity of channel margins can be an important factor influencing early rearing (Moore and Gregory 1988*b*); therefore, ecologically healthy streams possess complex margins that include backwaters, secondary channels, fallen trees, boulders, and other features that create areas of slowly moving water. Undercut streambanks with the root systems of riparian trees also provide excellent habitat for fry, as well as protection from terrestrial predators. Interactions between stream channels and riparian vegetation become critical to the maintenance of habitat for fry.

As fish grow larger they become better able to maintain feeding stations in swifter currents and more adept at avoiding predators. Preferred habitat often shifts from stream and river margins to deeper water, where the availability of drifting aquatic invertebrates may be greater (Chapman and Bjornn 1969). Some larger salmonids and cyprinids move from tributaries into the river mainstem with the onset of summer. In most cases such movements are timed to take advantage of seasonally abundant food. However, in certain instances summer movements are dictated by the thermal regime, with cold-adapted species seeking out groundwater seeps and congregating around the mouths of cooler tributaries or at the bottom of thermally stratified pools (Berman 1990). Beschta et al. (1987) point out that many fish stocks are adapted to local temperature regimes and that significant alteration of these regimes can lead to disruption of important life cycle events, such as the timing of migrations.

Segments of streams and rivers that provide productive rearing environments in summer may not be optimal winter rearing sites. Periods of high discharge and low food availability force some fishes to seek overwintering locations away from headwater streams and adjacent to, but not in, river mainstems (Peterson 1982*a*, Tschaplinski and Hartman 1983). Quite often these areas are located in seasonally flooded wetlands, beaver ponds, and spring-fed tributaries at the base of steep valley walls (Skeesick 1970, Cederholm and Reid 1987, Brown and Hartman 1988). Almost invariably they are characterized by pondlike conditions having relatively stable discharge and low current velocities. Although such sites serve as important refugia from high flows and heavy sediment loads, certain types of invertebrates can be abundant in them and winter growth rates of fish in off-channel ponds and swamps can be considerably greater than those overwintering in the mainstem (Peterson 1982*b*, Brown and Hartman 1988). Some species that do not emigrate from headwater streams or main channel habitats to seasonally flooded wetlands and spring-fed tributaries along the valley floor instead make use of the protection afforded by woody debris accumulations along the channel margin (Bustard and Narver 1975). Perhaps at no other time of the year are riparian vegetation and floodplain interactions more important to the maintenance of productive rearing habitat in watersheds than during winter.

Watershed-Scale Patterns in Habitat

Spatial patterns of physical habitat expressed at the scale of large watersheds (>100 km^2) in the coastal ecoregion are largely controlled by regional geology and geomorphology. The bedrock geology is the result of millions of years of lithologic and tectonic processes, whereas the geomorphology of individual watersheds has been heavily influenced by glaciation for at least 20,000 years. These factors combine to create unique spatial patterns of salmonid habitats for individual watersheds. For example, in the recently deglaciated South Fork of the Stillaguamish River, Washington, low gradient (<2%), pool-dominated habitats tend to be located on a 1,700 year old terrace adjacent to the main river, whereas riffle-dominated streams tend to occupy slightly higher-gradient streams (2–4%) incised into the older terraces (Benda et al. 1992, Beechie and Sibley 1990). Both types of stream channels are downcutting into glacial-age clay and outwash sand deposits, whereas bedrock channels in this valley are steeper and provide little habitat for anadromous salmonids.

Along a longitudinal gradient from the headwaters to the mouths of major river systems, fish communities are correlated with stream order and stream gradient (Platts 1974). Valley-wall or headwater streams, usually first- and second-order tributaries, are not accessible to anadromous salmonids and may be dominated by resident cutthroat trout and rainbow trout. When fish are absent, these tributaries remain an important part of the stream system, since they transport allochthonous nutrients such as leaf litter, sediments from the hillslope, and LWD to higher-order tributaries.

In moderate-gradient (2–5%) third- to fifth-order streams, anadromous salmonids tend to dominate when there are no barriers to upstream migration. Steelhead and cutthroat trout occupy the steepest streams in this range, whereas coho salmon tend to utilize all accessible, low-gradient tributaries. Small tributaries with gradients less than 2% are usually utilized by coho salmon during the spawning and summer rearing stages. Steelhead and cutthroat trout of several age classes may coexist with juvenile coho salmon when the habitat is diverse.

Chinook salmon tend to utilize larger tributaries and main rivers that are used little by steelhead and coho salmon. Chinook spawning preferences are for larger gravels than most other salmonid species, allowing them to avoid competition for spawning space. Chinook juveniles rear in deeper and faster water than either coho or steelhead, alleviating some of the competition for rearing space and food resources.

Influence of Woody Debris on Habitat Development

Large woody debris is an important part of salmonid habitats in streams, both as a structural element (Grette 1985, Bilby 1985, Sedell et al. 1988) and as cover or refugia from high flows (Bisson et al. 1982, Murphy et al. 1985). Furthermore, LWD tends to reinforce meanders (Mason and Koon

Table 6.4. Longitudinal patterns in channel roughness, effectiveness of LWD in controlling channel morphology, and habitat complexity in an ecologically healthy watershed.

Stream Order	Channel Roughness	Effect of LWD	Habitat Complexity
1	Very high	Very high	Moderate
3	High	Very high	High
5	Moderate	High	Very high
7	Low	Low to moderate	Moderate
9	Low	Low to moderate	Moderate

1985) and trap sediment (Bilby 1979) and smaller organic debris (Naiman and Sedell 1979*a*, Harmon et al. 1986) in stream channels. Historically, woody debris piles covered enormous areas of small streams and large rivers. For example, a driftwood jam on the Skagit River, Washington, was reported to have been 1.2 km long and 0.4 km wide (Sedell et al. 1988). Current estimates of woody debris biomass and volume in aquatic ecosystems are extremely variable. Biomass reports range from 18 Mg/ha to 550 Mg/ha and volumes from 45 m^3/ha to 1,400 m^3/ha (Harmon et al. 1986).

A habitat classification scheme to describe woody debris accumulations separates the debris piles by their typical geomorphic settings; in the main channel, in a side channel, along a cut bank, on an overflow bank, at the island head, or on a gravel bar (Mason and Koon 1985). The habitat functions of LWD vary along the longitudinal gradient from headwaters to mouth (Table 6.4). In low-order, high-gradient streams LWD has a reduced function as a structural element in pool formation because the roughness of a large log is small relative to the inherent roughness of a boulder and bedrock channel. As stream order increases and gradient decreases, LWD becomes increasingly important in creating salmonid habitats. In streams where LWD spans the width of the channel, LWD becomes a dominant roughness element relative to gravel and pebble substrates. Thus, in third- to fifth-order streams, LWD is a dominant channel-forming feature.

As roughness elements, LWD pieces deflect the flow of water and increase hydraulic diversity. Flow deflections create a number of pool types that serve as different habitats for juvenile salmonids in summer (Bisson et al. 1982). When LWD pieces are too small or located such that they do not create pools, they create local hydraulic diversity (i.e. localized low-velocity areas) that serve as refugia for juveniles at higher discharges (Murphy et al. 1985). LWD in third- to fifth-order streams also traps sediments and nutrients which often enhance the suitability of gravels for spawning and slows the transport of vital nutrients in the stream system. This also allows invertebrate communities to more fully utilize the allochthonous inputs to the stream.

Though LWD is less frequently a dominant channel-forming feature in larger rivers, it remains an important feature along the channel banks. LWD in rivers can influence meander cutoffs (Swanson and Lienkaemper 1982) and provide cover and increase invertebrate production for juvenile salmonids (Ward et al. 1982).

Additionally, in ecologically healthy watersheds, large estuaries and slough complexes historically provided significant rearing areas for salmonids. These largely have been eliminated to convert land to agricultural or residential uses (Maser et al. 1988). Much of the lower floodplain area in the Puget Lowland is now diked to contain floods, and large areas of wetlands and marshes no longer serve as habitat. These areas are critical for maintaining healthy anadromous fish populations, because the fish make physiological and metabolic adjustments to either marine or fresh waters in this transition zone.

Woody Debris Accumulations as Nodes of Ecological Organization

The functional role of large piles of debris deposited on riverbanks has received little investigation. Substantial amounts of LWD are associated with streams of old-growth Douglas-fir (*Pseudotsuga menziesii*), western hemlock (*Tsuga heterophylla*), and Sitka spruce (*Picea sitchensis*) forests. Lienkaemper and Swanson (1987) measured 92 Mg/ha to 300 Mg/ha in the wetted channel. In mid- and high-order streams substantial debris is deposited outside the channel but the mass has not been measured. These large accumulations of debris appear to provide critical nodes of biotic organization within the river-riparian corridor. By this we mean that woody debris accumulations act as key loci of habitat for small mammals and invertebrates, which in turn act as prey for larger predators. Without the woody debris accumulations, much of the biodiversity and productivity of the riparian zone would disappear.

In coastal Oregon at least 80 species of snag- or log-dependent wildlife frequent riparian forests (Cline and Phillips 1983). The importance of downed logs and standing snags for habitat complexity in forest ecosystems is well known (Raedeke 1988, Spies and Cline 1988; Franklin, this volume). Woody debris accumulations in close proximity to the channel add complexity to the habitat for terrestrial life and may be a critical resource for connecting upland and aquatic communities.

Twenty North American species of small mammals are known to use coarse woody debris for denning, feeding, and reproduction (Harmon et al. 1986). Doyle (1990) indicates that woody debris accumulations in riparian environments of montane areas provide superior habitat for several species of small mammals. In her study, several less commonly captured species of small mammals were collected only in riparian habitat. Deer mice (*Peromyscus maniculatus*) and chipmunk (*Tamias townsendii*) are often located in microhabitats that contain relatively large amounts of woody debris (Doyle

1990). Shrew (*Sorex trowbridgii*) and mole (*Neurotrichus gibbsii*) have also been found in association with decayed wood (Maguire 1983, Whitaker et al. 1979). These same woody debris accumulations may also provide unique habitat for invertebrates and decomposers (Anderson 1982, Shearer and von Bodman 1983). This is an aspect of watershed health requiring substantial investigation in the near future.

Habitat Alterations by Wildlife Communities

Wildlife communities are also sensitive indicators of ecological health at the watershed scale, provided a broad spatial and temporal perspective is taken. Wildlife affect ecological systems through feeding strategies and day-to-day activities (e.g., ponding water, burrowing). These are important control processes on the riparian forest and the stream channel which have reverberations throughout the entire ecological system. The fundamental features of the role of wildlife in ecologically healthy watersheds are related to the use of wildlife to detect broad-scale environmental change, the nature of long-term population cycles and their relation to environmental conditions, and the seasonal phenology of habitat use and migration to maximize individual fitness (e.g., connectivity).

The riparian zone provides an exceptional array of vegetative conditions that support diverse and productive wildlife communities (Thomas et al. 1979, Oakley et al. 1985). Whereas fish are usually incapable of modifying the physical environment of streams, some wildlife populations are quite capable of modifying the structure and dynamics of riparian zones (Kauffman 1988, Naiman 1988, Pastor et al. 1988). Large herbivores such as elk (*Cervus elaphus*) and deer (*Odocoileus hemionus* and *O. virginianus*) may alter the abundance of understory vegetation through browsing of herbaceous plants and by rubbing or trampling. These activities contribute to a patchy mosaic of plant communities in various successional stages, which in turn increase habitat and soil complexity (Figure 6.13).

Interactions of wildlife with riparian plants not only affect vegetative patchiness but may also alter habitat characteristics of stream channels themselves (Figure 6.13). Water quality parameters including temperature, light, nutrients, and sediment are all influenced by wildlife activities (Green and Kauffman 1989). Beaver have perhaps the most profound effects on streams and riparian habitat (Naiman et al. 1986, 1988). Beaver ponds provide suitable environments for lentic species as well as stream-dwelling forms preferring low current velocity. Beaver ponds are known to be important overwintering areas for some coastal fishes (Bisson et al. 1987). Beaver ponds also serve as important storage and processing sites for terrestrial plant materials entering the stream (Naiman and Melillo 1984), and thus play a major role in regulating nutrient availability downstream (Dahm et al. 1987). Beaver herbivory decreases tree density and basal area by as much as 43% within forage zones around beaver ponds (Johnston and Naiman 1990). Selective

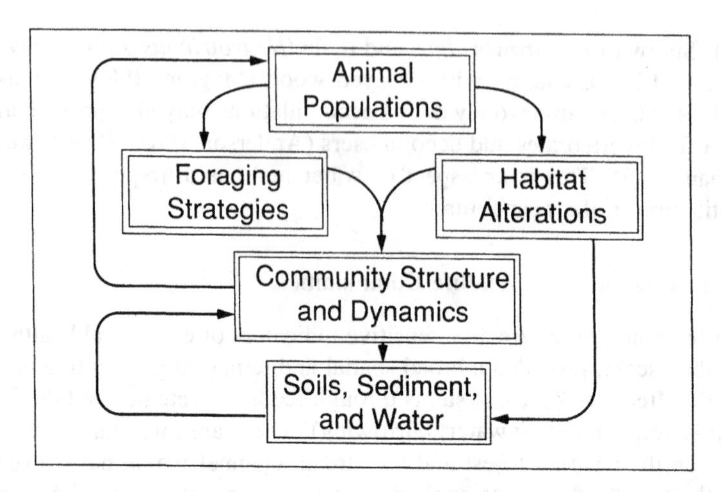

FIGURE 6.13. Wildlife influence ecosystem dynamics by their foraging strategies and by physical habitat alterations. These impacts are transmitted to the riparian forest community, resulting in long-term changes to biogeochemical cycles in soils, sediment, and water. (*Note.* After Naiman 1988. Copyright by American Institute of Biological Science. Used by permission.

foraging also alters tree species composition, which may affect habitat availability for other wildlife species. Because only a small fraction of the wood biomass felled by beaver is actually ingested, beaver contribute large amounts of woody debris to both riparian zones and stream channels. Flooding of alluvial terraces and valley floors by beaver dams causes long-term changes in forest stand succession, which in turn affects the location and dynamics of vegetation patches on a broad scale (Johnston and Naiman 1987).

Food Web Dynamics

Trophic processes in stream ecosystems are strongly influenced by channel morphology and the nature of riparian and upland vegetation (Vannote et al. 1980). In steep, cool headwater streams (first and second order), much of the organic matter processed by the aquatic community originates from riparian trees and is stored in stream channels by LWD (Naiman and Sedell 1979*a*, Cummins et al. 1982). Invertebrate communities are often dominated by detritivorous species that break down wood fragments, needles, leaves, and other debris particles into successively smaller pieces. Because small headwater streams in the coastal ecoregion are often heavily shaded and nutrient poor, aquatic plant production is limited to epilithic diatoms and a few green and blue-green algae. Large aquatic consumers are relatively rare and omnivory tends to be the rule (Anderson and Sedell 1979). Typical assemblages of large consumers in headwater Cascade Mountain streams include crayfish (*Pacifastacus leniusculus*), Pacific giant salamander (*Di-*

camptodon ensatus), sculpins (*Cottus* spp.), and coastal cutthroat trout (*Oncorhynchus clarki clarki*). Small streams draining the Coast Range usually have lower gradients, and anadromous fishes can penetrate far into watershed drainage networks. Assemblages of large consumers include, in addition to the taxa listed above, steelhead, coho salmon, sea-run cutthroat trout, lamprey (*Lampetra* spp.), and occasionally dace (*Rhinichthys* spp.). Other anadromous salmonids (chinook, chum) may be transitory residents of small coastal streams from several days to several months before migrating to the ocean.

Mid-order streams have a more even balance of allochthonous (terrestrial) and autochthonous (aquatic) sources of organic matter than headwater streams (Naiman and Sedell 1979*b*). A substantial fraction of the allochthonous material processed in third- to fifth-order streams is transported from headwater tributaries rather than entering laterally from riparian vegetation, although the latter source of organic matter remains important (Conners and Naiman 1984). Because much of it has already been consumed and excreted by aquatic invertebrates, fluvially transported organic matter has been reduced to fine particles, and invertebrate communities contain species that are adapted to processing this fine organic material. Mid-order stream channels are less heavily shaded than headwater streams, and periphyton production plays a greater role in community metabolism (Naiman and Sedell 1980, Gregory et al. 1987). Invertebrates specialized to consume algae are prominent members of the benthic community. In unconstrained valleys, composition of riparian vegetation along these streams also changes to a more even mixture of conifer and hardwood species. Deciduous trees contribute considerable amounts of easily decomposed and relatively nutrient-rich materials (leaves, catkins) to the streams on a seasonal basis.

In many respects, mid-order streams possess the greatest diversity of both trophic pathways and physical habitat conditions within the watershed. Pool-riffle sequences remain intact and are coupled with lateral habitat development in the form of backwaters and secondary channels on alluvial surfaces. These streams tend to support the greatest diversity of cold-water fishes, such as Salmonidae and Cottidae, and usually contain all of the large consumers found in headwater tributaries as well as species adapted to larger rivers, provided suitable temperatures exist. Salmonids typically found in third- to fifth-order streams include bull trout (*Salvelinus confluentus*; rarely in coastal streams), Dolly Varden (*S. malma*), and mountain whitefish (*Prosopium williamsoni*). Juvenile chinook salmon make greater use of these streams for rearing, and both chum and pink salmon use them to spawn. Because summer temperatures are warmer than in first- to second-order streams, mid-order streams also possess a more diverse community of minnows (*Rhinichthys* spp., *Richardsonius balteatus, Ptychocheilus oregonensis, Acrocheilus alutaceus*), and suckers (*Catostomus* spp.) than are represented in headwaters (Li et al. 1987). Included in this assemblage are species that consume algae and detritus.

Seasonal input of nutrients from the carcasses of spawned-out salmon can be an important source of nitrogen and phosphorus in these and downstream systems (Brickell and Goering 1970, Sugai and Burrell 1984), although carcasses can provide nutrients to small streams if they are retained in the channel by large woody debris (Cederholm and Peterson 1985).

In addition, riparian wildlife may benefit from nutrients transported to watersheds in the form of returning adult fish. Cederholm et al. (1989) documented the consumption of experimentally released coho salmon carcasses by 22 species of mammals and birds (51% of the total number of species observed in riparian zones) in small streams on the Olympic Peninsula, Washington. Cederholm and Peterson (1985) found that the presence of large woody debris served to hinder downstream transport of dead fish and increase carcass availability to terrestrial scavengers. Accumulations of salmon carcasses attract aggregations of scavenging birds to larger rivers (Stalmaster and Gessaman 1984). Long-term declines in the numbers of spawning salmon may result in a significant loss of seasonally abundant nutrients to some wildlife species (Cederholm et al. 1989, Spenser et al. 1991).

The trophic support of large streams (>sixth order) is dominated by autotrophic production of phytoplankton, periphyton, and rooted vascular plants, and by fine particulate organic matter transported into these larger systems from upstream sources and laterally from the floodplain (Vannote et al. 1980, Cummins et al. 1982). Although the direct input of wood and leaf litter from riparian forests is relatively unimportant compared with aquatic plant production and fluvially transported organic material, the food web of higher-order streams is still heavily dependent on resources, including nutrients, that originate from upstream areas (Naiman et al. 1987) as well as from periodic inundation of the floodplain (Pinay et al. 1990). Zooplankton and benthic detritivores are important invertebrate consumers in large streams, and fish communities reflect this shift in available food organisms. Both plankton feeders and benthic insectivores are well represented in the fishes of large rivers. Included among these are many of the thermally tolerant forms inhabiting mid-order streams as well as additional species of minnows (*Mylocheilus caurinus, Hybopsis crameri, Couesius plumbeus*) and other species that often inhabit marshes and sloughs (*Novumbra hubbsi, Percopsis transmontana*). Also represented among the fishes of larger streams are euryhaline species (*Acipenser* spp., *Thaleichthys pacificus, Gasterosteus aculeatus* [occasionally in smaller streams], several Cottidae, *Platichthys stellatus*) that inhabit brackish water ponds, the estuary, or ocean and move into rivers to feed or spawn. Introduced species (*Cyprinus carpio, Carassius auratus, Esox americanus, Gambusia affinis,* several Ictaluridae, many Centrarchidae, several Percidae) have become established in the lower reaches of many Pacific Northwest river basins. In general, nonnative fishes were imported from eastern North America and are associated with aquatic vegetation along the channel margin and floodplain. Some exotic species are piscivorous and in certain areas have displaced native forms from preferred habitat, either through predation or competition (Li et al. 1987). Overall, however, introduced fishes have tended to increase species richness in large streams.

An Ecologically Healthy Watershed

In reference to the original working hypothesis—that delivery and routing of water, sediment, and woody debris are the key processes regulating the characteristics of drainage networks in the Pacific Northwest coastal ecoregion—we offer several observations. The available evidence suggests that ecologically healthy watersheds are maintained by an active natural disturbance regime operating over a range of spatial and temporal scales. Ecologically healthy watersheds are dependent on the nature of the disturbance (e.g., fire, landslides, debris torrents, channel migration) and the ability of the system to adjust to constantly changing conditions. This natural disturbance regime imparts considerable spatial heterogeneity and temporal variability to the physical components of the system. In turn, this is reflected in the life history strategies, productivity, and biodiversity of the biotic community.

This natural disturbance regime produces a dynamic equilibrium for riparian forests, habitat, water storage, water quality, animal migration, and biodiversity resulting in resilient and productive ecological systems. The net result is an ecological system, at the watershed scale, which possesses a biotic integrity strongly valued for its long-term social, economic, and ecological characteristics.

The heart of an ecologically healthy watershed is the riparian forest (Décamps and Naiman 1989, Naiman and Décamps 1990). The riparian forest is shaped by channel geomorphology, hydrologic pattern, spatial position of the channel in the drainage network, and the inherent disturbance regimes. Yet the riparian forest affects, and is affected by, habitat dynamics, water quality, and the animal community. This strongly suggests that maintenance of riparian forests in their historic abundance and in healthy ecological condition is of fundamental importance for long-term ecological and socioeconomic vitality of watersheds in the Pacific Northwest coastal ecoregion.

Acknowledgments. We thank S.V. Gregory, J.R. Karr, and H.Décamps and the students and faculty associated with the Center for Streamside Studies for suggestions and comments which substantially improved the manuscript. Development of concepts was supported by research grants from the U.S. Forest Service, National Science Foundation, Environmental Protection Agency, Weyerhaeuser Company, and governmental and forest industry support of the Center for Streamside Studies. This is Contribution 56 from the University of Washington Center for Streamside Studies.

References

Agee, J.K. 1990. The historical role of fire in Pacific Northwest forests. Pages 25–38 *in* J. Walstad, S. Radosevich, and D. Sandberg, editors. Natural and prescribed fire in Pacific Northwest forests. Oregon State University Press, Corvallis, Oregon, USA.

American Public Health Association. 1989. Standard methods for the examination of water and wastewater. APHA, Washington, D.C., USA.

Anderson, N.H. 1982. A survey of aquatic insects associated with wood debris in New Zealand Streams. Mauri Ora **10**:21–23.

Anderson, N.H., and J.R. Sedell. 1979. Detritus processing by macroinvertebrates in stream ecosystems. Annual Review of Entomology **24**:351–377.

Aumen, N.G., T.J. Grizzard, and R.H. Hawkins. 1989. Water quality monitoring in the Bull Run watershed, Oregon, USA. Task Force Final Report to Bureau of Water Works, Portland, Oregon, USA.

Beaudry, P.G., D.L. Hogan, and J.W. Schwab. 1990. Hydrologic and geomorphic considerations for silvicultural investments on the lower Skeena River floodplain. British Columbia Ministry of Forests, Research Branch. Forest Research Development and Agreement Report 122.

Beechie, T.J., and T.H. Sibley. 1990. Evaluation of the TFW stream classification system: stratification of physical habitat area and distribution. Timber/Fish/Wildlife Final Report TFW-16B-89-006 to Timber, Fish and Wildlife Ambient Monitoring Steering Committee and State of Washington Department of Natural Resources, Olympia, Washington, USA.

Bencala, K.E. 1984. Interactions of solutes and stream bed sediment. 2. A dynamic analysis of coupled hydrologic and chemical processes that determine solute transport. Water Resources Research **20**:1804–1814.

Bencala, K.E., V.C. Kennedy, G.W. Zellweger, A.P. Jackman, and R.J. Avanzino. 1984. Interactions of solutes and streambed sediment. 1. An experimental analysis of cation and anion transport in a mountain stream. Water Resources Research **20**:1797–1803.

Benda, L.E. 1988. Debris flows in the Tyee sandstone foundation of the Oregon Coast Range. M.S. Thesis. Department of Geological Sciences, University of Washington, Seattle, Washington, USA.

Benda, L.E. 1990. The influence of debris flows on channels and valley floors in the Oregon Coast Range, USA. Earth Surface Processes and Landforms **15**:457–466.

Benda, L.E., T.J. Beechie, A. Johnson, and R.C. Wissmar. 1992. The geomorphic structure of salmonid habitats in a recently deglaciated river basin, Washington State. Canadian Journal of Fisheries and Aquatic Sciences, *in press*.

Benda, L.E., and T. Dunne. 1987. Sediment routing by debris flows. Pages 213–223 *in* R.L. Bestchta, T. Blinn, G.E. Grant, G.G. Ice, and F.J. Swanson, editors. Erosion and sedimentation in the Pacific Rim. International Association of Hydrological Sciences, Publication 165, Oxfordshire, United Kingdom.

Berman, C. 1990. The effect of elevated holding temperatures on adult spring chinook salmon reproductive success. M.S. Thesis. School of Fisheries, University of Washington, Seattle, Washington, USA.

Berris, S.N., and R.D. Harr. 1987. Comparative snow accumulation and melt during rainfall in forested and clear-cut plots in the western Cascades of Oregon, USA. Water Resources Research **23**:135–142.

Beschta, R.L., R.E. Bilby, G.W. Brown, L.B. Holtby, and T.D. Hofstra. 1987. Stream temperature and aquatic habitat: fisheries and forestry interactions. Pages 191–232 *in* E.O. Salo and T.W. Cundy, editors. Streamside management: forestry and fishery interactions. Contribution 57, Institute of Forest Resources, University of Washington, Seattle, Washington, USA.

Beschta, R.L., and W.S. Platts. 1986. Morphological features of small streams: significance and function. Water Resources Bulletin **22**:369–379.

Bilby, R.E. 1979. The function and distribution of organic debris dams in forest stream ecosystems. Dissertation. Cornell University, Ithaca, New York, USA.

Bilby, R.E. 1981. Role of organic debris dams in regulating the export of dissolved and particulate matter from a forested watershed. Ecology **62**:1234–1243.

Bilby, R.E. 1985. Influence of stream size on the function and characteristics of large organic debris. *In* Proceedings of the West Coast Meeting of the National Council of the Paper Industry for Air and Stream Improvement, Portland, Oregon, USA.

Bilby, R.E., and J.W. Ward. 1989. Changes in characteristics and function of woody debris with increasing size of streams in western Washington. Transactions of the American Fisheries Society **118**:368–378.

Bisson, P.A., R.E. Bilby, M.D. Bryant, C.A. Dolloff, G.B. Grette, R.A. House, M.L. Murphy, KV. Koski, and J.R. Sedell. 1987. Large woody debris in forested streams in the Pacific Northwest: past, present, and future. Pages 143–190 *in* E.O. Salo and T.W. Cundy, editors. Streamside management: forestry and fishery interactions. Contribution 57, Institute of Forest Resources, University of Washington, Seattle, Washington, USA.

Bisson, P.A., J.L. Nielsen, R.A. Palmason, and L.E. Grove. 1982. A system of naming habitat types in small streams, with examples of habitat utilization by salmonids during low streamflow. Pages 62–73 *in* N.B. Armantrout, editor. Acquisition and utilization of aquatic habitat inventory information. Western Division American Fisheries Society, Portland, Oregon, The Hague Publishing, Billings, Montana, USA.

Bisson, P.A., K. Sullivan, J.L. Nielsen. 1988. Channel hydraulics, habitat use, and body form of juvenile coho salmon, steelhead, and cutthroat trout in streams. Transactions of the American Fisheries Society **117**:262–273.

Bodhaine, G.L., and D.M. Thomas. 1964. Magnitude and frequency of floods in the United States. Part 12. Pacific Slope basins in Washington and upper Columbia River basin. United States Geological Survey, Water-Supply Paper 1687.

Brickell, D.C., and J.J. Goering. 1970. Chemical effects of salmon decomposition on aquatic ecosystems. Pages 125–138 *in* R.S. Murphy, editor. Proceedings of a Symposium on Water Pollution Control in Cold Climates. United States Government Printing Office, Washington, D.C., USA.

Brierley, G.J., and E.J. Hickin. 1985. The downstream gradation of particle sizes in the Squamish River, British Columbia. Earth Surface Processes and Landforms **10**:597–606.

Brown, G.W. 1983. Forestry and water quality. Book Stores, Inc., Oregon State University, Corvallis, Oregon, USA.
Brown, G.W., and J.T. Krygier. 1970. Effects of clearcutting on stream temperature. Water Resources Research **5**:1133–1139.
Brown, G.W., A.R. Gahler, and R.B. Marston. 1973. Nutrient losses after clearcut logging and slash burning in the Oregon Coast Range. Water Resources Research **9**:1450–1453.
Brown, T.G., and G.F. Hartman. 1988. Contribution of seasonally flooded lands and minor tributaries to the production of coho salmon in Carnation Creek, British Columbia. Transactions of the American Fisheries Society **117**:546–551.
Bruijnzeel, L.A. 1983. Evaluation of runoff sources in a forested basin in a wet monsoonal environment: a combined hydrological and hydrochemical approach. Pages 165–174 *in* R. Keller, editor. Hydrology of humid tropical regions. Proceedings of a Symposium of the International Union of Geodesy and Geophysics, Hamburg, West Germany. International Association of Hydrological Sciences, Publication 140, Oxfordshire, United Kingdom.
Burt, T.P. 1989. Storm runoff generation in small catchments in relation to the flood response of large basins. Pages 11–35 *in* K. Beven, editor. Floods: hydrological, sedimentological and geomorphological implications. John Wiley and Sons, New York, New York, USA.
Burt, T.P., and B.P. Arkell. 1986. Variable source areas of stream discharge and their relationship to point and non-point sources of nitrate pollution. Pages 155–164 *in* D. Lerner, editor. Monitoring to detect changes in water quality series. Proceedings of the Budapest Symposium, July 1986. International Association of Hydrological Sciences, Publication 157, Oxfordshire, United Kingdom.
Bustard, D.R., and D.W. Narver. 1975. Aspects of the winter ecology of juvenile coho salmon (*Oncorhynchus kisutch*) and steelhead trout (*Salmo gairdneri*). Journal of the Fisheries Research Board of Canada **32**:667–680.
Carter, V. 1986. An overview of the hydrologic concerns related to wetlands in the United States. Canadian Journal of Botany **64**:364–374.
Cederholm, C.J., D.B. Houston, D.L. Cole, and W.J. Scarlett. 1989. Fate of coho salmon (*Oncorhynchus kisutch*) carcasses in spawning streams. Canadian Journal of Fisheries and Aquatic Sciences **46**:1347–1355.
Cederholm, C.J., and N.P. Peterson. 1985. The retention of coho salmon (*Oncorhynchus kisutch*) carcasses by organic debris in small streams. Canadian Journal of Fisheries and Aquatic Sciences **42**:1222–1225.
Cederholm, C.J., and L.M. Reid. 1987. Impact of forest management on coho salmon (*Oncorhynchus kisutch*) populations of the Clearwater River, Washington: a project summary. Pages 373–398 *in* E.O. Salo and T.W. Cundy, editors. Streamside management: forestry and fishery interactions. Contribution 57, Institute of Forest Resources, University of Washington, Seattle, Washington, USA.
Chapman, D.W. 1988. Critical review of variables used to define effects of fines in redds of large salmonids. Transactions of the American Fisheries Society **117**:1–21.
Chapman, D.W., and T.C. Bjornn. 1969. Distribution of salmonids in streams, with special reference to food and feeding. Pages 153–176 *in* T.G. Northcote, editor. Symposium on Salmon and Trout in Streams. H.R. MacMillan Lectures in Fisheries, Institute of Fisheries, University of British Columbia, Vancouver, Canada.

Chapman, D.W., and K.P. McLeod. 1987. Development of criteria for fine sediment in the Northern Rockies Ecoregion. United States Environmental Protection Agency, Water Division, Report 910/9–87–162, Seattle, Washington, USA.

Cline, S.P., and C.A. Phillips. 1983. Coarse woody debris and debris-dependent wildlife in logged and natural riparian zone forests: a western Oregon example. Pages 33–39 *in* J.W. Davis, G.A. Goodwin, and R.A. Ockenfels, technical coordinators. Snag Habitat Management: Proceedings of the Symposium, Flagstaff, Arizona, USA. United States Forest Service General Technical Report RM-GTR99.

Compana, M.E., and R.L. Boone. 1986. Hydrologic monitoring of subsurface flow and groundwater recharge in a mountain watershed. Pages 263–274 *in* D.L. Kane, editor. Proceedings of the Symposium: Cold Regions Hydrology. American Water Resources Association, Bethesda, Maryland, USA.

Conners, E.M., and R.J. Naiman. 1984. Particulate allochthonous inputs: relationships with stream size in an undisturbed watershed. Canadian Journal of Fisheries and Aquatic Sciences **41**:1473–1484.

Crandell, D.R. 1965. The glacial history of western Washington and Oregon. Pages 241–353 *in* H.E. Wright and D.G. Fry, editors. The Quaternary of the United States. Princeton University Press, Princeton, New Jersey, USA.

Cummans, J.E., M.R. Collings, and E.G. Nassar. 1975. Magnitude and frequency of floods in Washington. United States Geological Survey, Open-File Report 74–336, Tacoma, Washington, USA.

Cummins, K.W., J.R. Sedell, F.J. Swanson, G.W. Minshall, S.G. Fisher, C.E. Cushing, R.C. Petersen, and R.L. Vannote. 1982. Organic matter budgets for stream ecosystems: problems in their evaluation. Pages 299–353 *in* G.W. Minshall and J.R. Barnes, editors. Stream ecology: application and testing of general ecological theory. Plenum, New York, New York, USA.

Cupp, C.E. 1989*a*. Stream corridor classification for forested lands of Washington. Report prepared for Washington Forest Protection Association, Olympia, Washington, USA.

Cupp, C.E. 1989*b*. Identifying spatial variability of stream characteristics through classification. M.S. Thesis. School of Fisheries, University of Washington, Seattle, Washington, USA.

Dahm, C.N., E.H. Trotter, and J.R. Sedell. 1987. The role of anaerobic zones and processes in stream ecosystem productivity. Pages 157–178 *in* R.C. Averett and D.M. McNight, editors. Chemical quality of water and the hydrologic cycle. Lewis Publishers, Chelsea, Michigan, USA.

Davis, G.H. 1988. Western alluvial valleys and the High Plains. Pages 283–300 *in* G.H. Davis editor. Hydrogeology. Geological Society of North America, Boulder, Colorado, USA.

DeAngelis, D.L., W.M. Post, and C.C. Travis. 1986. Positive feedback in natural systems. Springer-Verlag, New York, New York, USA.

Décamps, H., and R.J. Naiman. 1989. L'écologie des fleuves. La Recherche **20**:310–319.

Doyle, A.T. 1985. Small mammal micro- and macrohabitat selection in streamside ecosystems. Dissertation. Oregon State University, Corvallis, Oregon, USA.

Doyle, A.T. 1990. Use of riparian and upland habitat by small mammals. Journal of Mammology **71**:14–23.

Dunne, T. 1978. Field studies of hillslope processes. Pages 227–293 *in* M.J. Kirby, editor. Hillslope hydrology. John Wiley and Sons, London, England.

Dunne, T., and L.B. Leopold. 1978. Water in environmental planning. W.H. Freeman, San Francisco, California, USA.
Environmental Protection Agency. 1986. Quality criteria for water. 1986. United States Environmental Protection Agency EPA 440/5-86-001, Office of Water Regulations and Standards, Washington, D.C., USA.
Everest, F.H. 1987. Salmonids of western forested watersheds. Pages 3–8 *in* E.O. Salo and T.W. Cundy, editors. Streamside management: forestry and fishery interactions. Contribution 57, Institute of Forest Resources, University of Washington, Seattle, Washington, USA.
Everest, F.H., R.L. Beschta, J.C. Scrivener, KV. Koski, J.R. Sedell, and C.J. Cederholm. 1987. Fine sediment and salmonid production: a paradox. Pages 98–142 *in* E.O. Salo, and T.W. Cundy, editors. Streamside management: forestry and fishery interactions. Contribution 57, Institute of Forest Resources, University of Washington, Seattle, Washington, USA.
Feller, M.C., and J.P. Kimmins. 1979. Chemical characteristics of small streams near Haney in southwestern British Columbia. Water Resources Research **15**:247–258.
Fisher, S.G., and G.E. Likens. 1973. Energy flow in Bear Brook, New Hampshire: an integrative approach to stream ecosystem metabolism. Ecological Monographs **43**:421–439.
Ford, T.E., and R.J. Naiman. 1989. Groundwater-surface water relationships in boreal forest watersheds: dissolved organic carbon and inorganic nutrient dynamics. Canadian Journal of Fisheries and Aquatic Sciences **46**:41–49.
Freeze, R.A., and J.A. Cherry. 1979. Groundwater. Prentice-Hall, Englewood Cliffs, New Jersey, USA.
Frere, M.H., E.H. Seeley, and R.H. Leonard. 1982. Modeling the quality of water from agricultural land. Pages 381–405 *in* C.T. Haan, H.P. Johnson, and D.L. Brakensiek, editors. Modeling of small watersheds. American Society of Agricultural Engineering, St. Joseph, Michigan, USA.
Frissell, C.A., W.J. Liss, C.E. Warren, and M.D. Hurley. 1986. A hierarchical framework for stream habitat classification: viewing streams in a watershed context. Environmental Management **10**:199–214.
Fritsch, J.M., P.L. Dubreuil, and J.M. Sarrailh. 1987. From the plot to the watershed: scale effect in the Amazonian forest ecosystem. Pages 131–142 *in* R.H. Swanson, editor. Forest hydrology and watershed management. Proceedings of the Vancouver Symposium, August 1987. International Association of Hydrological Sciences, Publication 167, Oxfordshire, United Kingdom.
Gibert, J., M.J. Dole-Olivier, P. Marmonier, and P. Vervier. 1990. Surface water-groundwater ecotones. Pages 199–221 *in* R.J. Naiman and H. Décamps. The ecology and management of aquatic-terrestrial ecotones. UNESCO, Paris, and Parthenon Publishing Group, Carnforth, United Kingdom.
Grant, G.E. 1986. Downstream effects of timber harvest activity on the channel and valley floor morphology of western Cascade streams. Dissertation. Johns Hopkins University, Baltimore, Maryland, USA.
Grant, G.E., F.J. Swanson, and M.G. Wolman. 1990. Pattern and origin of stepped-bed morphology in high-gradient streams, western Cascades, Oregon. Geological Society of America Bulletin **102**:340–352.
Green, D.M., and J.B. Kauffman. 1989. Nutrient cycling at the land-water interface: the importance of the riparian zone. Pages 61–68 *in* R.E. Gresswell, B.A.

Barton, and J.L. Kerschner, editors. Practical approaches to riparian resource management. United States Department of the Interior, Bureau of Land Management, Billings, Montana, USA.
Gregory, S.V., G.A. Lamberti, D.C. Erman, KV. Koski, M.L. Murphy, and J.R. Sedell. 1987. Influence of forest practices on aquatic production. Pages 233–255 *in* E.O. Salo and T.W. Cundy, editors. Streamside management: forestry and fishery interactions. Contribution 57, Institute of Forest Resources, University of Washington, Seattle, Washington, USA.
Gregory, S.V., F.J. Swanson, W.A. McKee, and K.W. Cummins. 1991. An ecosystem perspective of riparian zones. BioScience, **41**:540–551.
Grette, G.B. 1985. The role of large organic debris in juvenile salmonid rearing habitat in small streams. M.S. Thesis. School of Fisheries, University of Washington, Seattle, Washington, USA.
Grimm, N.B., and S.G. Fisher. 1984. Exchange between interstitial and surface water: implications for stream metabolism and nutrient cycling. Hydrobiologia **111**:219–228.
Harmon, M.E., J.F. Franklin, F.J. Swanson, P. Sollins, S.V. Gregory, J.D. Lattin, N.H. Anderson, S.P. Cline, N.G. Aumen, J.R. Sedell, G.W. Lienkaemper, K. Cromack, Jr., and K.W. Cummins. 1986. Ecology of coarse woody debris in temperate ecosystems. Advances in Ecological Research **15**:133–302.
Harr, R.D. 1976. Hydrology of small forest streams in western Oregon. United States Forest Service General Technical Report PNW-55, Pacific Northwest Forest and Range Experiment Station, Portland, Oregon, USA.
Harr, R.D., and R.L. Fredriksen. 1988. Water quality after logging small watersheds within the Bull Run watershed, Oregon, USA. Water Resources Bulletin **24**:1103–1111.
Hartman, G., J.C. Scrivener, L.B. Holtby, and L. Powell. 1987. Some effects of different streamside treatments on physical conditions and fish population processes in Carnation Creek, a coastal rain forest stream in British Columbia. Pages 330–372 *in* E.O. Salo and T.W. Cundy, editors. Streamside management: forestry and fishery interactions. Contribution 57, Institute of Forest Resources, University of Washington, Seattle, Washington, USA.
Harvey, G.W. 1989. Technical review of sediment criteria. Water Quality Bureau, Idaho Department of Health and Welfare, Boise, Idaho, USA.
Heede, B.H. 1972. Influences of a forest on the hydraulic geometry of two mountain streams. Water Resource Bulletin **8**:523–529.
Hem, J.D. 1985. Study and interpretation of the chemical characteristics of natural water. United States Geological Survey, Water-Supply Paper 2254.
Hemstrom, M.A., and J.F. Franklin. 1982. Fire and other disturbances of the forests in Mount Rainier National Park. Quaternary Research **18**:32–51.
Hewlett, J.D., and A.R. Nutter. 1970. The varying source area of stream flow from upland basins. Pages 65–83 *in* Interdisciplinary aspects of watershed management. American Society of Civil Engineers, New York, New York, USA.
Higgins, C.G. 1984. Piping and sapping: development of landforms by groundwater outflow. Pages 18–58 *in* R.G. LaFleur, editor. Groundwater as a geomorphic agent. Proceedings of the 13th Annual Geomorphology Symposium, Rensselaer Polytechnic Institute, Troy, New York, USA.
Hilgert, J.W., and C.W. Slaughter. 1988. Water quality and streamflow in the Caribou-Poker Creeks Research Watershed, Central Alaska, 1979. United States Forest Service PNW Research Note 463, Portland, Oregon, USA.

Hixson, S.E., R.F. Walker, and C.M Skau. 1990. Soil denitrification rates in four subalpine plant communities of the Sierra Nevada. Journal of Environmental Quality **19**:617–620.

Holland, M.M., D.F. Whigham, and B. Gopal. 1990. The characteristics of wetland ecotones. Pages 171–198 *in* R.J. Naiman and H. Décamps, editors. The ecology and management of aquatic-terrestrial ecotones. UNESCO, Paris, and Parthenon Publishing Group, Carnforth, United Kingdom.

Hughes, D.A. 1980. Floodplain inundation: processes and relationships with channel discharge. Earth Surface Processes **5**:297–304.

Hynes, H.B.N. 1966. The biology of polluted waters. Liverpool University Press, Liverpool, England.

Hynes, H.B.N. 1970. The ecology of running waters. University of Toronto Press, Toronto, Ontario, Canada.

Hynes, H.B.N. 1983. Groundwater and stream ecology. Hydrobiologia **100**:93–99.

Hynes, H.B.N. 1985. The stream and its valley. Internationale Vereinigung für theoretische und angewandte Limnologie, Verhandlungen **19**:1–15.

Johnston, C.A., and R.J. Naiman. 1987. Boundary dynamics at the aquatic-terrestrial interface: the influence of beaver and geomorphology. Landscape Ecology **1**:47–57.

Johnston, C.A., and R.J. Naiman. 1990. Browse selection by beaver: effects on riparian forest composition. Canadian Journal of Forest Research **20**:1036–1043.

Karr, J.R. 1991. Biological integrity: a long-neglected aspect of water resource management. Ecological Applications **1**:66–84.

Karr, J.R., K.D. Fausch, P.L. Angermeier, P.R. Yant, and I.J. Schlosser. 1986. Assessing biological integrity in running waters: a method and its rationale. Illinois Natural History Survey, Special Publication 5, Champaign, Illinois, USA.

Karr, J.R., and I.J. Schlosser. 1978. Water resources and the land-water interface. Science **201**:229–234.

Kauffman, J.B. 1988. The status of riparian habitats in Pacific Northwest forests. Pages 45–55 *in* K.J. Raedeke, editor. Streamside management: riparian wildlife and forestry interactions. Contribution 59, Institute of Forest Resources, University of Washington, Seattle, Washington, USA.

Keller, E.A., and F.J. Swanson. 1979. Effects of large organic material on channel form and fluvial processes. Earth Surface Processes **4**:361–380.

Keller, H.M., F. Forster, and P. Weibel. 1986. Factors affecting stream water quality: results of a 15 year monitoring study in the Swiss prealps. Pages 215–225 *in* Monitoring to detect changes in water quality series. International Association of Hydrological Sciences Publication 157, Wallingford, Oxfordshire, United Kingdom.

Kunkle, S.H., and G.H. Comer. 1971. Estimating suspended sediment concentrations in streams by turbidity measurements. Journal of Soil and Water Conservation **26**:18–20.

Leopold, L.B., M.G. Wolman, and J.P. Miller. 1964. Fluvial processes in geomorphology. W.H. Freeman, San Francisco, California, USA.

Li, H.W., C.B. Schreck, C.E. Bond, and E. Rexstad. 1987. Factors influencing changes in fish assemblages of Pacific Northwest streams. Pages 193–202 *in* W.J. Matthews and D.C. Heins, editors. Community and evolutionary ecology of North American stream fishes. University of Oklahoma Press, Norman, Oklahoma, USA.

Lienkaemper, G.W., and F.J. Swanson. 1987. Dynamics of large woody debris in streams in old-growth Douglas-fir forests. Canadian Journal of Forest Research **17**:150–156.

Likens, G.E., editor. 1989. Long-term studies in ecology: approaches and alternatives. Springer-Verlag, New York, New York, USA.

Lisle, T.E. 1987. Overview: channel morphology and sediment transport in steepland channels. Pages 287–298 *in* R.L. Beschta, T. Blinn, G.E. Grant, G.G. Ice, and F.J. Swanson, editors. Erosion and sedimentation in the Pacific Rim. International Association of Hydrological Sciences, Publication 165, Oxfordshire, United Kingdom.

Lowrance, R.R., R.L. Todd, and L.E. Asmussen. 1984. Nutrient cycling in an agricultural watershed. I. Phreatic movement. Journal of Environmental Quality **13**:22–27.

Lum, W.E. 1984. Availability of ground water from the alluvial aquifer on the Nisqually Indian Reservation, Washington. United States Geological Survey, Water-Resources Investigations Report 83–4185.

MacDonald, L.H., A. Smart, and R.C. Wissmar. 1991. Monitoring guidelines to evaluate effects of forestry activities on streams in the Pacific Northwest and Alaska. United States Environmental Protection Agency, Technical Report EPA/910/9-91-001, Seattle, Washington, USA.

Maguire, C.C. 1983. First year responses of small mammals populations to clear-cutting in the Klamath Mountains at Northern California. Dissertation. Rutgers University, New Brunswick, New Jersey, USA.

Martin, C.W., and R.D. Harr. 1989. Logging of mature Douglas-fir in western Oregon has little effect on nutrient budget outputs. Canadian Journal of Forest Research **19**:35–43.

Martin, L.J. 1982. Ground water in Washington. Report 40, Washington Water Research Center, Washington State University, Pullman, Washington, USA.

Maser, C., R.F. Tarrant, J.M. Trappe, and J.F. Franklin. 1988. From the forest to the sea: a story of fallen trees. United States Forest Service. General Technical Report PNW-GTR-229, Portland, Oregon, USA.

Maser, C., and J.M. Trappe, editors. 1984. The seen and unseen world of the fallen tree. United States Forest Service General Technical Report PNW-164, Pacific Northwest Forest and Range Experiment Station, Portland, Oregon, USA.

Mason, D.T., and J. Koon. 1985. Habitat values of woody debris accumulations of the lower Stehekin River, with notes on disturbances of alluvial gravels. Final report to the National Park Service, Contract CX-9000–3-8066. Fairhaven College, Western Washington University, Bellingham, Washington, USA.

McKee, E.B. 1972. Cascadia: the geologic evolution of the Pacific Northwest. McGraw-Hill, New York, New York, USA.

McPhail, J.D. 1967. Distribution of freshwater fishes in western Washington. Northwest Science **41**:1–11.

Melillo, J.M., R.J. Naiman, J.D. Aber, and K.N. Eshleman. 1983. The influence of substrate quality and stream size on wood decomposition dynamics. Oecologia (Berlin) **58**:281–285.

Melillo, J.M., R.J. Naiman, J.D. Aber, and A.E. Linkins. 1984. Factors controlling mass loss and nitrogen dynamics of plant litter decaying in northern streams. Bulletin of Marine Science **35**:341–356.

Minshall, G.W. 1988. Stream ecosystem theory: a global perspective. Journal of the North American Benthological Society **7**:263–288.

Mitsch, W.J., and J.G. Gosselink. 1986. Wetlands. Van Nostrand Reinhold, New York, New York, USA.

Mohaupt, V. 1986. Nutrient-discharge relationships in a flatland river system and optimization of sampling. Pages 297–304 *in* Monitoring to detect changes in water quality series. International Association of Hydrological Sciences Publication 157, Oxfordshire, United Kingdom.

Moore, K.M.S., and S.V. Gregory. 1988*a*. Response of young-of-the-year cutthroat trout to manipulation of habitat structure in a small stream. Transactions of the American Fisheries Society **117**:162–170.

Moore, K.M.S., and S.V. Gregory. 1988*b*. Summer habitat utilization and ecology of cutthroat trout fry (*Salmo clarki*) in Cascade Mountain streams. Canadian Journal of Fisheries and Aquatic Sciences **45**:1921–1930.

Mosley, M.P. 1981. The influence of organic debris on channel morphology and bedload transport in a New Zealand forest stream. Earth Surface Processes and Landforms **6**:571–579.

Moyle, P.B., and B. Herbold. 1987. Life history patterns and community structure in stream fishes of western north America: comparisons with eastern north America and Europe. Pages 25–32 *in* W.J. Matthews and D.C. Heins, editors. Community and evolutionary ecology of North American stream fishes. University of Oklahoma Press, Norman, Oklahoma, USA.

Munter, J.A. 1986. Evidence of groundwater discharge through frozen soils at Anchorage, Alaska. Pages 245–252 *in* D.L. Kane, editor. Proceedings of the Symposium: Cold Regions Hydrology. American Water Resources Association, Bethesda, Maryland, USA.

Murphy, M.L., KV. Koski, J. Heifetz, S.W. Johnson, D. Kirchofer, and J.F. Thedinga. 1985. Role of large organic debris as winter habitat for juvenile salmonids in Alaska streams. Proceedings, Western Association of Fish and Wildlife Agencies **1984**:251–262.

Naiman, R.J. 1983. The annual pattern and spatial distribution of aquatic oxygen metabolism in boreal forest watersheds. Ecological Monographs **53**:73–94.

Naiman, R.J. 1990. Forest ecology: influence of forests on streams. Pages 151–153 *in* 1991 McGraw-Hill Yearbook of Science and Techology, McGraw-Hill, New York, New York, USA.

Naiman, R.J., editor. 1988. How animals shape their ecosystems. BioScience **38**:750–800.

Naiman, R.J., and H. Décamps, editors. 1990. The ecology and management of aquatic-terrestrial ecotones. UNESCO, Paris, and Parthenon Publishing Group, Carnforth, United Kingdom.

Naiman, R.J., H. Décamps, J. Pastor, and C.A. Johnston. 1988. The potential importance of boundaries to fluvial ecosystems. Journal of the North American Benthological Society **7**:289–306.

Naiman, R.J., C.A. Johnston, and J.C. Kelley. 1988. Alteration of North American streams by beaver. BioScience **38**:753–762.

Naiman, R.J., D.G. Lonzarich, T.J. Beechie, and S.C. Ralph. 1991. General principles of classification and the assessment of conservation potential in rivers. Pages 93–123 *in* P.J. Boon, P. Calow, and G.E. Petts, editors. River conservation and management. John Wiley and Sons, Chichester, England.

Naiman, R.J., and J.M. Melillo. 1984. Nitrogen budget of a subarctic stream altered by beaver (*Castor canadensis*). Oecologia (Berlin) **62**:150–155.

Naiman, R.J., J.M. Melillo, and J.E. Hobbie. 1986. Ecosystem alteration of boreal forest streams by beaver (*Castor canadensis*). Ecology **67**:1254–1269.

Naiman, R.J., J.M. Melillo, M.A. Lock, T.E. Ford, and S.R. Reice. 1987. Longitudinal patterns of ecosystem processes and community structure in a subarctic river continuum. Ecology **68**:1139–1156.

Naiman, R.J., and J.R. Sedell. 1979*a*. Benthic organic matter as a function of stream order in Oregon. Archiv für Hydrobiologie **87**:404–422.

Naiman, R.J., and J.R. Sedell. 1979*b*. Characteristics of particulate organic matter transported by some Cascade mountain streams. Journal of the Fisheries Research Board of Canada **36**:17–31.

Naiman, R.J., and J.R. Sedell. 1980. Relationships between metabolic parameters and stream order in Oregon. Canadian Journal of Fisheries and Aquatic Sciences **37**:834–847.

Nehlsen, W., J.E. Williams, and J.A. Lichatowich. 1991. Pacific salmon at the crossroads: stocks of salmon at risk from California, Oregon, Idaho, and Washington. Fisheries **16**:4–21.

Newbold, J.D., P.J. Mulholland, J.W. Elwood, and R.V. O'Neill. 1982. Organic spiraling in stream ecosystems. Oikos **38**:266–272.

Northcote, T.G. 1978. Migratory strategies and production in freshwater fishes. Pages 326–359 *in* S.D. Gerking, editor. Ecology of freshwater fish production. Blackwell Scientific Publications, Oxford, England.

Oakley, A.L., J.A. Collins, L.B. Everson, D.A. Heller, J.C. Howerton, and R.E. Vincent. 1985. Riparian zones and freshwater wetlands. Pages 58–79 *in* E.R. Brown, editor. Management of wildlife and fish habitats in forests of western Oregon and Washington. Part 1. United States Forest Service, Pacific Northwest Region, Portland, Oregon, USA.

Oliver, C.D., and T.M. Hinckley. 1987. Species, stand structures, and silvicultural manipulation patterns for the streamside zone. Pages 259–276 *in* E.O. Salo and T.W. Cundy, editors. Streamside management: forestry and fishery interactions. Contribution 57, Institute of Forest Resources, University of Washington, Seattle, Washington, USA.

O'Loughlin, E.M. 1986. Prediction of surface saturation zones in natural catchments by topographic analysis. Water Resources Research **22**:794–804.

Pastor, J., R.J. Naiman, B. Dewey, and P. McInnes. 1988. Moose, microbes, and the boreal forest. BioScience **38**:770–777.

Pennak, R.W., and J.V. Ward. 1986. Interstitial faunal communities of the hyporheic and adjacent groundwater biotopes of a Colorado mountain stream. Archiv für Hydrobiologie **74**:356–396.

Perkins, S.J. 1989. Interactions of landslide-supplied sediment with channel morphology in forested watersheds. M.S. Thesis. Department of Geology, University of Washington, Seattle, Washington, USA.

Peterjohn, W.T., and D.L. Correll. 1984. Nutrient dynamics in an agricultural watershed: observations on the role of a riparian forest. Ecology **65**:1466–1475.

Peterson, N.P. 1982*a*. Population characteristics of juvenile coho salmon (*Oncorhynchus kisutch*) overwintering in riverine ponds. Canadian Journal of Fisheries and Aquatic Sciences **39**:1303–1307.

Peterson, N.P. 1982*b*. Immigration of juvenile coho salmon (*Oncorhynchus kisutch*) into riverine ponds. Canadian Journal of Fisheries and Aquatic Sciences **39**:1308–1310.

Peterson, N.P., and L.M. Reid. 1984. Wall-base channels: their evolution, distribution, and use by juvenile coho salmon in the Clearwater River, Washington. Pages 215–226 *in* J.M. Walton and D.B. Houston, editors. Proceedings of the Olympic Wild Fish Conference, March 23–25, 1983, Port Angeles, Washington, USA.

Pickett, S.T.A., and P.S. White, editors. 1985. The ecology of natural disturbance and patch dynamics. Academic Press, New York, New York, USA.

Pierson, T.C. 1977. Factors controlling debris flow initiation on forested hillslopes in the Oregon Coast Range. Dissertation. University of Washington, Seattle, Washington, USA.

Pinay, G., H. Décamps, E. Chauvet, and E. Fustec. 1990. Functions of ecotones in fluvial systems. Pages 141–170 *in* R.J. Naiman and H. Décamps, editors. The ecology and management of aquatic-terrestrial ecotones. UNESCO, Paris, and Parthenon Publishing Group, Carnforth, United Kingdom.

Plamondon, A.P., A. Gonzalez, and Y. Thomassin. 1982. Effects of logging on water quality: comparison between two Quebec sites. Pages 49–70 *in* Canadian Hydrologic Symposium: 82, Fredericton, New Brunswick, Canada.

Platts, W.S. 1974. Geomorphic and aquatic conditions influencing salmonids and stream classification—with application to ecosystem management. United States Forest Service, Intermountain Forest and Range Experiment Station, Boise, Idaho, USA.

Ponce, V.M., and D.S. Lindquist. 1990. Management of baseflow augmentation: a review. Water Resources Bulletin **26**:259–268.

Potts, D.F., and B.K.M. Anderson. 1990. Organic debris and the management of small stream channels. Western Journal of Applied Forestry **5**:25–28.

Pringle, C.M., R.J. Naiman, G. Bretschko, J.R. Karr, M.W. Oswood, J.R. Webster, R.L. Welcomme, and M.J. Winterbourn. 1988. Patch dynamics in lotic systems: the stream as a mosaic. Journal of the North American Benthological Society **7**:503–524.

Quinn, T.P., and R.F. Tallman. 1987. Seasonal environmental predictability in riverine fishes. Environmental Biology of Fishes **18**:155–159.

Raedeke, K.J., editor. 1988. Streamside management: riparian wildlife and forestry interactions. Contribution 59, Institute of Forest Resources, University of Washington, Seattle, Washington, USA.

Reimers, P.E., and C.E. Bond. 1967. Distribution of fishes in tributaries of the lower Columbia River. Copeia **1967**:541–550.

Rhodes, J.J., C.M. Skau, and D.L. Greenlee. 1986. The role of snowcover on diurnal nitrate concentration patterns in streamflow from a forested watershed in the Sierra Nevada, Nevada, USA. Pages 157–166 *in* D.L. Kane, editor. Proceedings of the Symposium: Cold Regions Hydrology. American Water Resources Association, Bethesda, Maryland, USA.

Ringler, N.H., and J.D. Hall. 1975. Effects of logging on water temperature and dissolved oxygen in spawning beds. Transactions of the American Fisheries Society **104**:111–121.

Roberge, J., and A.P. Plamondon. 1987. Snowmelt runoff pathways in a boreal forest hillslope, the role of pipe throughflow. Journal of Hydrology **95**:39–54.

Rothacher, J. 1963. New precipitation under a Douglas-fir forest. Forest Science **9**:423–429.

Rutherford, J.E., and H.B.N. Hynes. 1987. Dissolved organic carbon in streams and groundwater. Hydrobiologia **154**:33–48.

Salo, E.O., and T.W. Cundy, editors. 1987. Streamside management: forestry and fishery interactions. Contribution 57. Institute of Forest Resources, University of Washington, Seattle, Washington, USA.

Satterlund, D.R. 1985. Variable source areas of watershed runoff in a small forest watershed: Phase I. Department of Forestry and Range Management, State of Washington Water Research Center Report A-128. Washington State University, Pullman, Washington, USA.

Scarlett, W.S., and C.J. Cederholm. 1984. Juvenile coho salmon fall-winter utilization of two small tributaries in the Clearwater River, Washington. Pages 227–242 *in* J.M. Walton and D.B. Houston, editors. Proceedings of the Olympic Wild Fish Conference, March 23–25, 1983, Port Angeles, Washington, USA.

Schlosser, I.J., and J.R. Karr. 1981*a*. Riparian vegetation and channel morphology impact on spatial patterns in water quality in agricultural watersheds. Environmental Management **5**:233–243.

Schlosser, I.J., and J.R. Karr. 1981*b*. Water quality in agricultural watersheds: impact of riparian vegetation during base flow. Water Resources Bulletin **17**:233–240.

Sedell, J.R., P.A. Bisson, F.J. Swanson, and S.V. Gregory. 1988. What we know about large trees that fall into streams and rivers. Pages 47–81 *in* C. Maser, R. F. Tarrant, J.M. Trappe, and J.F. Franklin, editors. From the forest to the sea: a story of fallen trees. United States Forest Service General Technical Report PNW-GTR-229, Portland, Oregon, USA.

Sedell, J.R., and J.L. Froggatt. 1984. Importance of streamside forests to large rivers: the isolation of the Willamette River, Oregon, USA, from its floodplain by snagging and streamside forest removal. Internationale Vereinigung für theoretische und angewandte Limnologie, Verhandlungen **22**:1828–1834.

Shearer, C.A., and S.B. von Bodman. 1983. Patterns of occurrence of ascomycetes associated with decomposing twigs in a Midwestern Stream. Mycologia **75**:518–530.

Sidle, R. 1986. Groundwater accretion in the unstable hillslopes of coastal Alaska. Pages 335–343 *in* S.M. Gorelick, editor. Conjunctive water use. Proceedings of the Budapest Symposium, July 1986. International Association of Hydrological Sciences Publication 156, Oxfordshire, United Kingdom.

Skeesick, D.G. 1970. The fall immigration of juvenile coho salmon into a small tributary. Research Reports of the Fish Commission of Oregon **2**:90–95.

Sklash, M.G., and R.N. Farvolden. 1979. The role of groundwater in storm runoff. Journal of Hydrology **43**:45–65.

Sloan P.G., and I.D. Moore. 1984. Modeling subsurface stormflow on steeply sloping forested watersheds. Water Resources Research **20**:1815–1822.

Smith, D.G. 1973. Aggradation of the Alexandra-Saskatchewan River, Banff Park, Alberta. Pages 201–219 *in* M. Morisawa, editor. Fluvial geomorphology. Binghamton Symposia in Geomorphology International Series **4**:314.

Sollins, P., C.C. Grier, F.M. McCorison, K. Cromack, Jr., R. Fogel, and R.L. Fredriksen. 1980. The internal element cycles of an old-growth Douglas-fir ecosystem in western Oregon. Ecological Management **50**:261–285.

Spenser, C.N., B.R. McClelland, and J.A. Stanford. 1991. Shrimp stocking, salmon collapse, and eagle displacement. BioScience **41**:14–21.

Spies, T.A., and S.P. Cline. 1988. Coarse woody debris in forests and plantations of coastal Oregon. Pages 5–24 *in* C. Maser, R.F. Tarrant, J.M. Trappe, and J.F. Franklin, editors. From the forest to the sea: a story of fallen trees. United States Forest Service General Technical Report PNW-GTR-229, Portland, Oregon, USA.

Stalmaster, M.V., and J.A. Gessaman. 1984. Ecological energetics and foraging behavior of overwintering bald eagles. Ecological Monographs **54**:407–428.

Stanford, J.A., and J.V. Ward. 1988. The hyporheic habitat of river ecosystems. Nature **335**:64–66.

Starfield, A.M., and A.L. Bleloch. 1986. Building models for conservation and wildlife management. Macmillan, New York, New York, USA.

Statzner, B., J.A. Gore, and V.H. Resh. 1988. Hydraulic stream ecology: observed patterns and potential applications. Journal of the North American Benthological Society **7**:307–360.

Stewart, J.E., and G.L. Bodhaine. 1961. Floods in the Skagit River Basin, Washington. U.S. Geological Survey, Water-Supply Paper 1688.

Strayer, D., J.S. Glitzenstein, C.G. Jones, J. Kolasa, G.E. Likens, M.J. McDonald, G.G. Parker, and S.T.A. Pickett. 1986. Long-term ecological studies: an illustrated account of their design, operation, and importance to ecology. Occasional Publication 2, Institute of Ecosystem Studies, New York Botanical Garden, Millbrook, New York, USA.

Sugai, S.F., and D.C. Burrell. 1984. Transport of dissolved organic carbon, nutrients and trace metals from the Wilson and Blossom Rivers to Smeaton Bay, southeast Alaska. Canadian Journal of Fisheries and Aquatic Sciences **41**:180–190.

Sunada, K., and T.F. Hong. 1988. Effects of slope conditions on direct runoff characteristics by the Interflow and Overland Flow Model. Journal of Hydrology **102**:381–406.

Swanson, F.J., R.L. Fredrickson, and F.M. McCorison. 1982. Material transfer in a western Oregon forested watershed. Pages 233–266 *in* R.L. Edmonds, editor. Analysis of coniferous forest ecosystems in the western United States. Hutchinson Ross, Stroudsburg, Pennsylvania, USA.

Swanson, F.J., R.L. Graham, and G.E. Grant. 1985. Some effects of slope movements on river channels. Pages 273–278 *in* International Symposium on Erosion, Debris Flow and Disaster Prevention. Tsukuba, Japan.

Swanson, F.J., S.V. Gregory, J.R. Sedell, and A.G. Campbell. 1982. Land-water interactions: the riparian zone. Pages 267–291 *in* R.L. Edmonds, editor. Analysis of coniferous forest ecosystems in the western United States. Hutchinson Ross, Stroudsburg, Pennsylvania, USA.

Swanson, F.J., and G.W. Lienkaemper. 1978. Physical consequences of large organic debris in Pacific Northwest streams. United States Forest Service General Technical Report, PNW-69, Pacific Northwest Forest and Range Experiment Station, Portland, Oregon, USA.

Swanson, F.J., and G.W. Lienkaemper. 1982. Interactions among fluvial processes, forest vegetation, and aquatic ecosystems, South Fork Hoh River, Olympic National Park. Pages 30–35 *in* E.E. Starkey, J.F. Franklin, and J.W. Matthews, editors. Ecological research in the national parks of the Pacific Northwest. Oregon State University Press, Corvallis, Oregon, USA.

Swanston, D.N., and F.J. Swanson. 1976. Timber harvesting, mass erosion, and steepland forest geomorphology in the Pacific Northwest. Pages 199–221 *in* D.R. Coates, editor. Geomorphology and engineering. Hutchinson Ross, Stroudsburg, Pennsylvania, USA.

Sweeney, B.W., and R.L. Vannote. 1978. Size variation and the distribution of hemimetabolous aquatic insects: two thermal equilibrium hypotheses. Science **200**:444–446.

Swistock, B.R., D.R. DeWalle, and W.E. Sharpe. 1989. Sources of acidic storm flow in an Appalachian headwater stream. Water Resources Research **25**:2139–2147.

Teensma, P.D.A. 1987. Fire history and fire regimes of the central western Cascades of Oregon. Dissertation. University of Oregon, Eugene, Oregon, USA.

Thomas, J.W., C. Maser, J.E. Rodiek. 1979. Riparian zones. Pages 40–47 *in* J.W. Thomas, editor. Wildlife habitats in managed forests: the Blue Mountains of Oregon and Washington. Agriculture Handbook 553. United States Forest Service, Washington, D.C., USA.

Toth, J. 1963. A theoretical analysis of groundwater flow in small drainage basins. Journal of Geophysical Research **68**:4795–5012.

Triska, F.J., V.C. Kennedy, R.J. Avanzino, G.W. Zellweger, and K.E. Bencala. 1989*a*. Retention and transport of nutrients in a third-order stream: channel processes. Ecology **70**:1877–1892.

Triska, F.J., V.C. Kennedy, R.J. Avanzino, G.W. Zellweger, and K.E. Bencala. 1989*b*. Retention and transport of nutrients in a third-order stream in northwestern California: hyporheic processes. Ecology **70**:1893–1905.

Troendle, C.A. 1985. Variable source area models. Pages 347–403 *in* M.G. Anderson and T.R. Burt, editors. Hydrological forecasting. John Wiley and Sons, New York, New York, USA.

Troendle, C.A. 1987. Effect of clearcutting on streamflow generating processes from a subalpine forest slope. Pages 545–552 *in* R.H. Swanson, editor. Forest hydrology and watershed management. Proceedings of the Vancouver Symposium, August 1987. International Association of Hydrological Sciences, Publication 167, Oxfordshire, United Kingdom.

Tschaplinski, P.J., and G.F. Hartman. 1983. Winter distribution of juvenile coho salmon (*Oncorhynchus kisutch*) before and after logging in Carnation Creek, British Columbia, and some implications for overwinter survival. Canadian Journal of Fisheries and Aquatic Sciences **40**:452–461.

Vannote, R.L., G.W. Minshall, K.W. Cummins, J.R. Sedell, and C.E. Cushing. 1980. The river continuum concept. Canadian Journal of Fisheries and Aquatic Sciences **37**:130–137.

Vervier, P., and R.J. Naiman. 1992. Spatial and temporal fluctuations of dissolved organic carbon in subsurface flow of the Stillaguamish River (Washington, USA). Archiv für Hydrobiologie, *in press*.

Wald, A.R., and M.G. Schaefer. 1986. Hydrologic functions of wetlands of the Pacific Northwest. Pages 17–24 *in* Wetland functions, rehabilitation, and creation in the Pacific Northwest: the state of our understanding. Olympia, Washington, USA.

Walker, A., editor. 1960. Water resources of the Nooksack River Basin and certain adjacent streams. Washington State Department of Ecology, Water Supply Bulletin 12, Olympia, Washington, USA.

Wallis, P.M., H.B.N. Hynes, and S.A. Telang. 1981. The importance of groundwater in the transportation of allochthonous dissolved organic matter to the streams draining a small mountain basin. Hydrobiologia **79**:77–90.

Walters, M.S., R.O. Teskey, and T.M. Hinckley. 1980. Impact of water level changes on woody riparian and wetland communities. Volume III. Pacific Northwest and Rocky Mountain Regions. United States Fish and Wildlife Service Report FWS/OBS-78/94, Washington, D.C., USA.

Ward, G.M., K.W. Cummins, R.W. Speaker, A.K. Ward, S.V. Gregory, and T.L. Dudley. 1982. Pages 9–14 *in* E.E. Starkey, J.F. Franklin, and J.W. Matthews, editors. Ecological research in national parks of the Pacific Northwest. Forest Research Laboratory, Oregon State University, Corvallis, Oregon, USA.

Waring, R.H., and W.H. Schlesinger. 1985. Forest ecosystems: concepts and management. Academic Press, Orlando, Florida, USA.

Whitaker, J.O., Jr., C. Maser, and R.J. Pederson. 1979. Food and ectoparasitic mites of Oregon moles. Northwest Science **53**:268–273.

Whittaker, J.G., and M.N.R. Jaeggi. 1982. Origins of step-pool systems in mountain streams. Journal of the Hydraulics Division, American Society of Civil Engineers **108**:758–773.

Winter, T.C. 1987. A conceptual framework for assessing cumulative impacts on the hydrology of nontidal wetlands. Environmental Management **12**:605–620.

Wolock, D.M., G.M. Hornberger, K.J. Bevan, and W.G. Campbell. 1989. Relationship of catchment topography and soil hydraulic characteristics to lake alkalinity in the Northeastern United States. Water Resources Research **25**:829–837.

7

Best Management Practices, Cumulative Effects, and Long-Term Trends in Fish Abundance in Pacific Northwest River Systems

PETER A. BISSON, THOMAS P. QUINN, GORDON H. REEVES, AND STANLEY V. GREGORY

Abstract

Although it is widely believed that forest management has degraded streams and rivers, quantitative relationships between long-term trends in fish abundance and forestry operations have not been successfully defined. In this article we review the difficulties in describing cumulative effects of forest management on fishes of the Pacific Northwest. Despite uncertainties in interpreting long-term trends from catch and escapement statistics as well as widespread programs of hatchery production, many local fish populations are declining. We suggest that trends in the abundance of *individual populations* are often of limited use in identifying the cumulative effects of forest management within a river system. Shifts in the composition and organization of *fish communities* may provide more comprehensive evidence of the extent of environmental alteration. Reduced stream habitat complexity has been one of the most pervasive cumulative effects of past forest practices and probably has contributed to significant changes in fish communities, particularly when accompanied by other land use activities that have led to straightened, confined channels. In simplified streams a few fish species have characteristically been favored while others have declined or disappeared completely. Likewise, fish culture practices have resulted in overall losses of genetic diversity among species. In order to protect channel complexity and biodiversity, best management practices (BMPs) should include measures to preserve physical and biological linkages between streams, riparian zones, and upland areas. Connections must include transfer processes that deliver woody debris, coarse sediment, and organic matter to streams, as these materials are largely responsible for creating and maintaining channel complexity and trophic diversity. Past forest practice regulations have required attainment of individual water quality standards, such as temperature or dissolved oxygen, and have been aimed at protecting certain life history stages of single species (e.g., salmon eggs in spawning gravels). This approach is inadequate to achieve the goal of restoring and maintaining natural levels of complexity at the level of a stream ecosystem. New BMPs are beginning to address this issue by prescribing riparian management zones with a greater range of vegetative species and structural diversity, thus providing for future sources of large woody debris, floodplain connections, and other linkages important to ecosystem function. Benefits of new BMPs in terms of improved habitat complexity and increased diversity of fishes on the scale of a river basin will require coordinated planning and extensive application, and will take years—perhaps decades—to become apparent.

Key words. Streams, habitat, cumulative effects, fish populations, biodiversity.

Introduction

For many years the cumulative effects of forest management activities on fish populations in river systems of the Pacific Northwest have been of concern (Hicks et al. 1991). Quite often the term "cumulative effects" has been implicitly or explicitly taken to mean the repeated, additive, or synergistic effects of forestry or other land use practices on various components of a stream's environment in time and space (Burns 1991). The term has considerable intuitive appeal, as it suggests that environmental impacts of specific management activities cannot properly be viewed in isolation from a broad perspective of land management at large spatial scales and long time scales. An underlying assumption has been that although individual management actions by themselves may not cause undue harm, taken collectively such land use activities may result in unacceptable stream habitat degradation and long-term declines in fish abundance, particularly when accompanied by heavy fishing pressure and competition with introduced species or hatchery stocks (Cederholm et al. 1981, Salo and Cederholm 1981).

As seemingly logical as this concept is, clear examples of cumulative effects of forest management on stream habitat have been difficult to demonstrate in all but the most severely degraded river systems (Platts and Megahan 1975, Coats and Miller 1981, Tripp and Poulin 1986*a*, *b*; Megahan et al., this volume). Furthermore, establishing unambiguous relationships between abundance of fish populations and cumulative environmental change has been equally difficult, if not more so (Pella and Myren 1974, Platts and Nelson 1988, Holtby and Scrivener 1989). This article reviews these difficulties and describes trends in habitat quality that appear to be common to river basins with histories of forest management and other types of land use activities. We discuss how the concept and implementation of best management practices (BMPs), a term generally taken to mean state-of-the-art environmental protection measures, have both succeeded and failed in attempting to (1) ameliorate adverse cumulative impacts, (2) protect natural interactions between streams and riparian zones, and (3) preserve the integrity of aquatic communities. Although we focus primarily on forest management, the conclusions apply equally to other land uses that cause similar types of environmental change.

Lack of Knowledge About Cumulative Effects and BMPs

Timber harvesting has been practiced in the Pacific Northwest since the early 19th century, but logging and reforestation techniques have changed dramatically. In general, new developments in logging systems have come at approximately 20–30 year intervals, while reforestation technology has evolved much more rapidly over the last 50 years (Figure 7.1). Many timber management techniques considered to be technologically advanced decades ago are viewed as outdated and environmentally destructive today (Franklin, this volume; Oliver et al., this volume). An example is the use of streams and rivers for log transport to mills (Sedell and Luchessa 1982, Sedell and Duval 1985). At one time, water-based log transport was the most practical means of moving very large logs from forested headwaters to downstream processing facilities. This management practice, considered the best in its day, caused a great deal of damage to streams and riparian zones, especially when it involved the use of splash dams (Bisson et al. 1987). Other technological advancements such as high lead and skyline yarding systems, as well as helicopter and balloon logging systems, were developed for harvesting in steep terrain but have proved to be far more environmentally sound on steep slopes than older methods that required skidding logs over forest soils.

The concept of best management practices with regard to environmental protection and restoration was essentially a post-World War II phenomenon and has been applied to land management activities both in Europe and North America (Petts 1990). Public pressure to protect stream habitat in western North America was at least partly responsible for an end to splash damming and log drives in the 1950s, and for some of the first restrictions on yarding across stream channels (Figure 7.1). The first comprehensive long-term forestry related aquatic research program in the Pacific Northwest—the Alsea watershed study—began in 1959 and continued uninterrupted until 1974 (Hall et al. 1987). As a result of findings in the Alsea watershed and elsewhere, most western states and provinces enacted forest practices regulations by the early 1970s. Many of the regulations addressed changes in temperature and fine sediment, two parameters shown to have been increased by logging activities. BMPs were therefore defined mostly in terms of water quality standards (i.e., temperature protection and erosion control).

New concerns about the effects of logging on peak flows and on the abundance of large woody debris in streams began to take shape in the 1970s and resulted in renewed research activity in the 1980s (Salo and Cundy 1987). Forest practice regulations have been revised to accommodate this new research information, yielding BMP regulations that have changed greatly within

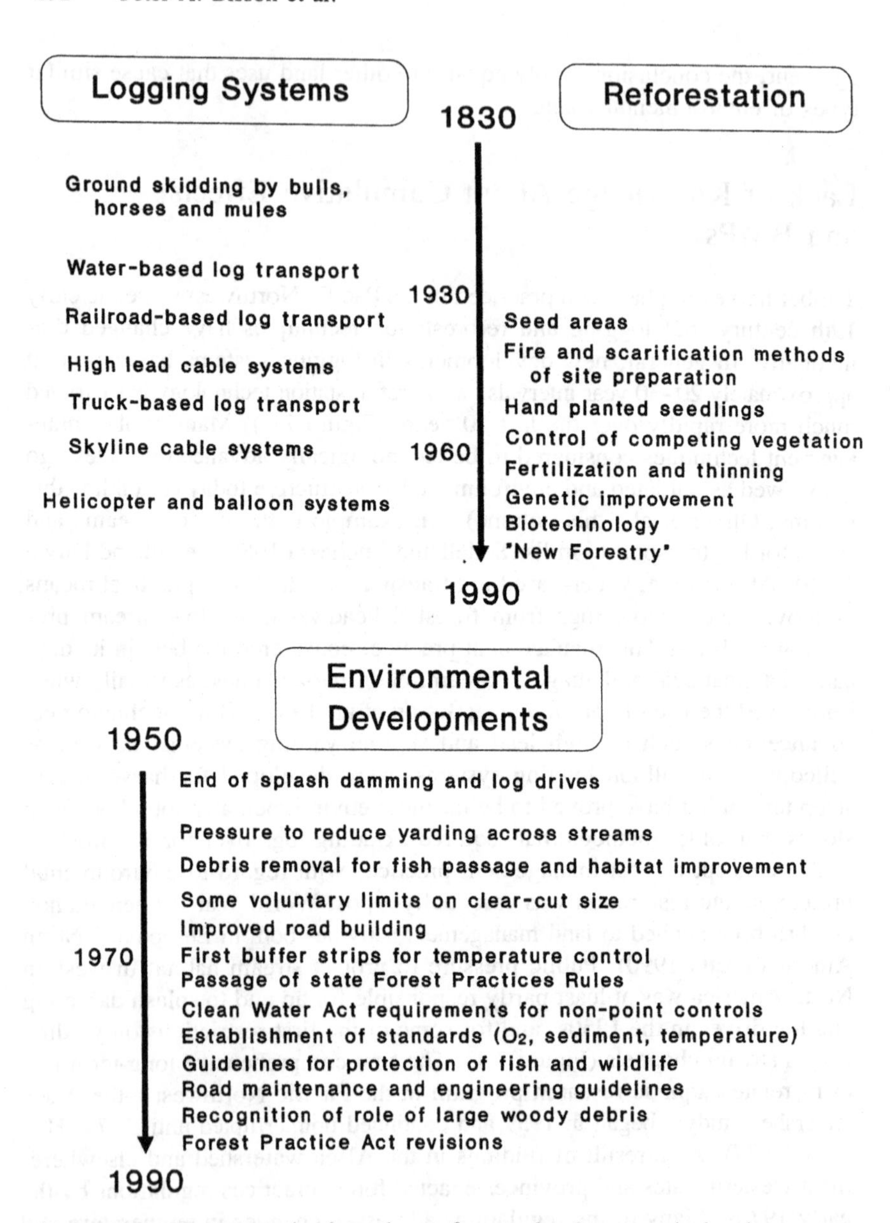

FIGURE 7.1. Diagrammatic chronology of some developments in logging systems, reforestation techniques, and environmental protection for streams and rivers in the Pacific Northwest.

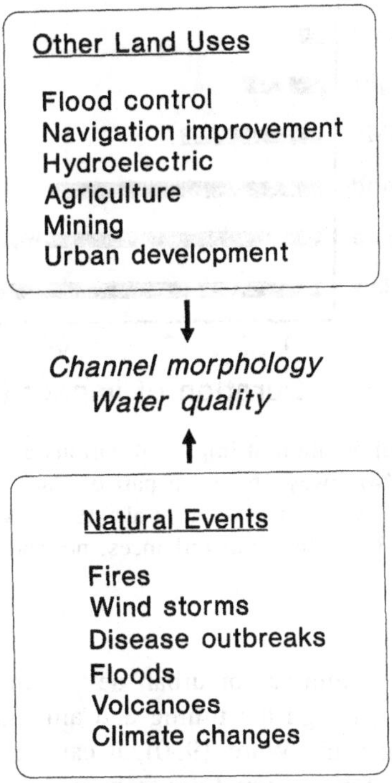

Other Land Uses

Flood control
Navigation improvement
Hydroelectric
Agriculture
Mining
Urban development

↓

Channel morphology
Water quality

↑

Natural Events

Fires
Wind storms
Disease outbreaks
Floods
Volcanoes
Climate changes

FIGURE 7.2. Some other important types of disturbances caused by various land management activities and natural events that can influence channel morphology and water quality in streams.

a single decade (Bilby and Wasserman 1989). Taken together, changes in logging systems, reforestation techniques, and environmental protection requirements have meant that our concepts of best forest management practices have always been evolving. Understanding the cumulative effects of forest management in light of changing BMPs has posed a difficult challenge.

Another reason for the poor understanding of cumulative effects of forest management on streams is that disturbances unrelated to forestry operations from both natural events and other land use activities have occurred concurrently with logging and reforestation (Figure 7.2). Naturally occurring events have taken place throughout drainage systems, although some (wildfires, windstorms) have probably had greater impact in forested headwaters than in nonforested lowlands (Keller and Swanson 1979). Many of the changes resulting from other types of land use have taken place in larger river systems, although agriculture, urban development, and mining have all affected small streams. Different types of disturbances can cause characteristic changes in stream habitat, but in some cases the environmental impacts of natural events or other land use activities may be relatively similar (Hicks et al. 1991, Schlosser 1991). For example, increased sediment deposition may

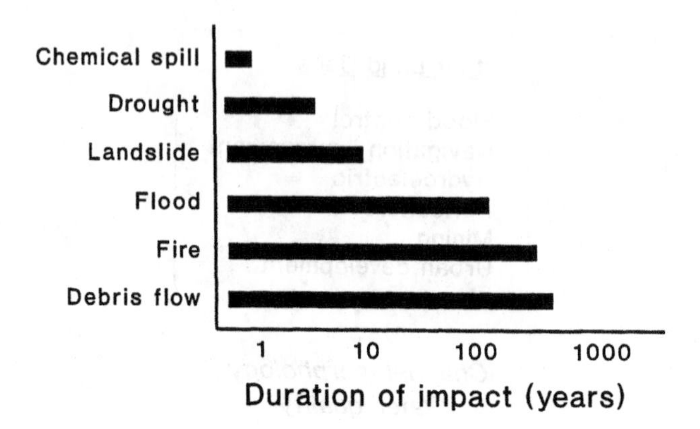

FIGURE 7.3. Hypothetical duration of impact of various disturbances on stream ecosystems in the Pacific Northwest, based in part on data from Swanson and Lienkaemper (1978), Grant (1988), and Niemi et al. (1990). Variation in the resilience of stream ecosystems after different disturbances, not shown in this figure, can be considerable.

accompany agriculture, mining, or urban development, as well as natural wildfires or floods. Although the timing and amount of inputs from these sources may vary (Poff and Ward 1990), it can be difficult to distinguish sediment produced by forestry practices from sediment produced from other types of disturbance (Everest et al. 1987). Effects of other disturbances may therefore obscure or mask cumulative environmental change attributable to forestry operations.

Recovery rates of the physical environment from different types of environmental disturbance are highly variable (Swanson and Lienkaemper 1978, Grant 1986). Likewise, the recovery of stream biota is highly variable, depending both on the nature of the disturbance (i.e., its spatial extent and temporal duration; Poff and Ward 1990) and the assemblages of plants and animals in the stream (Niemi et al. 1990). The duration of physical and water quality impacts from disturbance (Figure 7.3) may range from very short (a few days) to very long (several hundred years). Stream channels reflect disturbances that took place recently as well as changes that occurred decades or even centuries ago (Gregory et al. 1987, Grant et al. 1990). Reconstructing the disturbance history of a stream from surveys of existing conditions often requires subjective interpretation of cause and effect. Nevertheless, some recent procedural advances have improved our ability to interpret changes caused by past catastrophic disturbances (Grant 1988). Recovery rates of stream biota may depend, among other things, on the biota's "preadaptedness" to a particular type of disturbance (Poff and Ward 1990).

All of these factors make quantification of the effects of past disturbances based solely on recent physical and biological surveys very difficult. Thus long-term records of environmental conditions are essential if habitat change is to be associated with specific natural events and land use activities (Sedell and Luchessa 1982). Unfortunately, long-term habitat monitoring has not taken place in the majority of Pacific Northwest river systems. Very early habitat assessments were in the form of verbal descriptions of river valleys (Sedell et al. 1988, Gonor et al. 1988), and it was not until the early 20th century that biologists first attempted to quantify habitat conditions (Pacific Fishery Management Council 1979). By this time many rivers had been changed by logging and other land use practices (Chapman 1986).

Many of these original habitat surveys were poorly archived or never published in a widely available format; in most cases original data have been lost. Changing inventory methods and personnel transfers have also been a barrier to monitoring continuity. In rare instances where earlier habitat surveys could be related to modern conditions, we have been able to obtain a much more accurate picture of long-term trends in habitat quality (Sedell and Everest 1991). But these instances are rare, and may be limited to portions of river basins.

Long-Term Trends in Fish Abundance

Long-term trends in the abundance of Pacific Northwest fish populations, especially anadromous salmonids, have often been determined through examination of commercial, sport, and Native American catches, as well as estimates of the escapement of adult fish to spawn. Although catch and escapement records occasionally extend back into the late 19th or early 20th century, they are prone to measurement errors that can confound actual trends in stock abundance. Estimates of historical run size often require numerous assumptions and conversions, many of which cannot be verified but which nevertheless are necessary (Chapman 1986). Several different types of errors are associated with catch statistics. Unreported catches can be significant where catch monitoring efforts are small and where harvest records are supplied by fishermen themselves. Significant changes in fishing gear and other methods of harvest can result in significant increases or decreases in catch unrelated to stock abundance. Likewise, changes in size restrictions and fishing season openings can strongly affect the number of fish caught. Fishing pressure may change from year to year depending on weather, economic conditions, and a variety of political considerations. Finally, mixed-stock catches, which are the rule in most ocean fisheries, often prevent the separation of commercial and sport catches into component stocks from different drainage systems. Many of these problems, and efforts to circumvent them, are discussed by Healey (1982).

Although there are fewer types of errors associated with escapement estimates, the magnitude of the errors can be great. Where escapement is based on counts from viewing windows at fish ladders or on fish traps at impassable barriers, estimates can be relatively accurate, provided that no fishing or other significant mortality occurs between the counting location and spawning areas. However, these techniques require proper functioning of equipment at all times, including periods of high streamflow. But even ladder counts can be very inaccurate. For example, U.S. Army Corps of Engineers Annual Fish Passage Reports for the Columbia River often record substantially more anadromous shad (*Alosa sapidissima*) at The Dalles than at Bonneville Dam (ASACE 1989). In 1988, 2.01 million shad were counted at The Dalles Dam while only 1.16 million were counted from Bonneville Dam, which is downstream and would have had to pass at least as many fish as were counted at The Dalles.

Counts of adult fish on spawning grounds are often used to gauge run size, but such counts require frequent surveys during the period of spawning, and water quality must be conducive to fish viewing (Beidler and Nickelson 1980). Quite often these requirements are not met. Surveys of entire drainages may be logistically difficult, bad weather may hamper viewing, fish may be counted twice or more, and aerial surveys may be obscured by riparian vegetation (Neilson and Geen 1981). Changes in survey methods or locations of index areas often add unknown errors to escapement estimates. Finally, indices of adult abundance such as spawning counts from designated index sites or redd counts that assume a given number of redds per female—and in turn a ratio of females to males—require careful, local verification (Solazzi 1984, Nickelson et al. 1986); where these assumptions are not verified, escapement estimates may be subject to bias.

Catch statistics by themselves are often of limited use in determining the causes of advances or declines in stock abundance. Anadromous salmonids experience a variety of environmental conditions over the period of freshwater and marine rearing prior to being taken in a fishery. Because so many potentially limiting factors are encountered before capture (Figure 7.4), it is often impossible to determine the reasons for annual changes in run strength. For example, McDonald and Hume (1984) reported over tenfold variation in marine survival of Babine Lake sockeye salmon (*Oncorhynchus nerka*) between 1961 and 1977, and fivefold differences in marine survival between consecutive years. A variety of correlation approaches have been attempted in order to relate catch statistics to freshwater and marine habitat parameters such as streamflow and ocean upwelling (Neave 1949, Smoker 1955, Scarnecchia 1981, Nickelson 1986). While these authors claim varying degrees of success in predicting annual adult abundance at broad geographical scales, all have acknowledged that identification of specific environmental factors governing abundance using statistical correlation is very difficult. Furthermore, statistical design becomes crucial to the interpretation of correlation analyses and may lead to inaccurate conclusions if improper design is employed (Walters et al. 1988, Peterman 1989).

An illustration of the difficulty in relating catch statistics to land use is shown by a comparison of commercial catches (Mullen 1981) of coho salmon (*O. kisutch*) in two nearly adjacent drainages of similar characteristics on

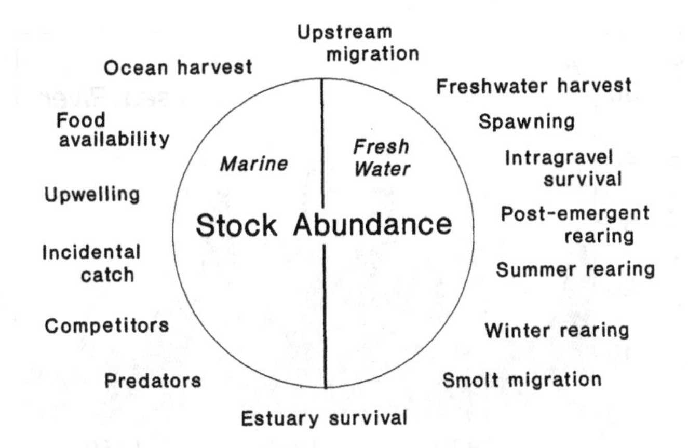

FIGURE 7.4. Some of the factors known to exert an influence on stock abundance in freshwater and marine environments.

the Oregon coast (Figure 7.5). The catch records cover a period of more than 30 years when both basins were being actively logged and relatively little protection was given to streams. The fishing methods in both cases were terminal nets located near the river mouths, and the records describe the catches of wild fish. Salmon hatcheries in this area began production in the mid-1950s. In the Alsea River Basin, catches of coho salmon generally increased to peaks in the 1930s and early 1940s, after which there was a decline. In the Siletz River Basin there was a steady decline over the entire period. Because these nearby basins were being logged at approximately the same time, it is difficult to explain why the catch record for the Alsea River appeared to be dome-shaped while that of the Siletz River trended consistently downward. In all likelihood, factors other than or in addition to habitat damage associated with timber harvest were influencing the commercial catch of coho salmon in the rivers. Determination of the causes of increased or decreased catch in such instances becomes statistically intractable without a comprehensive long-term knowledge of both freshwater and marine conditions, as well as thorough records of fishing methods and pressure.

Over the past several decades the output of smolts from hatcheries has risen along the Pacific Northwest coast. Nickelson (1986) notes that coho salmon smolt releases in the Oregon Production Area increased from fewer than 1 million during the 1950s to over 30 million by 1970. After stabilizing at approximately 35 million smolts in the mid-1970s, large-scale releases of coho from privately owned facilities increased total hatchery production to 62 million fish by 1981, although hatchery smolt releases have since declined. Light (1992) found that production of steelhead (*Oncorhynchus mykiss*) smolts throughout the Pacific coast rose by a factor of 10 from approximately 3 million in 1960 to 30 million by 1987 (Figure 7.6A). Similar increases in hatchery output of other anadromous salmonids have occurred

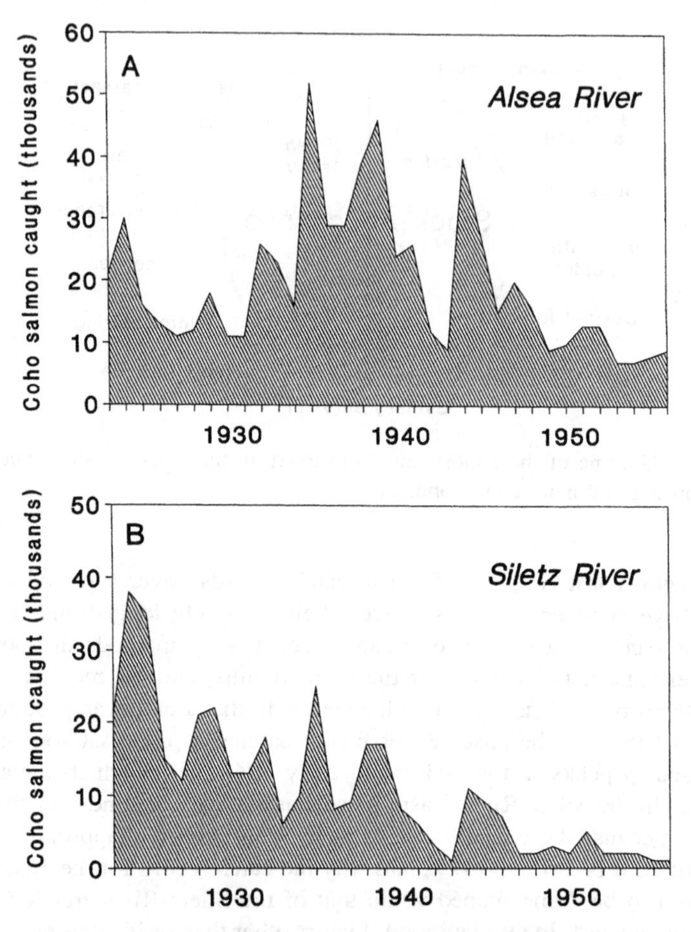

FIGURE 7.5. Number of adult coho salmon caught in (A) the Alsea River and (B) the Siletz River, Oregon, from 1923 to 1956. Data from Mullen (1981).

in response to greater demands by commercial, sport, and Native American fishermen, in addition to mitigation requirements for habitat loss. As a result of extensive hatchery production, overall catches may remain stable even though the abundance of naturally produced fish has declined (Figure 7.6B). It is possible too that hatchery fish have had a directly negative impact on wild stocks through competition for limited freshwater or marine resources (Nickelson et al. 1986, Lichatowich and McIntyre 1987, Hilborn 1992) or through genetic introgression of nonadaptive traits (Leider et al. 1984, Chilcote et al. 1986). These factors have often not been considered when transferring salmonid eggs and fry between river systems. Furthermore, an intense fishery targeting a large hatchery run can incidentally depress the escapement of wild fish (Nehlsen et al. 1991, Hilborn 1992).

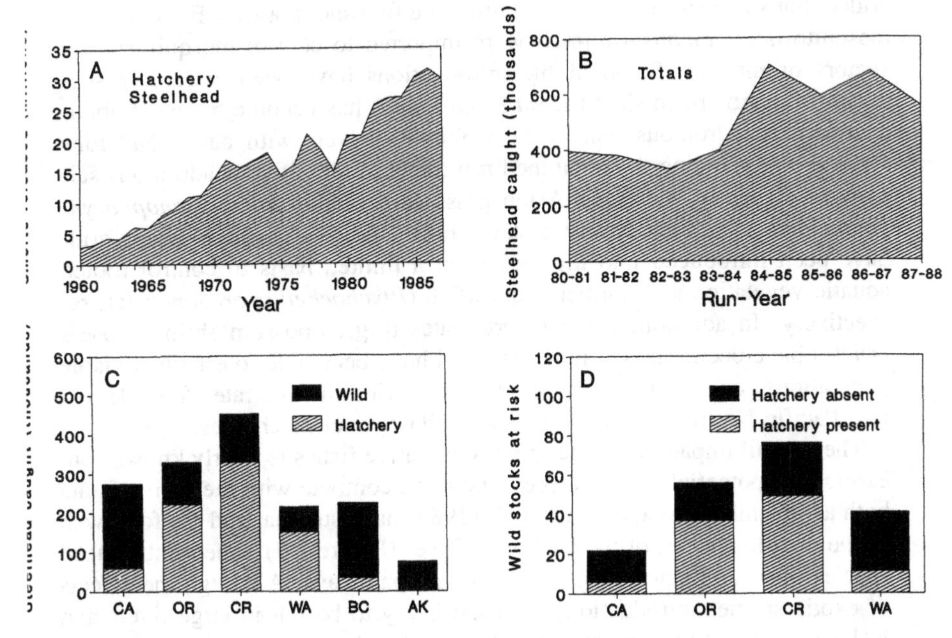

FIGURE 7.6. (A) Total number of steelhead smolts released from hatcheries along the Pacific coast from 1960 to 1987. (B) Total catch of adult steelhead on the Pacific coast from run-years 1980–81 through 1987–88. (C) Relative contribution of hatchery and wild steelhead to the total catch by region. CA=California, OR=Oregon coastal rivers, CR=Columbia River Basin, WA=Washington coastal and Puget Sound rivers, BC=British Columbia coastal rivers, AK=Alaska coastal rivers. (D) Number of wild stocks of *Oncorhynchus* considered to be at risk of extinction in river systems with and without hatcheries producing the same species. Locations as in C. Data for A-C from Light (1992); data for D from Nehlsen et al. (1991).

The heavy contribution of hatchery fish to total harvest for some species is illustrated by the relative catches of hatchery and wild steelhead over the range of steelhead distribution in western North America (Figure 7.6C). In the geographical center of the range, which includes coastal Washington, Oregon, and the Columbia River Basin, hatchery steelhead comprise about 70–80% of the total catch. Thus hatchery fish have largely supplanted wild steelhead in the most productive part of the range. Additionally, Nehlsen et al. (1991) note that substantial fractions of the anadromous salmonid stocks at risk of extinction in the Columbia River basin and in rivers on the Oregon coast occur in drainages where hatcheries are propagating the same species (Figure 7.6D).

Many of the nonnative fishes that have become established in Pacific Northwest river systems were introduced to provide a wider variety of sport fishing opportunities than existed with the native fauna. The majority of introduced species have belonged to the Centrarchidae, Percidae, and Ictaluridae, but some species have been imported for other reasons. For example, mosquitofish (*Gambusia affinis*) were imported to control mosquitoes, the vectors of malaria. Some of the introductions have been extremely successful; the American shad (*Alosa sapidissima*) has become the most abundant large anadromous fish in the Columbia River, with adult shad runs sometimes exceeding the combined run totals of all native anadromous salmonids. Recently, sterilized Asian grass carp (white amur, *Ctenopharyngodon idella*) and tiger musky (a sterile muskellunge x northern pike hybrid) have been introduced into the region on a limited basis to control rooted aquatic vegetation and northern squawfish (*Ptychocheilus oregonensis*), respectively. In addition, some invertebrates (e.g., opossum shrimp, *Mysis relicta*) have been released into lakes and have been able to establish viable populations, and many species of exotic marine invertebrates are released into Pacific Northwest waters through ballast water discharges.

The overall impact of exotic species on native fishes is poorly known, but exotics are potentially able to prey upon and compete with the native fauna both as juveniles and adults. Li et al. (1987) have summarized the food web of the middle reaches of the Columbia River (Figure 7.7), where introduced species now dominate many of the trophic pathways. Although the effects of exotic species introductions are most likely to be felt in large rivers and in lakes, many native species use these areas at some point in their freshwater life cycle. In an examination of the extinctions of North American freshwater fishes during the 20th century, Miller et al. (1989) found that 67% of the extinctions occurred in areas with established populations of nonnative species. It seems highly likely that exotic species have contributed to the decline of certain stocks of Pacific Northwest fishes, but the extent to which these negative interactions can be separated from those involving competition with hatchery fish or from the impacts of habitat damage and overfishing has not been quantitatively determined.

Because of their great importance to the region, the majority of research on fish population abundance in the Pacific Northwest has focused on salmon and trout. Very little is known about the effects of cumulative habitat changes on the abundance of most nonsalmonid species, particularly those that do not contribute directly to sport or commercial fisheries. Some of these species may be more sensitive to habitat change than anadromous salmonids because they spend their entire lives in freshwater and may be associated with a specific type of habitat.

We are aware of no studies that have attempted to assess the abundance of nonsalmonid populations at the scale of a drainage basin in the Pacific Northwest. Furthermore, there are few if any long-term records of nonsalmonids at index sites, where only records of salmonid abundance tend to be maintained. There are several reasons for the paucity of information on nonsalmonids, apart from their lack of commercial or recreational significance. Sampling gear for streams and rivers is usually designed to capture salmonids and other midwater species; it is often very inefficient at sampling

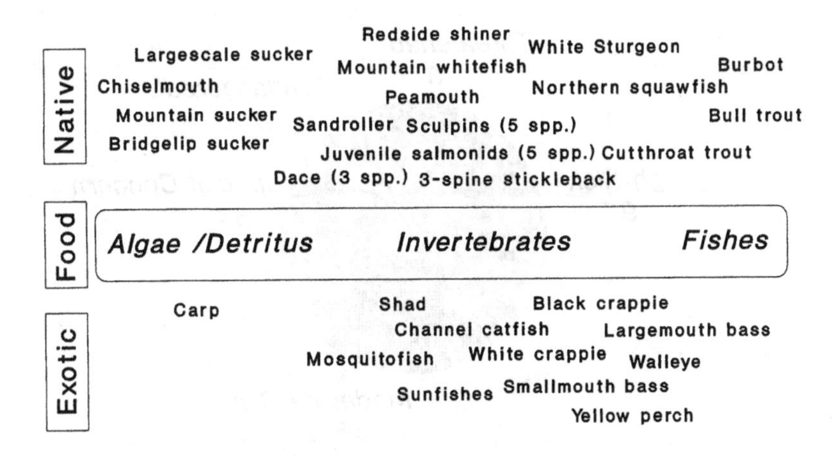

FIGURE 7.7. Native and exotic fishes of the middle and lower Columbia River, arranged according to their approximate preferences for different food types. Based on a diagram in Li et al. (1987). White sturgeon=*Acipenser transmontanus*, American shad=*Alosa sapidissima*, mountain whitefish=*Prosopium williamsoni*, juvenile salmonids=*Oncorhynchus* spp., cutthroat trout=*Oncorhynchus clarki*, bull trout=*Salvelinus confluentus*, chiselmouth=*Acrocheilus alutaceus*, carp=*Cyprinus carpio*, peamouth=*Mylocheilus caurinus*, northern squawfish=*Ptychocheilus oregonensis*, dace=*Rhinichthys* spp., redside shiner=*Richardsonius balteatus*, bridgelip sucker=*Catostomus columbianus*, largescale sucker=*Catostomus macrocheilus*, mountain sucker=*Catostomus platyrhynchus*, channel catfish=*Ictalurus punctatus*, sandroller=*Percopsis transmontana*, burbot=*Lota lota*, mosquitofish=*Gambusia affinis*, threespine stickleback=*Gasterosteus aculeatus*, sunfishes=*Lepomis* spp., smallmouth bass=*Micropterus dolomieui*, largemouth bass=*Micropterus salmoides*, white crappie=*Pomoxis annularis*, black crappie=*Pomoxis nigromaculatus*, yellow perch=*Perca flavescens*, walleye=*Stizostedion vitreum*, sculpins=*Cottus* spp.

fishes living on or in the benthos. Some taxa are difficult to identify and are often assigned simply to genera (e.g., *Cottus* spp.). Other species do not live in areas of the drainage inhabited by salmonids, and are likely to be overlooked in fish surveys. All of these factors have contributed to the general absence of knowledge of the status of nonsalmonids in the region.

Decline of Wild Salmonid Stocks

Recent analysis of the status of anadromous salmonid stocks (genetically distinct populations native to particular drainage systems) in the western United States has revealed over 200 stocks that are currently at some level of risk of becoming extinct (Figure 7.8). Nehlsen et al. (1991) have separated these stocks into three risk categories. Those with a high risk of extinction include stocks with consistent and significantly declining spawning escapements or those where the total adult population is believed to be less than 200 indi-

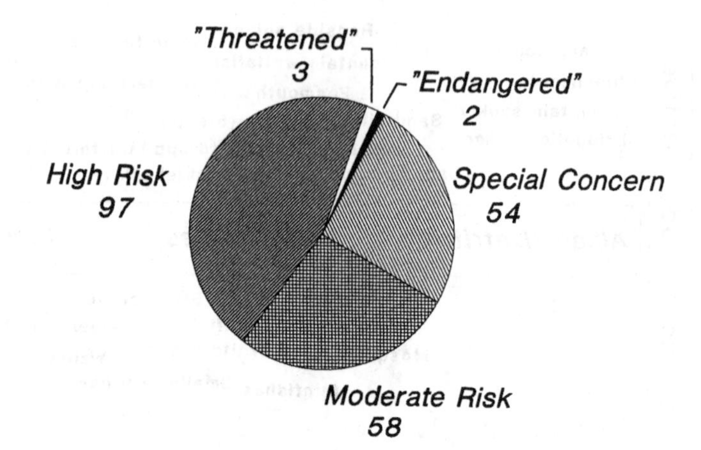

High Risk of Extinction	Moderate Risk of Extinction	Stocks of Special Concern
Spawning escapements significantly declining	Population stable after period of major decline	Vulnerable to minor disturbance
Fewer than 200 adults remain	More than 200 adults remain	Population trends suggest decline
		Interbreeding with non-native fish
		Unique character

FIGURE 7.8. Native stocks of anadromous Pacific salmon (*Oncorhynchus*) from California, Oregon, Idaho, and Washington considered by Nehlsen et al. (1991) to be at risk of extinction, and their criteria for classifying stocks in different risk categories. Some stocks have recently been classified as "Endangered" or "Threatened" under the U.S. Endangered Species Act.

viduals. Those with a moderate risk of extinction have relatively stable populations after a period of significant decline and have more than 200 spawning adults. Those stocks considered to be of special concern include populations that are vulnerable to minor disturbances, populations whose trends in abundance suggest a consistent pattern of decline, and populations in which there is significant interbreeding with nonnative fish.

Nearly half of the stocks considered to be at risk by Nehlsen et al. (1991) were placed in the high risk category (Figure 7.8). Of these, the greatest number occurred in the Columbia River Basin, although approximately 20 high risk stocks occurred respectively in Oregon coastal streams and in Washington coastal and Puget Sound streams (Figure 7.9A). Stocks that may already be extinct were identified from California, the Columbia River, and Washington. The greatest number of stocks classified at moderate risk of extinction occurred in Oregon. Without exception, habitat damage (Figure

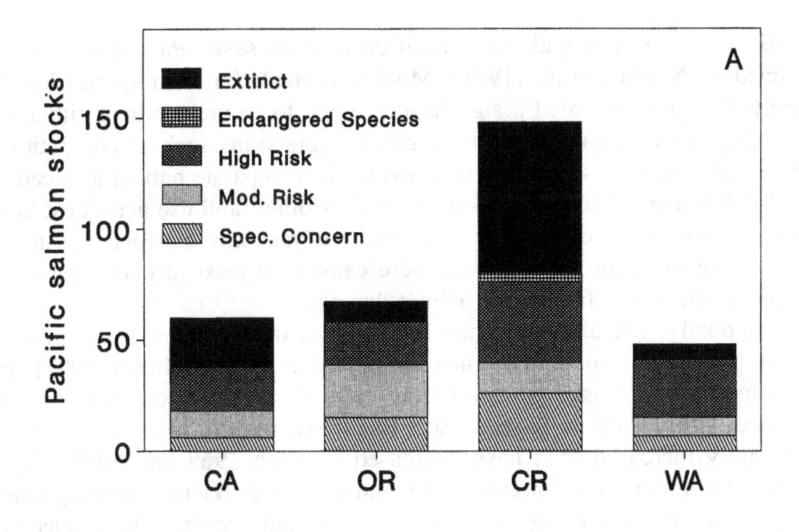

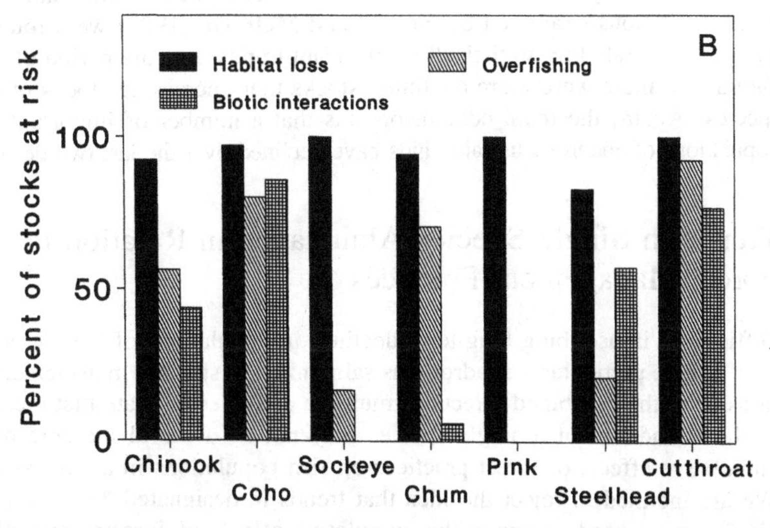

FIGURE 7.9. (A) Distribution of stocks of anadromous Pacific salmon (*Oncorhynchus*) in different extinction risk categories within various regions of the Pacific coast. (B) The percentage of stocks in which habitat damage, overfishing, and harmful biotic interactions have been implicated in declines of stock abundance. Data from Nehlsen et al. (1991).

7.9B) was associated with declines of each of the seven Pacific salmon examined by Nehlsen et al. (1991). More important, for each species habitat destruction was accorded a significance equal to or greater than either overfishing or the negative effects of biotic interactions such as competition, predation, and disease. No attempt was made to separate habitat loss caused by forest management from losses caused by other land use activities; however, an important conclusion from this analysis was that declines in the abundance of many native stocks were caused, at least in part, by the cumulative effects of freshwater habitat damage.

Regionally, not all species are declining at the same rate, and in some areas populations are actually increasing (Konkel and McIntyre 1987). For example, more Alaska populations of chinook salmon, coho salmon, and steelhead for which long-term escapement records are available have significantly increased than have decreased between 1968 and 1984 (Figure 7.10). In other regions, notably the Columbia River Basin, declining stocks far outnumber increasing stocks for all salmonid species. On a coastwide basis the majority of chinook, coho, chum (*Oncorhynchus keta*), and steelhead populations examined by Konkel and McIntyre (1987) were found to have demonstrated no statistically significant trends over the period of evaluation, but there were more declining stocks than increasing stocks of each species. Again, the main conclusion was that a number of important wild populations of anadromous salmonids have declined over the last two decades.

Trends in Single Species Abundance in Relation to Forest Management Practices

Difficulties in ascribing long-term declines in populations of Pacific Northwest fishes, particularly anadromous salmonids, to specific management actions or to the combined effects of multiple activities indicate that a change in the abundance of a single species may not be a useful measure of the cumulative effects of forest practices on fish populations in a river system. We are inclined to reject the idea that trends in designated "indicator species" can be used to gauge the cumulative effects of forestry operations, unless it can be convincingly demonstrated that those species are not impacted by other land use activities or overfishing. Even where it can be shown that stocks are undergoing severe declines, it is usually impossible to determine with reasonable certainty the relative effects of habitat degradation, fishing pressure, and biotic interactions such as competition, predation, and disease. For most populations, declines have resulted from a combination of several factors, the relative importance of which may change from year to year.

A potentially more powerful approach is to examine the relationship between forestry-related habitat changes and the structure of fish communities in streams and rivers. Community-level studies have been used successfully

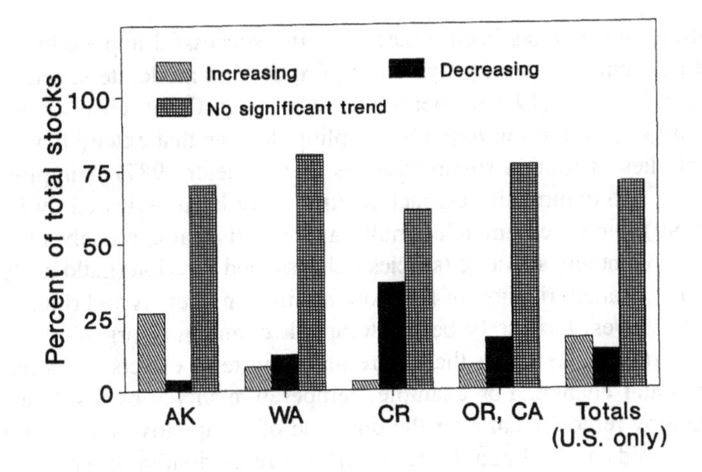

Region	Chinook UP	Chinook DN	Coho UP	Coho DN	Chum UP	Chum DN	Steelhead UP	Steelhead DN
Alaska	43%	1%	15%	11%	3%	13%	17%	0%
Coastal WA	12%	32%	9%	0%	6%	15%		
Columbia R. Basin	3%	39%	0%	45%	0%	33%	8%	25%
Coastal OR, CA	19%	12%	2%	17%	11%	11%	20%	40%
U.S. Total	20%	22%	6%	17%	4%	14%	11%	23%

FIGURE 7.10. Trends in the abundance of wild stocks of chinook salmon (*Oncorhynchus tshawytscha*), coho salmon (*O. kisutch*), chum salmon (*O. keta*), and steelhead (*O. mykiss*) from river systems along the Pacific coast. UP = percentage of stocks significantly increasing, DN = percentage of stocks significantly decreasing. Data from Konkel and McIntyre (1987).

elsewhere in North America to detect the effects of persistent environmental disturbances on fish assemblages (Karr 1981, Berkman and Rabeni 1987, several papers in Matthews and Heins 1987, Fausch et al. 1990) but there have been relatively few studies of this type in the Pacific Northwest (Li et al. 1987, Pearsons et al. 1992). One reason for lack of information on forestry-related environmental disturbance and fish communities in the region has been the almost complete absence of studies on the relationship between forest management and nonsalmonid species. A second reason is that fish assemblages are structured according to functional groups that include trophic, habitat, or reproductive guilds (Berkman and Rabeni 1987), but little is known about the food habits and spawning behavior of many native Pacific Northwest species. A third reason is that, unfortunately, comprehensive faunal surveys have not been carried out in most Pacific Northwest rivers. A fourth reason is that many Pacific Northwest river systems are faunally depauperate due to recent glaciation and lack of zoogeographic access (Moyle and Herbold 1987), and this has been a barrier to the successful application of indices of community integrity, which require at least moderate species richness (Fausch et al. 1990). Identification of significant changes in fish community structure often requires sampling designs that extend beyond intensive studies of limited stream reaches (Angermeier 1987). Sampling approaches to fish community characterization have been reviewed by Fausch et al. (1990), who recommended multivariate statistical approaches to characterizing community structure (species richness) and function (guild analysis).

Although characterization of fish communities in streams and rivers in the Pacific Northwest has rarely been attempted, community organization may yield important clues about the nature and long-term effects of cumulative environmental change. For example, temperature increases resulting from forest canopy removal can alter the outcome of competitive interactions between cyprinids and salmonids, resulting in a redistribution of species dominance along thermal gradients (Reeves et al. 1987). Deposition of fine sediment may reduce fish species diversity by eliminating spawning habitat and altering invertebrate food resources (Karr and Schlosser 1978, Karr et al. 1985, Berkman and Rabeni 1987). Separate age classes of many salmonids have particular habitat requirements (Bustard and Narver 1975, Bisson et al. 1988, Moore and Gregory 1988*a*) and can be considered functionally distinct members of stream communities (Schlosser 1991); therefore, they are often unequally impacted by habitat alteration. Stream fishes tend to be habitat specialists (Gorman and Karr 1978) and are affected by an increase or loss of preferred habitat types (Moore and Gregory 1988*b*), but a transformation of one habitat type to another (e.g., conversion of a pool to a riffle) may not lead to an overall reduction in fish density; rather, the effect may be expressed as a shift in species or age class composition (Schlosser 1991). Although factors other than anthropogenic habitat disturbance can influence community structure (Grossman et al. 1982, 1990; Herbold 1984, Schlosser 1985, Schlosser 1987, Power 1990), persistent changes in stream habitat are more likely to be detectable through analysis of fish assemblages than through interpretation of long-term trends in the abundance of individual species. In the Pacific Northwest, we conclude that the strong research emphasis on salmon and trout to the exclusion of other fishes has become a significant obstacle to defining the cumulative effects of forest management on stream ecosystems.

To fully understand how stream fish communities might be altered by cumulative habitat change, it is necessary to identify general environmental trends associated with forestry operations. In the following section, evidence that past forestry practices have led to simplification of stream channels and truncation of natural linkages between streams and riparian zones is reviewed. We conclude with conceptual recommendations for future BMPs that will provide a basis for protecting these complex linkages in a way that conserves the natural biodiversity of stream communities.

Past Forest Management Practices

Specific changes in stream environment caused by past forest practices in the Pacific Northwest vary according to logging and reforestation history, watershed geology, regional climate, and the degree of protection given to riparian zones during management activities. The one change that appears to be consistent over all areas in which the effects of forest management on streams have been studied is a trend toward simplification of stream channels and a loss of habitat complexity (Bisson and Sedell 1984, Grant 1986). This trend has resulted from a combination of management and regulatory actions, and represents a cumulative effect upon which there appears to be general scientific agreement (Hicks et al. 1991, Sedell and Beschta 1991). Simplification of stream channels involves loss of hydraulic complexity (i.e., caused by variation in current velocity and depth, and structural obstructions to flow), elimination of physical and biological interactions between a stream and its floodplain (see Naiman et al., this volume), reduction of structures that serve as cover from predators, an increase in the dominance of one particular substrate type, and loss of sediment and organic matter storage capacity (Sullivan et al. 1987).

Stream simplification and loss of complexity is perhaps most evident in the changes in frequency, size, and location of different types of habitat units within the channel. Over the last decade, physically based systems of channel unit classification (Bisson et al. 1982, Frissell et al. 1986, Sullivan 1986, Cupp 1989, Grant et al. 1990) have increased our ability to resolve stream morphology at a scale that is meaningful to understanding the distribution and abundance of fishes (Bisson et al. 1988). The most pervasive change has been a reduction in the frequency and size of pools, particularly large plunge and scour pools that constitute preferred habitat of certain species and age classes (Bisson and Sedell 1984, Sedell and Everest 1991). There have been two principal causes of pool reduction in Pacific Northwest streams: the filling of pools by sediment (Megahan 1982) and the loss of pool-forming structures such as boulders and large woody debris (Bryant 1980, Sullivan et al. 1987, Meehan 1991). Although both causes have been directly related to forest management activities, woody debris removal has been practiced in streams and rivers for navigation improvement, to aid water-based log transport, and to promote fish passage (Sedell and Luchessa 1982).

Examples of studies that have demonstrated relationships between forestry and pool frequency are shown in Figure 7.11. In ten Oregon coastal streams, Hicks (1990) found that the number of scour pools associated with large woody debris decreased in proportion to the percentage of the drainage basin that had been logged, and that the decrease occurred in drainages with both basalt and sandstone parent rock. Bilby and Ward (1991) found that streams in old-growth forests held more pools for a given channel width than streams in clearcuts or in second-growth forests. Bisson et al. (1987) cited numerous studies that have associated declines in fish abundance with loss of pools and woody debris in Pacific Northwest streams.

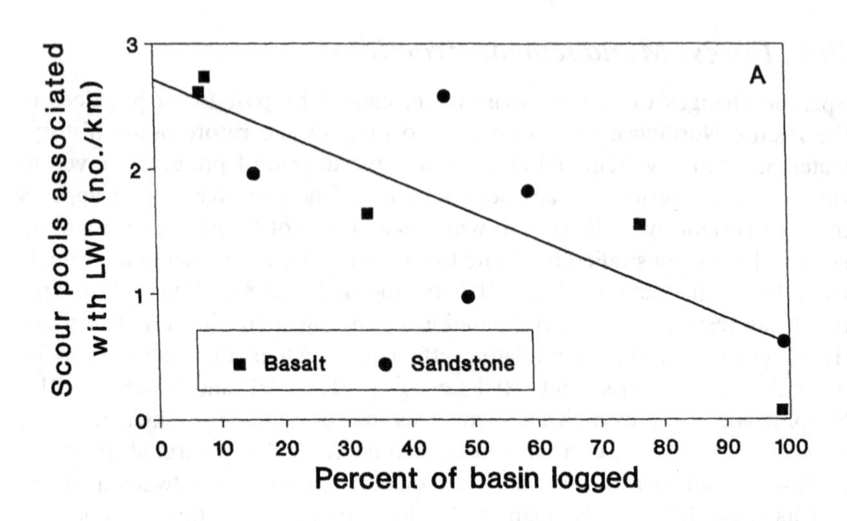

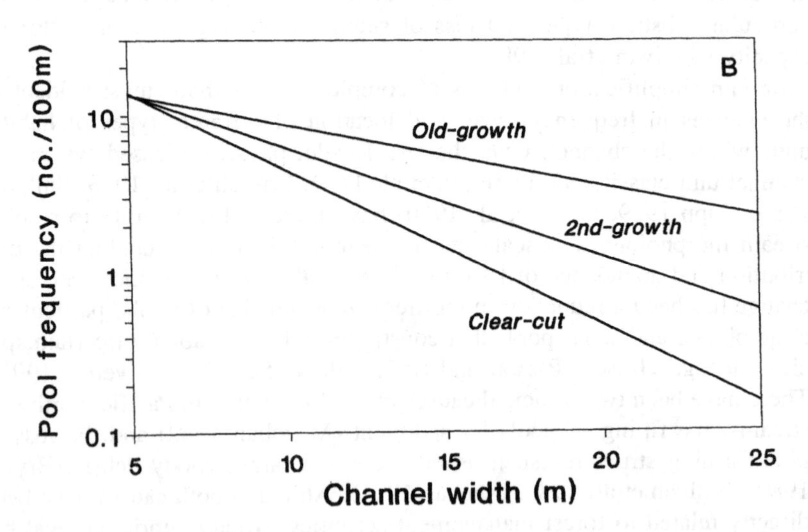

FIGURE 7.11. (A) Frequency of pools associated with large woody debris in ten Oregon coastal streams with different logging histories. Redrawn from Hicks (1990). (B) Relationship between pool frequency and stream channel width in old-growth, second-growth, and clearcut streams in western Washington. Redrawn from Bilby and Ward (1991) with permission of the *Canadian Journal of Fisheries and Aquatic Sciences*.

In addition to reductions in the number and size of large scour and plunge pools, forestry and other land use practices have led to stream channel simplification by eliminating edge habitat along stream margins. Complex channel margins are used extensively by underyearling fishes of many species, as these areas provide breaks in the stream current for resting and feeding. Edge habitat can be destroyed by reducing the abundance of large flow obstructions such as logs and boulders near the stream's margin and by isolating a stream from its floodplain through channelization, streambank stabilization, and other measures designed to confine the flow to a single channel. Experimental removal of eddy pools and backwater areas along the margin of a small stream greatly reduced the carrying capacity for young-of-the-year trout (Moore and Gregory 1988*b*). Complexity of channel margins is further reduced when management activities cause streambanks to collapse or when timber harvest in riparian zones results in destruction of root systems and eliminates future sources of large woody debris (Toews and Moore 1982, Bryant 1983).

Long-term reductions in the supply of large woody debris as the result of timber harvest have affected other important processes within stream ecosystems (Harmon et al. 1986). Small headwater streams serve as temporary storage sites for both sediment and fine particulate organic matter (FPOM) from the surrounding forest (Keller and Swanson 1979, Triska and Cromack 1980). Loss of sediment and FPOM storage capacity in small streams caused by reduced debris frequency greatly lessens the capacity of the streams to biologically process organic matter and ultimately make the energy of terrestrial plant materials available to fishes (Triska et al. 1982, 1984, Gregory et al. 1987). Because their storage and processing capacities are greatly diminished, streams with simplified channels route sediment and organic matter much more quickly downstream to larger streams (Naiman and Sedell 1980, Sedell and Beschta 1991). In some cases, rapid transport of sediment can overwhelm larger stream systems (Megahan and Nowlin 1976; Megahan, this volume), resulting in lower biological productivity (Platts and Megahan 1975) and reduced diversity of species requiring clean gravel substrate for spawning (Berkman and Rabeni 1987).

Large events of a catastrophic nature such as major floods or landslides can trigger debris flows that cause extensive scouring and simplification of headwater streams (Benda 1990, Lamberti et al. 1991). In the Queen Charlotte Islands, British Columbia, Rood (1984) found that the frequency of landslides increased over 30-fold after logging on geologically unstable hillslopes, and the frequency of debris flows in logged and roaded areas increased 40–76 times. Tripp and Poulin (1986*a*) investigated morphological changes in streams that had undergone massive debris flows and found that average pool depth was reduced by 20–24%, average pool area was reduced by 38–45%, large woody debris in the channel was reduced by 57%, and undercut bank cover was reduced by 76%. They found that riffle area was increased by an average of 47–57% and average channel width increased

48–77%. Large debris flows have also been found to strongly influence the species composition and structural characteristics of riparian vegetation in western Oregon (Gecy and Wilson 1990). Lamberti et al. (1991) have suggested that debris flows diminish stream ecosystem stability, leading to large annual fluctuations in fish populations.

Simplification of streams and rivers has also resulted from other types of land use activities, including agriculture (Schlosser 1982, Karr et al. 1983), grazing (Chapman and Knudsen 1980, Platts and Nelson 1985), and urban development (Leidy and Fiedler 1985). Habitat changes associated with these land uses are usually located downstream from forestry operations, but the net effect of combined management activities can be extensive loss of channel complexity from headwater streams to river mouths (Sedell et al. 1988, Gonor et al. 1988).

Simplified Streams and Biodiversity

The extent to which habitat simplification has led to an increase in the number of Pacific salmon stocks either extinct or at risk of extinction (Nehlsen et al. 1991) is not known. However, the large number of cases in which habitat degradation has been cited as a factor in stock declines (Figure 7.9B) suggests that loss of critical habitat has played an important role, particularly with species spending extended periods in fresh water and undertaking extensive seasonal movements within the drainage system. Severe reductions in stock abundance or outright extinctions have led to losses of genetic diversity within species, with one possible outcome being limited ability to maintain viable populations under unusual conditions. Potential consequences of the interactions between habitat simplification (Schlosser 1991) and genetic "simplification," whether due to loss of locally adapted stocks or to fish culture practices such as widespread planting of fry from a single hatchery population (Hilborn 1992), have not been adequately investigated.

At present there is little direct evidence that diversity of fishes has been reduced in simplified streams in the Pacific Northwest, because few studies have attempted to relate fish community composition to habitat characteristics. Most investigations have emphasized only salmon and trout; however, habitat simplification has been shown to alter the proportions of different salmonid species and age classes. Bisson and Sedell (1984) found that streams in western Washington in which logging debris had been removed had fewer pools and longer riffles than streams in old-growth forests. Although total salmonid biomass was greater in logged and cleaned streams than in old-growth sites, the communities were dominated by underyearling trout and there were proportionately fewer age one and older trout. Additional data of P. Bisson (cited by Sullivan et al. 1987, Hicks et al. 1991, and Naiman et al., this volume) indicate that conversion of pool to riffle habitat favors species and age classes that utilize riffles at the expense of those that prefer pools. The latter habitat type is usually preferred by juvenile coho salmon

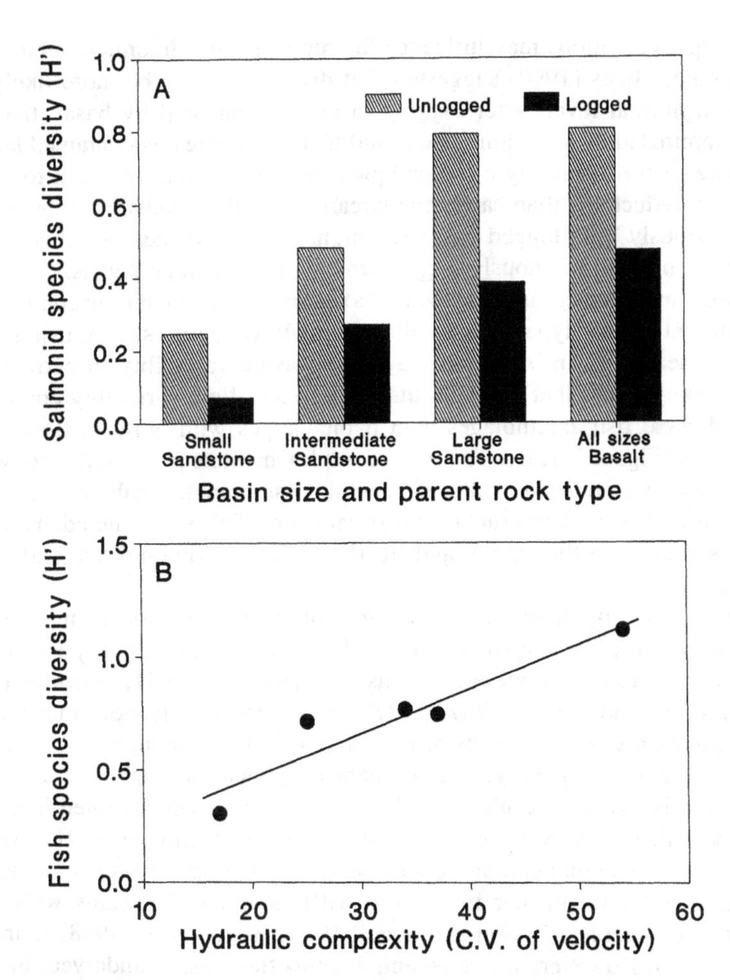

FIGURE 7.12. (A) Diversity of salmonid fishes expressed by H', the Shannon-Wiener index (Pielou 1969), in logged and unlogged Oregon coastal streams with different parent rock types. Unpublished data of G.H. Reeves, F.H. Everest, and J.R. Sedell. (B) Relationship between fish species diversity (H') and water current variability (coefficient of variation of velocity as measured by an inert dye tracer) in pools possessing different levels of hydraulic complexity in Huckleberry Creek, Washington. Different salmonid age classes were treated as separate taxa to reflect differences in habitat utilization (Bisson et al. 1988). Unpublished data of P.A. Bisson and B.R. Fransen.

and by older, larger trout. In a survey of Oregon coastal streams, G. Reeves, F. Everest, and J. Sedell (unpublished data) found that diversity of salmonids was lower in streams in which timber harvest had taken place than in similar streams in unlogged basins (Figure 7.12A). Diversity was lower in harvested sites regardless of drainage basin size and parent rock type.

Geologic conditions may influence the recovery of salmonid diversity after logging. Hicks (1990) suggested that diversity would be more likely to return to normal levels after logging in areas dominated by basalt than in areas dominated by sandstone. He found that basalt streams contained larger substrate, were inherently more complex, and were more resistant to summer flow reductions than sandstone streams. Basalt streams in Oregon that had previously been logged possessed higher salmonid species richness and evenness than did previously logged streams in sandstone basins.

Bisson and Fransen (unpublished data) estimated the hydraulic complexity of different habitat types in a small western Washington stream using controlled releases of an inert tracer dye to measure variability in current velocity. They found that habitat units with greater flow variability contained more diverse fish assemblages than habitat types with uniform flow characteristics (Figure 7.12B). The more complex habitats possessed woody debris, boulders, and undercut banks, while less complex habitat units possessed few structural obstructions to streamflow. This study included all fish species present in the stream and treated different salmonid age classes as separate taxa.

Several investigations in the Pacific Northwest have shown that timber harvest can result in increased salmonid productivity, chiefly by enhancing autotrophic production within streams (Murphy et al. 1981, Hawkins et al. 1983, Bilby and Bisson 1987, 1992). It is important to note that habitat simplification does not necessarily cause total fish production to decline. Rather, loss of complexity, if accompanied by increased light and dissolved nutrients, is likely to result in productivity increases concentrated in only a few taxa that directly benefit from the changes (Schlosser 1991). Where there are no potential competitors present, younger age classes of some salmonids appear to prosper in shallow, riffle-dominated streams with open vegetative canopies (Bisson and Sedell 1984, Bisson et al. 1988, Lamberti et al. 1991). However, where potential competitors exist, underyearling salmonids may be at a competitive disadvantage to species that are better adapted to warmer water that accompanies forest canopy removal (Reeves et al. 1987). The result may be loss of salmonid species and a large increase in the abundance of certain nonsalmonids, particularly cyprinids.

A similar pattern has been observed in the structure of aquatic invertebrate communities after logging. Erman et al. (1977) found that the density of benthic invertebrates was greater in northern California streams logged without buffer strips than in unlogged streams. Invertebrate species diversity, however, was greater in the unlogged streams (Figure 7.13). A few taxa (Ephemeroptera, Chironomidae) were much more abundant after logging while others disappeared from the streams. Results suggested that the pattern of increased production of a few taxa accompanied by a reduction in overall biodiversity may be common to all consumer trophic levels in streams where habitat has been simplified but light and nutrients are more plentiful (Gregory et al. 1987). Bilby and Bisson (1992) found that logging riparian veg-

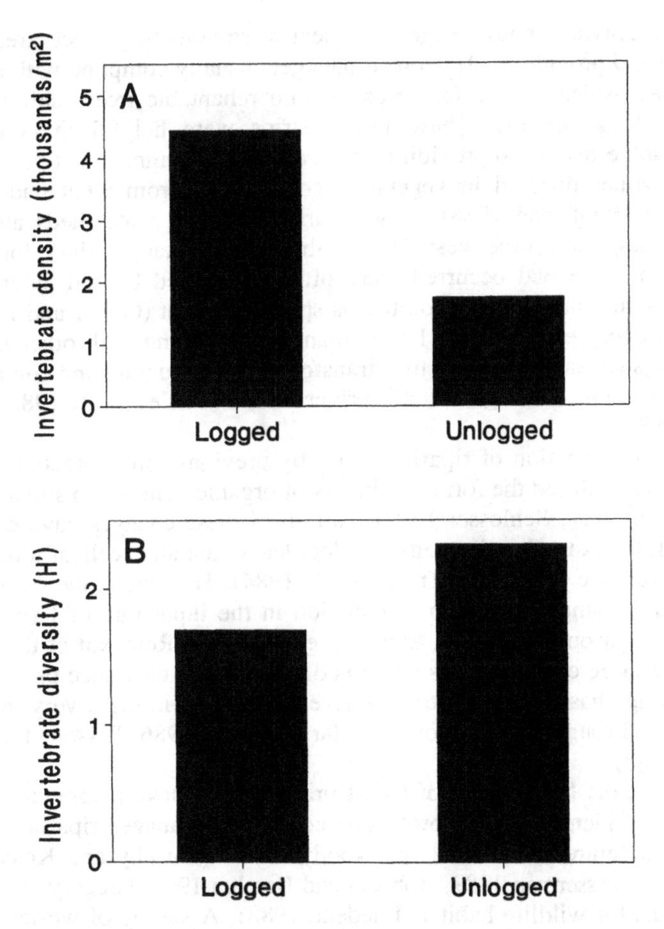

FIGURE 7.13. Density (A) and diversity (B) of aquatic invertebrates in logged and unlogged streams in northern California. Data from Erman et al. (1977).

etation also led to reduced diversity in the forms of terrestrial organic matter (leaves, needles, and other plant materials) entering a small stream, but that increased autotrophic production in the open channel led to greater availability of invertebrate species that were directly utilized by salmonids.

Restoration of Healthy, Productive River Systems

Managing for Large Woody Debris

Many streams in the Pacific Northwest were logged to the edge of the channel prior to enactment of the first state and provincial forest practices laws in the early 1970s. Because initial stream protection regulations were concerned primarily with temperature and erosion control, early BMP guidelines called for leaving enough vegetation next to streams to protect streambank integrity and provide shade. Forest managers usually complied with the new guidelines by leaving buffer strips of unmerchantable trees, chiefly hardwoods, along streams. These buffer strips were helpful in controlling streambank erosion and providing shade during the summer, but postlogging riparian zones differed in vegetative composition from what had existed originally (Kauffman 1988). New riparian zones in watersheds along the Pacific coast and on the west slope of the Cascade Range where logging to the stream edge had occurred were often dominated by red alder (*Alnus rubra*), with relatively few coniferous species present (Oliver and Hinckley 1987, Gregory et al. 1991). Forest management along with other land use activities such as grazing has thus transformed the structure and composition of riparian zones throughout entire river basins (Oakley et al. 1985, Kauffman 1988).

The transformation of riparian zones by previous forest practices has altered and simplified the form and inputs of organic material to streams (Gregory et al. 1987, Schlosser 1991). Not all of these changes have been detrimental; for example, nitrogen-rich alder leaves are an excellent food source for invertebrate shredders (Triska et al. 1984). However, one of the most significant changes has been a reduction in the input rate of large conifer debris (Swanson et al. 1976, Swanson et al. 1982). Removal of the sources of future large conifer debris combined with stream clearance programs for fish passage has left many streams severely lacking in these very important storage and roughness components (Harmon et al. 1986, Bisson et al. 1987, Bilby 1988).

Recent calls for revision of forest practice laws have recognized the importance of identifying and protecting conifers in managed riparian zones to provide a future source of large woody debris (Murphy and Koski 1989, Bilby and Wasserman 1989, Robison and Beschta 1990, Gregory et al. 1991) as well as for wildlife habitat (Raedeke 1988). A survey of western Washington riparian zones after enactment of the Timber, Fish and Wildlife agreement and incorporation of its provisions into the Washington Forest Practices Act in 1989 has shown that, while red alder remains the dominant tree species, about one-third of the remaining trees are now coniferous species (Figure 7.14). Similar measures have been adopted by other western states (Oregon Department of Forestry 1987) and in internal forestry planning and operating guidelines within the U.S. Forest Service (Hemstrom 1989), Bureau of Land Management (Oakley 1988), and Indian tribes (Bradley 1988). The new riparian zone prescriptions attempt to promote riparian communities like those in naturally (not anthropogenically) disturbed watersheds (Agee 1988) and will differ considerably from earlier managed riparian zones consisting of nonmerchantable timber.

Water Quality Standards

Attempts to regulate cumulative effects in forested watersheds have often relied on determining if water quality standards, here taken broadly to mean fixed levels of chemical constituents, temperature, water clarity, and both

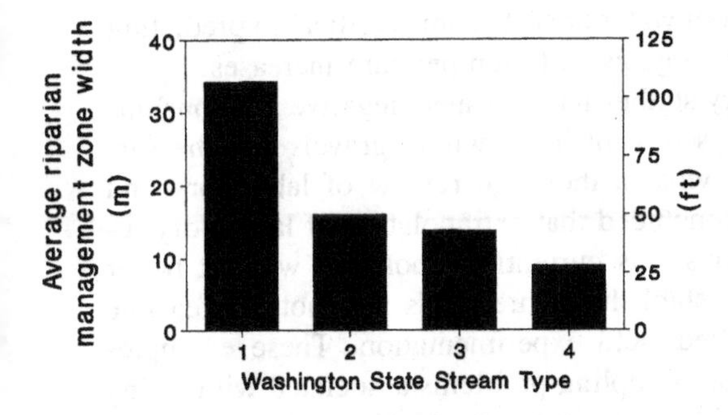

FIGURE 7.14. Average width and tree species composition of riparian vegetation left during logging after the revised 1989 Washington state riparian management requirements went into effect. The Washington State Stream Type refers to a regulatory classification system where Type 1 streams are large and Type 4 streams are small; generally, Type 1–3 streams contain fishes while Type 4 streams do not. Data from Quinault Indian Nation (1990).

suspended and deposited sediment, have been exceeded as the result of land management activities (Coats and Miller 1981). We wish to emphasize that while individual water quality standards, usually expressed as potentially harmful threshold levels, may serve useful functions as measures of relative risk to certain life history stages of individual species, their application to field situations in forested watersheds of the Pacific Northwest has been largely unsuccessful at either diagnosing or preventing cumulative environmental change. In many instances, difficulties have resulted from attempting to establish baseline levels of the parameter of interest, from attempting to extrapolate from laboratory experiments to field situations, or from attempting to extrapolate findings from one region to another (Burns 1991).

Although some water quality parameters are more easily quantified than others, ease of measurement does not guarantee predictable biological responses to cumulative disturbance. Quite often the characteristics of the stream system will influence the degree of impact. For example, Baltz et al. (1987) noted that cumulative temperature changes influenced microhabitat selection by trout in a California stream, but Modde et al. (1986) did not observe a clear impact of temperature modification on a trout population in a North Dakota stream. Berman and Quinn (1991) found that adult spring chinook salmon (*O. tshawytscha*) were able to locate cool water pockets for holding in a river during warm summer months. These cool water areas were generally undetectable from the streambank and would have gone unnoticed and unmeasured had the fish not been fitted with temperature-sensitive radio transmitters. The finding that adult salmon were capable of behavioral thermoregulation suggests that impacts of cumulative temperature change may be mediated by the presence of thermal refugia; therefore, understanding the distribution and abundance of cool water areas becomes critical to predicting the response of fishes and other organisms to temperature increases.

Although numerous laboratory studies have defined negative relationships between the percentage of fine sediment in spawning gravels and the survival of salmonid eggs and alevins, a thorough review of laboratory and field studies (Chapman 1988) concluded that extrapolation of laboratory results to natural stream conditions was currently impossible without better sampling techniques, and that establishing thresholds was not yet feasible without more carefully controlled field experimentation. These examples highlight a few of the formidable sampling problems associated with trying to relate measurements of individual parameters to the abundance of species of interest. Other problems encountered in applying fixed thresholds include adaptation of populations to local conditions (e.g., sediment rich glacially fed streams) and the possibility that harmful threshold levels may vary among species or even among life history stages of a single species (Davis 1975, Noggle 1978). We further believe that water quality standards and thresholds considered individually are not readily applicable to the goal of protecting biodiversity. On the other hand, we support the development of appropriate measures of ecosystem health focused on defining goals for the maintenance of important physical and biological processes that preserve the integrity of stream communities. With the exception of recent regulatory guidelines for maintaining trees in riparian zones for future recruitment of large woody debris, such goals remain largely unexplored.

BMPs and Stream Enhancement Programs

Restoration of stream habitat damaged by natural catastrophes, past forest practices, and other management activities has been undertaken by many federal, provincial, state, and tribal organizations. Some stream enhancement programs, including those of the Forest Service, have been extensively funded and cover large geographical areas, while others have been much more modest and local in scope. What have results from these programs taught us about the response of fish populations to habitat improvement, and what conclusions can we draw with regard to the prospects for recovery of altered stream systems with improved riparian management?

First, we have learned that correction of large and obvious problems can lead to measurable increases in population size, but recovery of stocks that are heavily fished is likely to take many years. In one of the longest continuous records of relative stock abundance on the Pacific coast, the abundance of sockeye salmon (*O. nerka*) in the Fraser River, British Columbia, declined precipitously when large rock slides created nearly impassable conditions in the Hell's Gate portion of the Fraser River canyon in 1913 and 1914, and splash dams in the Adams River prevented access to important spawning areas (Thompson 1945). Sockeye catches remained at only a small fraction of historical levels until completion of the Hell's Gate fishways in

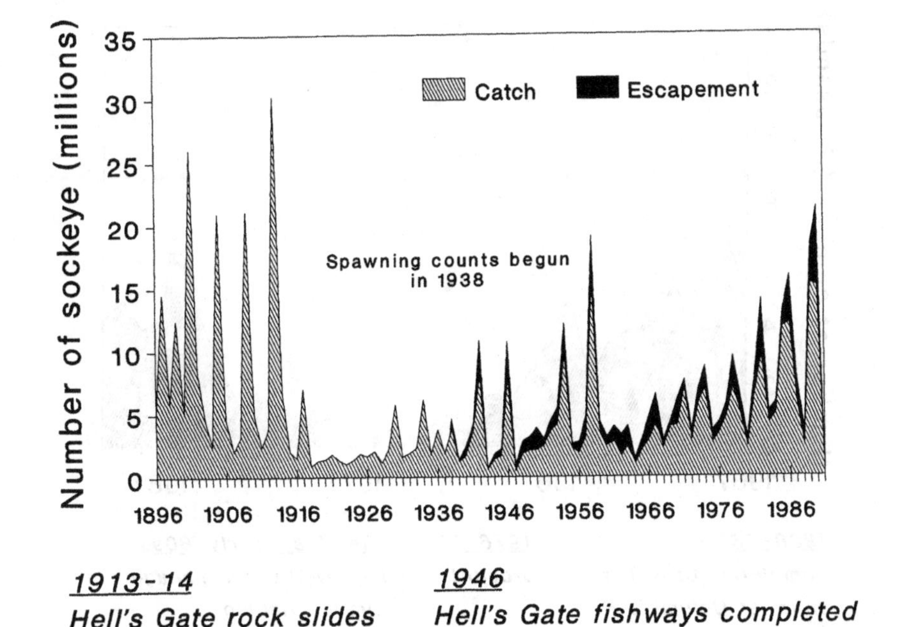

FIGURE 7.15. Catch and escapement of adult sockeye salmon in the Fraser River, British Columbia. Data of Thompson (1945), Ricker (1987), and annual reports of the International Pacific Salmon Fisheries Commission and the Pacific Salmon Commission.

1946, after which four-year cyclic peaks in the sockeye run gradually but steadily rose (Figure 7.15) despite a high exploitation rate. By 1989, sockeye catch plus escapement had almost reached the historically large run sizes of the early 1900s. This example demonstrated that a single restoration project could make a significant contribution toward rebuilding a depleted stock, but it has taken approximately 50 years for the population to recover.

Second, we have learned that carefully regulated fishing may be required to accompany habitat restoration in order to see an improvement in population abundance, particularly where populations are depressed to very low levels. In the Yakima River of central Washington, the population of spring chinook salmon declined from an estimated run of about 200,000 adults to 12,000 in 1920 and then to near extinction in the mid-1970s (Figure 7.16). There were a number of causes of the decline, but among the most important over the last 50 years have been the commercial and sport chinook salmon fisheries in the Columbia and Yakima rivers, completion of four hydroelectric dams downstream on the Columbia River between 1936 and 1968, and extensive water withdrawals from the Yakima River for crop irrigation (Fast et al. 1988). Droughts during the 1976 and 1977 summers further depleted the number of successfully spawning adults. When it became clear

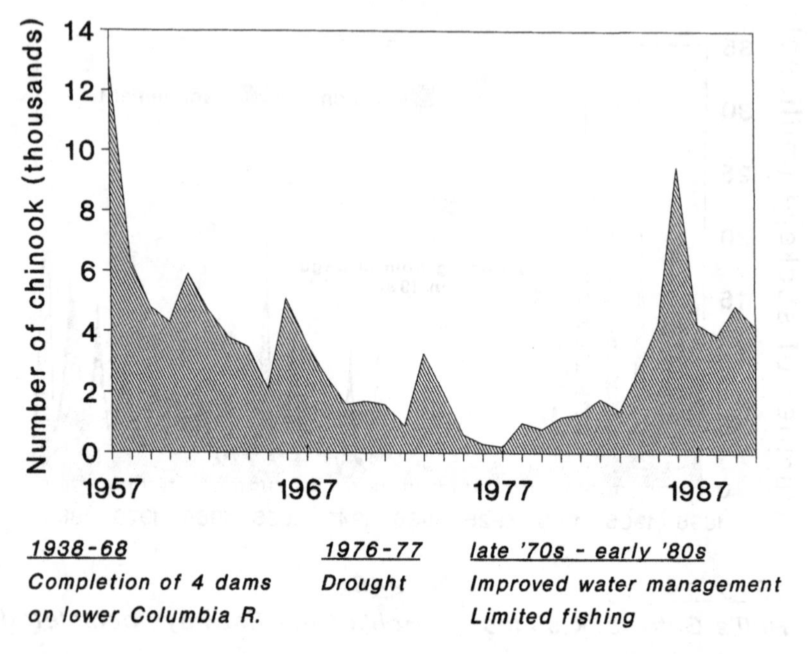

FIGURE 7.16. Estimated number of adult spring chinook salmon entering the Yakima River, Washington, from 1957 to 1990. Data from Fast et al. (1988) and S. Parker (pers. comm.).

that the population faced imminent extinction, severe limitations were placed on chinook salmon harvest in the river and an aggressive program of improved streamflow management was implemented. The population has since begun to increase but at a gradual rate. Fishery managers in the Yakima Basin believe that initial recovery of the run would not have been possible without both habitat improvement and reduced fishing intensity (Fast et al. 1988). This is a key point, because if survival in fresh water is reduced and marine survival remains fairly constant, fishing must be limited if stocks are to recover. However, even where freshwater survival has improved owing to habitat enhancement, continued fishing may lead to reductions in body size and, consequently, fecundity (Ricker 1980). Thus it is possible for exploited stocks to continue to decline due to reduced egg production per female in situations in which survival in streams and rivers has increased but average size of adults has become smaller. Complex interactions such as these between freshwater survival, marine survival, fishing intensity, and methods of harvest (i.e., seining, trolling, gillnetting) have surely contributed to the lengthy recovery period shown by several stocks in the Pacific Northwest.

Third, we have learned that adding large structural roughness elements such as logs or boulders to small streams can increase pool frequency and

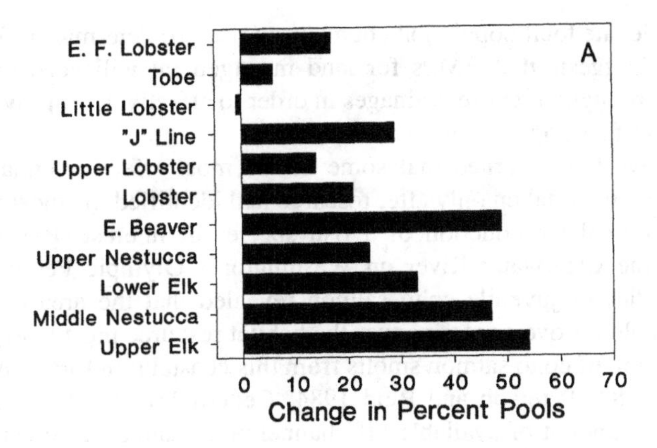

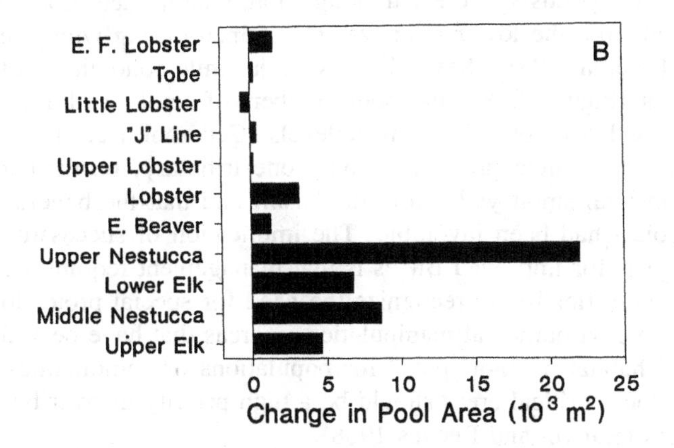

FIGURE 7.17. (A) Change in the percentage of stream area in pool habitat after addition of logs, boulders, and gabions in 11 Oregon coastal streams. (B) Change in the total area of pool habitat after enhancement. Data from House et al. (1989).

create additional habitat diversity (Figure 7.17) but the costs of such projects often limit habitat restoration to selected stream reaches, and would be prohibitively expensive at the scale of an entire drainage. Structure additions to small streams usually involve boulder, large woody debris, or gabion placement, and the purpose is usually to create large rearing pools or to trap gravel suitable for spawning (House and Boehne 1985). In terms of numbers of projects, structure additions constitute the most common form of stream enhancement in the Pacific Northwest. Yet the amount of stream habitat improved by these projects still represents only a tiny fraction of the total length of fish-bearing streams and rivers in the region (Sedell and Beschta 1991). Of course, not all streams need additional roughness elements, but the scale upon which structural enhancement of small streams has been practiced has been insufficient in most cases to determine if these activities significantly elevate total population abundance (also see Peterman 1990). This conclusion suggests that BMPs for land management will need to be implemented throughout entire drainages in order to effectively improve habitat and increase fish populations.

Finally, we have learned that some of the most effective enhancement projects were undertaken only after research had identified the most probable factor limiting the production of a fish species of interest (Reeves et al. 1989). In the Clearwater River on Washington's Olympic Peninsula, life history studies of juvenile coho salmon revealed that the area of riverine ponds suitable for overwintering was the habitat resource most likely to limit the production of coho salmon smolts from this coastal rain forest river basin (Peterson 1982, Peterson and Reid 1984, Cederholm and Reid 1987). To increase the amount of available off-channel pond habitat, a series of small interconnected ponds was created along an alluvial terrace tributary and access to and from the lower river was maintained during winter and spring (Cederholm et al. 1988). Several thousand juvenile coho entered the ponds each winter (Figure 7.18), and both numbers of smolts and average smolt size increased over preenhancement levels. Cederholm et al. (1988) estimated that this single project, on only one tributary, resulted in a 2.8% increase in total smolt yield from the basin, and that the benefit-cost ratio of the project had been favorable. The implication of successful enhancement projects for improved BMPs is that management requirements should be sufficiently flexible to recognize the need for special protection or possibly active environmental manipulation in areas that have been designated as critical habitat or "hot spots" for populations or communities. Identification of these critical areas should be a high priority in river basin habitat inventories (Hankin and Reeves 1988).

Patience, Patience, Patience

We conclude this review with the observation that restoration of naturally complex channels and unimpeded connections between streams and riparian zones is a formidable task requiring an unprecedented level of cooperation and willingness to alter current land use practices (Gregory et al. 1991, Sedell and Beschta 1991). Although we have freely used the term *best* management practice, we concur with Petts (1990) that management alternatives will continue to evolve as our knowledge increases. The BMPs of tomorrow will be better than those of today, and there is every reason to believe that there will always be room for environmental improvement. Recovery of habitat complexity and biological diversity in streams, even with the most benign of land use influences, will follow a trajectory dictated by landscape patterns and natural climatic regimes within the region (Schlosser 1991). Coupled with continued industrial development and fishing pressure, recovery rates of naturally produced aquatic resources to some desired but usually unspecified levels will not be rapid in any event. We wonder, for example,

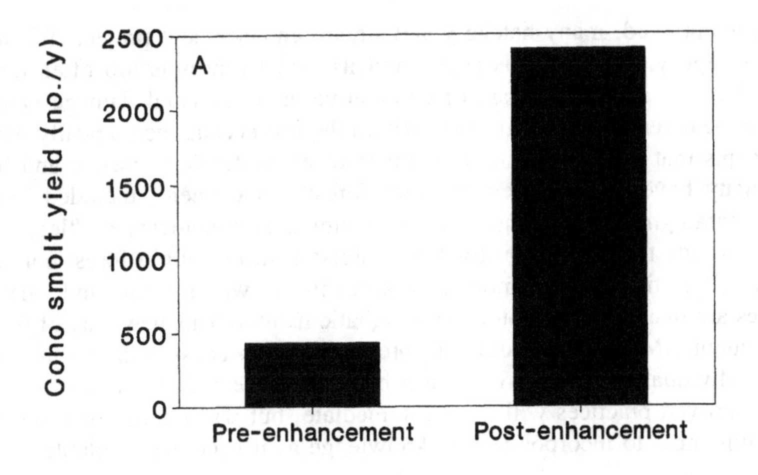

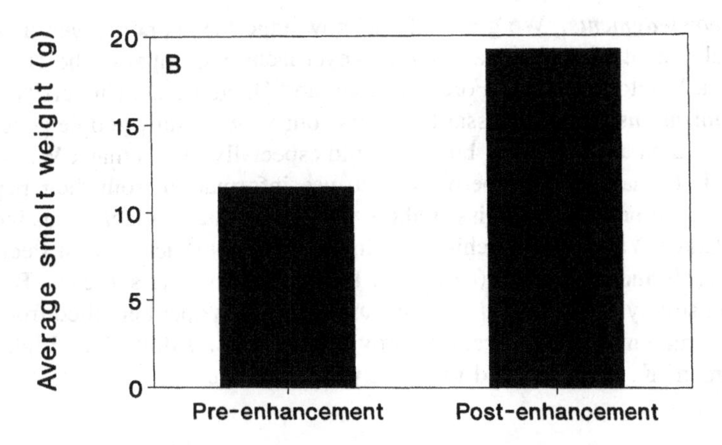

FIGURE 7.18. Average number of coho salmon smolts and average smolt weight after creation of additional winter rearing habitat in a small tributary of the Clearwater River, Washington. Data from Cederholm et al. (1988).

if programs intended to double the number of salmon returning to the Columbia River by the year 2000 are based on realistic assumptions and expectations.

But some of the most valuable aquatic resources in the Pacific Northwest are in jeopardy, and decisive action is needed. In many instances the need to take decisive action has led management organizations to adopt restoration approaches based more on *mitigation* of losses than on *protection or restoration* of natural ecological processes that have created and maintained diverse and productive stream habitat. Mitigation approaches seek rapid increases in numbers of harvestable fish rather than investments in long-term natural productivity that yield gradual but sustained improvement. However well-intentioned, many hatchery and stream enhancement projects fall into this category. Mitigation is costly and its effectiveness is too often questionable; the results of years of cumulative environmental damage cannot easily be reversed in a short time. Within the last decade, ecologically based concepts that emphasize the importance of ecosystem complexity and biodiversity have begun to revolutionize forest management. Included in the new paradigm are techniques for maintaining key elements or "legacies" that provide the foundation for the establishment of natural forest communities. It is time for a similar revolution in the way streams and riparian zones are managed for protection of aquatic habitat. This new way of thinking about BMPs as procedures to preserve stream ecosystem integrity and not individual fish populations must begin with the realization that benefits of improved practices will not be immediate, but will require patience and a willingness to incorporate new knowledge as it becomes available.

Acknowledgments. We gratefully acknowledge the support given us during development of these ideas by the Weyerhaeuser Company, the University of Washington, the U.S. Forest Service, and Oregon State University. These organizations do not necessarily endorse our views. Many people generously shared data and concepts, but we would especially like to thank W. Nehlsen and J. Lichatowich for permission to use information from their paper on Pacific salmon stocks at risk. Others who made data available to us included S. Parker (Yakima River chinook salmon), J. Light (Pacific coast steelhead), J. Sedell and F. Everest (Columbia River Basin habitat surveys), B. Hicks (pool surveys), and R. Bilby (pool surveys). This paper benefited from helpful comments and suggestions provided by R.E. Bilby, J.T. Light, R.J. Naiman, K. Sullivan, and two anonymous referees.

References

Agee, J.K. 1988. Successional dynamics in forest riparian zones. Pages 31–43 *in* K.J. Raedeke, editor. Streamside management: riparian wildlife and forestry interactions. Contribution 59, Institute of Forest Resources, University of Washington, Seattle, Washington, USA.

Angermeier, P.L. 1987. Spatiotemporal variation in habitat selection by fishes in small Illinois streams. Pages 52–60 *in* W.J. Matthews and D.C. Heins, editors. Community and evolutionary ecology of North American stream fishes. University of Oklahoma Press, Norman, Oklahoma, USA.

Baltz, D.M., B. Vondracek, L.R. Brown, and P.B. Moyle. 1987. Influence of temperature on microhabitat choices by fishes in a California stream. Transactions of the American Fisheries Society **116**:12–20.

Beidler, W.M., and T.E. Nickelson. 1980. An evaluation of the Oregon Department of Fish and Wildlife standard spawning fish survey system for coho salmon. Oregon Department of Fish and Wildlife, Fish Division Information Report Series, Fisheries Number 80-9, Portland, Oregon, USA.

Benda, L.E. 1990. The influence of debris flows on channels and valley floors in the Oregon Coast Range, U.S.A. Earth Surface Processes and Landforms **15**:457–466.

Berkman, H.E., and C.F. Rabeni. 1987. Effect of siltation on stream fish communities. Environmental Biology of Fishes **18**:285–294.

Berman, C., and T.P. Quinn. 1991. Behavioural thermoregulation and homing by spring chinook salmon, *Oncorhynchus tshawytscha* Walbaum, in the Yakima River. Journal of Fish Biology, **39**:301–312.

Bilby, R.E. 1988. Interactions between aquatic and terrestrial systems. Pages 13–29 *in* K.J. Raedeke, editor. Streamside management: riparian wildlife and forestry interactions. Contribution 59, Institute of Forest Resources, University of Washington, Seattle, Washington, USA.

Bilby, R.E., and P.A. Bisson. 1987. Emigration and production of hatchery coho salmon (*Oncorhynchus kisutch*) stocked in streams draining an old-growth and a clear-cut watershed. Canadian Journal of Fisheries and Aquatic Sciences **44**:1397–1407.

Bilby, R.E., and P.A. Bisson. 1992. Allochthonous vs. autochthonous organic matter contributions to the trophic support of fish populations in clear-cut and old-growth forested streams. Canadian Journal of Fisheries and Aquatic Sciences **49**, *in press*.

Bilby, R.E., and J.W. Ward. 1991. Large woody debris characteristics and function in streams draining old-growth, clear-cut, and second-growth forests in southwestern Washington. Canadian Journal of Fisheries and Aquatic Sciences **48**:2499–2508.

Bilby, R.E., and L.J. Wasserman. 1989. Forest practices and riparian management in Washington state: data based regulation development. Pages 87–94 *in* R.E. Gresswell, B.A. Barton, and J.L. Kerschner, editors. Practical approaches to riparian resource management. United States Bureau of Land Management, Billings, Montana, USA.

Bisson, P.A., R.E. Bilby, M.D. Bryant, C.A. Dolloff, G.B. Grette, R.A. House, M.L. Murphy, KV. Koski, and J.R. Sedell. 1987. Large woody debris in forested streams in the Pacific Northwest: past, present, and future. Pages 143–190 *in* E.O. Salo and T.W. Cundy, editors. Streamside management: forestry and fishery interactions. Contribution 57, Institute of Forest Resources, University of Washington, Seattle, Washington, USA.

Bisson, P.A., J.L. Nielsen, R.A. Palmason, and L.E. Grove. 1982. A system of naming habitat types in small streams, with examples of habitat utilization by salmonids during low streamflow. Pages 62–73 *in* N.B. Armantrout, editor. Acquisition and utilization of aquatic habitat inventory information. Western Division, American Fisheries Society, Portland, Oregon. The Hague Publishing, Billings, Montana, USA.

Bisson, P.A., and J.R. Sedell. 1984. Salmonid populations in streams in clearcut vs. old-growth forests of western Washington. Pages 121–129 *in* W.R. Meehan, T.R. Merrell, Jr., and T.A. Hanley, editors. Fish and wildlife relationships in old-growth forests. American Institute of Fishery Research Biologists, Juneau, Alaska, USA.

Bisson, P.A., K. Sullivan, and J.L. Nielsen. 1988. Channel hydraulics, habitat use, and body form of juvenile coho salmon, steelhead, and cutthroat trout in streams. Transactions of the American Fisheries Society **117**:262–273.

Bradley, W.P. 1988. Riparian management practices on Indian lands. Pages 201–206 *in* K.J. Raedeke, editor. Streamside management: riparian wildlife and forestry interactions. Contribution 59, Institute of Forest Resources, University of Washington, Seattle, Washington, USA.

Bryant, M.D. 1980. Evolution of large, organic debris after timber harvest: Maybeso Creek, 1949 to 1978. United States Forest Service General Technical Report PNW-101, Pacific Northwest Forest and Range Experiment Station, Portland, Oregon, USA.

Bryant, M.D. 1983. The role and management of woody debris in west coast salmonid nursery streams. North American Journal of Fisheries Management **3**:322–330.

Burns, D.C. 1991. Cumulative effects of small modifications to habitat. Fisheries **16**:12–17.

Bustard, D.R., and D.W. Narver. 1975. Preferences of juvenile coho salmon (*Oncorhynchus kisutch*) and cutthroat trout (*Salmo clarki*) relative to simulated alteration of winter habitat. Journal of the Fisheries Research Board of Canada **32**:681–687.

Cederholm, C.J., and L.M. Reid. 1987. Impact of forest management on coho salmon (*Oncorhynchus kisutch*) populations of the Clearwater River, Washington: a project summary. Pages 373–398 *in* E.O. Salo and T.W. Cundy, editors. Streamside management: forestry and fishery interactions. Contribution 57, Institute of Forest Resources, University of Washington, Seattle, Washington, USA.

Cederholm, C.J., L.M. Reid, and E.O. Salo. 1981. Cumulative effects of logging road sediment on salmonid populations in the Clearwater River, Jefferson County, Washington. Pages 38–74 *in* Proceedings of a Conference "Salmon Spawning Gravel: A Renewable Resource in the Pacific Northwest?" Report 39, State of Washington Water Resource Center, Washington State University, Pullman, Washington, USA.

Cederholm, C.J., W.J. Scarlett, and N.P. Peterson. 1988. Low-cost enhancement technique for winter habitat of juvenile coho salmon. North American Journal of Fisheries Management **8**:438–441.

Chapman, D.W. 1986. Salmon and steelhead abundance in the Columbia River in the nineteenth century. Transactions of the American Fisheries Society **115**:662–670.

Chapman, D.W. 1988. Critical review of variables used to define effects of fines in redds of large salmonids. Transactions of the American Fisheries Society **117**:1–21.

Chapman, D.W., and E. Knudsen. 1980. Channelization and livestock impacts on salmonid habitat and biomass in western Washington. Transactions of the American Fisheries Society **109**:357–363.

Chilcote, M.W., S.A. Leider, and J.J. Loch. 1986. Differential reproductive success of hatchery and wild summer-run steelhead under natural conditions. Transactions of the American Fisheries Society **115**:726–735.

Coats, R.N., and T.O. Miller. 1981. Cumulative silvicultural impacts on watersheds: a hydrologic and regulatory dilemma. Environmental Management **5**:147–160.

Cupp, C.E. 1989. Identifying spatial variablity of stream characteristics through classification. M.S. Thesis. School of Fisheries, University of Washington, Seattle, Washington, USA.

Davis, J.C. 1975. Minimal dissolved oxygen requirements of aquatic life with emphasis on Canadian species: a review. Journal of the Fisheries Research Board of Canada **32**:2295–2332.

Erman, D.C., J.D. Newbold, and K.B. Roby. 1977. Evaluation of streamside bufferstrips for protecting aquatic organisms. Contribution 165, California Water Resources Center, University of California, Davis, USA.

Everest, F.H., R.L. Beschta, J.C. Scrivener, KV. Koski, J.R. Sedell, and C.J. Cederholm. 1987. Fine sediment and salmonid production: a paradox. Pages 98–142 *in* E.O. Salo and T.W. Cundy, editors. Streamside management: forestry and fishery interactions. Contribution 57, Institute of Forest Resources, University of Washington, Seattle, Washington, USA.

Fast, D.E., J.D. Hubble, and B.D. Watson. 1988. Yakima River spring chinook: the decline and recovery of a mid-Columbia natural spawning stock. Pages 18–26 *in* B.G. Shepherd, editor. Proceedings of the 1988 Northeast Pacific Chinook and Coho Salmon Workshop. British Columbia Ministry of Environment, Penticton, British Columbia, Canada.

Fausch, K.D., J. Lyons, J.R. Karr, and P.L. Angermeier. 1990. Fish communities as indicators of environmental degradation. American Fisheries Society Symposium **8**:123–144.

Frissell, C.A., W.J. Liss, C.E. Warren, and M.D. Hurley. 1986. A hierarchical framework for stream habitat classification: viewing streams in a watershed context. Environmental Management **10**:199–214.

Gecy, J.L., and M.V. Wilson. 1990. Initial establishment of riparian vegetation after disturbance by debris flows in Oregon. American Midland Naturalist **123**:282–291.

Gonor, J.J., J.R. Sedell, and P.A. Benner. 1988. What we know about large trees in estuaries, in the sea, and on coastal beaches. Pages 83–112 *in* C. Maser, R.F. Tarrant, J.M. Trappe, and J.F. Franklin, editors. From the forest to the sea: a story of fallen trees. United States Forest Service General Technical Report PNW-GTR-229, Pacific Northwest Research Station, Portland, Oregon, USA.

Gorman, O.T., and J.R. Karr. 1978. Habitat structure and stream fish communities. Ecology **59**:507–515.

Grant, G.E. 1986. Downstream effects of timber harvest activity on the channel and valley floor morphology of western Cascade streams. Dissertation. Johns Hopkins University, Baltimore, Maryland, USA.

Grant, G.E. 1988. The RAPID technique: a new method for evaluating downstream effects of forest practices on riparian zones. United States Forest Service General Technical Report PNW-GTR-220, Pacific Northwest Research Station, Portland, Oregon, USA.

Grant, G.E., F.J. Swanson, and M.G. Wolman. 1990. Pattern and origin of stepped-bed morphology in high-gradient streams, Western Cascades, Oregon. Geological Society of America Bulletin **102**:340–352.

Gregory, S.V., G.A. Lamberti, D.C. Erman, KV. Koski, M.L. Murphy, and J.R. Sedell. 1987. Influence of forest practices on aquatic production. Pages 233–255 *in* E.O. Salo and T.W. Cundy, editors. Streamside management: forestry and fishery interactions. Contribution 57, Institute of Forest Resources, University of Washington, Seattle, Washington, USA.

Gregory, S.V., F.J. Swanson, W.A. McKee, and K.W. Cummins. 1991. An ecosystem perspective of riparian zones. BioScience **40**:540–551.

Grossman, G.D., J.F. Dowd, and M. Crawford. 1990. Assemblage stability in stream fishes: a review. Environmental Management **14**:661–671.

Grossman, G.D., P.B. Moyle, and J.O. Whittaker, Jr. 1982. Stochasticity in structural and functional characteristics of an Indiana stream fish assemblage: a test of community theory. American Naturalist **120**:423–454.

Hall, J.D., G.W. Brown, and R.L. Lantz. 1987. The Alsea watershed study: a retrospective. Pages 399–416 *in* E.O. Salo and T.W. Cundy, editors. Streamside management: forestry and fishery interactions. Contribution 57, Institute of Forest Resources, University of Washington, Seattle, Washington, USA.

Hankin, D.G., and G.H. Reeves. 1988. Estimating total fish abundance and total habitat area in small streams based on visual estimation methods. Canadian Journal of Fisheries and Aquatic Sciences **45**:834–844.

Harmon, M.E., J.F. Franklin, F.J. Swanson, P. Sollins, S.V. Gregory, J.D. Lattin, N.H. Anderson, S.P. Cline, N.G. Aumen, J.R. Sedell, G.W. Lienkaemper, K. Cromack, Jr., and K.W. Cummins. 1986. Ecology of coarse woody debris in temperate ecosystems. Advances in Ecological Research **15**:133–302.

Hawkins, C.P., M.L. Murphy, N.H. Anderson, and M.A. Wilzbach. 1983. Density of fish and salamanders in relation to riparian canopy and physical habitat in streams of the northwestern United States. Canadian Journal of Fisheries and Aquatic Sciences **40**:1173–1185.

Healey, M.C. 1982. Catch, escapement, and stock-recruitment for British Columbia chinook salmon since 1951. Canadian Technical Reports of Fisheries and Aquatic Sciences 1107, Ottawa, Ontario, Canada.

Hemstrom, M.A. 1989. Integration of riparian data in a geographic information system. Pages 17–22 *in* R.E. Gresswell, B.A. Barton, and J.L. Kerschner, editors. Practical approaches to riparian resource management. United States Bureau of Land Management, Billings, Montana, USA.

Herbold, B. 1984. Structure of an Indiana stream fish association: choosing an appropriate model. American Naturalist **124**:561–572.

Hicks, B.J. 1990. The influence of geology and timber harvest on channel morphology and salmonid populations in Oregon Coast Range streams. Dissertation. Oregon State University, Corvallis, Oregon, USA.

Hicks, B.J., J.D. Hall, P.A. Bisson, and J.R. Sedell. 1991. Response of salmonids to habitat changes. Pages 483–518 *in* W.R. Meehan, editor. Influences of forest and rangeland management on salmonid fishes and their habitats. American Fisheries Society Special Publication 19, Bethesda, Maryland, USA.

Hilborn, R. 1992. Hatcheries and the future of salmon in the Northwest. Fisheries **17**:5–8.

Holtby, L.B., and J.C. Scrivener. 1989. Observed and simulated effects of climatic variability, clear-cut logging and fishing on the numbers of chum salmon (*Oncorhynchus keta*) and coho salmon (*O. kisutch*) returning to Carnation Creek, British Columbia. Pages 62–81 *in* C.D. Levings, L.B. Holtby, and M.A. Henderson, editors. Proceedings of the National Workshop on Effects of Habitat Alterations on Salmonid Stocks. Canadian Special Publication of Fisheries and Aquatic Sciences 105, Ottawa, Ontario, Canada.

House, R.A., V. Crispin, and R. Monthey. 1989. Evaluation of stream rehabilitation projects: Salem District (1981–1988). United States Bureau of Land Management Technical Note BLM-OR-PT-90–10–6600.9, Salem, Oregon, USA.

Karr, J.R. 1981. Assessment of biotic integrity using fish communities. Fisheries **6**:21–27.

Karr, J.R., and I.J. Schlosser. 1978. Water resources and the land-water interface. Science **201**:229–234.

Karr, J.R., L.A. Toth, and D.R. Dudley. 1985. Fish communities of midwestern rivers: a history of degradation. BioScience **35**:90–95.

Karr, J.R., L.A. Toth, and G.D. Garman. 1983. Habitat preservation for midwest stream fishes: principles and guidelines. United States Environmental Protection Agency Report OR EPA-600/3–83–006, Corvallis, Oregon, USA.

Kauffman, J.B. 1988. The status of riparian habitats in Pacific Northwest forests. Pages 45–55 *in* K.J. Raedeke, editor. Streamside management: riparian wildlife and forestry interactions. Contribution 59, Institute of Forest Resources, University of Washington, Seattle, Washington, USA.

Keller, E.A., and F.J. Swanson. 1979. Effects of large organic material on channel form and fluvial processes. Earth Surface Processes **4**:361–380.

Konkel, G.W., and J.D. McIntyre. 1987. Trends in spawning populations of Pacific anadromous salmonids. United States Fish and Wildlife Service, Fish and Wildlife Technical Report 9, Washington, D.C., USA.

Lamberti, G.A., S.V. Gregory, L.R. Ashkenas, R.C. Wildman, and K.M.S. Moore. 1991. Stream ecosystem recovery following a catastrophic debris flow. Canadian Journal of Fisheries and Aquatic Sciences **48**:196–208.

Leider, S.A., M.W. Chilcote, and J.J. Loch. 1984. Spawning characteristics of sympatric populations of steelhead trout (*Salmo gairdneri*): evidence for partial reproductive isolation. Canadian Journal of Fisheries and Aquatic Sciences **41**:1454–1462.

Leidy, R.A., and P.L. Fiedler. 1985. Human disturbance and patterns of fish species diversity in the San Francisco Bay drainage, California. Biological Conservation **33**:247–267.

Li, H.W., C.B. Schreck, C.E. Bond, and E. Rexstad. 1987. Factors influencing changes in fish assemblages of Pacific Northwest streams. Pages 193–202 *in* W.J. Matthews and D.C. Heins, editors. Community and evolutionary ecology of North American stream fishes. University of Oklahoma Press, Norman, Oklahoma, USA.

Lichatowich, J.A., and J.D. McIntyre. 1987. Use of hatcheries in the management of Pacific anadromous salmonids. American Fisheries Society Symposium **1**:131–136.

Light, J.T. 1992. Distribution and origins of steelhead trout (*Oncorhynchus mykiss*) in offshore waters of the North Pacific Ocean. International North Pacific Fisheries Commission Report, Fisheries Research Institute, University of Washington, Seattle, USA, *in press*.

Matthews, W.J., and D.C. Heins, editors. 1987. Community and evolutionary ecology of North American stream fishes. University of Oklahoma Press, Norman, Oklahoma, USA.

McDonald, J., and J.M. Hume. 1984. Babine Lake sockeye salmon (*Oncorhynchus nerka*) enhancement program: testing some major assumptions. Canadian Journal of Fisheries and Aquatic Sciences **41**:70–92.

Meehan, W.R., editor. 1991. Influences of forest and rangeland management on salmonid fishes and their habitats. American Fisheries Society Special Publication 19, Bethesda, Maryland, USA.

Megahan, W.F. 1982. Channel sediment storage behind obstructions in forested drainage basins draining the granitic bedrock of the Idaho Batholith. Pages 114–121 in F.J. Swanson, R.J. Janda, T. Dunne, and D.N. Swanston, editors. Sediment budgets and routing in forested drainage basins. United States Forest Service Research Paper PNW-141, Pacific Northwest Forest and Range Experiment Station, Portland, Oregon, USA.

Megahan, W.F., and R.A. Nowlin. 1976. Sediment storage in channels draining small forested watersheds. Pages 4/115–4/126 *in* Proceedings of the Third Federal Interagency Sedimentation Conference, Denver, Colorado. Sedimentation Committee of the Water Resources Council, Washington, D.C., USA.

Miller, R.R., J.D. Williams, and J.E. Williams. 1989. Extinctions of North American fishes during the past century. Fisheries **14**:22–38.

Modde, T., H.G. Drewes, and M.A. Rumble. 1986. Effects of watershed alteration on the brook trout population of a small Black Hills stream. Great Basin Naturalist **46**:39–45.

Moore, K.M.S., and S.V. Gregory. 1988*a*. Summer habitat utilization and ecology of cutthroat trout fry (*Salmo clarki*) in Cascade Mountain streams. Canadian Journal of Fisheries and Aquatic Sciences **45**:1921–1930.

Moore, K.M.S., and S.V. Gregory. 1988*b*. Response of young-of-the-year cutthroat trout to manipulation of habitat structure in a small stream. Transactions of the American Fisheries Society **117**:162–170.

Moyle, P.B., and B. Herbold. 1987. Life history patterns and community structure in stream fishes of western North America: comparisons with eastern North America and Europe. Pages 25–32 *in* W.J. Matthews and D.C. Heins, editors. Community and evolutionary ecology of North American stream fishes. University of Oklahoma Press, Norman, Oklahoma, USA.

Mullen, R.E. 1981. Oregon's commercial harvest of coho salmon, *Oncorhynchus kisutch* (Walbaum), 1892–1960. Oregon Department of Fish and Wildlife, Information Report Series, Fisheries Number 81–3, Portland, Oregon, USA.

Murphy, M.L., C.P. Hawkins, and N.H. Anderson. 1981. Effects of canopy modification and accumulated sediment on stream communities. Transactions of the American Fisheries Society **110**:469–478.

Murphy, M.L., and KV. Koski. 1989. Input and depletion of woody debris in Alaska streams and implications for streamside management. North American Journal of Fisheries Management **9**:427–436.

Naiman, R.J., and J.R. Sedell. 1980. Relationships between metabolic parameters and stream order in Oregon. Canadian Journal of Fisheries and Aquatic Sciences **37**:834–847.

Neave, F. 1949. Game fish populations in the Cowichan River. Bulletin of the Fisheries Research Board of Canada 84, Ottawa, Ontario, Canada.

Nehlsen, W., J.E. Williams, and J.A. Lichatowich. 1991. Pacific salmon at the crossroads: stocks of salmon at risk from California, Oregon, Idaho, and Washington. Fisheries **16**:4–21.

Neilson, J.D., and G.H. Geen. 1981. Enumeration of spawning salmon from spawner residence time and aerial counts. Transactions of the American Fisheries Society **110**:554–556.

Nickelson, T.E. 1986. Influences of upwelling, ocean temperature, and smolt abundance on marine survival of coho salmon (*Oncorhynchus kisutch*) in the Oregon

Production Area. Canadian Journal of Fisheries and Aquatic Sciences **43**:527–535.

Nickelson, T.E., M.F. Solazzi, and S.L. Johnson. 1986. Use of hatchery coho salmon (*Oncorhynchus kisutch*) presmolts to rebuild wild populations in Oregon coastal streams. Canadian Journal of Fisheries and Aquatic Sciences **43**:2443–2449.

Niemi, G.J., P. DeVore, N. Detenbeck, D. Taylor, A. Lima, J. Pastor, J.D. Yount, and R.J. Naiman. 1990. Overview of case studies on recovery of aquatic systems from disturbance. Environmental Management **14**:571–587.

Noggle, C.C. 1978. Behavioral, physiological and lethal effects of suspended sediment on juvenile salmonids. M.S. Thesis. University of Washington, Seattle, Washington, USA.

Oakley, A.L. 1988. Riparian management practices of the Bureau of Land Management. Pages 191–196 *in* K.J. Raedeke, editor. Streamside management: riparian wildlife and forestry interactions. Contribution 59, Institute of Forest Resources, University of Washington, Seattle, Washington, USA.

Oakley, A.L., J.A. Collins, L.B. Everson, D.A. Heller, J.C. Howerton, and R.E. Vincent. 1985. Riparian zones and freshwater wetlands. Pages 58–79 *in* E.R. Brown, editor. Management of wildlife and fish habitats in forests of western Oregon and Washington. Part 1. United States Forest Service, Pacific Northwest Region, Portland, Oregon, USA.

Oliver, C.D., and T.M. Hinckley. 1987. Species, stand structures, and silvicultural manipulation patterns for the streamside zone. Pages 259–276 *in* E.O. Salo and T.W. Cundy, editors. Streamside management: forestry and fishery interactions. Contribution 57, Institute of Forest Resources, University of Washington, Seattle, USA.

Oregon Department of Forestry. 1987. Riparian protection. Forest Practices Section, Oregon Department of Forestry, Forest Practices Notes Number 6, Salem, Oregon, USA.

Pacific Fishery Management Council. 1979. Freshwater habitat, salmon produced, and escapements for natural spawning along the Pacific coast of the United States. Report of the Anadromous Salmonid Task Force, Portland, Oregon, USA.

Pearsons, T.R., H.W. Li, and G.A. Lamberti. 1992. Influence of habitat complexity on resistance and resilience of stream fish assemblages to flooding. Transactions of the American Fisheries Society **121**, *in press*.

Pella, J.J., and R.T. Myren. 1974. Caveats concerning evaluation of effects of logging on salmon production in southeastern Alaska from biological information. Northwest Science **48**:132–144.

Peterman, R.M. 1989. Application of statistical power analysis to the Oregon coho salmon (*Oncorhynchus kisutch*) problem. Canadian Journal of Fisheries and Aquatic Sciences **46**:1183–1187.

Peterman, R.M. 1990. Statistical power analysis can improve fisheries research and management. Canadian Journal of Fisheries and Aquatic Sciences **47**:2–15.

Peterson, N.P. 1982. Immigration of juvenile coho salmon (*Oncorhynchus kisutch*) into riverine ponds. Canadian Journal of Fisheries and Aquatic Sciences **39**:1308–1310.

Peterson, N.P., and L.M. Reid. 1984. Wall-base channels: their evolution, distribution, and use by juvenile coho salmon in the Clearwater River, Washington. Pages 215–226 *in* J.M. Walton and D.B. Houston, editors. Proceedings of the

Olympic Wild Fish Conference, March 23–25, 1983, Port Angeles, Washington, USA.

Petts, G.E. 1990. The role of ecotones in aquatic landscape management. Pages 227–261 *in* R.J. Naiman and H. Décamps, editors. The ecology and management of aquatic-terrestrial ecotones. UNESCO, Paris, and Parthenon Publishing Group, Carnforth, United Kingdom.

Pielou, E.C. 1969. An introduction to mathematical ecology. John Wiley and Sons, New York, New York, USA.

Platts, W.S., and W.F. Megahan. 1975. Time trends in riverbed sediment composition in salmon and steelhead spawning areas: South Fork Salmon River, Idaho. Transactions of the North American Wildlife and Natural Resources Conference, Washington, D.C. **40**:229–239.

Platts, W.S., and R.L. Nelson. 1985. Stream habitat and fisheries response to livestock grazing and instream improvement structures, Big Creek, Utah. Journal of Soil and Water Conservation **40**:374–379.

Platts, W.S., and R.L. Nelson. 1988. Fluctuations in trout populations and their implications for land-use evaluation. North American Journal of Fisheries Management **8**:333–345.

Poff, N.L., and J.V. Ward. 1990. Physical habitat template of lotic systems: recovery in the context of historical pattern of spatiotemporal heterogeneity. Environmental Management **14**:629–645.

Power, M.E. 1990. Effects of fish in river food webs. Science **250**:811–814.

Quinault Indian Nation. 1990. Interim riparian management zone study 1990 report. Division of Environmental Protection, Quinault Department of Natural Resources, Tahola, Washington, USA.

Raedeke, K.J., editor. 1988. Streamside management: riparian wildlife and forestry interactions. Contribution 59, Institute of Forest Resources, University of Washington, Seattle, Washington, USA.

Reeves, G.H., F.H. Everest, and J.D. Hall. 1987. Interactions between the redside shiner (*Richardsonius balteatus*) and the steelhead trout (*Salmo gairdneri*) in western Oregon: the influence of water temperature. Canadian Journal of Fisheries and Aquatic Sciences **44**:1602–1613.

Reeves, G.H., F.H. Everest, and T.E. Nickelson. 1989. Identification of physical habitats limiting the production of coho salmon in western Oregon and Washington. United States Forest Service General Technical Report PNW-GTR-245, Pacific Northwest Research Station, Portland, Oregon, USA.

Ricker, W.E. 1980. Causes of the decrease in age and size of chinook salmon (*Oncorhynchus tshawytscha*). Canadian Technical Report of Fisheries and Aquatic Sciences 944, Ottawa, Ontario, Canada.

Ricker, W.E. 1987. Effects of the fishery and of obstacles to migration on the abundance of Fraser River sockeye salmon (*Oncorhynchus nerka*). Canadian Technical Report of Fisheries and Aquatic Sciences 1522, Ottawa, Ontario, Canada.

Robison, E.G., and R.L. Beschta. 1990. Identifying trees in riparian areas that can provide coarse woody debris to streams. Forest Science **36**:790–801.

Rood, K.M. 1984. An aerial photograph inventory of the frequency and yield of mass wasting on the Queen Charlotte Islands, British Columbia. British Columbia Ministry of Forests, Land Management Report 34, Victoria, British Columbia, Canada.

Salo, E.O., and C.J. Cederholm. 1981. Cumulative effects of forest management on watersheds: some aquatic considerations. Pages 67–78 *in* The Edgebrook Conference: Cumulative Effects of Forest Management on California Watersheds. Division of Agricultural Sciences, University of California, Berkeley, California, USA.

Salo, E.O., and T.W. Cundy, editors. 1987. Streamside management: forestry and fishery interactions. Contribution 57, Institute of Forest Resources, University of Washington, Seattle, Washington, USA.

Scarnecchia, D.L. 1981. Effect of streamflow and upwelling on yield of wild coho salmon (*Oncorhynchus kisutch*) in Oregon. Canadian Journal of Fisheries and Aquatic Sciences **38**:471–475.

Schlosser, I.J. 1982. Fish community structure and function along two habitat gradients in a headwater stream. Ecological Monographs **52**:395–414.

Schlosser, I.J. 1985. Flow regime, juvenile abundance, and the assemblage structure of stream fishes. Ecology **66**:1484–1490.

Schlosser, I.J. 1987. A conceptual framework for fish communities in small warmwater streams. Pages 17–24 *in* W.J. Matthews and D.C. Heins, editors. Community and evolutionary ecology of North American stream fishes. University of Oklahoma Press, Norman, Oklahoma, USA.

Schlosser, I.J. 1991. Stream fish ecology: a landscape perspective. BioScience **41**:704–712.

Sedell, J.R., and R.L. Beschta. 1991. Bringing back the "bio" in bioengineering. American Fisheries Society Symposium **10**:160–175.

Sedell, J.R., P.A. Bisson, F.J. Swanson, and S.V. Gregory. 1988. What we know about large trees that fall into streams and rivers. Pages 47–81 *in* C. Maser, R.F. Tarrant, J.M. Trappe, and J.F. Franklin, editors. From the forest to the sea: a story of fallen trees. United States Forest Service General Technical Report PNW-GTR-229.

Sedell, J.R., and W.S. Duval. 1985. Water transportation and storage of logs. Pages 1–68 *in* W.R. Meehan, editor. Influence of forest and rangeland management on anadromous fish habitat in western North America. United States Forest Service General Technical Report PNW-186, Portland, Oregon, USA.

Sedell, J.R., and F.H. Everest. 1991. Historic changes in pool habitat for Columbia River Basin salmon under study for TES listing. United States Forest Service General Technical Report, Pacific Northwest Research Station, Portland, Oregon, USA.

Sedell, J.R., and K.J. Luchessa. 1982. Using the historical record as an aid to salmonid habitat enhancement. Pages 210–223 *in* N.B. Armantrout, editor. Acquisition and utilization of aquatic habitat inventory information. Western Division, American Fisheries Society, Portland, Oregon. The Hague Publishing, Billings, Montana, USA.

Smoker, W.A. 1955. Effects of streamflow on silver salmon production in western Washington. Dissertation. University of Washington, Seattle, Washington, USA.

Solazzi, M.F. 1984. Relationships between visual counts of coho, chinook, and chum salmon from spawning surveys and the actual number of fish present. Oregon Department of Fish and Wildlife, Fish Division Information Report Series, Fisheries Number 84–7, Portland, Oregon, USA.

Sullivan, K. 1986. Hydraulics and fish habitat in relation to channel morphology. Dissertation. Johns Hopkins University, Baltimore, Maryland, USA.

Sullivan, K., T.E. Lisle, C.A. Dolloff, G.E. Grant, and L.M. Reid. 1987. Stream channels: the link between forests and fishes. Pages 39–97 *in* E.O. Salo and T.W. Cundy, editors. Streamside management: forestry and fishery interactions. Contribution 57, Institute of Forest Resources, University of Washington, Seattle, Washington, USA.

Swanson, F.J., S.V. Gregory, J.R. Sedell, and A.G. Campbell. 1982. Land-water interactions: the riparian zone. Pages 267–291 *in* R.L. Edmonds, editor. Analysis of coniferous forest ecosystems in the western United States. Hutchinson Ross, Stroudsburg, Pennsylvania, USA.

Swanson, F.J., and G.W. Lienkaemper. 1978. Physical consequences of large organic debris in Pacific Northwest streams. United States Forest Service General Technical Report PNW-69, Pacific Northwest Forest and Range Experiment Station, Portland, Oregon, USA.

Swanson, F.J., G.W. Lienkaemper, and J.R. Sedell. 1976. History, physical effects, and management implications of large organic debris in western Oregon streams. United States Forest Service General Technical Report PNW-56, Pacific Northwest Forest and Range Experiment Station, Portland, Oregon, USA.

Thompson, W.F. 1945. Effects of the obstruction at Hell's Gate on the sockeye salmon of the Fraser River. International Pacific Salmon Fisheries Commission, Bulletin 1, New Westminster, British Columbia, Canada.

Toews, D.A.A., and M.K. Moore. 1982. The effects of streamside logging on large organic debris in Carnation Creek. Province of British Columbia, Ministry of Forests, Land Management Report 11, Vancouver, British Columbia, Canada.

Tripp, D.B., and V.A. Poulin. 1986*a*. The effects of mass wasting on juvenile fish habitats in streams on the Queen Charlotte Islands. Land Management Report 45, British Columbia Ministry of Forestry and Lands, Victoria, British Columbia, Canada.

Tripp, D.B., and V.A. Poulin. 1986*b*. The effects of logging and mass wasting on salmonid spawning habitat in streams on the Queen Charlotte Islands. Land Management Report 50, British Columbia Ministry of Forestry and Lands, Victoria, British Columbia, Canada.

Triska, F.J., and K. Cromack, Jr. 1980. The role of wood debris in forests and streams. Pages 171–190 *in* R.H. Waring, editor. Forests: fresh perspectives from ecosystem analysis. Proceedings of the 40th Annual Biology Colloquium. Oregon State University Press, Corvallis, Oregon, USA.

Triska, F.J., J.R. Sedell, K. Cromack, Jr., S.V. Gregory, and F.M. McCorison. 1984. Nitrogen budget for a small coniferous forest stream. Ecological Monographs **54**:119–140.

Triska, F.J., J.R. Sedell, and S.V. Gregory. 1982. Coniferous forest streams. Pages 292–332 *in* R.L. Edmonds, editor. Analysis of coniferous forest ecosystems in the western United States. Hutchinson Ross, Stroudsburg, Pennsylvania, USA.

United States Army Corps of Engineers. 1989. Annual fish passage report, 1988. Columbia River projects and Snake River projects. USACE, Portland and Walla Walla Districts, Portland, Oregon, USA.

Walters, C.J., J.S. Collie, and T. Webb. 1988. Experimental designs for estimating transient responses to management disturbances. Canadian Journal of Fisheries and Aquatic Sciences **45**:530–538.

8

Sensitivity of the Regional Water Balance in the Columbia River Basin to Climate Variability: Application of a Spatially Distributed Water Balance Model

JAYNE DOLPH, DANNY MARKS, AND GEORGE A. KING

Abstract

A one-dimensional water balance model was developed and used to simulate the water balance for the Columbia River Basin. The model was run over a 10 km digital elevation grid representing the U.S. portion of the basin. The regional water balance was calculated using a monthly time step for a relatively wet year (1972 water year), a relatively dry year (1977 water year), and a double ($2xCO_2$) climate scenario. Input data, spatially distributed over the grid, included precipitation, maximum soil moisture storage capacity, potential evapotranspiration (PET), and threshold baseflow. The model output provides spatially distributed surfaces of actual evapotranspiration (ET), runoff, and soil storage. Model performance was assessed by comparing modeled ET and runoff with the input precipitation data, and by comparing modeled runoff with measured runoff. The model reasonably partitions incoming precipitation to evapotranspiration and runoff. However, modeled total annual runoff was significantly less than measured runoff, primarily because precipitation is underestimated by the network of measurement stations and because of limitations associated with the interpolation procedure used to distribute the precipitation across the grid. Estimated precipitation is less than measured runoff, a physical impossibility. Under warmer $2xCO_2$ climate conditions (January 4.0°K warmer, July 6.5°K warmer), the model predicts that PET increases by about 80%, ET increases, and runoff and soil moisture decrease. Under these climate conditions, the distribution and composition of forests in the region would change dramatically, and water resources would become more limited.

Key words. Regional water balance, runoff, evapotranspiration, soil moisture, climate change.

Introduction

Large changes may occur in global climate during the next 50 to 100 years, driven by increases in atmospheric carbon dioxide (CO_2) and other radiatively important trace gases (Keeling 1973, Manabe and Wetherald 1975, 1980; Keeling and Bacastow 1977, Ramanathan et al. 1985, Dickinson and Cicerone 1986, Hansen et al. 1988, Houghton et al. 1990). Global climate change would affect the terrestrial biosphere through changes in the regional energy balance (Dickinson 1983) and associated changes in the regional water balance (Strain 1985, Eagleson 1978, 1982, 1986, Lettenmaier and Burges 1978, Smith and Tirpak 1989, Lettenmaier and Gan 1990, Lettenmaier and Sheer 1991). In particular, changes in soil moisture and evapotranspiration could have a considerable impact on water and forest resources, since the distribution and abundance of these resources are controlled largely by the seasonality of moisture availability (Whittaker 1975, Mather and Yoshioka 1968, Neilson et al. 1989, Stephenson 1990). Major shifts in vegetation patterns and condition could occur as a result of climate change (Perrier 1982, Eagleson and Segarra 1985, Solomon 1986, Prentice and Fung 1990).

Changes in the regional water cycle could also influence feedback between vegetation and climate as described in detail by Hansen et al. (1984), Dickinson and Hanson (1984), and Rind (1984). These would influence both magnitude and timing of climate change by altering the surface albedo and radiation balance, soil moisture storage, and evapotranspiration. In sum, an altered regional water balance will play a key role in future climate-biosphere interactions.

Existing hydrologic models designed to operate over large river basins or regions are not adequate to analyze the impacts of climate change on the regional water balance. These models are intended to predict riverflow in response to index precipitation data (e.g., Anderson 1973, Burnash et al. 1973). They are spatially lumped models that do not predict the distribution of soil moisture, evaporation, or runoff over the basin. Model parameterization and calibration are not easily extended from catchment to river basin to large regional scales (Dooge 1986), even for prediction of runoff from the basin.

Consequently, new hydrologic models must be developed which simulate the spatial magnitude and extent of hydrologic processes and properties, such as precipitation, soil moisture, and evaporation, in response to changing climate conditions. This type of modeling is required particularly for simulating regional vegetation response to climate change at broad spatial scales, since vegetation distribution is controlled in large part by soil moisture.

In the experiment described in this article, we develop a spatially distributed water balance model over the Columbia River Basin in the Pacific Northwest, USA (Figure 8.1). Water balance calculations are made over a regular digital elevation grid with a 10 km spacing representing the U.S. portion of the basin. The specific objectives of the experiment are (1) to evaluate our ability to estimate the distribution of precipitation, evaporative

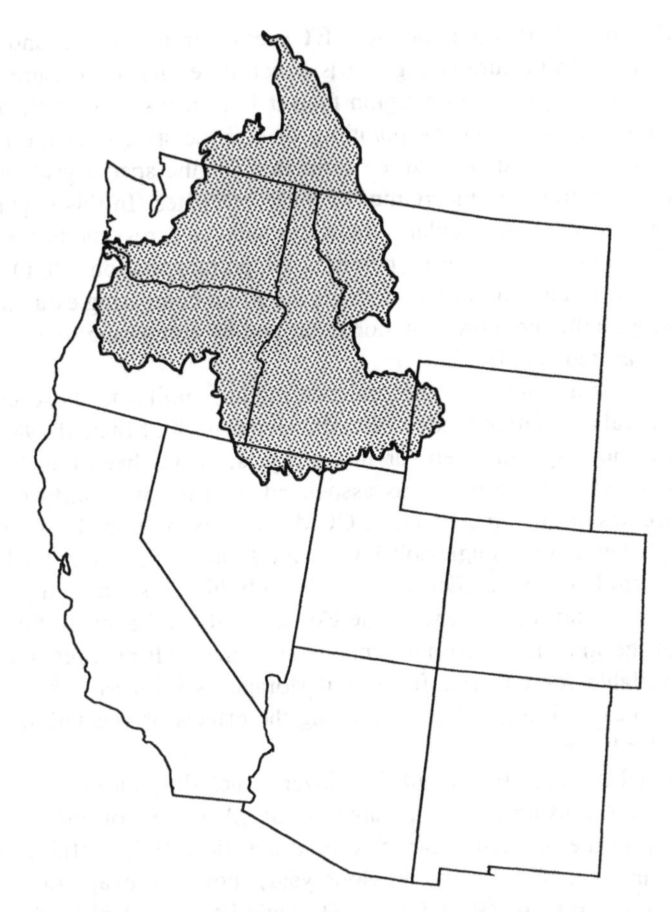

FIGURE 8.1. Location of Columbia River Basin in the Pacific Northwest, USA. The outflow of the basin is at the Washington-Oregon border at the Pacific coast.

demand, and soil moisture at the regional scale, (2) to estimate the sensitivity of runoff, soil moisture, and evapotranspiration to variations in both the magnitude and distribution of these parameters, during a wet year and a dry one, (3) to evaluate the sensitivity of soil moisture, evapotranspiration, and runoff over the Columbia River Basin to climate change conditions as simulated by a general circulation model (GCM) for double CO_2 ($2xCO_2$) concentrations, and (4) to discuss the potential effects of climate change on water and forest resources in the river basin.

Methods

A Simple One-Dimensional Water Balance Model

The water balance at a point can be represented as:

$$RO = P + ET + \Delta St \quad (1)$$

where RO is runoff, P is precipitation, ET is evapotranspiration, and ΔSt is the change in soil moisture storage. It is generally easier to measure runoff or streamflow from a basin or region than it is to measure or estimate the distribution of precipitation, evaporation, or soil moisture over the region. However, if P, ET, and ΔSt can be estimated for the spatial grid, we can predict RO to within the uncertainty of these estimates. In this experiment we develop a model that calculates the water balance across the basin, using precipitation and an estimate of potential evapotranspiration (PET) as the model drivers. Evaluation of the model is based on comparing estimated ET and runoff with the input precipitation data, and by comparing modeled runoff with measured runoff.

The model is a "bucket" water balance model, similar to those used by several general circulation models (GCMs) (Jenne 1990, Mitchell 1983), and is run at a monthly time step for this analysis. It has been developed to overcome several of the problems associated with the way surface hydrologic processes are parameterized in GCMs (Eagleson 1986) by simulating the water balance for a single soil layer with drainage (a "leaky bucket" or "bucket" with baseflow model) over a relatively high resolution digital grid (10 km grid spacing) representing the Columbia River Basin. In this initial version of the model, we do not simulate snowmelt; all precipitation is assumed available for ET, runoff, or soil storage as soon as it falls. Also, there are no explicit algorithms estimating the effects of vegetation canopy on the water balance.

The model requires five input data layers, including an estimate of the maximum soil moisture retention capacity (St_{max}), initial soil moisture storage (St_i) from the previous year, threshold baseflow (Bf_{th}, defined here as the September runoff from the previous year), potential evapotranspiration (PET), and precipitation (P). Of these, St_{max} and Bf_{th} are established as initial conditions, and held constant for the model run. PET and P must be estimated for each time step in the model run, and St_i is recalculated at each time step and passed to the next time step.

Input Data Surfaces for Current Climate Conditions

The data used for this analysis were organized into a geographic information system (GIS; Campbell et al. 1990). This system is compatible with a variety of both raster and vector GIS software formats, but this analysis was done using only raster data and processing tools. Most of the processing was done using Image Processing Workbench (IPW; developed by Frew 1990). Data projection, management, and display were done using the GRASS GIS software system, Version 3.2, which was acquired from the U.S. Army Corps of Engineers (USACE 1988). The model was run initially for only the U.S. portion of the basin, because, at the time of this analysis, digital data sets were unavailable for the Canadian portion.

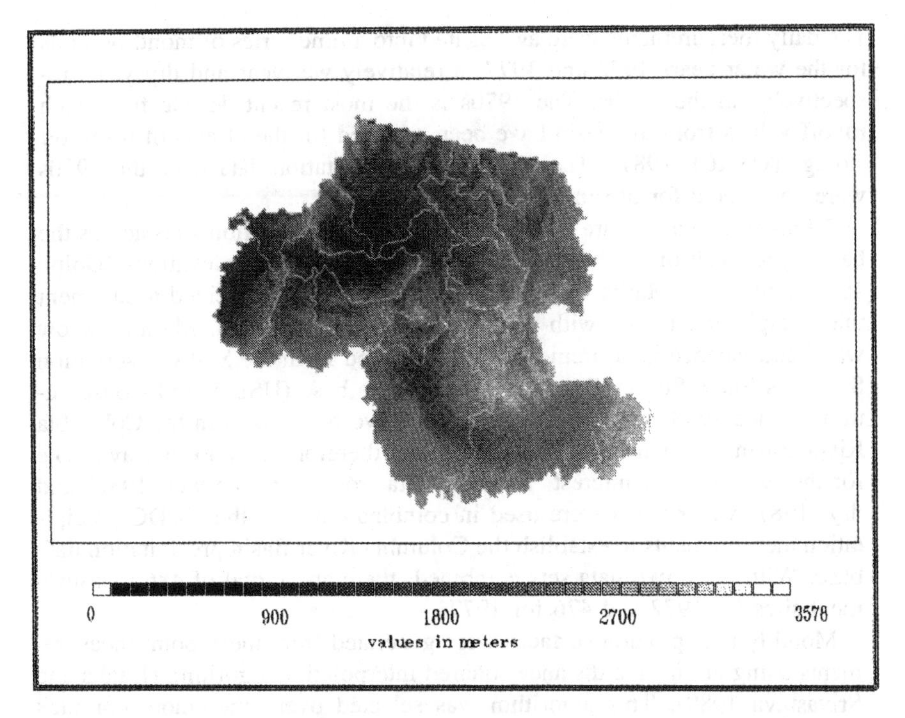

FIGURE 8.2. Relief map of Columbia River Basin sampled at a five-minute latitude-longitude grid spacing (approximately 10 km) (NGDC 1989). Highest elevations are lightest shading. Major subbasins are outlined in white.

Digital Elevation Grid

The base layer used in the analysis is a digital elevation model (DEM) of the continental United States sampled at a five-minute latitude-longitude spacing. It was acquired from the NOAA National Geophysical Data Center in Boulder, Colorado, on CD-ROM (NOAA 1989). This data base was re-projected into an Albers Equal-Area Conic projection (Snyder and Voxland 1989) at a uniform 10 km grid spacing. This is a grid of 110 rows by 120 columns, with 13,200 grid points. When the Columbia River Basin is used to mask the file, 5,701 grid points fall within the area used for this analysis. The minimum elevation is sea level, at the outflow of the basin, and the maximum elevation is 3,578 m, located in the northern Cascades near the summit of Mount Rainier. Figure 8.2 is a shaded relief map of this data layer, showing the distribution of relief and topography across the Columbia River Basin. The DEM was used to create the elevation-corrected temperature and humidity surfaces used in this analysis.

Precipitation Data

Daily precipitation measurements were extracted from the EarthInfo climate data base on CD-ROM, which is a compilation of historical climatological data from the National Climatological Data Center (NCDC) (EarthInfo 1988).

The daily measurements were aggregated into a time series of monthly totals for the water years 1972 and 1977, a relatively wet year and dry year, respectively, in the 1970s. The 1970s is the most recent decade for which runoff values from the basin have been adjusted for the effects of reservoir storage (USACE 1981). Therefore, only precipitation data from the 1970s were considered for use in this analysis.

Precipitation gauges are primarily located at low elevation sites across the basin, yet much of the basin's precipitation falls at high elevations (Dolph 1990, Dolph and Marks 1991). For this reason, it was decided to augment the precipitation record with snow water equivalence data. Monthly snow water equivalence measurements were obtained from the Soil Conservation Service's Snow Survey Program (Snotel) data base (USDA Soil Conservation Service 1988). Snotel measurements have been taken in the Columbia River Basin only since the late 1970s, and therefore data are not available for the two years of interest. Monthly data from surrogate wet (1986) and dry (1987) water years were used in combination with the NCDC precipitation measurements to establish the Columbia River Basin precipitation data base. With these two data sets combined, there are a total of 486 measurement sites for 1972 and 476 for 1977.

Monthly precipitation surfaces were generated from these point measurements using an inverse distance squared interpolation algorithm (Isaaks and Srivastava 1989). This algorithm was selected over other more complex methods such as spline interpolation (Eubank 1988) because it is numerically simple, relatively fast computationally, and an exact interpolation method (the interpolated surfaces are constrained to go through the input data points).

Calculating Potential Evapotranspiration

The energy balance of the earth's surface is expressed as

$$\Delta Q = R_n + H + L_v E + G + M \tag{2}$$

where ΔQ is change in surface energy, and R_n, H, $L_v E$, G, and M are net radiative, sensible, latent, conductive, and advective energy fluxes, respectively. In temperature equilibrium, $\Delta Q = 0$; a negative energy balance will cool the surface, decreasing its temperature, while a positive energy balance will warm the surface. The sign convention used in this analysis is that a flux of energy or mass is negative if it is away from the surface, and positive if it is toward the surface.

In most terrestrial environments, conductive (G) and advective (M) fluxes are relatively small. Turbulent energy and mass flux at the earth's surface is second only to radiation in importance for the energy balance of terrestrial ecosystems. Over land, net radiation (R_n) is of approximately equal magnitude to sensible (H) and latent heat exchange ($L_v E$) (Budyko 1974, Baumgartner and Reichel 1975, Korzun 1978). However, because $L_v E$ is usually negative (a heat loss), and H is usually positive (a heat gain), the sum of

$H + L_vE$ is generally much smaller than R_n when integrated over a day or longer (Marks and Dozier 1991).

The turbulent transfers of momentum, heat, and water vapor described by H and L_vE are the most complicated forms of energy exchange, and are not easily measured in a natural environment. The data required to calculate them are difficult to measure at a point, and they have a highly variable distribution over a topographic surface. Not only does significant energy transfer occur by turbulent exchange, but in most environments significant water loss can occur from sublimation, direct evaporation, or transpiration by vegetative cover (Budyko 1974, Baumgartner and Reichel 1975, Korzun 1978, Beaty 1975, Stewart 1982, Davis et al. 1984).

Tractable approaches to calculating sensible and latent fluxes have been summarized (Fleagle and Businger 1980, Brutsaert 1982). The method used to calculate potential evapotranspiration over the Columbia River Basin is described in detail by Marks (1990), who used it to estimate potential evapotranspiration over the continental United States. It was adapted from Brutsaert (1982) by Marks (1988) and is similar to the approach used by Martin et al. (1990) to estimate the sensitivity of ET to climate change over several types of vegetation in North America.

A surface roughness length characteristic of a grassland was used for all grid points for these calculations. The method requires inputs of elevation-corrected temperature, humidity, and wind surfaces as described in Marks (1990).

Water Holding Capacity

Maximum soil moisture storage capacity (St_{max}), the water retained between field capacity (33 kPa) and wilting point (1,500 kPa), was estimated using texture and organic carbon data from the Soil Map of the World (SMW) (United Nations Food and Agriculture Organization 1975, Kern 1991). A digital version of the SMW provided a geographical framework to distribute the results spatially. Each soil layer was converted to an estimated water holding capacity using the relationships described by Verheye (1989) for texture and De Jong et al. (1983) for organic carbon. The average soil moisture holding capacity was calculated by weighting the soil map unit components by the percent composition. The water holding capacity of the dominant soil in each map unit was corrected for the phase, if present. Stony phases were assumed to have 35% rock fragments by volume, lithic phases to have 100 cm depth, and duripan phases to have 130 cm depth. The upper 50 cm, and from 50 to a maximum of 200 cm, were calculated as separate layers. The water holding capacity of the upper 50 cm was calculated from both texture and carbon because this layer typically has most of the carbon. The water holding capacity of the second layer was estimated from texture and assumed to extend to 200 cm depth unless otherwise indicated by phase. Stmax was set equal to the sum of the water holding capacity of both soil layers.

Input Data Surfaces for Predicted 2xCO_2 Conditions

The objective of the climate change analysis was to determine the sensitivity of the regional water balance to potential future climate conditions, and to make a preliminary assessment of whether these conditions are similar to extreme years in the recent past or are outside our range of experience. The potential regional water balance under 2xCO_2 climate conditions was compared with the regional water balance simulated for the relatively wet and dry years as a first analysis.

The challenge for this and other climate change studies is to produce reasonable estimates of future climate for the region of interest. Currently, the only way to produce quantitative and spatially distributed estimates of key climate variables under 2xCO_2 concentrations is to use output from GCMs (Dickinson 1986). GCMs are complex numerical models that simulate the fundamental physical relationships of the earth's ocean-atmosphere-land surface system (Dickinson 1986, Houghton et al. 1990).

However, regional estimates of climate change produced by GCMs are still highly uncertain and variable (Houghton et al. 1990). Several GCMs have been formulated, but differ in their parameterization of important atmospheric processes. These differences in model formulations lead to differences in their simulations of future climate conditions and their geographic distribution of predicted change. These inconsistencies are due, in part, to the unrealistic parameterization of land surface processes and coarse grid cell resolution. The difficulty in both improving model performance and making future climate estimates stems from the inherent complexity of the earth's climate system and our incomplete knowledge of how that system works (Dickenson 1986, Houghton et al. 1990).

Consequently, researchers conducting climate impact studies have focused on generating climate scenarios of plausible future climates to test the sensitivity of the resource under study to global climate change. If a large climate change produces little change in resource quality or quantity, then any actual future climate change would probably not place that resource at risk. If resource changes are large under several climate change scenarios, future climate change could place the resource at risk. Resource managers might then be advised to consider mitigation and adaptation steps to decrease the potential impacts.

Generating a climate scenario from GCM output is complicated by the fact that the actual values predicted by the 1xCO_2 GCM runs are inconsistent with historical data distribution in both space and time. This is caused by the coarse grid cell resolution of the GCMs, the lack of a realistic topography which influences parameters such as temperature, wind, humidity, and precipitation, and the relatively crude parameterization of surface properties within GCMs (Rosenzweig and Dickinson 1986, Schlesinger 1988). To deal with this problem, ratios between the 2xCO_2 values and 1xCO_2 values are used to modify historical averages of each climate variable required in the analysis. This approach to using GCM scenario data was presented by Parry et al. (1987) and Parry and Carter (1989), and was used for precipitation analysis in the EPA study of the potential impacts of climate change on the United States (Smith and Tirpak 1989).

Data from the Geophysical Fluid Dynamics Laboratory (GFDL) GCM (Manabe and Wetherald 1987), developed at NOAA, were used for the initial $2xCO_2$ sensitivity analysis. Output from the GFDL model was chosen for this test because (1) it has been used in a case study of climate change impacts on water resources in California (Smith and Tirpak 1989, Lettenmaier and Gan 1990), (2) data on projected future changes in humidity and wind speed were readily available for the model, and (3) the resolution of the model is greater than that of the GISS (Goddard Institute for Space Studies) model output (Hansen et al. 1983) used in the California case study. The GFDL resolution is 4.5° by 7.5° latitude-longitude. A detailed description of the GCM data is given by Jenne (1990).

The $2xCO_2$ PET surface calculated using the GFDL data by Marks (1990) was used in the scenario. This surface was created by modifying the long-term average temperatures for the basin (1948–87) by the ratio of $2xCO_2$/ $1xCO_2$ conditions. (The GCM ratios were not applied to the 1972 or 1977 years individually because the GCM data are from simulations of average climate conditions under double CO_2 concentrations. Applying the ratios to individual years under the current climate regime would thus be inappropriate.) The 1972 (wet year) and 1977 (dry year) precipitation data were averaged together to approximate "average" precipitation over the basin. These data were not modified for the climate change scenario, as the GFDL model predicts essentially unchanged annual precipitation for the region (Manabe and Wetherald 1987). Soil moisture storage (St_i) values for 1972 and 1977 were averaged to create the input soil moisture storage surface, and baseflow (Bfth) values for 1972 and 1977 were averaged to create the baseflow threshold surface.

Model Algorithm

In the model, all units are in depth of water (mm), and fluxes to the surface are positive, while fluxes away from the surface are negative. For example, ET is negative while P is positive, so that (P + ET) is nearly always less than P alone. (Refer to Table 8.1 for a listing of model variables.)

For a given time step, the model first estimates the available water (Av_w), and the relative soil saturation (S_s):

$$Av_w = MAX(St_i + P, 0.0) \tag{3}$$

$$S_s = Av_w / St_{max} \tag{4}$$

Actual evapotranspiration (ET) is then calculated as a function of potential evapotranspiration (PET) and relative soil saturation (S_s):

$$ET = (PET)[\exp(S_s \times K_e) - K_e] \tag{5}$$

Table 8.1. Key to water balance model variables.

Av_w	=	available water
Bf	=	actual baseflow
Bf_{th}	=	threshhold baseflow
ET	=	evapotranspiration
K_{bf}	=	baseflow decay factor (1.25)
K_e	=	evapotranspiration decay factor (-2)
Ov	=	overflow
P	=	precipitation
PET	=	potential evapotranspiration
RO	=	runoff
S_s	=	soil saturation
St_i	=	initial soil moisture storage
St_{max}	=	maximum soil moisture retention capacity (or water holding capacity)
ΔSt	=	change in soil moisture storage

The evapotranspiration decay factor K_e is -2. S_s can be greater than 1, but ET cannot exceed PET. Equation 5 (Figure 8.3) is a simple exponential decay function fashioned after the relative transpiration efficiency curves reported by Denmead and Shaw (1962). It includes effects of soil matrix suction, stomatal resistance, and root density in relation to percentage of soil moisture in the region of water available between field capacity and the wilting point (roughly between 0.3 and 15 atmospheres) (Brady 1974).

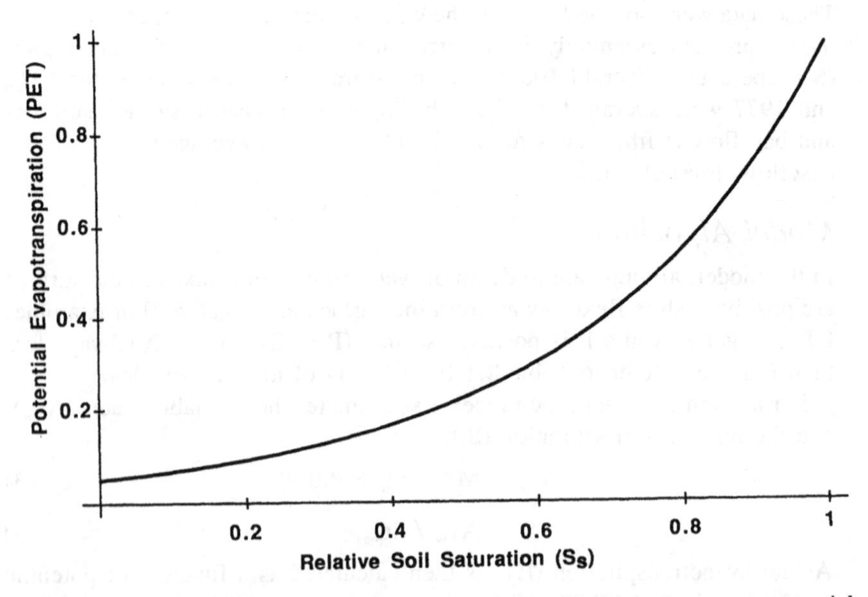

FIGURE 8.3. Relationship between percent soil saturation and percent potential evapotranspiration used in the water balance model algorithm (see text for equation).

If ET is greater than Av_w, then

$$ET = -Av_w \tag{6}$$

since ET cannot be greater than the water available for evapotranspiration.

Once calculated, ET is then removed from Av_w:

$$Av_w + ET = Av_w \tag{7}$$

and S_s is recalculated:

$$S_s = Av_w/St_{max} \tag{8}$$

Actual baseflow Bf is then calculated as a function of the threshold baseflow Bf_{th} and S_s. If S_s is greater than 1.0 (greater than maximum soil storage), Bf is allowed to increase above Bf_{th} by

$$Bf = (K_{bf})(Bf_{th}) \tag{9}$$

where the baseflow decay factor (K_{bf}) is set to 1.25. This allows actual baseflow (Bf) to vary linearly between zero and up to 125% of the threshold value.

Otherwise,

$$Bf = (S_s)(K_{bf})(Bf_{th}) \tag{10}$$

Bf cannot exceed either St_{max} or Av_w, so:

$$Bf = MIN[Av_w, MIN(Bf, St_{max})] \tag{11}$$

Overflow, Ov, is calculated as:

$$Ov = MAX[Av_w - (Bf + St_{max}), 0.0] \tag{12}$$

Runoff RO is then:

$$RO = Bf + Ov \tag{13}$$

The final soil moisture storage is set as St_i for the next iteration:

$$St_i = MAX(Av_w - RO, 0.0) \tag{14}$$

and RO, St_i, and ET are written to the output file.

Results

The model inputs and outputs are continuous data surfaces. The values for each surface have been categorized into distinct classes for display in this manuscript (e.g., Figures 8.5, 8.6). We first discuss the input data surfaces, followed by descriptions of the model output.

The detail of topographic structure shown in the shaded relief map of the DEM (Figure 8.2) illustrates the significant variability of relief and topography over the Columbia Basin. The 10 km grid spacing of the DEM cap-

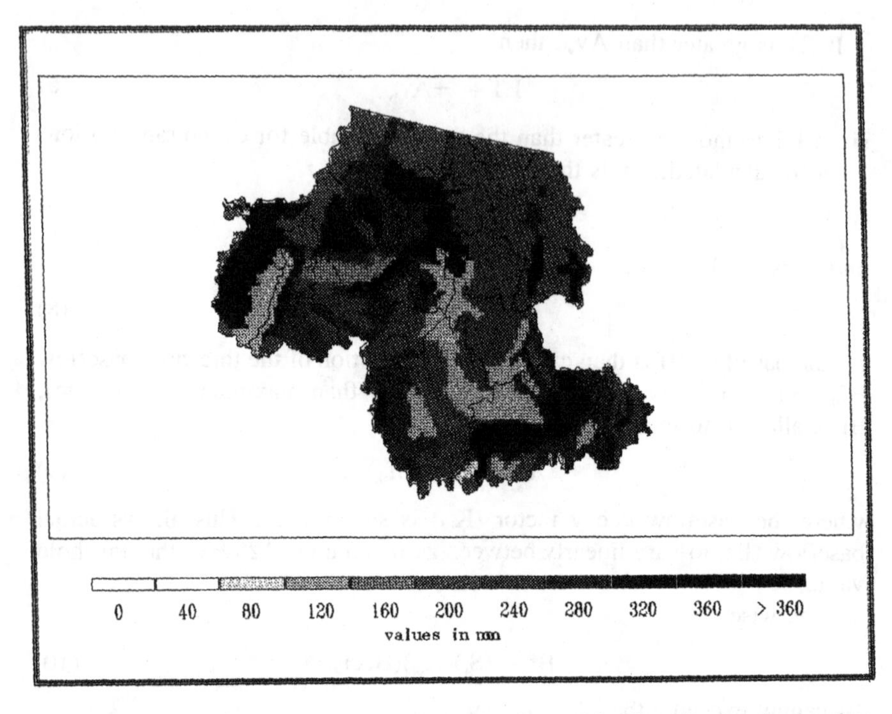

FIGURE 8.4. Soil water holding capacity for the Columbia River Basin calculated at a five-minute resolution.

tures considerable topographic detail for this large-scale analysis, including the major mountain ranges and valleys of the basin, while keeping the number of grid points computationally acceptable.

The spatial variability of soil water holding capacity (St_{max}) for the region is depicted in Figure 8.4. The average St_{max} for the basin is 236 mm, with a minimum value of 60 mm and a maximum value of 375 mm. Estimates of soil water holding capacity are greatest in the Willamette Valley, central Washington, and southern Idaho. The smallest estimates of St_{max} occur just east of the Cascade Mountains in eastern Oregon and Washington, and in central Idaho.

Threshold baseflow for both years averages 29 mm, with a range from 1 to 224 mm for 1972 and 1 to 230 mm for 1977. Threshold baseflow (Bf_{th}) is assumed to be equal to the September runoff value from the previous year, and therefore may be underestimated due to the limitations associated with the interpolation of point values of runoff.

Monthly precipitation data for 1972 and 1977 are depicted in Figures 8.5 and 8.6. The characteristic seasonality of precipitation in the Northwest is evident for both years, with the bulk of the region's precipitation falling in the winter months, and very little input from June through September. Within

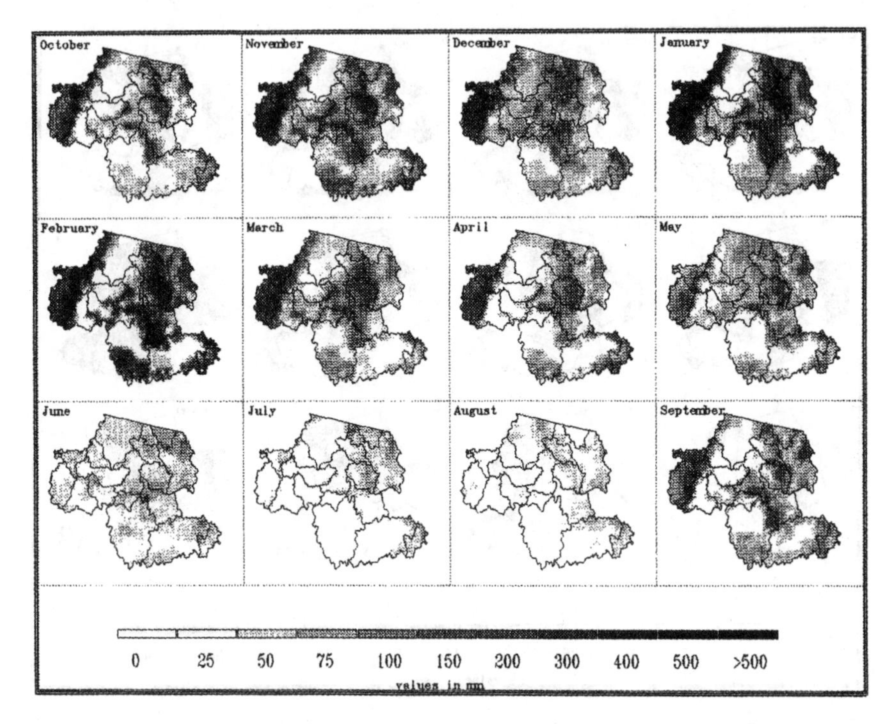

FIGURE 8.5. Monthly measured precipitation for the 1972 water year, interpolated at a five-minute resolution using an inverse distance squared interpolation algorithm. Major subbasins are outlined in black.

each month the spatial distribution is generally the same for both years. Winter months in 1972 were wetter in the basin than in 1977, particularly in January and February. However, 1977 had a wetter summer than 1972, most notably in parts of the Willamette Valley.

The annual surfaces of precipitation for 1972 and 1977 (Figure 8.7) readily illustrate the substantial differences in estimates of precipitation volume for the wet year and the dry year. Precipitation values in 1972 are higher across much of the basin. A comparison of unit runoff values from Table 8.2 reinforces this point, with measured runoff at the basin outflow approximately three times greater in 1972 than in 1977. The highest precipitation estimates occur in the Willamette Valley, just west of the Cascade Mountains. The rain shadow effect of the Cascades is a notable feature, with precipitation estimates just east of the Cascades among the lowest in the basin.

PET appears generally unchanged in distribution between wet and dry years, and in total is little different between years (Figure 8.8; Table 8.2). This is not surprising, as the same average wind and humidity data were used in both years, and summer temperatures were not very different between years (Table 8.3). Differences between the modeled values of ET for

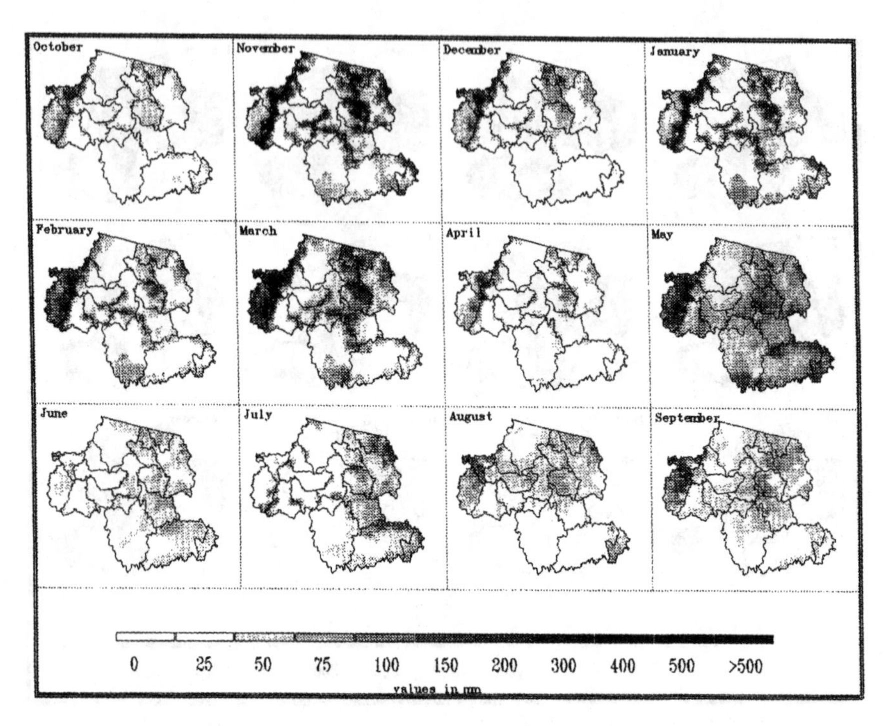

FIGURE 8.6. Monthly measured precipitation for the 1977 water year, interpolated at a five-minute resolution using an inverse distance squared interpolation algorithm. Major subbasins are outlined in black.

the wet and dry years are therefore primarily a function of differences in the magnitude of precipitation input, which determines the amount of water available to meet the evaporative demand. Precipitation input for the 1972 model run was greater than input for the 1977 run during most months.

The GFDL model predicts significant regional warming under $2xCO_2$ conditions, with January temperatures an average 4.0°K warmer, and July temperatures 6.5°K warmer. PET increases significantly by about 80% over the 1972 and 1977 values (Table 8.2).

Output from the model simulations for 1972, 1977, and $2xCO_2$ 1977 conditions are displayed in Figures 8.9 through 8.13. Relevant summary statistics are presented in Table 8.2. The water balance model predicts greatest runoff where precipitation is greatest, and less runoff in the dry year (1977) than in the wet year (1972) (Figure 8.10; Table 8.2). Evapotranspiration is similarly greatest where precipitation is greatest, and lower in the dry year (Figure 8.9; Table 8.2). Soil moisture storage increased slightly during 1972, and decreased slightly in 1977.

One aspect of the climate change scenario used in this analysis that deserves mention is that both 1972 and 1977 were 2°C cooler than the long-

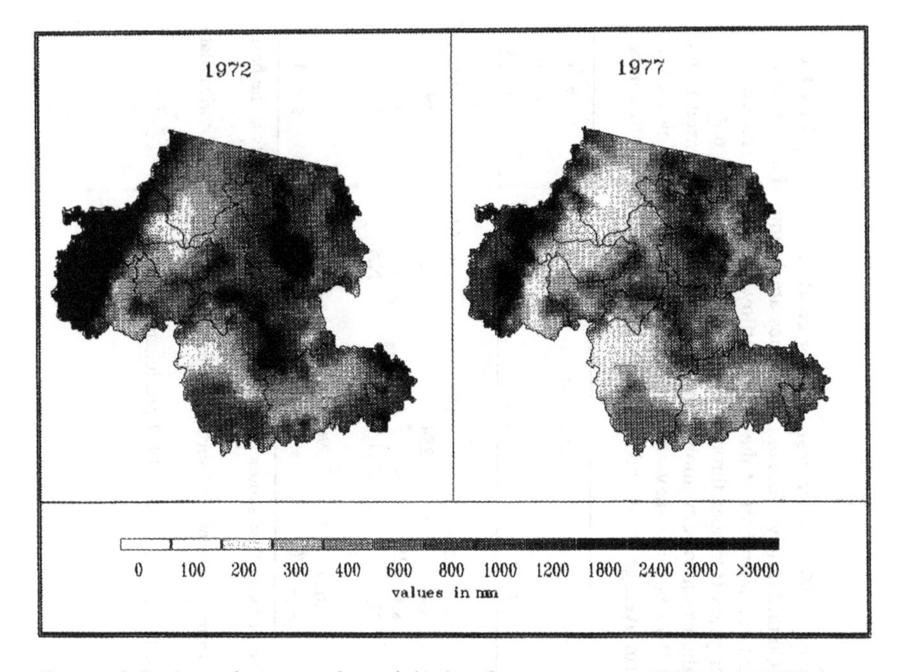

FIGURE 8.7. Annual measured precipitation for water years 1972 and 1977, interpolated at a five-minute resolution using an inverse distance squared interpolation algorithm.

term mean annual temperature (Table 8.3) to which the GCM $2xCO_2/1xCO_2$ temperature rates were applied. Thus the $2xCO_2$ scenario is even warmer compared to 1972 and 1977 than to the long-term average temperature for the basin.

PET is the principal climate driver controlling potential future changes in ET and runoff in this climate scenario. As mentioned, PET increases by about 80% over the 1972 and 1977 values (Figure 8.11; Table 8.2). The increased atmospheric demand for moisture results in higher ET (Figure 8.13), with ET becoming a larger fraction of precipitation (60% versus 40% for 1972 and 50% for 1977). Runoff is about the same as the dry year (Figure 8.12), while soil moisture drops by over 50% compared to the 1972 and 1977 values. In sum, $2xCO_2$ climate conditions as simulated here would be considerably drier across the basin than today.

Discussion

Evaluating Model Performance

In evaluating model performance we compare model output with the input precipitation data to see if the model is internally consistent, and we compare modeled and measured data. Given the precipitation input volume and spa-

Table 8.2. Annual water balance results for a very wet (1972) and a very dry (1977) water year, and for conditions predicted by the GFDL GCM for 2 × CO_2 climate conditions for the U.S. portion of the Columbia River Basin. All values are in mm H_2O per unit area, so they represent an average depth of water over the basin. Measured annual runoff at the basin outflow has been corrected to reflect only discharge from the U.S. portion. Annual values refer to water years (October through September). NA = data not available or not applicable. The standard deviation(s) is used to indicate the extent of deviation from the basin average reported in the table. No standard deviation is given for measured runoff from 1972 and 1977 because they are derived from single values measured at the basin outflow. A standard deviation has been given for the long-term average measured runoff because it is based on 40 annual values.

Year	Measured Annual Precipitation	Measured Annual Runoff	Modeled Annual Runoff	Annual PET	Modeled Annual ET	Soil Initial Storage	Soil Final Storage
Wet year	776	1,447*	437	878	311	65	93
1972	s = 547		s = 475	s = 315	s = 151	s = 61	s = 60
Dry year	507	332*	259	898	254	65	59
1977	s = 377		s = 295	s = 325	s = 150	s = 57	s = 63
Long-term	NA	741[†]	NA	NA	NA	NA	NA
average		s = 490					
GFDL $2xCO_2$	636[‡]	NA	276	1,627	396	63	27
scenario	s = 543		s = 319	s = 470	s = 215	s = 57	s = 37

* Annual runoff over the U.S. portion of the Columbia River Basin (Canadian portion of basin flow subtracted out) from gauge measurements at the basin outflow, adjusted for storage effects (USACE 1981).

† Forty-year average unit runoff for the U.S. portion of the Columbia River Basin, from an interpolated surface of runoff calculated using historical data from Wallis et al. (1991).

‡ Average precipitation for the U.S. portion of the Columbia River Basin calculated from the 1972 and 1977 precipitation data.

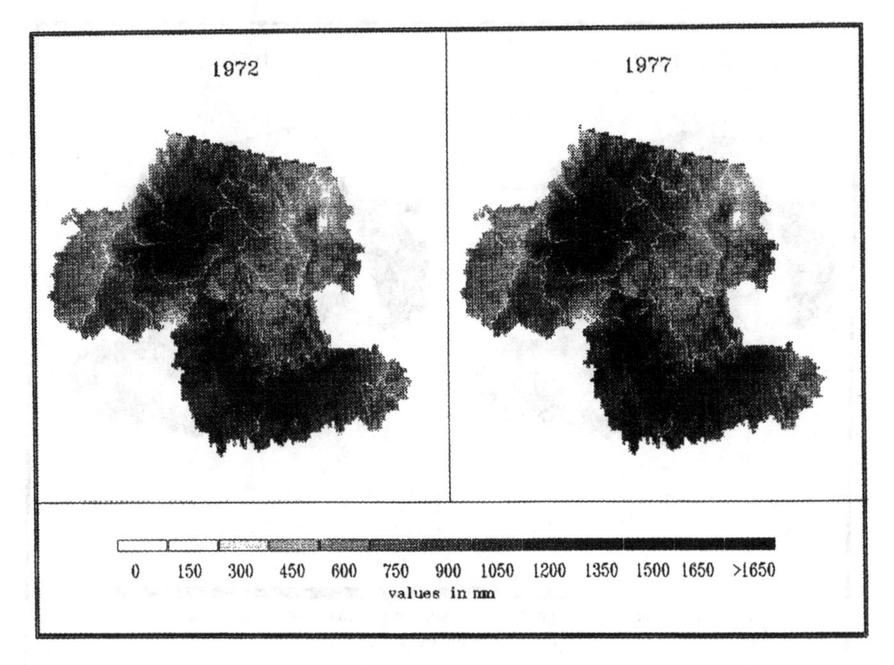

FIGURE 8.8. Annual potential evapotranspiration for water years 1972 and 1977 calculated using a turbulent transfer model and interpolated at a five-minute resolution using an inverse distance squared interpolation algorithm.

tial distribution, the model makes a reasonable partitioning of that available water into runoff (56% in 1972, 48% in 1977) and evapotranspiration (40% in 1972, 50% in 1977). These values fall into the ranges estimated for temperate systems (Eagleson 1986, McNaughton 1986).

Comparing modeled versus measured ET is problematic because measur-

Table 8.3. Elevation-corrected average temperature data for the U.S. portion of the Columbia River Basin. Values in degrees Celsius per unit area. Procedure for elevation correction described in Marks (1990). Temperature data from 85 Historical Climatology Network stations in the Columbia River Basin (Quinlan et al. 1987, Karl et al. 1990).

Year	January	July	Mean Annual Temperature*
1972	-8.8	13.4	2.0
1977	-9.7	13.4	2.0
40-year average (1948–87)	NA	NA	4.0

* Average of monthly mean temperatures.
NA = data not available.

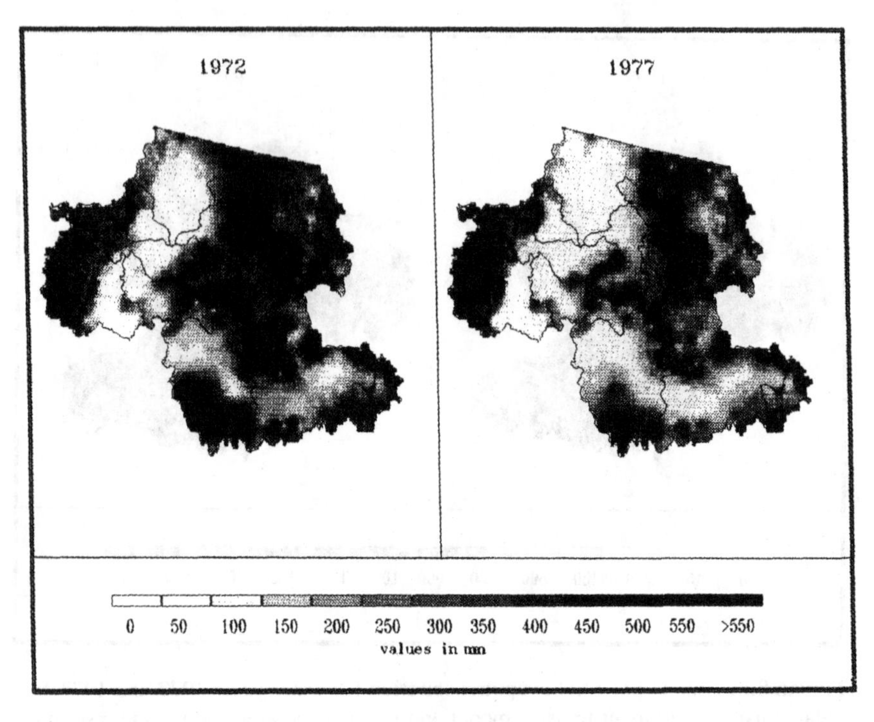

FIGURE 8.9. Annual modeled evapotranspiration for water years 1972 and 1977.

ing ET is very difficult, especially for a region or large river basin. However, a comparison can be made with other modeled values of ET for parts of the basin. Running et al. (1989) simulated ET in a 1,540 km^2 region surrounding Flathead Lake in northwestern Montana, which is in the Columbia River Basin. Evapotranspiration was estimated for two gauged watersheds in their study area as 28% and 40% of measured precipitation. Their modeled values for ET range from 25 to >55 cm/yr, with most of the region between 40 and 55 cm/yr. Our values range from 22 to 43 cm/yr in 1972, and 33 to 44 cm/yr in 1977 for the same region within the Columbia River Basin, indicating that our estimates tend to be somewhat lower than those of Running et al. (1989).

In comparing modeled and measured data, the water balance model predicts considerably less runoff than that measured at the outlet of the basin after accounting for the runoff originating in the Canadian portion of the watershed (Table 8.2). On average, the Canadian side of the basin supplies about 30% of the total basin runoff. Our estimate of total precipitation volume over the basin is greater than the total volume of measured runoff for the U.S. portion of the basin in the dry year (1977). However, in the wet year (1972) more water flows out of the basin than is accounted for by our

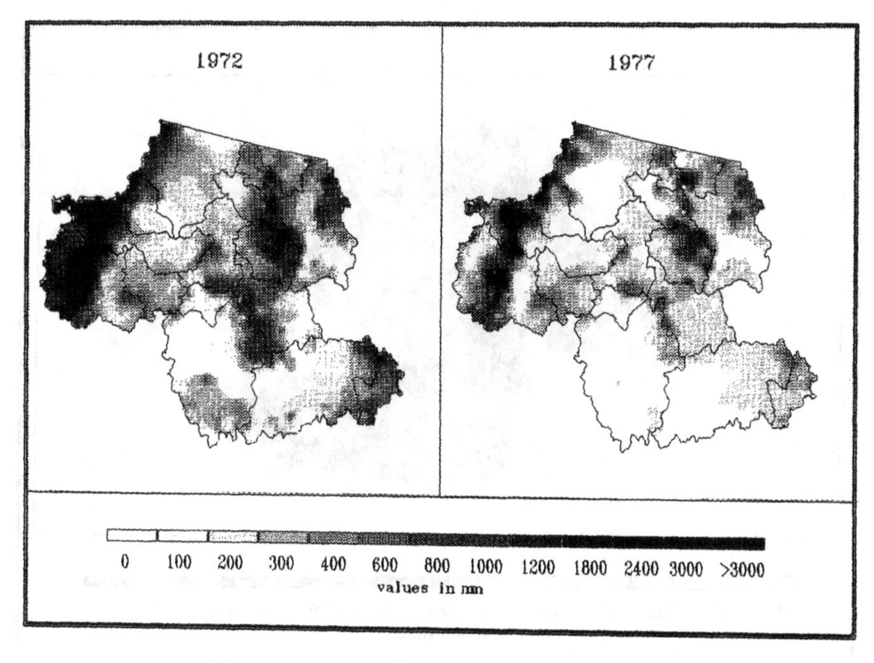

FIGURE 8.10. Annual modeled runoff for water years 1972 and 1977.

precipitation estimates (Table 8.2). It is important to note that for neither 1972 nor 1977 are our estimates of precipitation large enough to account for both runoff and the evaporative demand.

The most critical factor accounting for the low values of modeled runoff is the underestimation of precipitation input, due to the sparse spatial distribution and density of measurement sites and the associated limitations in the methods used for interpolating the point precipitation measurements across the basin (Dolph 1990, Dolph and Marks 1991). Precipitation measurements at high elevations, which receive a disproportionate amount of the regional precipitation, are especially lacking. This limits the application of the model in regions of high topographic diversity such as the Columbia River Basin. While the inclusion of information on snow water equivalence from high elevation Snotel sites improves precipitation estimates by increasing the elevational range and spatial distribution of sites, the simple interpolation technique cannot capture orographic influences on precipitation distribution across the large, topographically diverse Columbia River Basin. Thus far, techniques that are computationally feasible have not been developed to interpolate precipitation based on precipitation-elevation relationships across regions this large (Phillips et al. 1991). Ultimately, spatially distributed precipitation estimates that account for the spatial arrangement of measurement sites, precipitation-elevation relationships, and topographic constraints

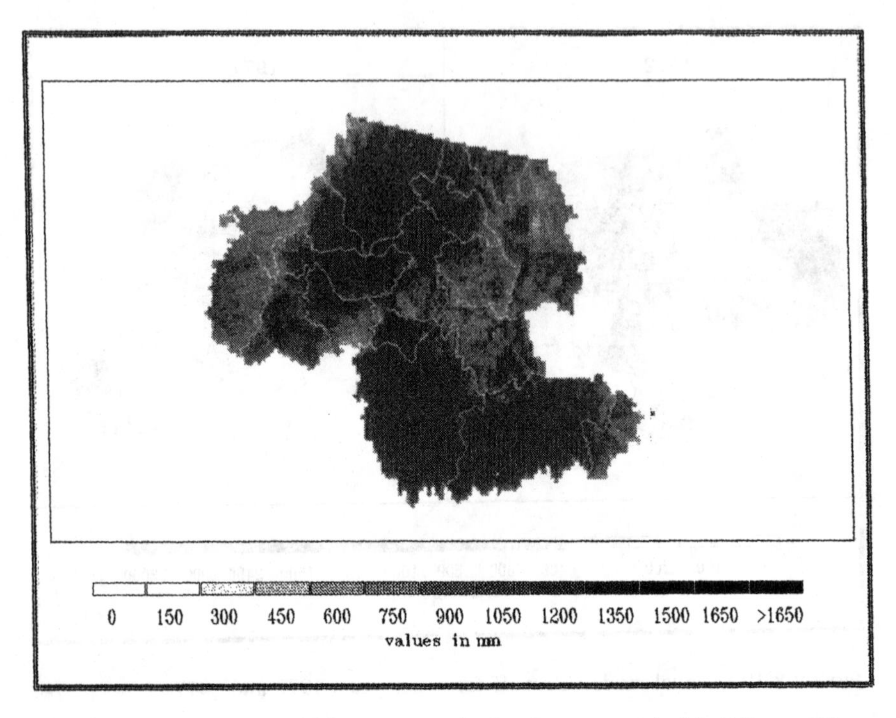

FIGURE 8.11. Annual potential evapotranspiration for average double CO_2 conditions as predicted by the GFDL model (Manabe and Wetherald 1987), calculated using a turbulent transfer model.

on precipitation distribution will be required (i.e., an orographic precipitation model).

In areas of relatively low relief (e.g., eastern Washington and Oregon), orographic influences on precipitation are minimal and the precipitation data should be more representative of the moisture that actually reaches the ground. Therefore, the model results should be less biased by the input data, and differences between modeled runoff and measured runoff should be primarily the result of model limitations. Detailed comparisons of measured and modeled runoff in these areas will be made in subsequent analyses.

Model Limitations and Future Enhancements

There are several assumptions inherent in the water balance model as it has been developed for this experiment. Resulting limitations for the application of this model and future improvements to model structure are discussed below.

The water balance model does not partition precipitation into snow and rain, and thus does not simulate snowmelt. It cannot resolve the temporal variability of precipitation-runoff relationships. The model assumes that pre-

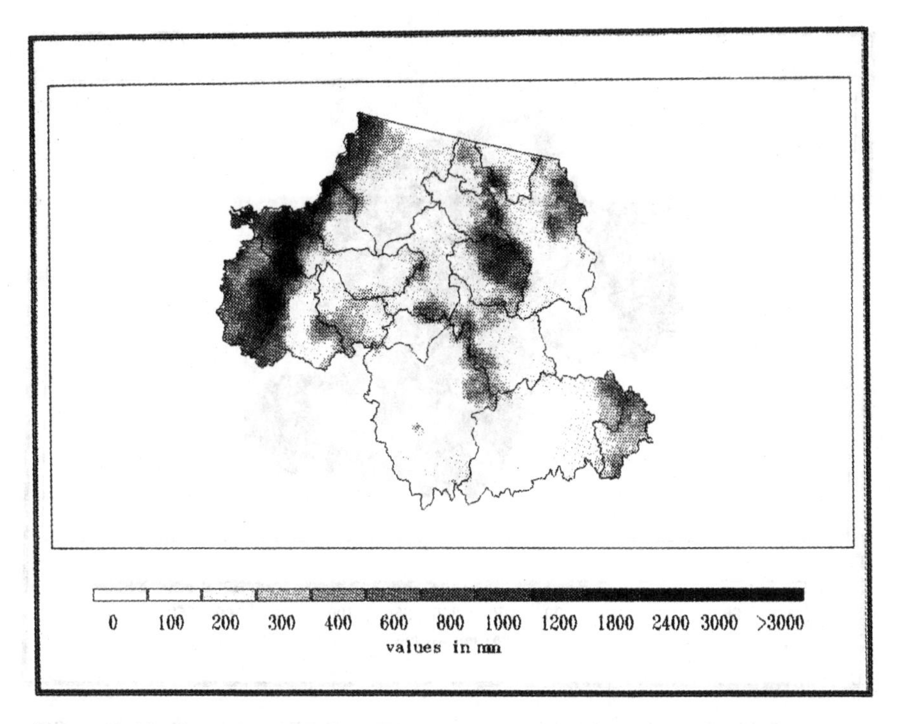

FIGURE 8.12. Annual modeled runoff calculated using the double CO_2 GFDL potential evapotranspiration estimates and average precipitation estimates.

cipitation may contribute to runoff immediately after the evaporative demand for the month is satisfied. For this reason, we have presented only annual results, although the model is run at a monthly time step. While we did include snow data in our precipitation data base, this will not in itself account for snow accumulation and melt, nor resolve the temporal variability of precipitation-runoff relationships. We will need to develop a snow accumulation/depletion model to supplement precipitation data availability and increase our spatial estimates of precipitation. In future work, a snowmelt algorithm will be added and the model will be run at a daily time step to allow weekly, monthly, and seasonal analyses. Initially, we will account for snowmelt runoff using a degree-day method, where temperature will be used as an index for snowmelt (Linsley et al. 1982). Ideally, we would like to modify and incorporate a spatially distributed, more physically based snowmelt model such as that described by Marks (1988).

Soil moisture holding capacity has been calculated for soil profiles extending to a depth of 200 cm, and the model assumes no partitioning of water holding capacity into layers. In future work, we will develop a multilayer model which will include water holding capacities calculated for each layer of the soil and variable ET depletion functions based on the soil char-

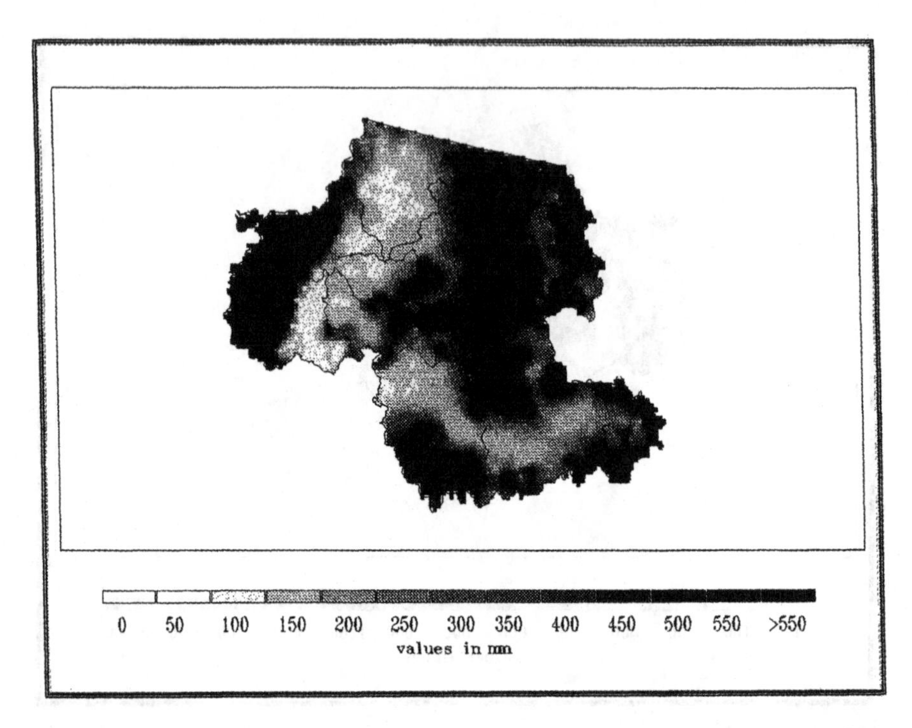

FIGURE 8.13. Annual modeled evapotranspiration calculated using the double CO_2 GFDL potential evapotranspiration estimates and average precipitation estimates.

acteristics and vegetation cover type. The upper layer will essentially function as the current single-bucket model. The lower layers will include a deep rooting zone where vegetation canopy type will affect moisture depletion by damping the response to runoff. The baseflow function will also be adapted to the multilayer model such that baseflow may increase, decrease, or become zero in the upper layer, but will be relatively constant in the lower layer, feeding spring and summer runoff when precipitation is lowest.

It will be essential to include a distributed vegetation layer in the water balance model because vegetation rooting depths vary with vegetation type, and actual evapotranspiration is modulated by vegetation type, in part through variations in stomatal conductance and canopy density (Woodward 1987, Running et al. 1989). Spatially distributed estimates of vegetation cover will help determine the way in which each layer of the multilayer model functions, and thus affect predicted runoff. The distributed vegetation canopy will be estimated from satellite data coupled with digital vegetation maps to characterize vegetative type. Estimation of the seasonal variation of roughness length and canopy density will be possible with the use of leaf area index (LAI) derived from remotely sensed data. This will improve our estimates of PET, which may be underestimated for the basin in this study because a characteristic grassland surface roughness length was used in the PET calculation.

Simulating soil moisture status in several soil layers is important for predicting the effects of climate change on vegetation. Major plant life-forms (e.g., deciduous trees, shrubs, grasses) have different overall moisture requirements. They also vary in rooting depths, imparting on them certain adaptive abilities and liabilities in relation to obtaining sufficient moisture under different types of precipitation regimes. These abilities and liabilities to a large degree control vegetation patterns in the continental United States (e.g., Daubenmire 1968, Neilson et al. 1989, Stephenson 1990).

Limitations in precipitation data availability due to sparse gauging networks and the underestimation of precipitation have already been discussed. In the absence of an orographic model, as a first approximation to accounting for topographic effects, precipitation will be enhanced by first accounting for snow storage and melting, and then by increasing the volume of precipitation deposition using a simple orographic lapse rate to increase deposition volume at higher elevations. More important, the model should be run over a relatively flat basin of roughly the same size as the Columbia River Basin, to assess the improvement in model performance in a region of minimal terrain and orographic influence on the spatial distribution of precipitation. Dolph (1990) and Dolph and Marks (1991) showed that precipitation estimates from the Historical Climatology Network stations (HCN) (Quinlan et al. 1987, Karl et al. 1990) for much of the eastern and midwestern United States better characterize both the spatial distribution and volume of total precipitation input than do estimates from HCN stations in the intermountain west.

The model results using the 2xCO_2 scenario should be viewed as a sensitivity analysis rather than a prediction of future water balance in the Northwest with global climatic change. There are numerous caveats that must be included with any climate change impact study, including the limitations of the GCMs used to generate the climate scenarios (Dickinson 1986, Smith and Tirpak 1989, Houghton et al. 1990). Other limitations pertain to the early stage of development of the water balance model, including the lack of a snowmelt component and vegetation feedbacks. Climate change will induce changes in vegetation cover which will in turn affect regional PET and ET and hence the regional water balance and potentially the regional climate. An additional limitation of this study is that output from only one GCM was used. Scenarios from several GCMs should be used to generate a range of possible future conditions (e.g., Gleick 1987*a*, *b*; Lettenmaier and Gan 1990). The GCM run chosen for this study predicts one of the more severe temperature and wind increases, and humidity decreases, for the Northwest and thus results in a more severe impact on the regional water balance. Other GCM model scenarios would produce different and probably more moderate decreases in soil moisture and runoff for the region.

Potential Impacts of Climate Change on Natural Resources

Even though the water balance model described here is in an early stage of development and only a limited 2xCO_2 sensitivity analysis was conducted, some qualitative statements of the impacts of climate change on natural resources can be made based on the preliminary output from the model and our general understanding of how climate controls the quality and quantity of water and forest resources in the basin.

Generally, GCMs predict warmer temperatures in the Northwest under 2xCO_2 conditions, and unchanged or slightly greater precipitation: GFDL (Manabe and Wetherald 1987); Oregon State University (Schlesinger and Zhao 1989); Goddard Institute for Space Studies (Hansen et al. 1983). Regional estimates of climate change from GCMs need to be used with care, but the fact that several GCMs consistently predict warmer temperatures for the Northwest allows one to be more confident that the direction of climate change in the Northwest will indeed be toward warmer temperatures throughout the year.

Warmer temperatures, combined with increases in wind and decreases in humidity, will increase PET over the region (Table 8.2). This change will increase evaporative losses over the region (assuming there is no change in precipitation). If precipitation does not increase to satisfy the increased atmospheric demand for water, soil moisture and runoff will decrease, as was simulated by our model. The decrease in annual runoff will obviously have a negative impact on available water resources for irrigation, urban and industrial uses, fisheries, power generation, and navigation (see Simenstad et al., this volume).

An additional concern is the potential change in seasonality of runoff caused by the warmer temperatures. Warmer temperatures will raise the elevation of the 0°C isotherm, decreasing snow-covered area and storage of moisture in the winter snowpack (Lettenmaier and Gan 1990). Consequently, the seasonality of runoff will change, with more occurring in the winter months. This shift in seasonality was not calculated for this paper because of the lack of a snowmelt algorithm in the water balance model. However, such a shift in seasonality has been predicted for California under climate warming (Gleick 1987*a*, *b*; Lettenmaier and Gan 1990).

Even if total annual runoff is unchanged, a decrease in spring runoff caused by a reduced snowpack could reduce growing season water supplies because of the dual role of reservoirs for flood control and water storage (Sheer and Randall 1989). Currently in winter, water levels in reservoirs are kept below capacity to capture infrequent but potentially high flows during flood events. After the probability of flooding decreases in the spring, reservoirs are filled with runoff from the melting snowpack. With warmer temperatures, the reservoirs would have to be operated in the same fashion in the winter to prevent lowland flooding, but would not fill during the spring because of the reduced runoff. In California, reservoir operating rules could not be changed enough to compensate for the reduced spring flow (Sheer and Randall 1989). Finally, future changes in extreme events (droughts or floods) are highly uncertain, but will be important to consider in analyzing the water resource impacts of climate change.

If water supplies decline in the future (or as demand for a limited supply increases) and flood frequencies change, efficient management of the resource will become more critical. Several potential management strategies have been identified by Williams (1989), including reallocation of water resources, stricter accounting of water resource costs, development of incentives for alternative agricultural production, increased energy conservation, improved management of soils and vegetation to control floods, and redesigning water resource facilities. Determining the combination of strategies most effective for the Columbia River Basin is an important task for regional water policy development.

Climate change could have a significant affect on vegetation patterns in the Northwest, although only limited modeling of climate change impacts on vegetation has been conducted in the region to date (Leverenz and Lev 1987, Franklin et al. 1991; Urban, unpublished, University of Virginia).

A general statement about potential vegetation change can be formulated considering that the seasonality of precipitation, PET, and soil moisture is a primary control on the distribution of vegetation in the Northwest (Daubenmire 1968, Running et al. 1975, Waring and Franklin 1979, Neilson et al. 1989). Higher PET and unchanged precipitation combined with a higher snow line will increase drought stress on regional vegetation. Drought stress and changes in thermal regime are likely to result in upward movement of all vegetation zones, reduce forest productivity, and increase disturbance probabilities (Leverenz and Lev 1987, Neilson et al. 1989, Franklin et al. 1991). Whether more drought tolerant vegetation currently restricted to the east slopes of the Cascades (e.g., *Pinus ponderosa* forests) would be favored at low elevations on the west slope is uncertain, but warrants further consideration (Leverenz and Lev 1987). In sum, total forest cover in the region could decrease. Alpine vegetation may be particularly at risk, since temperature regimes favoring it may be eliminated from all but the highest elevations in the region.

The Role of Regional Water Balance Models in Natural Resource Management

A key component in natural resource management (forest and water resources management in particular) is understanding how resources respond to seasonal and annual variations in climate and then predicting future supplies. The potentially large change in regional climate underscores the need for understanding how climate affects resource abundance and distribution, and the need for tools with which to evaluate the potential effects of climate change on natural resources. The development of regional-scale water balance models will enable integrated assessments of climate change impacts throughout the river basin as well as a better understanding of how interannual shifts in precipitation patterns can affect local-scale runoff and soil moisture.

Specifically, the preliminary assessments of climate change impacts on forests in the Pacific Northwest are not at a broad enough spatial scale to enable regional quantification of changes in forest composition and location, or are too imprecise in terms of the vegetation types simulated to enable calculations of changes in species composition and productivity (Franklin et al. 1991, Prentice and Fung 1990). In addition, the models do not adequately simulate current vegetation patterns (Prentice and Fung 1990; Urban, unpublished, University of Virginia). Since seasonality of soil moisture controls much of the vegetation in the region, developing a technique to adequately simulate seasonal dynamics of soil moisture will be the most important step in developing an accurate vegetation model for the region.

The modeling approach taken here is a first step in developing an improved class of vegetation models. The water balance approach, incorporating a physically based model of potential evapotranspiration and explicit calculation of soil water holding capacity, will improve our ability to simulate soil moisture under current and future climate conditions. Additional model refinements as discussed above should significantly improve model performance.

Also of fundamental value for making regional assessments of climate change impacts is the implementation of the model on a relatively fine spatial grid over the basin. Since the climatic controls on vegetation are spatially distributed, modeling vegetation change using spatially distributed estimates of the driving climate variables will produce the highest resolution and precision possible given existing data bases. Managers can assess geographical aspects of forest vegetation change, as well as site-specific changes in composition. Modeling vegetation in a spatially distributed framework is especially important when the arrangement of vegetation types on the landscape is of interest, particularly when the goal of the simulation is to predict changes in habitat type and location for biodiversity analyses. Simulating the transient dynamics of vegetation change in response to climate change, especially migration and disturbance, will require spatially distributed modeling.

Similar statements can be made for assessing the impact of changes in climate on water resources in the area. As the wet and dry year analysis showed, precipitation variation over the region is not uniform. For a given year, some areas of the basin can be wetter than normal, while others are drier. This is to be expected given the size and complex topographic structure of the basin. Shifts in seasonal weather patterns could affect different sections of the basin in an opposite manner. Moreover, future changes in regional climate associated with trace-gas induced global warming could produce new patterns of precipitation in the basin relative to those experienced historically. Increases in the elevation of the 0°C isotherm will reduce snow-covered area and alter runoff patterns even if regional patterns of precipitation are largely unchanged. In order to predict the effects of these shifts in precipitation on water resources, a spatially distributed modeling approach using physically based calculations of the regional water balance is required. Regression-based approaches for predicting runoff using historical data will not suffice under future climate conditions that do not have historical analogues.

The spatially distributed modeling approach adopted here will also be useful in the broader geographic context to improve our simulations of global climate patterns and thus our predictions of future climate change. Current GCMs have relatively simple parameterizations of land surface interactions with the atmosphere at coarse spatial scales. This limits their ability to simulate current climate patterns and conditions accurately at spatial scales that are ecologically meaningful (Houghton et al. 1990). Feedbacks between the atmosphere and the land surface (vegetation in particular) play a large role in determining regional climate (e.g., Shukla et al. 1990, Houghton et al. 1990). Vegetation-climate feedbacks are largely mediated via changes in energy and water balance. Thus one of the keys to improving the climate models is to improve the simulations of energy and water balance at a spatial resolution fine enough to capture the majority of the natural variability that determines the character of the regional climate. Because of limitations of computer power, perhaps the finer spatial resolution of energy and water balance processes can be first simulated independently of the GCM runs to determine the appropriate aggregation of key parameters for each GCM grid cell. Likewise, GCM output can be disaggregated across a grid cell using the same spatially distributed modeling techniques to generate finer resolution climate scenarios for the region, as was done in this analysis.

Acknowledgments. The authors would like to thank Jeff Kern for supplying the water holding capacity data, two anonymous reviewers, and Dr. Jonathan Istok for helpful comments on an earlier version of this manuscript. The research described in this chapter has been funded by the U.S. Environmental Protection Agency. This document has been prepared at the EPA Environmental Research Laboratory in Corvallis, Oregon, through contract 68-C8-0006 with ManTech Environmental Technology Incorporated. It has been subjected to the Agency's peer and administrative review and approved for publication. Mention of trade names or commercial products does not constitute endorsement or recommendation for use.

References

Anderson, E.A. 1973. National Weather Service river forecast system: snow accumulation and ablation model. Technical Memorandum NWS HYDRO-17, National Oceanic and Atmospheric Administration, Silver Springs, Maryland, USA.

Baumgartner, A., and E. Reichel. 1975. The world water balance. Elsevier Scientific Publications, Amsterdam, The Netherlands.

Beaty, C.B. 1975. Sublimation or melting: observations from the White Mountains, California and Nevada, USA. Journal of Glaciology **14**:275–286.

Brady, N.C. 1974. The nature and properties of soils. Macmillan, New York, New York, USA.

Brutsaert, W. 1982. Evaporation into the atmosphere. D. Reidel, Dordrecht, The Netherlands.

Budyko, M.I. 1974. Climate and life. Academic Press, New York, New York, USA.

Burnash, R., R. Ferral, and R. McGuire. 1973. A generalized streamflow simulation system. Technical Report, Joint Federal-State River Forecast Center, Sacramento, California, USA.

Campbell, W.G., D. Marks, J.J. Kineman, and R.T. Lozar. 1990. A geographic data base for modeling the role of the biosphere in climate change. Chapter 2 *in* H. Gucinski, D. Marks, and D.P. Turner, editors. Biospheric feedbacks to climate change: the sensitivity of regional trace gas emissions, evapotranspiration, and energy balance to vegetation redistribution; status of ongoing research. United States Environmental Protection Agency EPA/600/3–90/078.

Daubenmire, R. 1968. Soil moisture in relation to vegetation distribution in the mountains of Northern Idaho. Ecology **49**:431–438.

Davis, R.E., J. Dozier, and D. Marks. 1984. Micrometeorological measurements and instrumentation in support of remote sensing observations of an alpine snow cover. Pages 161–164 *in* Proceedings of the Western Snow Conference, 52nd Annual Meeting, April 17–19, 1984, Colorado State University, Fort Collins, Colorado, USA.

De Jong, R., C.A. Campbell, and W. Nicholaichuk. 1983. Water retention equations and their relationship to soil organic matter and particle size distribution for disturbed samples. Canadian Journal of Soil Science **63**:291–302.

Denmead, O.T., and R.H. Shaw. 1962. Availability of soil water to plants as affected by soil moisture content and meteorological conditions. Agronomy Journal **45**:385–390.

Dickinson, R.E. 1983. Land surface processes and climate: surface albedos and energy balance. Advances in Geophysics **25**:305–353.

Dickinson, R.E. 1986. How will climate change? The climate system and modelling of future climate. Pages 207–270 *in* B. Bolin, B.R. Doos, J. Jager, and R.A. Warrick, editors. The greenhouse effect, climatic change, and ecosystems. SCOPE 29. John Wiley and Sons, Chichester, England.

Dickinson, R.E., and R.J. Cicerone. 1986. Future global warming from atmospheric trace gasses. Nature **319**:109–115.

Dickinson, R.E., and B. Hanson. 1984. Vegetation-albedo feedbacks. Pages 180–187 *in* J.E. Hansen and T. Takahashi, editors. Climate processes and climate sensitivity. Geophysical Monograph 5. American Geophysical Union, Washington, D.C., USA.

Dolph, J. 1990. An analysis of the distribution of precipitation and runoff from the historical record for the continental United States. M.A. Thesis. Department of Geoscience, Oregon State University, Corvallis, Oregon, USA.

Dolph, J., and D. Marks. 1991. Characterizing the distribution of precipitation over the continental United States using historical data. Climate Change, *in press*.

Dooge, J.C. 1986. Looking for hydrologic laws. Water Resources Research **22**:46S-58S.

Eagleson, P.S. 1978. Climate, soil, and vegetation. I. Introduction to water balance dynamics. Water Resources Research **14**:705–712.

Eagleson, P.S. 1982. Land surface processes in atmospheric general circulation models. Page 560 *in* Proceedings of Greenbelt Study Conference. Cambridge University Press, New York, New York, USA.

Eagleson, P.S. 1986. The emergence of global-scale hydrology. Water Resources Research **22**:6S-14S.

Eagleson, P.S., and R.I. Segarra. 1985. Water-limited equilibrium of savanna vegetation systems. Water Resources Research **21**:1483–1493.

EarthInfo, Inc. 1988. Climate data users manual. Denver, Colorado, USA.

Eubank, R.L. 1988. Spline smoothing and nonparametric regression. Statistics: Textbooks and Monographs 90. Marcel-Dekker, New York, New York, USA.

Fleagle, R.G., and J.A. Businger. 1980. An introduction to atmospheric physics. Second edition. Academic Press, New York, New York, USA.

Franklin, J.F., F.J. Swanson, M.E. Harmon, D.A. Perry, T.A. Spies, V.H. Dale, A. McKee, W.K. Ferrell, S.V. Gregory, J.D. Lattin, T.D. Schowalter, D. Larsen, and J.E. Means. 1991. Effects of global climate change on forests in northwestern North America. *In* R.L. Peters and T.E. Lovejoy, editors. Global warming and biological diversity. Yale University Press, New Haven, Connecticut, USA, *in press*.

Frew, J.E. 1990. The Image Processing Workbench. Dissertation. Department of Geography, University of California, Santa Barbara, California, USA.

Gleick, P.H. 1987*a*. Regional hydrologic consequences of increases in atmospheric CO_2 and other trace gases. Climatic Change **10**:137–161.

Gleick, P.H. 1987*b*. The development and testing of a water balance model for climate impact assessment: modeling the Sacramento Basin. Water Resources Research **23**:1049–1061.

Hansen, J., I. Fung, A. Lacis, D. Rind, S. Lebedeff, R. Ruedy, and G. Russell. 1988. Global climate changes as forecast by Goddard Institute for Space Studies three dimensional model. Journal of Geophysical Research **93**:9341–9364.

Hansen, J., A. Lacis, D. Rind, G. Russell, P. Stone, I. Fung, R. Ruedy, and J. Lerner. 1984. Climate sensitivity, analysis of feedback mechanisms. Pages 130–163 *in* J.E. Hansen and T. Takahashi, editors. Climate processes and climate sensitivity. Geophysical Monograph 5. American Geophysical Union, Washington, D.C., USA.

Houghton, J.T., G.J. Jenkins, and J.J. Ephraums, editors. 1990. Climate change: the IPCC scientific assessment. Cambridge University Press, Cambridge, England.

Isaaks, E.H., and R.M. Srivastava. 1989. Applied geostatistics. Oxford University Press, New York, New York, USA.

Jenne, R.L. 1990. Data from climate models, the CO_2 warming. National Center for Atmospheric Research, Boulder, Colorado, USA.

Karl, T.R., C.N. Williams, F.T. Quinlan, and T.A. Boden. 1990. United States Historical Climatology Network (HCN) serial temperature and precipitation data. United States Department of Energy, Carbon Dioxide Information Analysis Center, Oak Ridge, Tennessee, USA.

Keeling, C.D. 1973. Industrial production of carbon dioxide from fossil fuels and limestone. Tellus **25**:174–198.

Keeling, C.D., and R.B. Bacastow. 1977. Impact of industrial gasses on climate. Pages 72–95 *in* Energy and climate. National Academy of Sciences, Washington, D.C., USA.

Kern, J.S. 1991. The geography of soil carbon in the contiguous United States. Soil Science Society of America Journal, *submitted*.

Korzun, V.I. 1978. World water balance and water resources of the earth. USSR National Committee for the International Decade. UNESCO Press, Paris, France.

Lettenmaier, D.P., and S.J. Burges. 1978. Climate change: detection and its impact on hydrologic design. Water Resources Research **14**:679–687.

Lettenmaier, D.P., and T.Y. Gan. 1990. Hydrologic sensitivities of the Sacramento-San Joaquin River Basin, California, to global warming. Water Resources Research **26**:69–87.

Lettenmaier, D.P., and D.P. Sheer. 1991. Assessing the effect of global warming on water resources systems: an application to the Sacramento-San Joaquin River Basin, California. Journal of Water Resources Planning and Management **117**:108–125.

Leverenz, J.W., and D.J. Lev. 1987. Effects of carbon dioxide-induced climate changes on the natural ranges of six major commercial tree species in the western United States. Pages 123–155 *in* W.E. Shands and J.S. Hoffman, editors. The greenhouse effect, climate change, and U.S. forests. Conservation Foundation, Washington, D.C., USA.

Linsley, R.K., M.A. Kohler, and J.L. Paulhus. 1982. Hydrology for engineers. McGraw-Hill, New York, New York, USA.

Manabe, S., and R.T. Wetherald. 1975. The effects of doubling the concentration on the climate of a general circulation model. Journal of Atmospheric Sciences **32**:3–15.

Manabe, S., and R.T. Wetherald. 1980. On the distribution of climate change resulting from an increase in CO_2 content of the atmosphere. Journal of Atmospheric Sciences **37**:99–118.

Manabe, S., and R.T. Wetherald. 1987. Large-scale changes in soil wetness induced by an increase in carbon dioxide. Journal of Atmospheric Sciences **44**:1211–1235.

Marks, D. 1988. Climate, energy exchange, and snowmelt in Emerald Lake watershed, Sierra Nevada. Dissertation. Departments of Geography and Mechanical Engineering, University of California, Santa Barbara, California, USA.

Marks, D. 1990. A continental-scale simulation of potential evapotranspiration for historical and projected doubled-CO_2 climate conditions. Chapter 3 *in* H. Gucinski, D. Marks, and D.P. Turner, editors. Biospheric feedbacks to climate change: the sensitivity of regional trace gas emissions, evapotranspiration, and energy balance to vegetation redistribution; status of ongoing research. United States Environmental Protection Agency EPA/600/3–90/078.

Marks, D., and J. Dozier. 1991. Climate and energy exchange of the snow surface in the alpine region of the Sierra Nevada. II. Snow cover energy balance. Water Resources Research, *in press*.

Martin, P., N.J. Rosenberg, and M.S. McKenney. 1990. Sensitivity of evapotranspiration in a wheat field, a forest, and a grassland to changes in climate and direct effects of carbon dioxide. Climatic Change **14**:111–117.

Mather, J.R., and G.A. Yoshioka. 1968. The role of climate in the distribution of vegetation. Annals of the American Association of Geographers **58**:29–41.

McNaughton, K.G. 1986. Regional evaporation models. Pages 103–106 *in* C. Rosenzweig and R. Dickinson, editors. Climate-vegetation interactions. Proceedings of a Workshop held January 27–29, 1986, NASA/Goddard Space Flight Center, Greenbelt, Maryland. Office for Interdisciplinary Earth Studies (OIES) and University Corporation for Atmospheric Research (UCAR), Boulder, Colorado, USA.

Mitchell, J.F.B. 1983. The hydrological cycle as simulated by an atmospheric general circulation model. Pages 429–446 *in* A. Street-Perrott, M. Beran, and R. Ratcliffe, editors. Variations in the global water budget. D. Riedel, Dordrecht, The Netherlands.

NOAA, National Geophysical Data Center. 1989. Geophysics of North America: users guide. Unites States Department of Commerce, Boulder, Colorado, USA.

Neilson, R.P., G.A. King, R.L. DeVelice, J. Lenihan, D. Marks, J. Dolph, B. Campbell, and G. Glick. 1989. Sensitivity of ecological landscapes and regions to global climate change. United States Environmental Protection Agency EPA/600/3–89/073, Environmental Research Laboratory, Corvallis, Oregon, USA.

Parry, M.L., and T.R. Carter. 1989. An assessment of the effects of climatic change on agriculture. Climatic Change **15**:95–116.

Parry, M.T., T.R. Carter, N. Konijin, and J. Lockwood. 1987. The impact of climatic variations on agriculture. *In* Introduction to the IIASA/UNEP case studies in semi-arid regions. International Institute for Applied Systems Analysis, Laxenburg, Austria.

Perrier, A. 1982. Land surface processes: vegetation. *In* Land surface processes in atmospheric general circulation models. Proceedings of Greenbelt Study Conference. Cambridge University Press, New York, New York, USA.

Phillips, D.L., J. Dolph, and D. Marks. 1991. A comparison of geostatistical procedures for spatial analysis of precipitation in mountain terrain. Agricultural and Forest Meteorology, *in press*.

Prentice, J.C., and I.Y. Fung. 1990. The sensitivity of terrestrial carbon storage to climatic change. Nature **346**:48–51.

Quinlan, F.T., T.R. Karl, and C.N. Williams, Jr. 1987. United States Historical Climatology Network (HCN) serial temperature and precipitation data. NDP-019, Carbon Dioxide Information Analysis Center, Oak Ridge National Laboratory, Oak Ridge, Tennessee, USA.

Ramanathan, V.R., R.J. Cicerone, H.B. Singh, and J.T. Kiehl. 1985. Trace gas trends and their potential role in climate change. Journal of Geophysical Research **90**:5547–5566.

Rind, D. 1984. The influence of vegetation on the hydrologic cycle in a global climate model. Pages 73–92 *in* J.E. Hansen and T. Takahashi, editors. Climate processes and climate sensitivity. Geophysical Monograph 5. American Geophysical Union, Washington, D.C., USA.

Rosenzweig, C., and R. Dickinson. 1986. Climate-vegetation interactions. Workshop Proceedings, NASA/Goddard Space Flight Center, Greenbelt, Maryland, USA.

Running, S.W., R.R. Nemani, D.L. Peterson, L.E. Band, D.F. Potts, L.L. Pierce, and M.A. Spanner. 1989. Mapping regional forest evapotranspiration and photosynthesis by coupling satellite data with ecosystem simulation. Ecology **70**:1090–1101.

Running, S.W., R.H. Waring, and R.A. Rydell. 1975. Physiological control of water flux in conifers: a computer simulation model. Oecologia (Berlin) **18**:1–16.

Schlesinger, M.E. 1988. Quantitative analysis of feedbacks in climate model simulations of CO_2-induced warming. Pages 653–737 *in* M.E. Schlesinger, editor. Physically-based modelling and simulation of climate and climatic change. Volume C. Mathematical and physical sciences. NATO ASI Series 243. Kluwer, Boston, Massachusetts, USA.

Schlesinger, M.E., and Z.C. Zhao. 1989. Seasonal climatic change introduced by doubled CO_2 as simulated by the OSU atmospheric GCM/mixed-layer ocean model. Journal of Climate **2**:429–495.

Sheer, D.P., and D. Randall. 1989. Methods for evaluating the potential impacts of global climate change: case studies of the state of California and Atlanta, Georgia. Chapter 2 *in* J.B. Smith and D.A. Tirpak, editors. The potential effects of global climate change on the United States. Volume A. Water resources. United States Environmental Protection Agency EPA/230/05–89/051, Washington, D.C., USA.

Shukla, J., C. Nobre, and P. Sellers. 1990. Amazon deforestation and climate change. Science **247**:1322–1325.

Smith, J.B., and D.A. Tirpak. 1989. The potential effects of global climate change on the United States. United States Environmental Protection Agency EPA/230/05–89/051, Washington, D.C., USA.

Snyder, J.P., and P.M. Voxland. 1989. An album of map projections. United States Geological Survey, Professional Paper.

Solomon, A.M. 1986. Transient response of forests to CO_2-induced climate change: simulation modeling experiments in eastern North America. Oecologia (Berlin) **68**:567–579.

Stephenson, N.L. 1990. Climatic control of vegetation distribution: the role of the water balance. American Naturalist **135**:649–670.

Stewart, B.J. 1982. Sensitivity and significance of turbulent energy exchange over an alpine snow surface. M.A. Thesis. Department of Geography, University of California, Santa Barbara, California, USA.

Strain, B.R. 1985. Physiological and ecological controls on carbon sequestering in terrestrial ecosystems. Biogeochemistry **1**:219–232.

United Nations Food and Agriculture Organization. 1975. Soil map of the world. Volume 2. North America, 1:5,000,000 digital map with text. UNESCO, Paris, France.

United States Army Corps of Engineers (USACE). 1981. Adjusted streamflow and storage, 1928–1978, Columbia River and coastal basins. USACE Columbia River Water Management Group, Portland, Oregon, USA.

United States Army Corps of Engineers (USACE). 1988. GRASS users and programmers manual. USACE, Construction Engineering Research Laboratory, Champaign, Illinois, USA.

United States Department of Agriculture, Soil Conservation Service (SCS). 1988. Snow survey and water supply products reference. USDA-SCS Western National Technical Center, Portland, Oregon, USA.

Verheye, W.H. 1989. Le régime hydrique des sols d'Europe, basé sur des données pédologiques et climatologiques courantes: principes et approche méthodologique. Science du sol **27**:117–130.

Wallis, J.R., D.P. Lettenmaier, and E.F. Wood. 1991. A daily hydro-climatological data set for the continental United States. Water Resources Research **27**:1657–1664.

Waring, R.H., and J.F. Franklin. 1979. Evergreen coniferous forests of the Pacific Northwest. Science **204**:1380–1386.

Whittaker, R.H. 1975. Communities and ecosystems. Second edition. Macmillan, New York, New York, USA.

Williams, P. 1989. Adapting water resources management to global climate change. Climatic Change **15**:83–93.

Woodward, F.I. 1987. Climate and plant distribution. Cambridge University Press, London, England.

9

Impacts of Watershed Management on Land-Margin Ecosystems: The Columbia River Estuary

CHARLES A. SIMENSTAD, DAVID A. JAY, AND
CHRISTOPHER R. SHERWOOD

Abstract

Patterns of land use development that have arisen in the Columbia River Basin over the last century are occurring in large river basins worldwide. The consequent modifications of river flow, physical properties, and discharge of sediment and other constituents appear as cumulative effects in land-margin ecosystems, where estuarine processes intercept, entrap, and transform both riverine and oceanic material. These watershed changes alter both the input to the estuary and the fundamental estuarine processes. Our studies of the Columbia River estuary indicate that these human alterations to watersheds can affect the interaction between river flow and the tides, modifying circulation patterns important to estuarine food webs. Mean river flow has decreased approximately 20% since the 19th century; probably 6 to 8% is due to irrigation withdrawal, the remaining 12 to 14% to climate variability. Regulation of river flow has reduced spring freshet flows to about 50% of the natural level, and has increased fall minimum flows by 10 to 50%. The reduction in spring freshets has lowered modern-day sediment input to the estuary to ~25% of that recorded in the latter part of the 19th century. Navigation structures and filling and diking in the lower river and estuary have decreased the tidal prism by about 15%, increased sediment residence time and shoaling, simplified the channel network, and concentrated flow in the navigation channel. In addition to sediment, temperature, organic matter, nutrients, pollutants, and biotic influxes at the estuarine interface, changes in the river discharge regime have modified estuarine stratification, mixing, and residence time. Such modifications have profound effects on sensitive estuarine processes such as those that occur in the estuarine turbidity maximum (ETM), where trapping of suspended material occurs, organic matter is incorporated in a dynamic microbial loop, and important food web linkages to higher level consumers occur. A landscape perspective on the impacts of watershed alterations needs to be included in our emerging regional and global approaches to ecosystem management if land-margin impacts are to be predicted and mediated.

Key words. Estuary, river flow, watershed management, land margin ecosystems, estuarine turbidity maximum.

Introduction

Watershed Impacts on Land-Margin Ecosystems

Watershed management policies have tended to ignore or deny the ultimate consequences of a basic hydrological principle: *water flows downhill.* Despite a growing awareness that diverse alterations to watersheds can have major, long-term effects on the landscape—especially the riparian and riverine elements thereof—seldom have our perspectives extended downriver to higher stream order segments of a watershed, much less to the confluence of the watershed with the land margin (i.e., to the estuary and to coastal areas measurably influenced by riverine inflow). In fact, most of the evidence describing watershed impacts on land-margin ecosystems has been published only in the last 15 years (Skreslet 1986). In many respects, this recent interest mirrors a new environmental perspective brought about by irrefutable evidence of regional and global impacts from radioactive fallout, acid rain, and organochloride pesticides, and strong indications of significant changes in global carbon and ozone balances (Risser, this volume; Dolph et al., this volume). It is impossible to ignore the deterioration in land-margin ecosystems any longer. Increasing demands for water resources, food production augmented by irrigation, fertilization and herbicide/pesticide addition, and the prominent aggregation of human populations on coastal margins demonstrate the need for watershed management at the appropriate landscape scale.

Estuaries and near-shore coastal waters constituting land-margin ecosystems are arguably the most vulnerable segments of watershed landscapes; most alterations to watersheds are invariably manifested, and often amplified, at the land margin. Because of the unique structure and processes of land-margin ecosystems, the effects of watershed manipulations are typically focused along steep environmental gradients. The interaction of oceanic and river physics over short spatial scales mediates material inflow at the river-ocean interface; geo- and biochemical processes promote the transformation of both inorganic and organic constituents along the same gradients; and highly productive estuarine communities are adapted to, and dependent on, these strong gradient systems.

We describe in this synthesis how the structure and dynamics of land-margin ecosystems, specifically estuaries, are affected by watershed modifications. Processes considered include river flow, circulation, water characteristics and constituents, sediment transport, and important estuarine processes. (We have adopted the term *river flow* to distinguish between the river's discharge of water and that of sediment, the latter of which we have called *sediment discharge* or *transport*.) We do not focus on the effects of nutrient loading and its role in eutrophication, which can dominante land-margin effects in other estuaries. In illustrating how watershed manipulations affect land-margin ecosystems by altering these factors and interactions, we utilize the Columbia River Basin and its estuary as a case study. We have selected the Columbia River because: (1) the watershed dominates our regional landscape; (2) its input into the eastern North Pacific is of primary importance to coastal circulation and ecology; (3) historical changes are well documented, and the effects at the land margin are the focus of our recent research; and (4) historical modifications of the system are typical of many temperate land-margin ecosystems, even if the scale is larger in the Columbia River than in many other systems. Impacts on the structure and dynamics of the Columbia River land-margin ecosystem have, aside from large-scale losses of anadromous fish runs, been moderate compared with those occurring in more heavily developed or less resilient land-margin ecosystems elsewhere. Nonetheless, we suggest that the Columbia River estuary is a system likely to be at the edge of more profound impacts if the basin's development continues without due consideration of interactions between the watershed and the land margin.

Land-Margin Ecosystems Illustrative of Watershed Impacts

Deficiencies in understanding and attention to interactions between river basin modifications and land-margin ecosystem processes have resulted in some phenomenal impacts (Cross and Williams 1981). River flow regulation and diversion have caused major ecosystem changes in south-central Asian republics of the Soviet Union, from the Sea of Azov in the Ukraine adjoining the north border of the Black Sea to the Caspian Sea region of Kazakhstan, including the Aral Sea. In the Sea of Azov, 30 years of river flow diversion (>60% of spring flow) of the Donets and Kuban rivers and a resulting increase in saline waters from the Black Sea have brought radical changes in the endemic biota and reduced productivity of planktonic and benthic communities (Bronfman 1977, Remisova 1984*a*, *b*; Volovik 1986, Rozengurt et al. 1987). The reasons for modifying river flow entering the Sea of Azov parallel those affecting the Columbia Basin (except for hydropower uses): flood control and increased water diversion for consumption and irrigation. And both areas have experienced qualitative and quantitative changes in nutrient loading. Changes in the Columbia River, however, have been less radical.

Extensive hydroelectric power and irrigation development of the Volga River watershed over the last 20 years has resulted in a 38% (normal years) to >50% (dry years) reduction in the spring discharge to the Caspian Sea. This has caused increased salinity, a compression of the mixing zone, reduced nutrient and sediment loading, reduced phytoplankton, zooplankton, and benthos standing stocks, and loss of a substantial area of fish nursery grounds (Rozengurt and Hedgpeth 1989). The consequence has been progressive deterioration of natural recruitment, stock, and commercial catches of anadromous fishes dependent on the land-margin ecosystem.

As a terminal lake, the Aral Sea is particularly vulnerable to disruption of the balance between river and groundwater inflow, precipitation, and evaporation. It once ranked as one of the world's largest lakes in area, but due to excessive consumption of the Amu Darya River water, the Aral Sea has decreased 66% in volume and has dropped to sixth in surface area among the world's lakes (Micklin 1988, Ellis 1990). Modifications of regional climate have occurred in the form of giant dust storms and greater temperature extremes, depression of the groundwater table and accompanying deterioration of domestic water sources, loss of productive fisheries, and destruction of major areas of native wetlands.

Estuarine and coastal fisheries often collapse as a result of severe watershed alteration. For example, while bringing an abundance of electrical energy and irrigation water to Egypt, the Aswan High Dam has contributed to a 95% loss in the sardine (*Sardinella aurita, S. madarensis*) catch in the eastern Mediterranean (Turner 1971, Mancy 1979). But even the agricultural benefit to balance this economic loss may be evaporating, because ~0.6 Mg (metric tons) ha^{-1} yr^{-1} of carbonates may be precipitating on the irrigated soils of Upper Egypt (Kempe 1983). Unfortunately, similar watershed impacts at the land margin may be found in North America. Annual freshwater flows into San Francisco Bay have decreased by 20 to 63%, and diversion of the spring freshet has exceeded 80%, over the last three decades (Rozengurt et al. 1987, Nichols et al. 1986). Dramatic declines in fish populations, including striped bass (*Morone saxatilis*) and chinook salmon (*Oncorhynchus tshawytscha*), have been correlated with the decline in river flow. These declines have been caused primarily by diversion and mortality associated with irrigation, loss of primary and secondary production associated with disruption of the turbidity maximum zone of fresh- and saltwater convergence, and increased concentrations of toxic wastes from agriculture and industry. A worse, though poorly documented, example is the almost total disappearance of the Colorado River estuary (USA-Mexico).

Water diversion and disruption in normal flow cycles do not account for all major alterations at land margins. Increased organic waste loading, nutrient enrichment, and toxic waste contamination from adjoining watersheds have caused deterioration of water quality (e.g., anoxia), reduction of seagrasses and other submerged aquatic vegetation, blooms of both phytoplankton (including toxic "red tide" forms) and macroalgae, and fish and shellfish contamination or mortality in other land-margin ecosystems such as Chesapeake Bay (Orth and Moore 1983, Sanders et al. 1985, Seliger et al. 1985, Officer et al. 1984). This review, however, focuses on systems where water flow modification, water withdrawal, and associated processes are the dominant human impacts, using the Columbia River estuary as an example.

Table 9.1. River characteristics affected by watershed management altering the structure and dynamics of land-margin ecosystems.

RIVER FLOW DYNAMICS
Short-term changes
diel
monthly-seasonal
Long-term changes
annual (hydroperiod)
extreme flows (periods of decades to centuries)
WATER CHARACTERISTICS AND CONSTITUENTS
Sediment
suspended load
bedload
Temperature
Chemistry
nutrients
alkalinity
pH
Organic matter
detritus
dissolved organics
large organic debris
Pollutants
herbicides and pesticides
heavy metals
organochlorides
metals
petroleum hydrocarbons
radionuclides
Biota
anadromous fishes
migratory waterfowl
exotic species

River Characteristics and the Influence of Watershed Alterations

The influences of watershed management practices on fluvial and land-margin processes fall within two broad categories: (1) river flow dynamics and (2) river water constituents and properties (Table 9.1).

River Flow Dynamics

The physical process of river flow that regulates freshwater input to land-margin ecosystems is fairly simple dynamically: water moves downhill under the influence of gravity. Potential energy (energy of position) is converted to kinetic energy and thence—via turbulence, sediment transport, and frictional dissipation—to heat. In the seaward portions of large rivers (what we consider the landward boundary of the land-margin ecosystem), but still landward of the estuarine boundary defined by the presence of salt, the downstream motion of river water is modulated by tidal effects. Effects of storm flows at the boundary of the land margin are dependent on the geographic scale, topography, and climate of the watershed; while small landscapes such as the river drainages on the western slopes of the Cascades can be dominated entirely by storm events, storms are less important in the Columbia River Basin because the scale of most storms is less than the size of the basin (Dolph et al., this volume). Temporal variations in river input to land-margin ecosystems have a variety of causes and time scales. Important processes include: seasonal variations caused by variations in atmospheric temperature and precipitation, climatic changes in these properties over decades to centuries, changes in land use which may alter both groundwater and surface flow over a wide variety of frequencies, flow regulation for flood control and power generation, and consumptive water use (e.g., irrigation).

Water Characteristics and Constituents

Water acquires certain characteristics and constituents as rivers pass through their watersheds (Table 9.1) and ultimately through the land margin to the ocean (if constituents are not transformed or deposited in estuarine or coastal areas). Sediment is the dominant river-borne component, and one of the most vulnerable to changes in watershed management. The total sediment load consists of two components: the material traveling in suspension (suspended sediment) and that which rolls and bounces (saltates) along the riverbed (bedload). These two fractions have distinct responses to changes in the watershed. Suspended sediment is generally composed of silts and clays. Once in suspension, these may be transported by all but the lowest river flows; and fine sediment transport tends, in a largely sand-bedded system like the Columbia River, to be a function of sediment supply in the watershed (supply limitation). Bedload transport of the larger fractions of the sediment particle size spectrum (sands and gravels) is limited by flow velocities (transport limitation). Major sediment transport events occur when high river flows provide a large amount of silt and bring fine sands, normally carried as bedload, into suspension. The few days during the year having the highest river flows usually account for most of the total annual sediment transport. Even a small reduction in these peak flows can cause a major decrease in sediment transport because of the nonlinear relationship between sediment transport and river flow.

Temperature regimes of highly manipulated watersheds and rivers tend toward greater maximums and stronger fluctuations. These can originate from watershed landscape changes such as increased sunlight due to deforestation (Holtby 1988), changes in stream morphology and diversion through irrigation, and reduction of stratification in flow in riverine impoundments behind hydropower and irrigation diversion dams (Petts 1984).

Alterations in river chemistry can originate from inorganic compounds introduced by excessive land use practices (e.g., drainage of crop fertilizers and nutrients from intensive animal husbandry), from domestic sewage and industrial discharge, and from geochemical and biological processes within the river itself. Under certain conditions, riverine (as well as groundwater) nutrient loadings can alter basic nutrient cycling and cause excessive primary production (i.e., eutrophication) in estuarine and near-coastal waters (Kemp et al. 1982, Nixon et al. 1986, Valiela and Costa 1988). This decreases water quality and can be extreme enough to cause fish kills. For many land-margin ecosystems, nutrients may continue to be the greatest influence in altering the natural biological processes (Rosenberg 1985, Turner and Rabalais 1991). Also, in rare instances, large-scale introductions from outside the watershed can alter riverine chemistry, such as the atmospheric input of acid rain in basins that do not have natural buffering capacity.

Transport of organic matter, especially fine particulate organic carbon detritus (FPOC; <1 mm diameter), is typical of natural watersheds. There is little documentation of changes in detritus loads in manipulated watersheds, and it is presumed to vary depending on the types of vegetation cover and soils, surface water flow, and so forth. Dissolved organics, however, can increase significantly in regulated rivers as a function of phytoplankton production in enlarged euphotic zones of impoundments. Transport of large particulate organic carbon (LPOC; >1 mm diameter), however, has probably declined precipitously with loss of riparian vegetation and natural forest maturation in the watershed (Sedell et al. 1988, Gonor et al. 1988).

Anthropogenic changes to anadromous fish and migratory waterfowl populations within the watershed are commonly manifested at the land margin. In addition to using the land margin as a migratory corridor between terrestrial and marine habitats, many species use estuaries or near-shore coastal regions as rearing ("nursery") habitats during critical periods in their life histories. Although not initially considered deleterious to watersheds, exotic biota have often expanded in distribution and influence into brackish and estuarine habitats, altering community and food web relationships at the land margin.

In summary, changes in river flow, the dynamics of the flow cycle, and the characteristics and constituents of rivers can affect, either directly or indirectly, at least five processes that structure land-margin ecosystems: (1) physical interaction between river flow, tidal propagation, and salinity intrusion; (2) transport and deposition of inorganic and organic sediment; (3) primary and secondary productivity; (4) ecological interactions (e.g., competition, predation); and (5) trophic (food web) linkages. These processes are not discrete. They are closely linked by their effects on, or use of, material and biota that are borne into, transformed, or exported from estuaries to near-shore coastal waters.

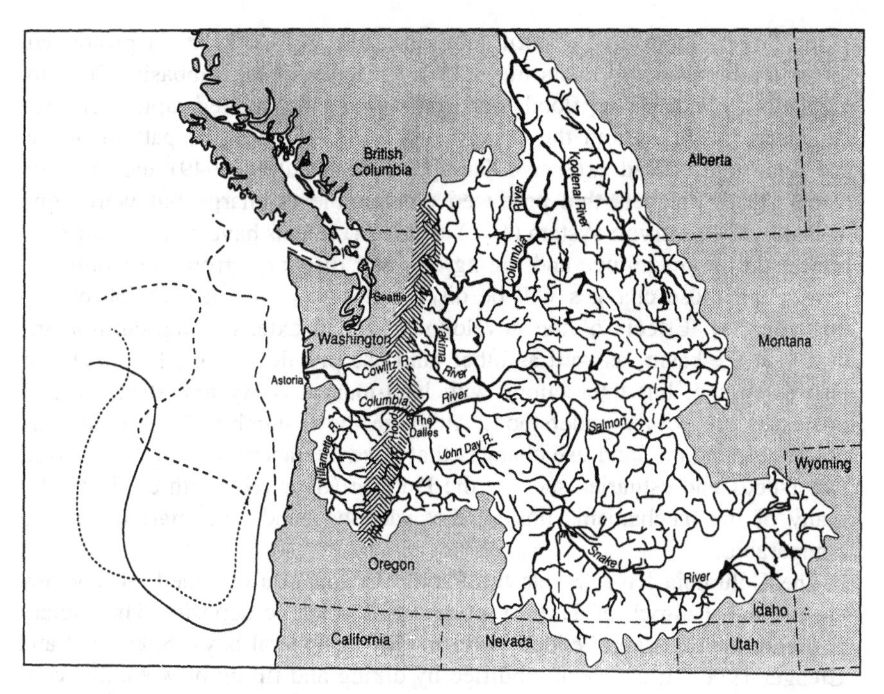

FIGURE 9.1. Columbia River Basin and average seasonal extent of the Columbia River discharge plume, as illustrated by the 32.5 psu (‰) salinity isopleth (spring ---, summer ——, autumn ·····, winter — —); modified from Barnes et al. (1972).

Historic Alterations to the Columbia River from the Watershed to the Land Margin

The Columbia River Basin

The Columbia River dominates the Pacific Northwest landscape and the coastal environs of the eastern North Pacific south of Vancouver Island (Figure 9.1). It contributes 60% (winter) to 90% (summer) of the freshwater input into the Pacific Ocean between San Francisco Bay and the Strait of Juan de Fuca (Barnes et al. 1972). In terms of mean water discharge, the Columbia is the second largest river on the west coast of the Americas. Only the Yukon River is larger. The Columbia River Basin encompasses 660,480 km^2 in seven U.S. states and two Canadian provinces. It can be subdivided into two portions, an eastern subbasin and a coastal subbasin. The coastal subbasin contains only about 8% of the total surface area, but provides about 24% of the total runoff (Orem 1968, Jay and Good 1978). It has a moist, maritime climate. Most of the river flow from the coastal subbasin occurs during winter as short, intense freshets caused by subtropical weather systems that bring heavy rains and rapid snowmelt. Most of the water originates in the larger, more arid eastern subbasin, and arrives during a prolonged spring freshet as snow melts in the upper reaches of the subbasin. Prior to regulation, major spring floods occurred when snow melted rapidly and simultaneously throughout the eastern subbasin. This was the pattern of the last flood that caused extensive damage (1948; Paulsen 1949) and of more recent spring freshets that produced comparable discharge but were controlled by the reservoir system (1972 and 1974). Dams have had a significant impact on the magnitude and timing of water discharge from the Columbia River, and those changes will be discussed below (see also Dolph et al., this volume). Effects of the development of an extensive agricultural industry in the basin, as well as other land use practices (e.g., forestry), are also discussed, but interactions with land-margin ecosystem structure and processes are poorly understood. Although the Columbia River Basin has not undergone the land use changes that have characterized more urbanized watersheds and estuaries such as the Hudson River (Howarth et al. 1991), many of the mechanisms linking the watershed and land margin may be comparable.

The Columbia River meets the Pacific Ocean in a drowned river valley cut through bedrock of Tertiary marine and volcanic deposits. The estuary is predominantly sand-bedded, with muddy peripheral bays (Sherwood and Creager 1990). It has been modified by diking and filling of wetlands, construction of solid jetties at the mouth and pile dikes along the channels, and extensive dredging of the deep-draft (13 m) navigation channel. The circulation of the estuary results from a complex interplay of tides, river flow, and topography; and historical changes in the latter two have affected estuarine circulation and sedimentary processes (Sherwood et al. 1990). These changes will be outlined in the following sections.

River and Watershed Alterations

The Columbia River has been highly developed by man despite the formidable dimensions of its basin and water discharge (Figure 9.1). The river is now free-flowing along only 4% (71 of the 1,931 km) of its length. The remainder of the river channel has been converted to a system of tributaries and lakes, with reservoirs holding 77.7×10^9 m^3 of water behind 28 major dams (Hunt 1988). Most dams constructed in the Columbia River Basin since the 1930s are multipurpose dams for power generation, flood control, irrigation water diversion, industrial- and municipal-use water diversion, and recreation (Table 9.2). Since 1933, the storage capacity of the reservoir system has risen and dams with longer residence times (defined below) have been built (Table 9.2). Water resource managers now modify the natural flow cycle, using the reservoir system for power production, irrigation water, and flood control, even to the point of transferring some of the runoff between water years. Since one of the goals of this chapter is to show that modifications of river flow directly affect physical and ecological processes

Table 9.2. Chronology of construction and characteristics of major Columbia River dams and reservoirs.

River	Year Completed	Reservoir (Dam)	Active Storage ($m^3 \times 10^3$)	Average Flow (m^3/s)	Residence Time (days)	Cumulative Active Storage ($m^3 \times 10^6$)
Spokane	1906	Coeur d'Alene Lake (Post Falls)	277.5	174	18.2	0.3
Boise	1914	Arrowrock	354.0	70	58.2	0.6
Spokane	1915	Lake Spokane (Long Lake)	128.3	218	6.7	0.8
Snake	1916	Jackson Lake	1,044.8	38	313.8	1.8
Snake	1926	American Falls	2,096.9	180	133.3	3.9
Chelan	1927	Lake Chelan	833.8	57	168.4	4.7
Payette	1930	Deadwood	199.8	7	347.6	4.9
Kootenai	1932	Kootenay Lake (Corra Linn)	970.7	772	14.4	5.9
Owyhee	1932	Owyhee	881.9	45	224.7	6.8
Yakima	1932	Five Reservoirs	1,314.9	8.1		
Columbia, Lower	1937	Bonneville	170.2	5,129	0.4	8.3
Columbia, Upper	1938	F.D.R. Lake (Grand Coulee)	6,453.6	3,013	24.5	14.7
Flathead	1938	Flathead Lake (Kerr)	1,503.6	3,234	53.2	16.2
Long Tom	1941	Fern Ridge	135.7	14	108.3	16.3
Willamette, Coast Fork	1942	Cottage Grove	37.0	7	57.3	16.4
Boise	1945	Anderson Ranch	521.8	28	213.3	16.8
Payette	1947	Cascade	805.5	29	322.1	17.6
Row	1949	Dorena	87.6	20	50.6	17.7
Flathead, South Fork	1951	Hungry Horse	3,706.6	98	430.8	21.4

Table 9.2. Continued.

River	Year Completed	Reservoir (Dam)	Active Storage ($m^3 \times 10^3$)	Average Flow (m^3/s)	Residence Time (days)	Cumulative Active Storage ($m^3 \times 10^6$)
Pend Oreille	1952	Pend Oreille Lake (Albeni Falls)	1,424.7	710	23.0	22.8
Columbia, Lower	1953	Lake Wallula (McNary)	228.2	4,754	0.6	23.0
Santiam, North Fork	1953	Detroit	419.4	44	109.4	23.5
Lookout Point	1955	Lookout Point	430.5	81	60.7	23.9
Boise	1956	Lucky Peak	342.9	72	54.8	24.2
Snake	1956	Palisades	1,506.1	169	102.1	25.7
Columbia, Lower	1957	Lake Celilo (The Dalles)	65.4	4,981	0.2	25.8
Columbia, Mid	1958	Rufus Woods Lake (Chief Joseph)	143.1	3,024	0.5	25.9
Snake	1958	Brownlee	1,208.8	463	29.9	27.2
Clark Fork	1959	Noxon	284.9	542	6.0	27.4
Columbia, Mid	1961	Priest Rapids	54.3	3,315	0.2	27.5
Willamette, Middle Fork	1961	Hills Creek	307.1	30	115.5	27.8
Columbia, Mid	1962	Lake Entiat (Rocky Reach)	44.4	3,212	0.2	27.9
McKenzie, South Fork	1963	Cougar	203.5	22	106.9	28.1
Columbia, Mid	1964	Wanapum	198.6	3,312	0.7	28.3
Fall Creek	1965	Fall Creek	141.9	15	105.8	28.4
Columbia, Mid	1967	Lake Pateros (Wells)	91.3	3,150	0.3	28.5
Duncan	1967	Duncan Lake	1,751.5	99	202.6	30.2

Table 9.2. Continued.

River	Year Completed	Reservoir (Dam)	Active Storage ($m^3 \times 10^3$)	Average Flow (m^3/s)	Residence Time (days)	Cumulative Active Storage ($m^3 \times 10^6$)
Santiam, Middle Fork	1967	Green Peter	410.7	60	78.4	30.6
Santiam, South Fork	1967	Foster	41.9	60	8.0	30.7
Columbia, Lower	1968	Lake Umatilla	659.9	4,827	1.6	31.5
Columbia, Lower	1968	Lake Umatilla (John Day)	659.9	4,827	1.6	31.5
Blue	1968	Blue River	104.8	12	100.6	30.8
Columbia, Upper	1968	Arrow Lakes (Hugh Keenleyside)	9,053.7	1,123	92.3	40.5
Cowlitz	1968	Davisson Lake	1,723.2	143	137.9	42.2
Clearwater, North Fork	1972	Dworshak	2,486.7	156	182.0	44.7
Kootenai	1972	Lake Koocanusa (Libby)	6,116.8	318	220.3	50.8
Columbia, Upper	1973	McNaughton Lake (Mica)	14,801.7	574	295.0	65.6
Columbia, Upper	1983	Revelstoke	5,304.1	840	72.3	70.9

in the estuary, a brief summary of historical changes in the river flow is provided here.

Alterations of the River Flow Frequency Spectrum

The first significant control over river flow was afforded by construction of Grand Coulee Dam, initiated by the U.S. Bureau of Reclamation (USBR) in 1933. The Bonneville Power Administration (BPA) was formed in 1935 to supervise the generation and distribution of Columbia River hydroelectric power. Following World War II, large dams were built by the U.S. Army Corps of Engineers (USACE) at McNary, Albeni Falls, Chief Joseph, and The Dalles and by the USBR at Hungry Horse, Anderson Ranch, and Palisades. Funding for federal water resources development decreased in the mid- to late 1950s, but dams with extensive generating capacities were completed thereafter on the mainstem of the Columbia River and major tributaries at Arrow Lakes, Dworshak, and McNaughton. A large dam at Revelstoke, built in 1983, also created a significant expansion in storage capacity.

The dams built along the lower and mid-Columbia River cannot affect average river flow over periods longer than a few days because their active storage capacity is small relative to the river discharge. Since their residence time (defined as active storage capacity/average discharge; Table 9.2) is generally less than one day, they must pass on whatever flow approaches them. The large reservoirs on the upper reaches of the Columbia River system have much longer residence times, and therefore have the capacity to alter flow cycles across seasons and even from year to year. Control over the timing of river discharge is related to the total active storage capacity of the reservoir system and to the residence times of the reservoirs. The active storage capacity has more than doubled in the last 25 years and is now about 74×10^9 m^3, nearly one-third of the total annual runoff for the basin (223.31×10^9 m^3). Many of the large reservoirs completed since 1967 have long residence times (Table 9.2). The operation of these reservoirs has resulted in long-term changes in the annual discharge of water and sediment, and altered the timing of river flow on scales ranging from hours to months and even years.

Changes to seasonal river flow patterns are evident in a comparison of the estimated adjusted long-term monthly mean runoff (Figure 9.2, top) at the mouth (eastern plus coastal basin flow) and the actual observed flow (Figure 9.2, bottom) for the same location. The adjusted flow is an estimate of the runoff that would have occurred in the absence of flow regulation (Orem 1968). (No correction is made, however, for consumptive loss to irrigation, which is discussed later in this section.) Although some flow regulation existed in the Columbia Basin before 1849, the effects of regulation on river flow have been most dramatic since 1968 (Figure 9.3; Sherwood et al. 1990). The flow record may be divided into a preregulation period (prior to 1969) and a modern period (1969 to date). Interannual transfers of

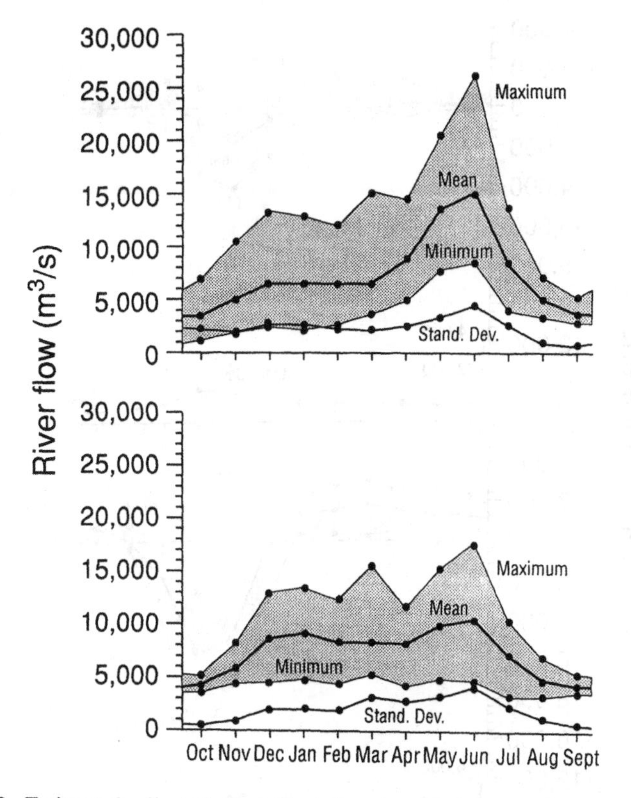

FIGURE 9.2. Estimated adjusted river flow (maximum, mean, minimum, and standard deviation) of the Columbia River at the mouth, averaged by month over the period 1928–82 (top), and modern monthly estimated observed river flow, 1969–82 (bottom). Estimated adjusted river flow is the river flow that would have occurred in the absence of storage. The large difference between long-term adjusted and modern observed river flow reflects the influence of regulation. (*Note*. From Sherwood et al. 1990. Copyright by Pergamon Press PLC. Used by permission.)

flow, for example, increased dramatically in the early 1970s at the beginning of the modern period (Figure 9.4).

Comparison of the actual and adjusted flows (Figures 9.2–9.4) and the seasonal contribution to the annual river flow over 108 years (Figure 9.5) demonstrates that spring and early summer flows (May through July) have been consistently and substantially reduced. Flows in April have also decreased in most years. Regulation has augmented flow in all other months except, on rare occasion, during November and August. The greatest flow reduction has occurred at the peak of the spring freshet (usually in June, but occasionally in May). In the period 1969–82, June river flow was reduced by an average of about 40%. A further reduction has occurred since then because of increased regulation and generally dry weather. In some years,

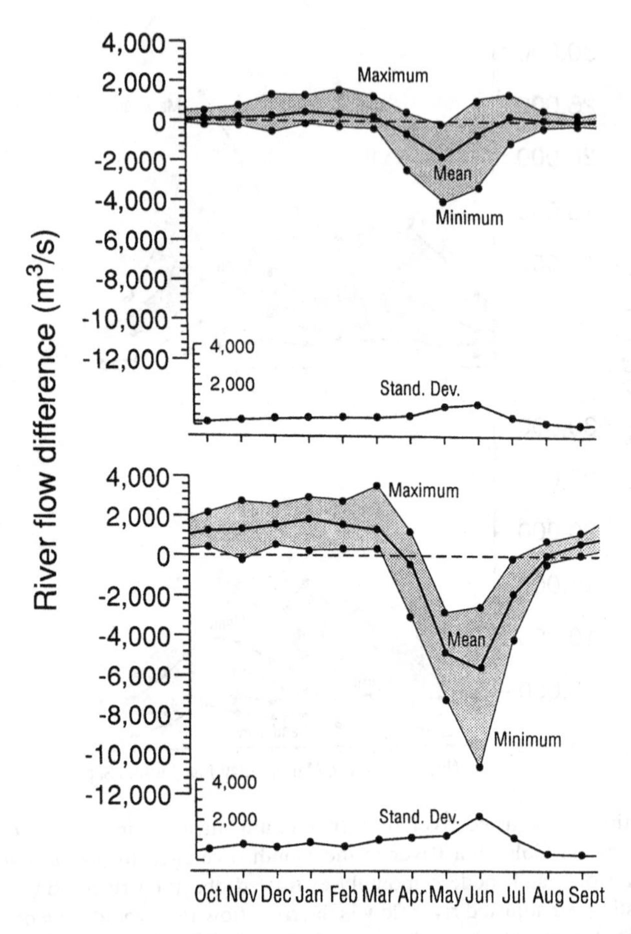

FIGURE 9.3. Monthly differences (maximum, mean, minimum, standard deviation) between observed and adjusted (storage effects removed) flow of the Columbia River at the mouth averaged over 1928–68 (top) and 1969–82 (bottom). (*Note*. From Sherwood et al. 1990. Copyright by Pergamon Press PLC. Used by permission.)

a combination of regulation of the spring freshet and large winter storms causes the largest monthly mean flow during one of the winter months. Because winter flood flows are of shorter duration than the spring freshet, the largest daily flows now usually occur during the winter.

Historical changes in the river flow frequency spectrum can be examined using spectral analysis. Analysis of the daily discharge at The Dalles confirms changes noted above and provides additional information (Sherwood et al. 1990). Flow regulation has reduced discharge variations at low frequency (periods of months to years) while increasing variations at the high frequency end of the spectrum (weeks to days). The reduction at the low frequencies has occurred because of the curtailment of spring freshets, and

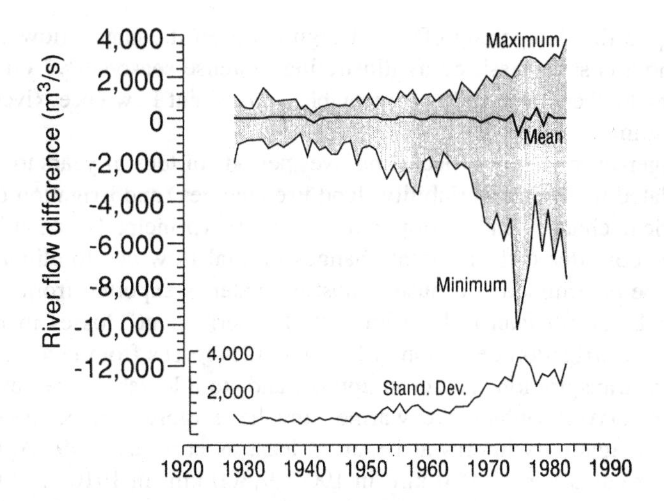

FIGURE 9.4. Differences between the estimated observed and estimated adjusted river flow (maximum, mean, minimum, standard deviation) of the Columbia River at the mouth for 1928–82.

because flow from high runoff years is stored and released in low runoff years. Large differences between the modern and historical high frequency variations reflect the close relationship between river flow and domestic and industrial power demand. River flow is now typically lower on weekends and holidays, and significant variations are evident not only at a seven-day period but at even multiples, or harmonics, of this frequency (periods of 3.5 d, 1.75 d, etc.) where none occurred historically.

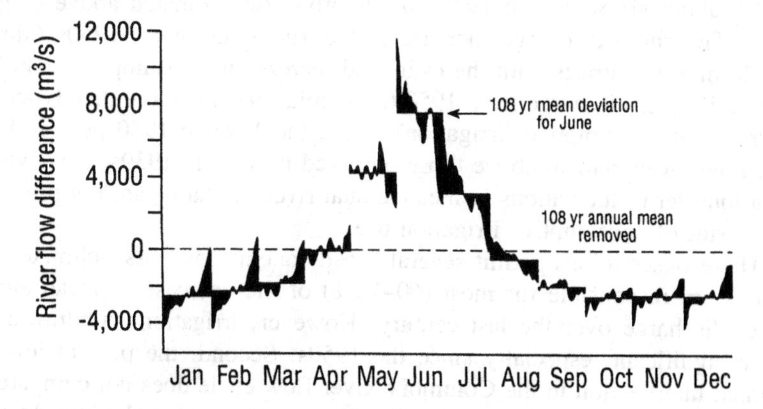

FIGURE 9.5. Historical changes in river flow for each month at The Dalles (1878–1985). The 108-year trends are shown as deviations from the long-term means. The 108-year mean has been removed.

Despite the dampening effect of regulation on its annual flow cycle, the Columbia is still classified as illustrating "intense seasonality" on the basis of its hydrochemistry, and is comparable to the Saint Lawrence River (Kempe and Lammerz 1983).

Longer-term changes (occurring over periods of tens of years to centuries) are related to climate variability, land use changes, and irrigation depletion. Historical changes in consumptive water use (principally irrigation) must also be considered if historical changes in total flow due to climate change are to be determined. Because almost no water is exported from the system (USACE, North Pacific Division 1984), "permanent" losses in river flow are due to irrigation depletion (which includes losses from land-based evaporation, transpiration, and crop uptake) and, to a lesser degree, evaporation from reservoirs (which are warmer and have more surface area than the river). Irrigation withdrawal began at least as early as 1849. Approximate areas irrigated were 2,000 km^2 in 1900, 9,300 km^2 in 1910, 14,600 km^2 in 1928, and 31,600 km^2 in 1980 (Depletions Task Force 1983). Withdrawal is typically greatest in spring during the freshet, with some return flow later in the year. The total net loss is uncertain because of the large number of irrigation projects, but estimates for the early 1980s ranged from 7 to 10% of the mean annual discharge, or about 450 to 680 m^3/s (Sherwood et al. 1990). It is not necessarily true that consumptive water loss is directly proportional to area irrigated, but this assumption, along with the estimate of a 10% reduction, permits calculation of consumptive irrigation-water use over time. Net consumption would have been about 44 m^3/s in 1900, 200 m^3/s in 1910, and 310 m^3/s in 1928. Other consumptive uses in industry and increased enhanced evaporation cannot be evaluated, but are likely to be less than that for agriculture.

The long-term trends in river discharge at The Dalles, smoothed with a 15-year filter, show a decrease of almost 2,000 m^3/s between 1880 and 1940, about six times the 1928 consumptive use estimated above (Figure 9.6). The sharp discharge increase in the 1940s and early 1950s (almost 1,000 m^3/s) contrasts with the estimated increase in consumption. The decrease in river discharge from 1950 to the mid-1980s is more than twice the increase in consumptive (irrigation) use in the 1928 to 1980 period. Even so, flow levels remain above those observed in the late 1930s. It is evident that long-term fluctuations in mean annual river discharge are not primarily the result of consumptive irrigation use.

These observations permit several broad conclusions. First, climate variations are responsible for most (60–70%) of the decrease in mean annual river discharge over the last century. However, irrigation withdrawal has been significant, especially since the 1950s. Second, the present level of human intervention in the Columbia River flow cycle does not compare (in terms of discharge) with the truly disastrous cases cited in the introduction. Caution for the future is definitely warranted, however, because any major increase in consumptive water use would make such use the dominant factor

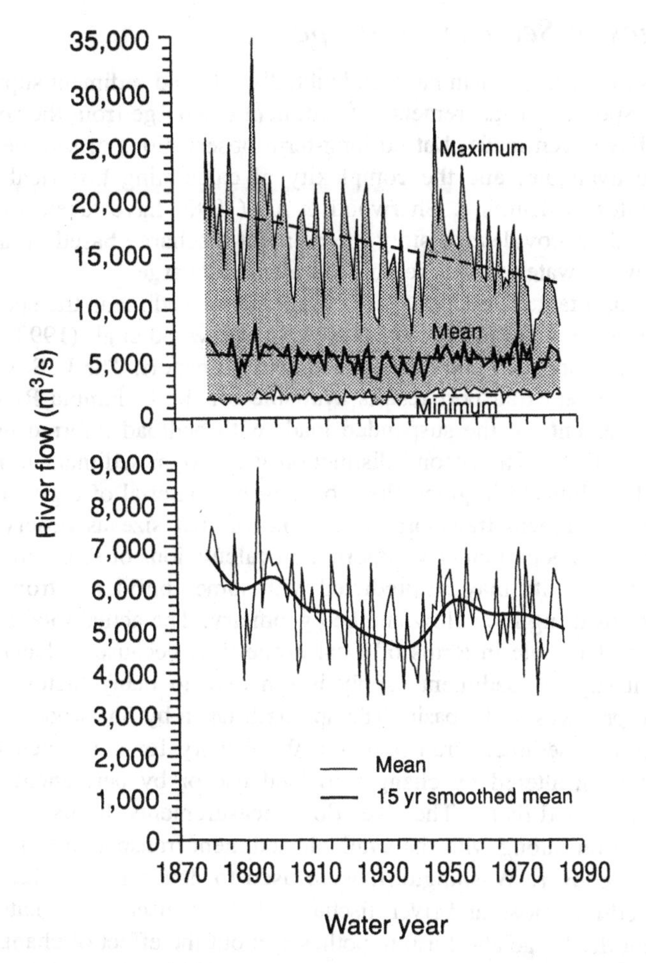

FIGURE 9.6. Top: annual mean, maximum, and minimum of daily river flows, Columbia River at The Dalles, Oregon, for water years 1878–1985. Long-dashed line indicates the least-squares sets to the data. Bottom: annual mean river flow (light line) of the top figure, plotted on an expanded scale. The smoothed curve (heavy line) averages over about 15 years and highlights low flows that occurred during drought years in the late 1930s. (*Note*. From Sherwood et al. 1990. Copyright by Pergamon Press PLC. Used by permision.)

in long-term changes in river discharge. Third, if global warming due to the greenhouse effect were to cause higher temperatures, increased evaporation, greater demand for water, or a reduction in precipitation, the sum total of human interventions in the system would likely become the dominant factor in changes in long-term mean discharge, as is already the case with the annual flow cycle (Dolph et al., this volume).

Changes in Sediment Transport

Changes in the river basin have undoubtedly affected sediment supply to the estuary. Sporadic measurements of sediment discharge from the river to the estuary have been made, but no long-term observations of sediment supply rates are available, and the complexity of calculating historical sediment supply rates is daunting. Sherwood et al. (1990) have taken an heuristic approach that provides a history of sediment discharge based on a relationship between water discharge and sediment discharge.

Two important distinctions regarding fluvial sediment transport are required to understand the approach used by Sherwood et al. (1990). The first is between suspended and bedload sediment transport. The U.S. Geological Survey sediment transport results available for the Columbia River consist of measurements of the suspended load, with bedload inferred using standard corrections. The second distinction is between sediment transport capacity, the ability of a given flow to transport material of a given size, and the actual sediment transport of material of that size as observed at any moment. This is particularly important in calculations of sediment transport. Transport capacity may be predicted with some confidence from hydraulic calculations using river flow and bed geometry. The actual total load is not, however, definable in terms of local properties, because it depends on the sediment supply. Sediment supply is sensitive to many factors, including land use practices in the basin. The approach taken by Sherwood et al. (1990) assumes that sediment transport into the estuary for any given flow level has not been altered by changes in land use or by permanent storage of sediment behind dams. The river flow measurements discussed in the previous section, along with the available sediment transport information (data from Vancouver, Washington), were used to determine a relationship between sediment load and river discharge that was used to estimate historical sediment discharge and form hypotheses about the effect of changes in river flow on sediment discharge (Sherwood et al. 1990). These estimates should be understood as the sediment transport that would have occurred at the time in question given the land use patterns of the 1960–74 period, during which the sediment transport observations were made.

The relationship between river flow and sediment transport is exponential. Because of this nonlinear character, most sediment transport in any year occurs on the few days having maximum flow during that year. Moreover, the percentage of sand increases much more rapidly than the fine sediment load, becoming the dominant fraction above 10,000 m^3/s. There are several reasons for these features. First, transport capacity for each size class of material (be it sand, silt, or clay) increases as a power greater than two of river discharge. Second, while fine materials (silts and clays) always travel in suspension, sand moves at low discharge levels principally as bedload. Only at higher flow levels is sand able to move in suspension, which is a far more efficient means to move sediment. Third, the fine sediment load

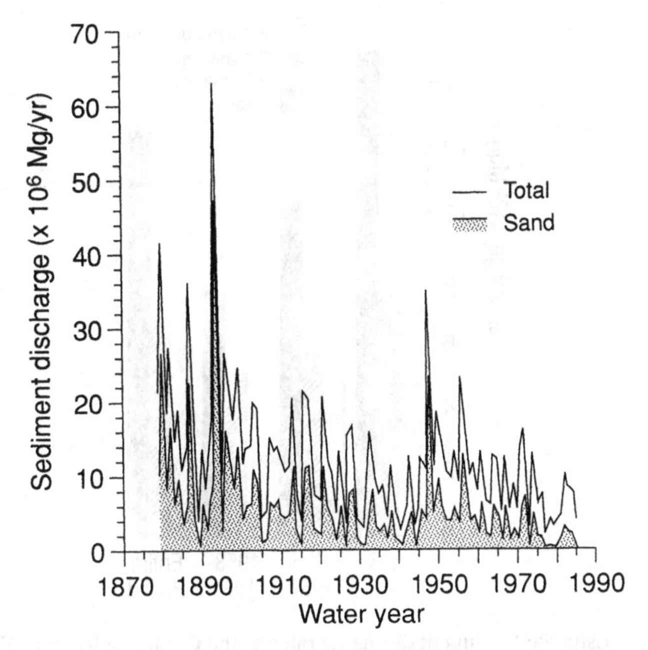

FIGURE 9.7. Estimated annual sediment discharge of the Columbia River at Vancouver, Washington, USA, resulting from the sediment discharge-river flow relationship. (*Note*. From Sherwood et al. 1990. Copyright by Pergamon Press PLC. Used by permission.

is supply limited under most conditions, so that transport of fine sediment rarely reaches transport capacity. In contrast, sand is always available on the bed and moves whenever there is sufficient transport capacity. Sand transport is, and probably always has been, capacity limited. Thus the fine sediment component of the total load is most difficult to calculate. It is also the fraction whose input to the river is most likely to be affected by changes in land use. For purposes of estimating total load, however, it is encouraging that sand transport is relatively more predictable and is also dominant under freshet conditions. It is also the fraction most subject to trapping in the estuary.

Sediment transport estimates for the period 1868–1981 (Figures 9.7 and 9.8) illustrate several obvious features: the very large transports associated with a few freshets (most notably those of 1880, 1887, 1894, and 1948), the dramatic decrease in transport after 1900 and again after about 1972, and the increased percentage of fine material. Just as the annual sediment load is dominated by the contribution from a few days of high flow during each year, major freshets are the dominant feature of the long-term sediment transport record. The average annual load for the period 1868–1934 was 14.9 x 10^6 Mg (metric tons) per year; this decreased to 7.6 x 10^6 Mg/yr during 1958–81. The percentage of sand decreased from >50% before 1900

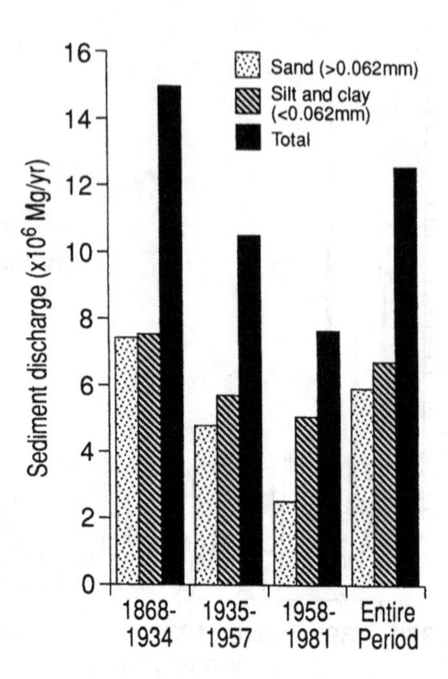

FIGURE 9.8. Estimated sediment discharge rates of the Columbia River at Vancouver, Washington, USA, by historical period, based on time series presented in Figure 9.7.

to about 33% for 1958–81. Thus, while the total input of fine sediment has decreased by about a third, sand input (the dominant size class retained by the estuary) has decreased by a factor of three. Calculation of sediment transport for the last decade would probably show an even more dramatic decrease in sand input, and anecdotal evidence suggests a decrease in dredging requirements that would be consistent with such a decrease.

Most of the change in calculated sediment supply is due to flow regulation. Flow regulation increases the length of the spring freshet but decreases its intensity. Reduced daily flows caused greatly reduced sediment transport. Long-term climate variations and irrigation losses have decreased the total river flow slightly, but flow regulation has severely reduced freshet flows and thus disproportionately decreased sediment discharge. Andrews (1986) concluded that regulation of river flow was responsible for an analogous drop in sediment transport in the Green River in Colorado and Utah.

There are several possible counterarguments to the hypothesis that sediment input to the river and estuary has decreased. The eruption of Mount St. Helens was estimated to have introduced about 100 x 10^6 Mg (metric tons) of sediment into the system in 1980 alone and as much as 5 x 10^6 Mg/yr for some years thereafter (Dunne and Leopold 1981, Schuster 1981, Fairchild and Wigmosta 1983). However, a large but unknown fraction of this material was immediately removed from the system by dredging. A second possibility is that large-scale agriculture and logging have increased the sediment supplied to the system. Given that fine sediment transport is usually supply limited, this would cause an increase in fine sediment transport at any given flow level. This hypothesis cannot be evaluated quantitatively, but, as discussed below, only about 30% of the fine sediment supplied to the estuary remains there, and most sediment retained by the estuary is sand. Sand transport is capacity limited, and the sediment transport-discharge relationship for sand should not be affected by changes in land use. There is also a question concerning the issue of the fate of the sediment allegedly not transported to the estuary under reduced flow conditions. It should be accumulating in storage reservoirs or elsewhere in the system. No systematic evaluation of this problem has been made, and 200 x 10^6 Mg of sediments (the apparent decrease in sediment transport over the last 40 years, given a 50% decrease in sediment transport) could easily have accumulated in reservoirs and along riverbanks between reservoirs without having any appreciable effect on storage capacity (which is about 780 times the volume of sediment in question). Deltas of rivers emptying into reservoirs would be a good place to look for sediment accumulation. Finally, it may be argued that navigational structures have increased the capacity of the system to transport sand at any given flow level. The counterargument is that this increased transport capacity is available only within the navigation channel, and that transport capacity outside the channel has decreased commensurately.

To summarize the changes in sediment transport, available evidence suggests that there has been a dramatic, threefold decline in sand input to the estuary and a 33% decrease in input of fine sediment, primarily as a result of flow regulation. Unlike river flow, where both human intervention and climate change have played a role in changes over the last century, human intervention is the dominant factor in the change in sediment supply. While the volume of sediment not carried to the estuary over the last 40 years is an insignificant fraction of the total upriver storage capacity, it will be shown below to be an important factor in the estuarine sediment budget.

Effects of Modifications to Watershed on Land-Margin Ecosystem Processes

Modification of Circulation and the Dynamics of the Estuarine Turbidity Maximum

Historical changes in estuarine circulation and dynamics of the estuarine turbidity maximum (ETM) in the Columbia River estuary may be inferred by comparing numerical circulation simulations for historical (1868) and modern (1958) bathymetry (Hamilton 1990, Sherwood et al. 1990). The clearest conclusions are those concerning the vertically averaged (barotropic) circulation, as this can be modeled with considerable confidence. A loss of tidal prism of about 15%—caused by shoaling, diking, and filling of peripheral wetlands and construction of navigation structures near the entrance—has brought about a reduction in energy of the tidal circulation. Construction of pile dikes and sand barriers has redirected the river flow into the navigation channel. The south channel through the estuary coincides with the navigation channel and is now more ebb dominant than before, while the north channel is more flood dominant because of loss of freshwater flow. Concentration of flow in the navigation channel has resulted in deepening of the navigation channel and shoaling of peripheral channels. Particularly in the river upstream of the estuary, widespread filling with dredged material has simplified the channel network and substantially narrowed the river by removing peripheral wetlands.

The above changes in the barotropic flow have in turn caused a variety of alterations in the density-driven and residual (tidal-cycle averaged) circulations and in sediment transport processes. In general, less confidence may be placed in evaluations of these changes than in those for the vertically averaged circulation, because these processes are less well understood. Nonetheless, we know at least the direction of the change in almost every instance. The decrease in energy of the tidal circulation has increased residence times of water and particles, and stratification. An increase in the salinity intrusion length on neap tides for any given river flow level must be balanced against a considerable increase in low-flow season minimum river flows. On the whole, it is likely that maximum salinity intrusion length has actually decreased from approximately 65 to 50 km. Minimum salinity intrusion has substantially increased because of the curtailment of freshets. As late as 1959, all salt was expelled from the estuary at the end of ebb on high-flow spring tides (USACE 1960), but this has probably not occurred on a regular basis since the mid-1970s.

A subtle but potentially very important change in the stratification regime has also occurred. Jay and Smith (1990) documented the occurrence of prominent tidal-monthly changes in salinity intrusion length and residual flow patterns associated with changes in stratification. On spring tides, strong tidal mixing destroys most of the stratification. The estuary is weakly stratified on the flood, and salinity intrusion length is at a minimum despite tidal advection, because strong vertical mixing renders impossible an advance of the density field in the form of a salt wedge. Two-layer residual flow is weak or absent. On neap tides, salinity intrusion increases dramatically because vertical mixing is weak enough to allow such a salt-wedge advance. This salt-wedge type of tidal flow causes a prominent two-layer residual circulation. At present, neap-spring transitions are most prominent during low-flow periods. Model results suggest that in the 19th century they were most prominent during the high-flow season (Hamilton 1990). At that time, the energy level of the tidal circulation was sufficient to cause strong mixing under all low-flow tidal conditions. Neap-spring transitions could occur only at intermediate river flow levels where river flow effects were strong enough to overcome tidal mixing. The decrease in tidal mixing has thus caused a shift in tidal-monthly transitions to lower river flow levels.

The importance of these changes in density-driven circulation and residual flow lies primarily in their effect on estuarine sediment transport and biological processes. Both bedload transport of sand and trapping of suspended particles by the estuarine turbidity maximum have been altered. The decrease in energy level in the tidal circulation and the decrease in peak river flows have lessened the ability of the estuary to transport sand seaward. This decrease must be balanced against the decrease in supply and increased removal by dredging. This will be discussed below, under sediment budget considerations.

Physical processes associated with the estuarine turbidity maximum can be thought of as a balancing of three processes: trapping of particles by circulation, settling and resuspension at the bed, and supply of particulates from the river. For example, the prominent two-layer circulation during low-flow neap tides provides ample opportunity for trapping of particles near the upstream limits of salinity intrusion (Figure 9.9, top). However, weak shear stress near the bed under these conditions allows settling of a large part of the suspended particle inventory to the bottom. Thus the present turbidity maximum during the low-flow season is best developed on spring tides, despite relatively weak trapping by the flow. It is likely, in fact, that most export of fine particulates occurs on spring tides, because particle trapping is inefficient under these circumstances. Both sediment supply and trapping capacity are at annual maximum levels during high-flow season because of increased river flow, which also brings increased stratification and a more salt-wedgelike flood tide flow pattern (Figure 9.9, bottom). Because stratification varies less strongly over the tidal month during high-flow season than during the fall (low-flow) season, turbidity maximum processes are believed to exhibit less tidal-monthly variability during this period.

The predevelopment estuary had a stronger tidal circulation in shallower, broader channels. This means that the particle trapping capacity of the flow (which is associated with shear and stratification) was less and bed stresses were stronger. Because minimum river flow levels were lower in the fall than at present, the supply of sediment was also less during this season. Moreover, periods of strong stratification that would have allowed settling of suspended material to the bottom were apparently absent during the low-flow season. This would suggest that the turbidity maximum (at least as measured by the presence of large amounts of mostly inorganic suspended sediment) may have been weak or nearly absent during periods of low flow. It is not clear, however, how closely associated biological activity is with inorganic sediment particles within the estuarine turbidity maximum. It is conceivable that trapping of organic detritus is little influenced by the presence or absence of inorganic sediment fractions and that there was a biologically active turbidity maximum in the predevelopment estuary during the low-flow season despite small total suspended sediment concentrations.

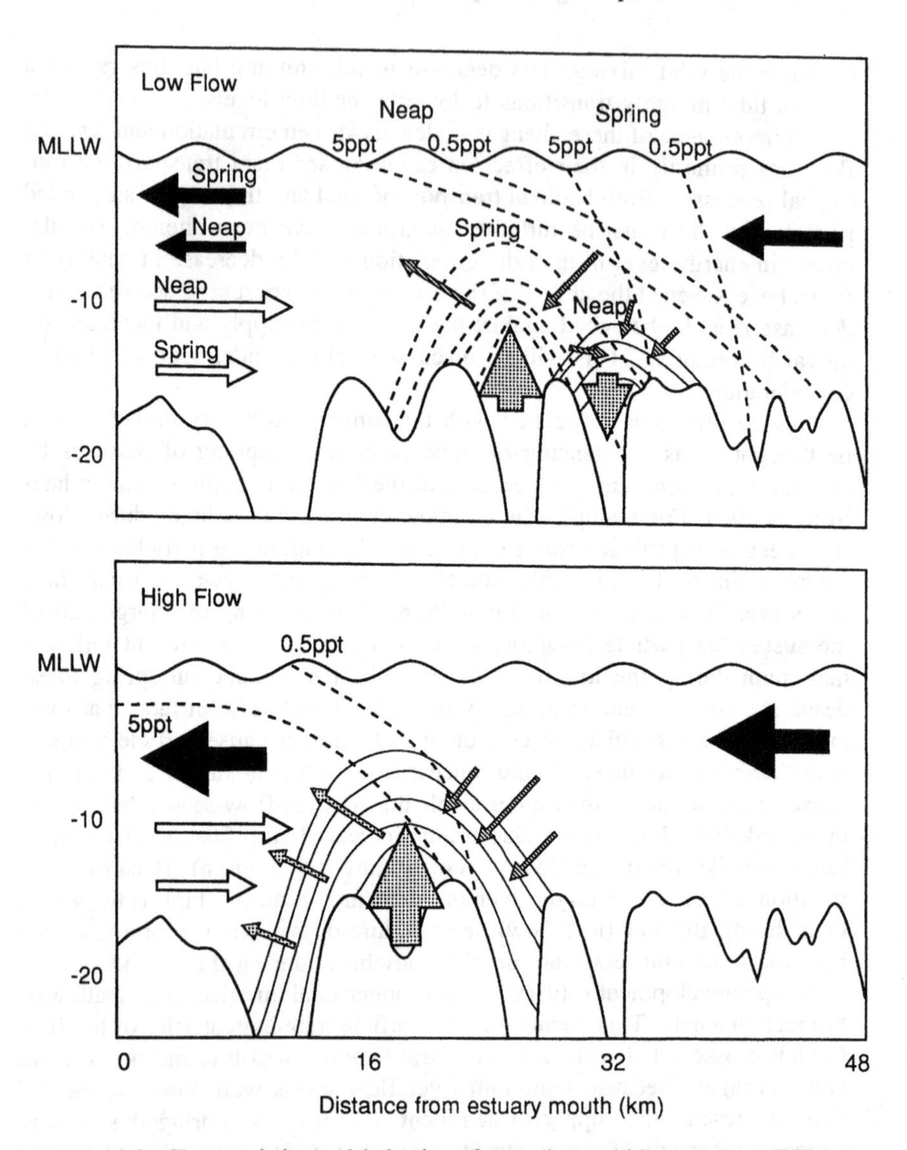

FIGURE 9.9. Characteristics and behavior of estuarine turbidity maxima (ETM) in the Columbia River estuary during low flow (top) and high flow (bottom) situations. Salinity is indicated in parts per thousand (ppt) for neap and spring tide conditions. Freshwater flow is indicated by dark arrows, saltwater intrusion by open arrows, and turbidity transport by stippled arrows. Vertical axis shows meters below mean lower low water (MLLW).

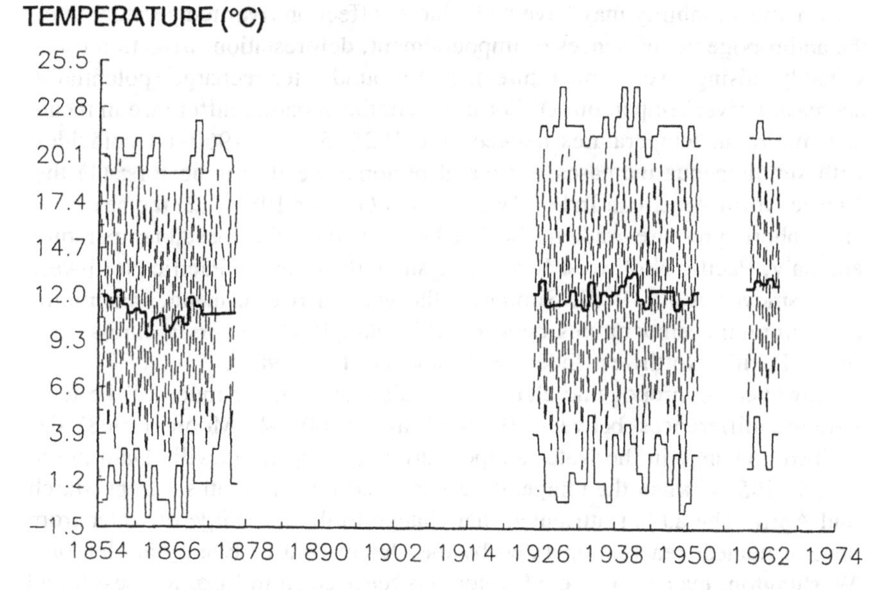

FIGURE 9.10. Three historical periods of surface water temperature recordings from the Columbia River estuary at Astoria, Oregon, illustrating maxima, means, and minima; based on USCGS (1952, 1967) data.

Changes in the Estuarine Heat Budget

Changes in water temperatures at the land margin have not been evaluated previously for the Columbia River estuary. As a preliminary examination of the potential historic changes in river temperature as manifested at the land margin, we utilized the U.S. Geological Survey data for surface water temperature at Astoria (USCGS 1952, 1967) to compare the three periods of continuous recordings: (1) 1854–76, (2) 1925–51, and (3) 1960–64. We assume that these represent equivalent conditions in terms of tidal mixing, such that the only differences could probably be attributed to change in river temperature and river flow. Examination of the monthly mean, maximum, and minimum temperatures (Figure 9.10) suggests differences between the three periods. Estuarine surface temperatures in the earliest period have annual means between 9.6° and 12.2°C, with extreme excursions below freezing (to -1°C) and to 21°C during four years; there may have been a slight downward trend in the mean temperature over the 23-year period. Between 1925 and 1951, the mean was approximately 1°C higher, between 11.1° and 13.2°C, and both the maxima (20.6° to 23.9°C) and minima (-0.6° to 3.9°C) were higher than the previous period. In the last, short period of record, temperatures averaged between 12.3° and 12.9°C, with the maxima between 21° and 22°C, but the minima were all above 2°C.

Climatic variability may have had a larger effect on river temperature than the anthropogenic influences of impoundment, deforestation, irrigation (presumably raising river temperatures), and groundwater recharge (potentially decreasing river temperatures). For instance, the apparent difference in mean and minimum temperatures between the 1925–51 and 1960–64 coincides with simultaneous increases in several regional weather indicators: (1) the Pacific North American Index (Wallace and Gutzler 1981), (2) the sea level atmospheric pressure gradient (Seckel 1988), and (3) the coastal temperature anomaly, Pacific Northwest Index, and snow depth in the Pacific Northwest (Ebbesmeyer et al. 1989). Similarly, the general rise in mean, maximum, and minimum temperatures between 1935 and 1941 mirrors a similar rise in the Pacific Northwest Index (Ebbesmeyer et al. 1989).

However, anthropogenic change may also account for much of the temperature differences between 1925–51 and 1960–64. Moore (1968) described a change in the water-temperature regime at Bonneville Dam in the middle 1950s, when the temperature increased for all months except March and April. The author attributed that change to the discharge of water from several nuclear reactors operated by the Department of Energy at Hanford, Washington, even though cold water was released from Lake Roosevelt and the Brownlee Reservoir to compensate for warming caused by the Hanford operations. Davidson (1964) indicated that the greater part of a 1.1° to 1.7°C rise in the river temperature between Priest Rapids (above Hanford) and Richland (~40 km below Hanford) was attributable to the Hanford operations. In his analysis of the effects of impoundment behind the Wanapum and Priest Rapids dams and the Wells and Rocky Reach dams, Raphael (1961 and 1962, respectively) estimated that surface water temperatures rise markedly in these impoundments during periods of low to moderate river flow and exceptionally warm weather, on the order of 1°C per 32 km of the river. Davidson (1969) also calculated that temperatures at Rock Island Dam averaged 10.3°C during low-flow periods in October 1913 to 1940, compared to 10.5°C between 1941 and 1968. Obviously, riverine impoundments behind dams have, in association with climate variability, a potential to contribute to cumulative increases in river temperatures. While the contribution of dams to the temperature regime downstream depends in part on how the dams are operated, increased regional temperatures would certainly increase the demand for water, further exacerbating impacts on the fluvial and estuarine heat budget.

Organic Carbon Input and Food Web Processes

Both the quantitative sources of production and the qualitative linkages between consumers and their food resources in the Columbia River estuary are affected by changes in the riverine inputs and the dynamic processes within the estuary. Annual phytoplankton production accounts for 57% of the annual primary production in the estuary; at only 100 g C m^{-2} yr^{-1}, the Columbia system is low among North American estuaries in water column production (Lara-Lara et al. 1990*b*). This production has been shown by several estimation methods, including the effects of the passage of the turbidity load from the Mount St. Helens eruption through the estuary, to be limited primarily by light attenuation (Lara-Lara et al. 1990*a, b*; Small et al. 1990). Thus any decrease in surface water turbidity that may have occurred as a function of decreased discharge of suspended fine sediments since 1868 (Figures 9.7 and 9.8) would probably have caused increased phytoplankton production. The estuary's rapid flushing rate is another significant contributor to the low phytoplankton production (Lara-Lara et al. 1990*b*). Increased residence time of particles in the estuary as a result of the modifications in estuarine circulation and ETM dynamics may also have enhanced phytoplankton production. The magnitude of these increases cannot at present be estimated.

Primary production by freshwater phytoplankton decreases in the portion of the estuary within the influence of salinity intrusion because the cells lyse upon encountering saline waters of $0^{\circ}/_{\infty}$ to $5^{\circ}/_{\infty}$ (Small et al. 1990). Changes in salinity intrusion length associated with regulation of the seasonal river flow cycle may have decreased production in the spring season and increased it in the summer and fall. In any event, the lysation of fluvial phytoplankton appears to add labile carbon (FPOC) to the detritus pool in the estuary, and this FPOC is likely to become trapped in the estuarine turbidity maximum, where it may contribute (via a microbial food web link) to the production of detritivorous consumers such as epibenthic zooplankters. Since the advent of flow regulation, the dominant source of fluvial phytoplankton biomass transported to the estuary is thought to be enhanced production from upriver reservoir impoundments (Simenstad et al. 1990*a*). Phytoplankton biomass, as measured by fluorosensor, has been observed to increase 4x from the middle reaches of the Snake (Lower Granite Dam reservoir) to the mouth of the river (Bristow et al. 1985; Figure 9.11, top); dissolved organic matter does not appear to be similarly enhanced, and is constant through much of the lower river (Figure 9.11, bottom).

Sherwood et al. (1990) argued on the basis of a hindcast estuarine carbon budget (Figure 9.12) that the organic base of the estuary's detritus-based food web has shifted from predominantly endogenous wetland vegetation (i.e., macrodetritus) to exogenous riverine phytoplankton, detrital (e.g., microdetritus) and dissolved carbon. The wetland-generated macrodetritus has decreased significantly with the extensive diking and filling of forested wetlands and emergent marshes (Thomas 1983, Sherwood et al. 1990). However, the impoundment of the river compensates the estuarine food web by providing freshwater phytoplankton production for the estuarine detritus pool. In comparison, the additional decline in LPOC input from the watershed (Gregory et al. 1987) has not likely contributed to a significant change in macrodetritus. However, the role of LPOC in enhancing the structural het-

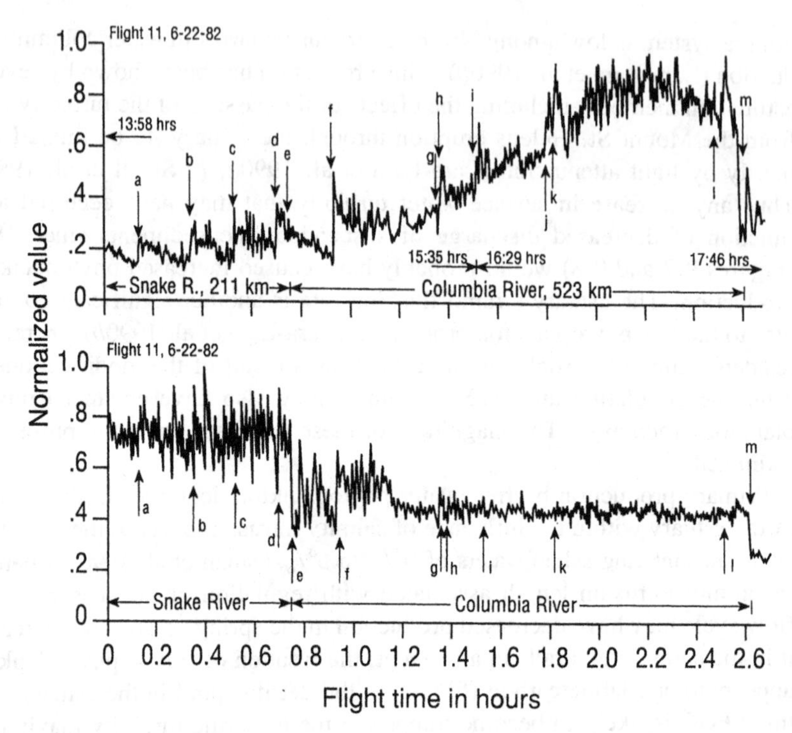

FIGURE 9.11. Profiles of chlorophyll-*a* fluorescence (top) and dissolved organic matter fluorescence (bottom) as a function of aircraft time during airborne laser fluorosensor survey of Snake and Columbia rivers from Lower Granite Dam reservoir to mouth; from Bristow et al. (1985). Key to location symbols: a=Lower Granite Dam; b=Little Goose Dam; c=Lower Monunmental Dam; d=Ice Harbor Dam; e=confluence of Snake and Columbia; f=McNary Dam; g=John Day River; h=John Day Dam; i=The Dalles Dam; j=Cascade Rapids; k=Bonneville Dam; l=groundtruthing sampling site; and m=mouth of Columbia River.

erogeneity and fish and wildlife habitat complexity in estuarine habitats has undoubtedly declined (Gonor et al. 1988).

Consumer Populations and Ecology

Watershed influences on the production, structure, and ecology of consumers in the Columbia River estuary include: (1) modification of anadromous populations that regularly dominate predator-prey interactions in the estuary during their migrations, (2) introduction of exotic species that infiltrate and modify estuarine communities, and (3) alteration of estuarine processes that control either trophic or competition-predation interactions among consumer organisms in the estuary.

One of the greatest historic changes in the role of consumer organisms in the Columbia River estuary has been the population and demographic changes

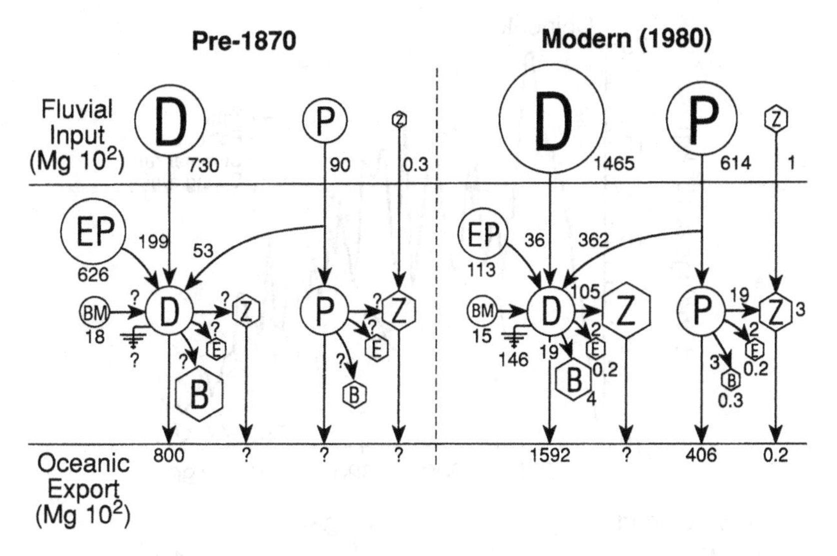

FIGURE 9.12. Hypothetical historical (left) and modern (right) carbon sources (in Mg) and pathways of the detritus-based food web in the Columbia River estuary. (*Note*. From Sherwood et al. 1990. Copyright by Pergamon Press PLC. Used by permission.)

in the abundances and assemblage structure of anadromous fishes migrating through the watershed and estuary. From the standpoint of the linkages between watershed and land margin, migrating juvenile salmon and steelhead (*Oncorhynchus* spp.) are extremely prominent estuarine predators and prey. Adults migrating back through the estuaries from their oceanic feeding grounds may also provide an important input of nutrients to the tributaries to the estuary and in spawning streams (Bisson et al., this volume). The diverse life histories of the myriad salmon and steelhead stocks in the Columbia River catchment produce a complex pattern of migrations through the estuary in various pulses throughout the year.

Although the initial decline in salmon populations between 1880 and 1930 has been attributed principally to overexploitation, modifications to freshwater spawning and rearing habitats in the Columbia River catchment undoubtedly account for decline or a lack of recovery in more modern time (Chapman 1986, Nehlsen et al. 1991). Of the over 200 stocks of anadromous salmonids on the West Coast considered to be at some risk of becoming extinct, most of those stocks in the high risk category originate in the Columbia River Basin.

Historical changes in estimated abundances of spring and summer chinook salmon (*O. tshawytscha*) returning to the Snake River and mid-Columbia River above Priest Rapids Dam since the 1930s (Figure 9.13) suggest that these populations have declined as a result of the hydroelectric development

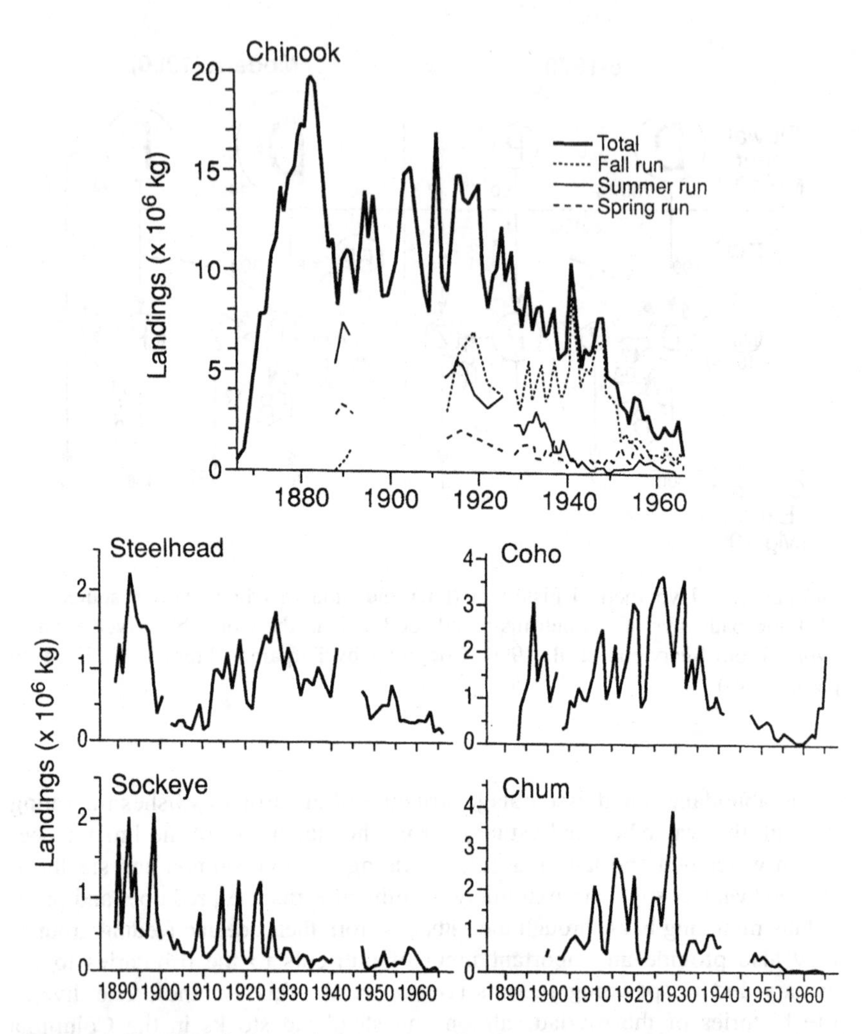

FIGURE 9.13. Historical trends in total catch of chinook salmon (1868–1966) and steelhead, coho salmon, sockeye salmon, and chum salmon (1888–1940, 1947–66) in the Columbia River; from Beiningen (1976) and Chapman (1986).

of the river (Raymond 1988) and other factors associated with watershed management practices (Bisson et al., this volume). One important factor has been increased mortality of juveniles migrating downstream through the dams and impoundments. Potential mechanisms of the cumulative mortality include gas supersaturation causing gas bubble disease, physical damage from passing through turbines, physiological stress and increased disease due to higher temperatures and pollutants, loss of genetic diversity with dilution by hatchery stocks, and increased predation by expanding populations of piscivorous fishes, such as the northern squawfish (*Ptychocheilus oregonensis*), in the reservoirs (Nehlsen et al. 1991); but the contribution of these interacting factors is poorly understood.

Habitat loss and degradation are also major contributors to the decline of Pacific salmon in the Columbia River Basin. Over 55% of the drainage basin has been blocked to upstream migrant salmon. Spawning areas for summer and spring chinook runs have virtually disappeared above Grand Coulee, Dexter, and Lewiston dams, and spawning areas for fall chinook have been lost in the Cowlitz, Lewis, Willamette, Wind, Little White Salmon, and Hood rivers (Fulton 1968). Mainstem spawning, principally by fall chinook salmon, has been curtailed by the inundation behind Chief Joseph, John Day, and Hells Canyon dams. Sockeye salmon (*O. nerka*) production in the Columbia River Basin is now marginal; of eight naturally productive lake systems, only three (Wenatchee, Osoyoos, and Redfish) support remnant populations (Fulton 1970). Chum salmon (*O. keta*) are not dependent on spawning areas in the upper and middle Columbia River Basin; only the reservoir behind Bonneville Dam inundated their spawning habitat. Although they are thus comparatively unaffected by the reservoir and dam migration gauntlet, they have not recovered from the major declines imposed by overfishing and habitat degradation at the turn of the century. Habitat degradation tends to be a persistent factor. Deleterious changes in the stream environments required for salmonid spawning and rearing have been produced by forestry and other intensive land use practices, such as simplification of stream channels and loss of habitat complexity (Holtby 1988; Bisson et al., this volume). It is likely that continued degradation of spawning habitats in the lower river's tributaries will inhibit recovery despite curtailed commercial fishing on chum salmon stocks. Only coho salmon (*O. kisutch*) and steelhead (*O. mykiss*) populations appear to be recovering pre-1940s population levels, principally due to exhaustive enhancement by hatcheries. Hatchery production, however, does not necessarily compensate lost natural salmonid production, and may actually impair wild stocks through a variety of mechanisms (Bisson et al., this volume).

Other endemic, anadromous species that occur in the Columbia River system, and play important functional roles in the estuary, are eulachon (*Thaleichthys pacificus*) and longfin smelt (*Spirinchus thaleichthys*) (Bottom and Jones 1990). Although their interactions in the estuary may have, like Pacific salmon, been strongly affected by watershed alterations, historic data on their populations are not sufficient to allow such an analysis.

Exotic species have also altered the nature of the Columbia River estuary (Bisson et al., this volume). Some species of fish, such as common carp (*Cyprinus carpio*), have become established in the system from introductions in the watershed. Other populations of anadromous exotics, such as American shad (*Alosa sapidissima*), were introduced intentionally into watersheds farther south and expanded into the Columbia River estuary and watershed. All of these species are prominent in the estuary's fish fauna, and American shad have become a prominent component of the estuary's pelagic zooplanktivores. Another introduced species, nutria (*Myocastor coypus*), has become a common and relatively important herbivore on emergent wetland plants in the estuary's marshes and swamps.

Recent studies in the Columbia River estuary have also documented the prominence of an exotic calanoid copepod, *Pseudodiaptomus inopinus* (Cordell et al. 1992), that is endemic to tidal freshwater and brackish waters of Asia (Shen 1979). This species was not found at all during intensive studies in the estuary in 1979–80 (Simenstad and Cordell 1985, Jones et al. 1990). The organisms were most likely introduced into the river in ballast water dumped by ships that had visited Asian ports. Although shipping traffic to the Columbia River from Asia has increased twofold during the last decade, the practice of pumping ballast water has been occurring for many decades. Three other possible explanations for the sudden establishment and numerical prominence of *P. inopinus* have been put forward: (1) numerous, repeated introductions may have been required before a critical density could be achieved; (2) undetectable, small populations (low genetic diversity) may have been present but required several decades to adapt to the conditions in the estuary before becoming reproductively viable; and (3) conditions may have been eliminated that had previously inhibited survival and reproduction (Carlton 1985). The species had not been reported previously on the West Coast of the United States, but two conspecifics, *P. marinus* and *P. forbesi*, have been reported in San Francisco Bay (Orsi and Walter 1991). A third Asian species of calanoid, *Sinocalanus doerrii*, also has increased rapidly in abundance and distribution throughout that estuary (Orsi et al. 1983). That San Francisco Bay and the Sacramento-San Joaquin Bay estuary have been invaded by these calanoids, and in an even more spectacular way by the clam *Potamocorbula amurensis* (Carlton et al. 1990), suggests that land-margin ecosystems stressed by water diversion and/or drought may be exceptionally vulnerable to establishment of invasive species. These invasions may be mediated by changes in limiting factors such as thermal and salinity regimes, competitors, and predators (Roberts 1990).

The consequences of historic changes to the Columbia River estuary's biotic communities are unclear. Even today, despite the potential extinction of some wild salmonid stocks in the Columbia River Basin, juvenile salmon are a significant component of the estuary's fish fauna during peak migration periods (Bottom and Jones 1990). Thus it is likely that they historically accounted for a considerably greater proportion of the estuary's aquatic secondary consumption by their active foraging on benthic organisms. Qualitative changes in the estuarine function of juvenile salmon have also occurred. Changes in the demography (temporal occurrence, density, size distributions, and residence time) of estuarine populations have most likely altered the effects of their predation on prey organisms. Whereas stocks that migrate through the estuary as fry or small, subyearling fish (e.g., chum, fall and summer chinook) occupy estuarine habitats the longest, and contribute most significantly to the estuarine food web, those stocks that migrate as yearlings and smolts (e.g., spring chinook, coho, steelhead) tend to pass rapidly through the estuary. Populations that have undergone the greatest and most persistent declines have historically served as important predator-prey linkages in the estuary. Hatchery enhancement of salmonid stocks intended to supplement population declines in the "wild" stocks has instead, by artificial feeding, produced larger juveniles ("smolts") rather than fry or subyearlings. Finally, compared to the historically diffuse migration of juveniles through the system, hatchery enhancement practices of large-scale, punctuated releases of fish undoubtedly produce more concentrated pulses of fish; the ecological effects on the salmon and their estuarine prey resources are unknown.

In summary, while the biotic communities of the Columbia River estuary do not exhibit striking land-margin impacts from intensive urbanization such as characterize the watersheds of Chesapeake Bay and the New York Bight (Orth and Moore 1983, Seliger et al. 1985), significant alteration of the estuary's basic physical, geochemical, and ecological processes by anthropogenic manipulation of the watershed has modified estuarine communities and food web dynamics. These modifications are representative of estuaries that have experienced significant water withdrawal and flow manipulation, such as San Francisco Bay (Nichols et al. 1986).

Discussion

Both improvements in some and continued degradation of other aspects of land use practices in the Columbia River Basin can be expected in the future. For instance, inputs of nutrients, fine sediments, and pollutants are likely to be further curtailed, and human influences on stream temperatures may moderate somewhat. Modern forestry and agricultural practices, if systematically employed, could considerably reduce these impacts. Similarly, water conservation measures could greatly decrease consumptive use. Recovery of other degraded processes, such as sediment transport, is much more problematic. Restoration of at least short-term freshet flows would be required, but it is dubious that sufficient discharge could be obtained to affect sediment deposition and accretion in the estuary without severely impacting the Columbia River's energy and water users. Moreover, future modifications to the watershed and estuary and continuing intensive land use cannot be ignored in anticipating impacts at the land margin. For example, deepening the dredged navigation channel from 13 to 16–18 m has been proposed. This would cause a major increase in dredge material disposal, increased salinity intrusion and stratification, and probably increase the trapping of particles in the estuarine turbidity maximum (Hamilton 1990). Any additional filling and diking of intertidal and other wetland habitats would decrease the tidal prism, causing further declines in flushing and increases in residence time.

Some restoration of freshet flows may, however, occur for a very different reason. Stocks of wild salmon are so severely depressed that classification of summer and fall chinook and sockeye salmon as endangered under the Endangered Species Act is imminent. This may require managers of the Columbia River water system (i.e., Bonneville Power Administration) to increase spring flows to enhance the migration of juveniles downstream. Additional watershed management practices may be imposed to enhance natural salmonid production and demography by restoring natural spawning and rearing habitat characteristics such as flow, temperature regimes, and input of LPOC debris. In the same light, hatchery practices should be examined with the objective of promoting salmon survival and passage through estuaries, rather than presuming that rapid downstream movement and estuarine residence are ultimately an advantage.

If we are to apply landscape and watershed management practices on an ecosystem scale, we need a better understanding of flow manipulation effects on estuarine circulation, estuarine turbidity dynamics, sedimentation, and biology. Detailed descriptive studies, such as the Columbia River Estuary Data Development Program (CREDDP; Simenstad et al. 1990*b*), or intensive process studies, such as the National Science Foundation's Land-Margin Ecosystem Research (LMER) studies initiated recently in the estuary, can provide critical information on the structure and processes of the estuary. Such studies can generate but not often test hypotheses concerning watershed and land-margin interactions. Ultimately, testing of such hypotheses and development of corresponding management strategies would require both manipulative experiments and extensive numerical modeling of ecosystem processes. Experiments might involve short-term river flow and storage manipulations to examine responses of estuarine circulatory, sediment transport, and biological processes. These experiments should be used to test the numerical model, which could then be utilized to investigate alternative scenarios of watershed, regional, and climate change.

Obviously, global trends may eclipse anthropogenic changes in flow characteristics, physical properties, and constituents of the Columbia River that alter processes at the land margin. For example, interannual transfers of flow are small relative to changes in annual river flow caused by natural, climatic variability over decades (Figure 9.6). Most global change scenarios involve decreased precipitation and increased temperatures that would exacerbate the historic watershed manipulations. For example, a 1° to 2°C warming coupled with 10% less precipitation could reduce runoff an average of 40 to 70% across the country (Houghton et al. 1990), although the effect may be somewhat less in a watershed as diverse as the Columbia (Dolph et al., this volume). This decreased runoff would be coupled with increased biotic productivity in impoundments and increased estuarine microdetritus input. Present policies of flow redistribution for hydropower production and irrigation in the Columbia River Basin would significantly lower river flows in the winter and spring, causing major changes in estuarine circulation. Employment of even more drastic management strategies to cope with dramatic climate changes would dictate consideration of responses from the land-margin ecosystem. Meeting the challenge of climate change may even prompt further large-scale interconnections among reservoirs and watersheds (Waggoner 1991; Stanford and Ward, this volume). Such an approach could critically exacerbate already stressed land-margin ecosystems such as San Francisco Bay/ Sacramento-San Joaquin River estuary and stress such systems as the Columbia River estuary.

Given the continuing degradation of land-margin ecosystems and the impending threat of global climate change, there is a need to instill an ecosystem perspective in our studies and in our management of watersheds. If we persist in ignoring downstream, land-margin effects, ecosystems such as the Columbia River estuary will be thrust in the same adverse direction as more stressed or failing systems such as San Francisco Bay and the Sacramento-San Joaquin estuary. Just as in highly urban-suburban and agricultural-based ecosystems, the challenge of managing extensive estuarine ecosystems is to a large degree one of managing their watersheds (Howarth et al. 1991; Stanford and Ward, this volume).

Acknowledgments. Preparation of this manuscript was supported under NSF LMER Grant OCE-8907118 to C. Simenstad and D. Jay at the University of Washington; C. Sherwood acknowledges the support of Battelle Pacific Northwest Laboratories, Marine Sciences Laboratory, Sequim, Washington. The authors wish to express their appreciation to Ms. Stephanie Martin for her editing efforts on the manuscript and to two anonymous reviewers for their constructive comments. Figures from Sherwood et al. (1990) have been reprinted with permission from Pergamon Press PLC. This is Contribution 835 of Fisheries Research Institute, School of Fisheries, and 1904 of the School of Oceanography, University of Washington.

References

Andrews, E.D. 1986. Downstream effects of Flaming Gorge Reservoir on the Green River, Colorado and Utah. Geological Society of America Bulletin **97**:1012–1023.

Barnes, C.A., A.C. Duxbury, and B.A. Morse. 1972. Circulation and selected properties of the Columbia River effluent at sea. Pages 41–80 *in* A.T. Pruter and D.L. Alverson, editors. The Columbia River estuary and adjacent ocean waters. University of Washington Press, Seattle, Washington, USA.

Beiningen, K.T. 1976. Fish runs. Pages E1-E65 *in* Investigative reports of the Columbia River fisheries project. Pacific Northwest Regional Commission, Vancouver, Washington, USA.

Bottom, D.L., and K.K. Jones. 1990. Species composition, distribution, and invertebrate prey of fish assemblages in the Columbia River estuary. Progress in Oceanography **25**:243–270.

Bristow, M.P.F., D.H. Bundy, C.M. Edmonds, P.E. Ponto, B.E. Frey, and L.F. Small. 1985. Airborne laser fluorosensor survey of the Columbia and Snake rivers: simultaneous measurements of chlorophyll, dissolved organics and optical attenuation. International Journal of Remote Sensing **6**:1707–1734.

Bronfman, A.M. 1977. The Azov Sea water economy and ecological problems: investigation and possible solutions. Pages 39–58 *in* G.F. White, editor. Environmental effects of complex river development. Westview Press, Boulder, Colorado, USA.

Carlton, J.T. 1985. Transoceanic and interoceanic dispersal of coastal marine organisms: the biology of ballast water. Oceanography and Marine Biology Review **23**:313–371.

Carlton, J.T., J.K. Thompson, L.E. Schemel, and F.H. Nichols. 1990. Remarkable invasion of San Francisco Bay (California, USA) by the Asian clam *Potamocorbula amurensis*. I. Introduction and dispersal. Marine Ecology Progress Series **66**:81–94.

Chapman, D.W. 1986. Salmon and steelhead abundance in the Columbia River in the nineteenth century. Transactions of the American Fisheries Society **115**:662–670.

Cordell, J.R., C.A. Morgan, and C.A. Simenstad. 1992. Establishment of the Asian calanoid copepod *Pseudodiaptomus inopinus* and its relationship with *Eurytemora affinis* (Copepoda: Calanoida) and *Scottolana canadensis* (Copepoda: Harpacticoida) in the Columbia River estuary. Journal of Crustacean Biology, *in press*.

Cross, R., and D. Williams, editors. 1981. Proceedings of the National Symposium on Freshwater Inflow to Estuaries. United States Fish and Wildlife Service FWS/OBS-81/04, Office of Biological Services, Washington, D.C., USA.

Davidson, F.A. 1964. The temperature regime of the Columbia River from Priest Rapids, Washington to the Arrow Lakes in British Columbia. Public Utility District 2, Grant County, Ephrata, Washington, USA.

Davidson, F.A. 1969. Columbia River temperature at Rock Island Dam from 1913 to 1968. Public Utility District 2, Grant County, Ephrata, Washington, USA.

Depletions Task Force. 1983. 1980 level modified streamflow, 1928–1978: Columbia River and coastal basins. Columbia River Water Management Group, Portland, Oregon, USA.

Dunne, T., and L.B. Leopold. 1981. Flood and sedimentation hazard in the Toutle and Cowlitz River system as a result of the Mt St Helens eruptions. Report to Federal Emergency Management Agency, Region X, January 1981.

Ebbesmeyer, C.C., C.A. Coomes, G.A. Cannon, and D.E. Bretschneider. 1989. Linkage of ocean and fjord dynamics at decadal period. Geophysics Monographs **55**:399–417.

Ellis, W.S. 1990. The Aral: a Soviet sea lies dying. National Geographic, February: 73–92.

Fairchild, L.H., and M. Wigmosta. 1983. Dynamic and volumetric characteristics at the 18 May 1980 lahars on the Toutle River, Washington. *In* Proceedings, Symposium on Erosion Control in Volcanic Areas, July 6–9, 1982. Technical Memorandum 1908, Public Works Research Institute, Ministry of Construction, Government of Japan.

Fulton, L.A. 1968. Spawning areas and abundance of chinook salmon (*Oncorhynchus tshawytscha*) in the Columbia River Basin—past and present. United States Fish and Wildlife Service Special Scientific Report 571.

Fulton, L.A. 1970. Spawning areas and abundance of steelhead trout and coho, sockeye, and chum salmon in the Columbia River Basin—past and present. United States Fish and Wildlife Service Special Scientific Report 618.

Gonor, J.J., J.R. Sedell, and P.A. Benner. 1988. What we know about large trees in estuaries, in the sea, and on coastal beaches. Pages 83–112 *in* C. Maser, R.F. Tarrant, J.M. Trappe, and J.F. Franklin, editors. From the forest to the sea: a story of fallen trees. United States Forest Service General Technical Report PNW-GTR-229, Pacific Northwest Forest and Range Experiment Station, Portland, Oregon, USA.

Gregory, S.V., G.A. Lamberti, D.C. Erman, KV. Koski, M.L. Murphy, and J.R. Sedell. 1987. Influence of forest practices on aquatic production. Pages 233–255 *in* E.O. Salo and T.W. Cundy, editors. Streamside management: forestry and fishery interactions. Contribution 57, Institute of Forest Resources, University of Washington, Seattle, Washington, USA.

Hamilton, P. 1990. Modeling salinity and circulation for the Columbia River estuary. Progress in Oceanography **25**:113–156.

Holtby, L.B. 1988. Effects of logging on stream temperatures in Carnation Creek, British Columbia, and resultant impacts on the coho salmon (*Oncorhynchus kisutch*). Canadian Journal of Fisheries and Aquatic Sciences **45**:502–515.

Houghton, J.T., G.J. Jenkins, and J.J. Ephraums, editors. 1990. Climate change: the IPCC scientific assessment. Cambridge University Press, Cambridge, England.

Howarth, R.W., J.R. Fruci, and D. Sherman. 1991. Inputs of sediment and carbon to an estuarine ecosystem: influence of land use. Ecological Applications **1**:27–39.

Hunt, C.E. 1988. Down by the river. Island Press, Washington, D.C., USA.

Jay, D.A., B.S. Giese, and C.R. Sherwood. 1990. Columbia River estuary: energetics and sedimentary processes. Progress in Oceanography **25**:157–174.

Jay, D.A., and J.W. Good. 1978. Columbia River estuary flushing characteristics. *In* M.H. Seaman, editor. Columbia River estuary inventory of physical, biological and cultural characteristics. Columbia River Estuary Study Taskforce, Astoria, Oregon, USA.

Jay, D.A., and J.D. Smith. 1990. Circulation, density distribution and neap-spring transitions in the Columbia River estuary. Progress in Oceanography **25**:81–112.

Jones, K.K., C.A. Simenstad, D.L. Higley, and D.L. Bottom. 1990. Community structure, distribution, and standing stock of benthos, epibenthos, and plankton in the Columbia River estuary. Progress in Oceanography **25**:211–241.

Kemp, M.W., R.L. Wetzel, W.R. Boynton, C.F. D'Elia, and J.C. Stevenson. 1982. Nitrogen cycling and estuarine interfaces: some current concepts and research directions. Pages 209–230 *in* V.S. Kennedy, editor. Estuarine comparisons. Academic Press, New York, New York, USA.

Kempe, S. 1983. Impact of Aswan High Dam on water chemistry of the Nile. SCOPE/UNEP Sonderband **55**:401–423.

Kempe, S., and U. Lammerz. 1983. Statistical interpretation of hydrochemical data from major world rivers. SCOPE/UNEP Sonderband **55**:39–54.

Lara-Lara, J.R., B.E. Frey, and L.F. Small. 1990*a*. Primary production in the Columbia River estuary. I. Spatial and temporal variability of properties. Pacific Science **44**:17–37.

Lara-Lara, J.R., B.E. Frey, and L.F. Small. 1990*b*. Primary production in the Columbia River estuary. II. Grazing losses, transport, and a phytoplankton carbon budget. Pacific Science **44**:38–50.

Mancy, K.H. 1979. The Aswan High Dam and its environmental implications. Socita Internationalis Limnologiae Workshop on Limnology of African Lakes, Nairobi, Kenya.

Micklin, P.P. 1988. Desiccation of the Aral Sea: a water management disaster in the Soviet Union. Science **241**:1170–1176.

Moore, A.M. 1968. Water temperatures in the lower Columbia River. United States Geological Survey Circular 551, Washington, D.C., USA.

Nehlsen, W., J.E. Williams, and J.A. Lichatowich. 1991. Pacific salmon at the crossroads: stocks of salmon at risk from California, Oregon, Idaho, and Washington. Fisheries **16**:4–21.

Nichols, F.H., J.E. Cloern, S.N. Luoma, and D.H. Peterson. 1986. The modification of an estuary. Science **231**:567–573.

Nixon, S.W., C.A. Oviatt, J. Frithsen, and B. Sullivan. 1986. Nutrients and the productivity of estuarine and coastal marine ecosystems. Journal of Limnological Society of South Africa **12**:43–71.

Officer, C.B., R.B. Biggs, J.L. Taft, L.E. Cronin, M.A. Tyler, and W.R. Boynton. 1984. Chesapeake Bay anoxia: origin, development, and significance. Science **223**:22–27.

Orem, H.M. 1968. Discharge in the lower Columbia River Basin, 1928–1965. United States Geological Survey Circular 550, Washington, D.C., USA.

Orsi, J.J., T.E. Bowman, D.C. Marelli, and A. Hutchinson. 1983. Recent introduction of the planktonic calanoid copepod *Sinocalanus doerri* (Centropagidae) from mainland China to the Sacramento-San Joaquin estuary of California. Journal of Plankton Research **5**:357–375.

Orsi, J.J., and T.C. Walter. 1991. *Pseudodiaptomus forbesi* and *P. marinus* (Copepoda: Calanoida), the latest copepod immigrants to California's Sacramento-San Joaquin estuary. Proceedings of Fourth International Conference on Copepodology, *in press*.

Orth, R.J., and K.A. Moore. 1983. Chesapeake Bay: an unprecedented decline in submerged aquatic vegetation. Science **222**:51–53.

Paulsen, C.G. 1949. Floods of May-June 1948 in the Columbia River Basin. United States Geological Survey Water Supply Paper 1080, Washington, D.C., USA.

Petts, G.E. 1984. Impounded rivers: perspectives for ecological management. John Wiley and Sons, Chichester, England.

Pruter, A.T., and D.L. Alverson, editors. 1972. The Columbia River estuary and adjacent ocean waters. University of Washington Press, Seattle, Washington, USA.

Raphael, J.M. 1961. The effect of Wanapum and Priest Rapids dams on the temperature of the Columbia River. Public Utility District 2, Grant County, Ephrata, Washington, USA.

Raphael, J.M. 1962. The effect of Wells and Rocky Reach dams on the temperature of the Columbia River. Public Utility District 2, Grant County, Washington, Ephrata, Washington, USA.

Raymond, H.L. 1988. Effects of hydroelectric development and fisheries enhancement on spring and summer chinook salmon and steelhead in the Columbia River Basin. North American Journal of Fisheries Management **8**:1–24.

Remisova, S.S. 1984*a*. Water balance of the Sea of Azov. Journal of Water Research **1**:109–121.

Remisova, S.S. 1984*b*. Salt balance of the Sea of Azov. Journal of Water Research **3**:9–14.

Roberts, L. 1990. Why do some invasions succeed? Science **249**:1371.

Rosenberg, R. 1985. Eutrophication—the future marine coastal nuisance? Marine Pollution Bulletin **16**:227–231.

Rozengurt, M.A., and J.W. Hedgpeth. 1989. The impact of altered river flow on the ecosystem of the Caspian Sea. Review of Aquatic Sciences **1**:337–362.

Rozengurt, M.A., M.J. Herz, and M. Josselyn. 1987. The impact of water diversions on the river-delta-estuary-sea ecosystems of San Francisco Bay and the Sea of Azov. *In* D.M. Goodrich, editor. San Francisco Bay: issues, resources status and management. NOAA Estuary of the Month, Seminar Series 6, NOAA, Washington, D.C., USA.

Sanders, B.M., K.D. Jenkins, W.G. Sunda, and J.D. Costlow. 1985. Chesapeake Bay: an unprecedented decline in submerged aquatic vegetation. Science **222**:51–54.

Schuster, R.L. 1981. Effects of the eruption on civil works and operations in Pacific Northwest. Pages 701–718 *in* P.W. Lipman and D.R. Mullineaux, editors. The 1980 eruption of Mt. St. Helens, Washington. United States Geological Survey Professional Paper 1250, Washington, D.C., USA.

Seckel, G.R. 1988. Indices for mid-latitude North Pacific winter wind systems: an exploratory investigation. Geojournal **16.1**:97–111.

Sedell, J.R., P.A. Bisson, F.J. Swanson, and S.V. Gregory. 1988. What we know about large trees that fall into streams and rivers. Pages 47–81 *in* C. Maser, R.F. Tarrant, J.M. Trappe, and J.F. Franklin, editors. From the forest to the sea: a story of fallen trees. United States Forest Service General Technical Report PNW-GTR-229.

Seliger, H.H., J.A. Boggs, and W.H. Biggley. 1985. Catastrophic anoxia in the Chesapeake Bay in 1984. Science **228**:70–73.

Shen, C.J. 1979. Fauna Sinica. Page 410 *in* Crustacea freshwater Copepoda. Fauna Editorial Committee, Academica Sinica, Science Press, Peking, China.

Sherwood, C.R., and J.S. Creager. 1990. Sedimentary geology of the Columbia River estuary. Progress in Oceanography **25**:15–79.

Sherwood, C.R., D.A. Jay, R.B. Harvey, P. Hamilton, and C.A. Simenstad. 1990. Historical changes in the Columbia River estuary. Progress in Oceanography **25**:299–352.

Simenstad, C.A., and J.R. Cordell. 1985. Structural dynamics of epibenthic zooplankton in the Columbia River delta. Internationale Vereinigung für theoretische und angewandte Limnologie, Verhandlungen **22**:2173–2182.

Simenstad, C.A., L.F. Small, and C.D. McIntire. 1990*a*. Consumption processes and food web structure in the Columbia River estuary. Progress in Oceanography **25**:271–298.

Simenstad, C.A., L.F. Small, C.D. McIntire, D.A. Jay, and C.R. Sherwood. 1990*b*. Columbia River estuary studies: an introduction to the estuary, a brief history, and prior studies. Progress in Oceanography **25**:1–14.

Skreslet, S. 1986. The role of freshwater outflow in coastal marine ecosystems. Ecological Sciences. Volume 7. Springer-Verlag, Berlin, Germany.

Small, L.F., C.D. McIntire, K.B. Macdonald, J.R. Lara-Lara, B.E. Frey, M.C. Amspoker, and T. Winfield. 1990. Primary production, plant and detrital biomass, and particle transport in the Columbia River estuary. Progress in Oceanography **25**:175–210.

Thomas, D.W. 1983. Changes in the Columbia River estuary habitat types over the past century. Columbia River Estuary Data Development Program, Astoria, Oregon, USA.

Turner, D.I. 1971. Dams and ecology. Civil Engineering **41**:76–80.

Turner, R.E., and N.N. Rabalais. 1991. Changes in Mississippi River water quality this century. BioScience **41**:140–147.

United States Army Corps of Engineers (USACE). 1960. 1959 current measurement program, Columbia River at mouth, Oregon and Washington. Volume 4. Portland, Oregon, USA.

United States Army Corps of Engineers (USACE). North Pacific Division. 1984. Columbia River Basin: master water control manual. Portland, Oregon, USA.

United States Coast and Geodetic Survey. 1952. Surface water temperatures at tide stations, Pacific Coast, North and South American and Pacific Ocean Islands. Special Publication 280, Washington, D.C., USA.

United States Coast and Geodetic Survey. 1967. Surface water temperature and density, Pacific Coast, North and South American and Pacific Ocean Islands. Publication 31–3, Rockville, Maryland, USA.

Valiela, I., and J.E. Costa. 1988. Eutrophication of Buttermilk Bay, a Cape Cod coastal embayment: concentration of nutrients and watershed nutrient budgets. Environmental Management **12**:539–553.

Volovik, S.P. 1986. Changes in the ecosystem of the Azov Sea in relation to economic development in the basin. Journal of Ichthyology **26**:1–15.

Waggoner, P.E. 1991. U.S. water resources versus an announced but uncertain climate change. Science **251**:1002.

Wallace, J.M., and D.S. Gutzler. 1981. Teleconnections in the geopotential height field during the northern hemisphere winter. Monthly Weather Review **109**:784–812.

10

Some Emerging Issues in Watershed Management: Landscape Patterns, Species Conservation, and Climate Change

F.J. SWANSON, R.P. NEILSON, AND G.E. GRANT

Abstract

Emerging issues in watershed management include the need to assess the effects of management activities on a time scale of several cutting rotations (>100 years) and on spatial scales that encompass influences from beyond watershed boundaries. Long-range analysis indicates that today's activities will have strong, long-lasting effects, though the ecological consequences may not be visible when the analysis horizon spans only a few decades. Land use decisions within watersheds are increasingly influenced by broader social, economic, and biological factors (e.g., wildlife management plans, such as the Northern Spotted Owl Conservation Strategy). Global climate change poses an even greater potential for altering watershed management. Consequently, improved social and technical tools are needed for planning management of multiple resources in an increasingly uncertain world.

Key words. Climate change, cumulative effects, watershed management.

Introduction

Management of natural resources has become increasingly complex and uncertain because of shifts in demography, political power bases, public expectations, and understanding of natural systems. The fields of watershed science and management are likely to undergo rapid change for some years to come. Uncertainties include (1) cumulative, long-term effects of current forest management practices, (2) effects of changing societal values and expectations in regard to natural resource management, especially when it comes to decisions about species of particular concern, and (3) effects of a changing global environment. Consequently, watershed managers are faced with several major challenges, including the sheer size and complexity of 5,000 to

50,000 hectare watersheds. Furthermore, managers must balance all the many demands made by increasingly diverse users of watershed resources.

In this article we discuss three emerging issues in watershed management and science as examples of these three classes of uncertainties. The first concerns the effects of forest management activities on forest patterns over a time scale of several centuries and the hydrological and ecological consequences of alternative patterns. Second, we describe two proposed wildlife conservation strategies that would affect watershed management. Third, the prospects of global climate change are considered in several respects that could impinge on watershed management. Climate change and future wildlife management systems are difficult to predict and will be difficult to manage, especially with current analytical tools and social institutions. However, in the near future they will increasingly influence watershed management decisions. Since the three issues we will discuss are all in an early stage of development, we will focus mainly on needed concepts, models, and data.

In this discussion we consider examples from federal forest lands in the Pacific Northwest, USA, recognizing that many of the issues are distinctly regional. Management of forest watersheds in this region is undergoing rapid change, as is our understanding of the prospective effects of climate change. Thus many of the concepts discussed here are similarly evolving. However, we believe that these examples argue strongly for broadening the traditional temporal and spatial scope of analysis used in watershed management, in this region and elsewhere.

Long-Term Cumulative Effects of Present Management

Much analysis of cumulative watershed effects has considered planning horizons of only a few years to a few decades. However, for the federal land designated for timber production in the Pacific Northwest, a planned century-long conversion of natural forests to managed forests is under way; thus analysis of a few decades considers only a portion of the transition period. Spatial patterns created by past disturbances and present cutting will have a strong influence on the pattern of forest age classes distributed across the landscape far into the future.

As of 1991 much of the federal forest land dedicated to timber production in the Pacific Northwest was only partly converted from natural to plantation (previously cut and replanted) forests. This conversion on federal lands is now 20 to over 50% complete (T. Spies, U.S. Forest Service, Corvallis, Oregon, pers. comm.), whereas private forest lands are almost completely converted. The rotation length in federal timber production areas is planned to be 70 to 120 years, with approximately equal areas cut each decade.

Assessing the effects of this conversion of natural forests to tree plantations is a crucial element in analyzing cumulative watershed effects and requires a long-term perspective. It can be argued that planning for more than one to two decades is foolish, considering the rapid pace of change in scientific understanding, technology, and, particularly, societal expectations concerning natural resources. However, present practices create landscape patterns that limit future management options for many decades to centuries. Magnuson's (1990) concept of the "invisible present" is very relevant to this issue. He argues that we "typically underestimate the degree of change that does occur" during slow processes that last for decades.

This is exemplified by analysis of forest cutting patterns used in the Pacific Northwest. Dispersion of 10 to 20 ha cutting units has been widely used in federal forestry in the region to meet a series of objectives (Franklin and Forman 1987). Initial objectives included: (1) dispersion of the hydrologic and sediment-production effects of forest cutting and road construction, (2) establishment of a road network for fire protection and other purposes, (3) creation of edge and early seral vegetation habitat for game species, and (4) creation of small openings where trees might regenerate naturally from seed from the adjacent stands. But by the time the dispersed cutting system is only 30% implemented, many of the initial objectives have been met or superseded by alternative practices (such as switching from natural regeneration to planting nursery-grown seedlings). In the meantime, other issues and objectives (such as the provision of large blocks of interior forest habitat for wildlife) have emerged. Therefore, it is appropriate to reevaluate the future cutting pattern.

An analysis of cutting patterns by Franklin and Forman (1987) shows the importance of taking the longer range view. They compared effects of dispersed, aggregated, and other cutting patterns on disturbance processes, wildlife habitat, and other properties of landscapes. A critical aspect of their analysis is that the eventual consequences of any long-term strategy, such as the dispersed cutting system, may not be evident for decades. Even very simple modeling exercises can be useful for examining patterns generated by different management systems. This concept has encouraged modeling of alternative pathways of landscape pattern development.

Analysis and Management of Within-Rotation Patterns

Using geographic information systems (GISs), several studies have projected alternative future vegetation patterns resulting from forest cutting in real rather than hypothetical grid landscapes (Cissel 1990, Hemstrom 1990). These studies have lead to several important conclusions: (1) Consideration of a full cycle of cutting is essential to planning future landscape management. (2) Patterns of stand age classes across a landscape can take very different courses, even within current standards and guidelines for U.S. Forest Service management. (3) There is inertia in the rate of landscape pattern change, in the sense that it may require a long time for a particular managed pattern to become established or for an established pattern to be altered once a change has been made in the rules governing pattern development—that is, rate of

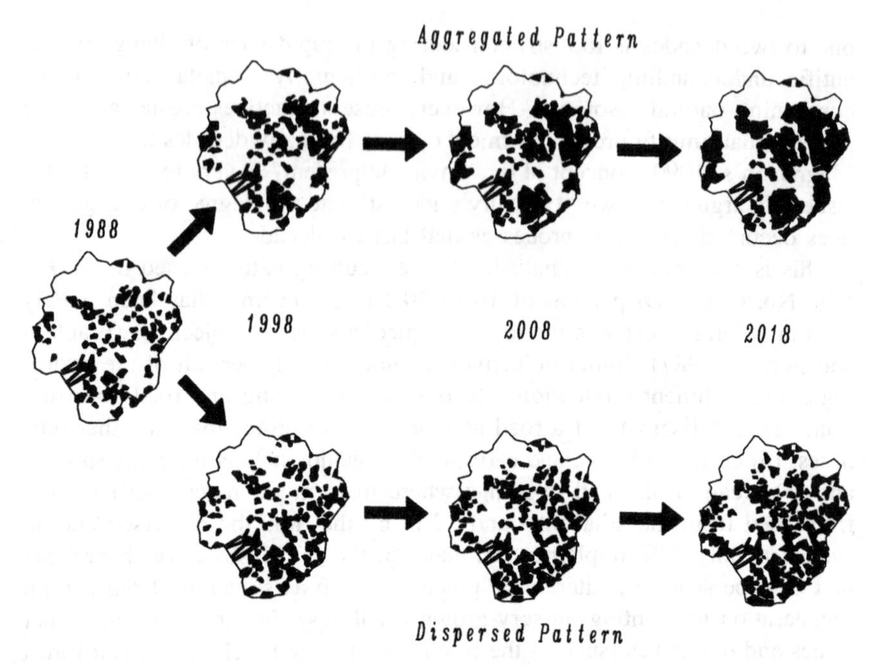

FIGURE 10.1. Example of rate of landscape pattern change after rules of pattern formation are changed. This example of the 5,000 ha Cook and Quentin creek drainages, Blue River Ranger District, Willamette National Forest, USA, starts with approximately 20% cut in dispersed pattern (condition existing in 1988). Using continued dispersed cutting at 12% area cut by decade, cutting patterns are projected for the next three decades. Beginning with the pattern in 1988, but using an aggregated cutting pattern, the pattern diverges over the next three decades.

cutting, size distribution of cutting units, age of units at time of cutting, and arrangement of cutting units. These points are expressed in a landscape modeling exercise contrasting dispersed and partial aggregation of cutting units in the 5,000 ha Cook-Quentin drainages in the Blue River Ranger District, Willamette National Forest, western Cascade Range, Oregon, USA (Figure 10.1) (Hemstrom 1990). By 1988 about 20% of this area had been cut in a dispersed pattern. The district planning staff then projected the next three decades of harvest under two different regimes: (1) continued dispersed cutting and (2) aggregated cutting with the objective of maintaining large blocks of contiguous interior forest habitat. In the two alternatives the same area would be cut each decade, the area cut would be distributed similarly by 305 m elevation bands so that there would be no differential effects of rain-on-snow hydrology (Harr 1981), and regional Forest Service standards and guidelines would be followed, including the rule that a patch of forest cannot be cut next to a plantation until the trees in the plantation are 1.5 m tall.

This model displays a substantial range of possible future vegetation patterns even under the quite restricted difference in rules governing pattern development. Assuming an edge effect of two tree lengths into forested areas, Hemstrom (1990) observed 10% greater area in interior forest habitat for the aggregated pattern in the year 2018. The density of forest-plantation edge (km of edge per km^2 of landscape area) would be 42% greater in the dispersed pattern at 2018. Landscapes with higher edge density may be more susceptible to windthrow (i.e., uprooted trees). These are just two of many possible points of contrast between the pair of vegetation patterns shown in Figure 10.1.

The Cook-Quentin example of landscape analysis has drawn criticism on several points. First, critics say that since most landscapes where future cutting will occur have already been more thoroughly cut, options are more restricted than in the Cook-Quentin case. Second, the year 2018 in Figure 10.1 is the point in landscape development when the patterns of residual forest are most different between the two cutting patterns. After 2018 the vegetation patterns (expressed only as residual forest and plantations) will converge as the scenarios progress toward the end of the first rotation. As discussed below, the resulting vegetation patterns in these two scenarios are quite different in future rotations when the distributions of age classes are considered.

Both of these concerns are valid, but it is important to look ahead into subsequent rotations. The two landscape management systems shown in Figure 10.1 are creating landscape patterns of very different grain size (i.e., scale of pattern), which will project into future rotations. The traditional dispersed cutting pattern creates a landscape with a 15 ha grain. The aggregated patterns used in the Cook-Quentin example have a dominant texture of >500 hectares, consisting of aggregates of cutting units harvested within a few decades and then left to grow into mature forest. Different future landscape structures are created, and their effects on wildlife habitat, watershed, and other aspects of ecosystems await analysis.

An interesting aspect of the Cook-Quentin exercise is the effect of inertia in changing landscape patterns. Landscape pattern inertia refers to the propagation far into the future of patterns of vegetation patches created by past and present events. Consider, for example, creation of landscape pattern under rule set A (e.g., maximum dispersion constrained by simultaneous development of a road network). Application of rule set A to a hypothetical or real landscape creates the pattern resulting from that rule set. If at some point we convert to rule set B, which produces another pattern (such as aggregated cutting from nuclei with a particular spacing), another pattern is created, but it will take some time to convert the landscape from type A to type B.

Factors that control the rate of vegetation pattern change include: (1) rates of cutting (percentage of area cut per unit of time) and vegetation regrowth, (2) magnitude of difference between the two pattern types, (3) extent of development of type A before imposition of rule set B, and (4) the flexibility of rule set B in overriding the pattern created by rule set A (i.e., there could be application of an initially modified form of rule set B to hasten the transition from type A to type B.

The effects and the rate of pattern change are evident in the Cook-Quentin study (Figure 10.1). Here we see that after 20% of the area was cut with a dispersed cutting pattern, it took another three decades to create the aggregated pattern after changing the rules for pattern development. The extent of aggregation was determined largely by the initial conditions (i.e., extent of previous dispersed cutting); and the rate of change is greatly constrained by the cutting rate of 12% of area per decade.

Considerations at the Multirotation Time Scale

Analysis of long-term consequences of pattern management on Pacific Northwest forests should include: (1) distinction of the transition rotation (i.e., conversion of natural forest to plantations) from subsequent rotations of fully regulated forest and (2) the inertia of patterns created in the first (transition) rotation which may carry through into the second and subsequent rotations.

A multirotation analysis is essential for two main reasons: (1) An analysis of a few decades of management activities, only a fraction of the rotation length, simply leaves invisible the long-range effects on forest age-class patterns and associated system properties. Examples of relevant system properties include wildlife habitat and vegetation effects on peak streamflows (Franklin and Forman 1987). (2) Activities in the middle of the transitional rotation may have different effects than the same activities in later rotations. In the middle of the transitional rotation, for example, about half of the landscape is in residual, yet-to-be-cut mature and old-growth stands. In subsequent rotations these areas are much younger plantations. Therefore, residual forest remaining during the first cutting rotation may buffer certain environmental effects of cutting that would be observed in subsequent rotations.

We can portray aspects of the contrast between the transition and subsequent rotations by a schematic representation of the percentage of a watershed area in forest stands over 100 years old (Figure 10.2). To assess cumulative effects, we would prefer to present the variation in a watershed aspect of interest (e.g., percentage variation in peak flows generated by a particular precipitation event) rather than the extent of a single age class, as depicted in Figure 10.2, but our analytical techniques are not that sophisticated yet.

Past analyses of long-range effects of cutting have focused on the transition rotation (Franklin and Forman 1987, Li 1989, Cissel 1990, Hemstrom 1990, Hansen et al. 1991). We are in the early stages of extending this analysis through two rotations. Ideally we would examine the effects of historic natural disturbances as well as management activities and consider the

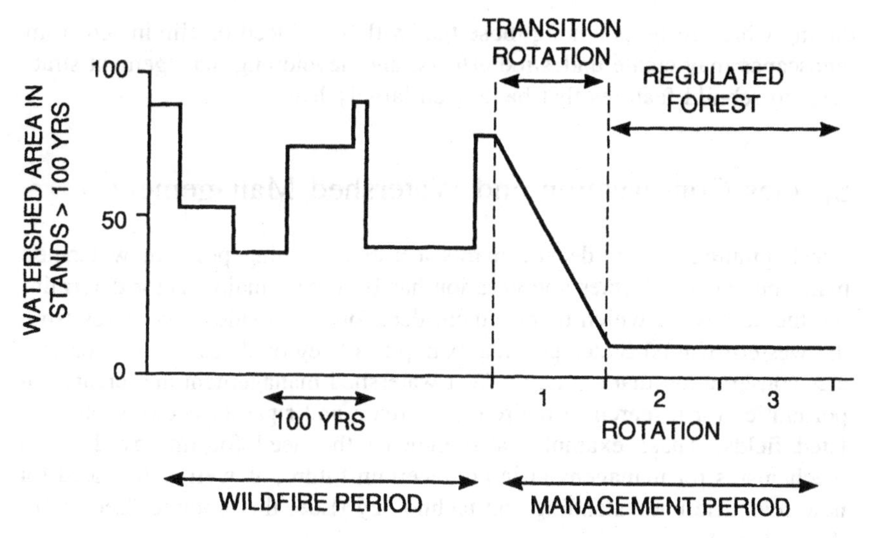

FIGURE 10.2. Hypothetical change in the extent (percent area) of forest age classes exceeding 100 years in age in a large watershed. Time scale depicts a period of natural disturbance (wildfire) through the first (transitional) rotation of forest management as natural forest is converted to plantations and into the subsequent, "fully regulated" rotations of cutting in the fully developed forest, where the only patches of natural forest are those dedicated to remain as such.

temporal variance of landscape pattern under the conditions of natural disturbance, transition, and fully developed states. For example, we might hypothesize that there was much greater decade-to-decade variation in the distribution of forest age classes and the grain size of patches in the prehistoric, wildfire-dominated disturbance regime than in the fully managed forest landscape (Naiman et al., this volume). This would provide a frame of reference for analysis of long-term effects of management on ecosystem attributes such as streamflow regime and biological diversity.

Modeling efforts to date have emphasized degree of aggregation of cutting units as the key variable (Franklin and Forman 1987, Li 1989, Cissel, 1990, Hemstrom, 1990). These exercises have used rather limited ranges of values for two landscape pattern variables: size distribution of cutting units and rotation length. These two variables have potentially great effects on wildlife habitat, hydrologic regimes, and other landscape functions. An initial step in a broader analysis could take the form of a sensitivity analysis of computer simulated vegetation patterns to evaluate the effects of varying unit size and rotation length, as well as arrangement of cutting units. Habitat assessment and process models, such as hydrologic models, could then be used to estimate effects of these patterns on key landscape functions.

The rates of vegetation pattern change have implications for management, including maintaining desirable system features as long as possible, predicting which features among those that will be reduced or eliminated from landscapes may create undesired effects, and developing management strategies to rebuild features that have been largely lost.

Species Conservation and Watershed Management

The beginning of this decade marks a critical turning point in watershed management, for species conservation has become a major factor determining the context in which management decisions are made. Two cases from the western United States provide examples of regional scale plans for species conservation that strongly affect watershed management and create opportunities for research in hydrology, stream and riparian ecology, and related fields. These examples also indicate the need for improved social mechanisms for management in an uncertain future, as well as the need for new scientific understanding and technology (Lee, this volume; Lee et al., this volume).

Northern Spotted Owl

The conservation strategy for the northern spotted owl (*Strix occidentalis caurina)* (Thomas et al. 1990) proposes establishment of a series of habitat conservation areas (HCAs) distributed across northern California and western Oregon and Washington. A majority of HCAs range in size from 20,000 to 40,000 ha, and each is intended to encompass habitat for approximately 20 pairs of spotted owls. The strategy also prescribes limits on forest cutting in lands between HCAs, with the intent of providing dispersal paths between HCAs. The geographic pattern of HCAs is designed to provide for owl dispersal and to use lands where timber harvest is already prohibited (e.g., legislated wilderness).

Consequently, HCAs assume a variety of configurations with respect to watershed boundaries and areas, including some that partly coincide with wilderness areas along the crest of the Cascade Range and therefore result in protection of large (50,000+ ha), complete headwater basins (Figure 10.3). In the central part of the western Cascades in areas removed from large wildernesses, individual HCAs typically straddle drainage divides, and therefore cover multiple, smaller (2,000–10,000 ha) watersheds. Complex patterns of HCA boundaries exist where several ownerships are involved.

Furbearers in the Sierras of California

Another system for species conservation is contained in the proposal for "Sustainable Resource Management for Forest Landscapes" developed by several national forests in California (M. Chappel, Tahoe National Forest, Nevada City, California, pers. comm.). This proposal derives from a desire to better protect furbearers and other special interest species in the Sierra

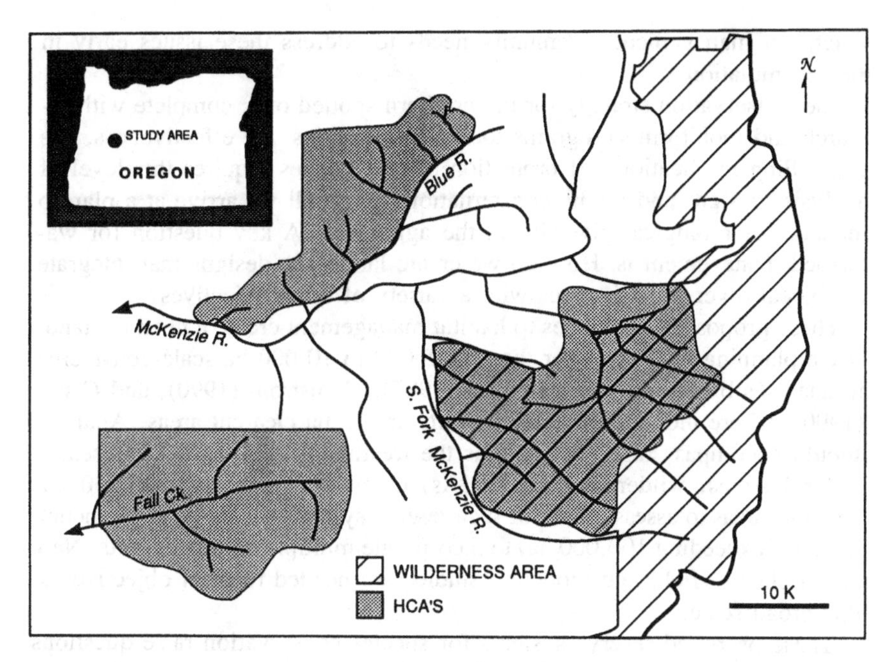

FIGURE 10.3. Portion of the habitat conservation area (HCA) system in the conservation strategy for northern spotted owls (modified after Thomas et al. 1990). Note that HCA O-7 (top) encompasses a series of small watersheds and borders mainstem Blue River and McKenzie River; O-8 (middle) encompasses lower elevation portions of larger watersheds, the upper reaches of which are outside the HCA but within Three Sisters Wilderness; O-9 (lower left) encompasses the full, single watershed of upper Fall Creek.

Nevada Range of California. This plan calls for management of (1) "old forest zones" in valley bottoms and steep lower valley walls of major river drainages in the central Sierras and (2) "bridges" of continuous forest cover over drainage divides to connect the old forest zones. In this scheme the old forest zone would receive low levels of timber harvest, and aerial logging systems would be used so that natural forest habitat could be maintained with minimal human disturbance. The bridges are intended to provide dispersal corridors between watersheds in case of wildfire, climate change, or other major disturbance. "Young forest zones" are located on the gentler, upper slopes where habitat value for furbearers is fairly low, timber harvest can be emphasized, and relatively inexpensive cable and ground-based yarding systems can be employed.

Implications for Watershed Management and Research

These two proposals reflect a trend emphasizing management for biological diversity as a driving force in landscape planning and management. More generally, new issues are setting the context of watershed management. The watershed management community needs to address these issues early in their formulation.

The conservation strategy for the northern spotted owl, complete with research and monitoring programs to test assumptions and effectiveness, has compelling implications. If protection of one species requires this level of analysis, design, and public consternation, how will we arrive at a plan to manage for biological diversity in the aggregate? A key question for watershed management is: How do we create landscape designs that integrate biological diversity objectives with a variety of other objectives?

These proposed approaches to habitat management create a need for landscape planning much broader than the 5,000 to 10,000 ha scale considered in analyses by Franklin and Forman (1987), Hemstrom (1990), and Cissel (1990), where individual treatment units are 15 ha clearcut areas. Analysis should encompass reserved areas as the treatment units, such as Research Natural Areas, wilderness, and HCAs, many of which are >20,000 ha. Analysis areas to assess effects of such reserve systems should be large enough (perhaps exceeding 100,000 ha) to incorporate multiple treatment units. New approaches to field and modeling studies are needed to meet objectives at this broad scale.

These proposed reserve systems for species conservation raise questions concerning the integration of stand- and landscape-scale management practices. For example, many proposed HCAs contain extensive young plantations of Douglas-fir (*Pseudotsuga menziesii*). How should these stands be managed to meet wildlife as well as hydrologic objectives, such as minimizing peak streamflows generated during rain-on-snow events (Harr 1981)? Some of these young stands will stagnate if left unattended, thereby slowing the rate at which they might reach desired habitat conditions. How should woody fuel loads be managed in reserve and nonreserve areas to balance objectives of maintaining woody debris-related habitat with the desired frequency and extent of wildfire across landscapes? Because large-scale disturbances by processes such as wildfire are highly probable in many parts of the Pacific Northwest on the time scale relevant to this land use planning, such events should be an explicit consideration in landscape design.

The proposed system of HCAs, if enacted, will create an important new set of large-scale experimental treatments that enhance the opportunity for understanding effects of various management systems. Forest lands in many HCAs have been substantially cut in the past few decades. Since establishment of HCAs will eliminate further harvest for some time, vegetation in HCAs will undergo succession without interruption, resulting in a net increase in age of vegetation.

HCAs would add an interesting and important new element to the matrix of landscape management treatments defined in terms of the trends of past and future cutting. The federal land management system without HCAs includes three treatments: (1) past and future cutting conducted in the forest land available for timber cutting, (2) no past or future cutting in wilderness areas and Research Natural Areas, and (3) no past cutting, but future cutting in roadless areas released for future forest cutting through the federal forest planning process. The HCA system would be made up of areas of past cutting and no future cutting. The importance of the HCA landscape treatment is that it sends large pieces of landscape on a trajectory of vegetation change unlike anything existing previously.

Given the large-scale, complex nature of these land management designations, the best opportunity to capitalize on this important, ad hoc landscape experiment may be through an adaptive management approach (Walters 1986). The essential theme of adaptive management is to acknowledge our present imperfect understanding of natural resource systems and learn answers to crucial questions by conducting experiments through management. This approach is a proposed part of the conservation strategy for the northern spotted owl (Thomas et al. 1990), and is equally relevant for landscape management. Monitoring of hydrologic, ecologic, and other system responses to these treatments would prove useful in evaluating the cumulative watershed effects of management activities (such as road building, harvesting) and in testing models developed to simulate them.

Climate Change

Past management activities and those being considered for the future largely assume a static climate (see Thomas et al. 1990). In this section, we will relax that assumption and address a few of the potential combined impacts of current land use in light of a rapidly changing climate.

Effects of change in the physical and chemical properties of the atmosphere present many uncertainties for future watershed management (Risser, this volume; Dolph et al., this volume). Clearly the concentrations of radiatively active gases in the atmosphere are rising (Houghton and Woodwell 1989, Keeling et al. 1989), but the effects on important ecosystem functions, such as biotic regulation of hydrology, are poorly known. Hydrologic effects of climate change could include increased evapotranspiration in response to warming and change in vegetation cover type and amount, change in amount and type of precipitation (e.g., snow versus rain), change in the water use efficiency of vegetation in response to elevated CO_2 levels in the atmosphere, and change in soil properties resulting from altered disturbance regimes and vegetation cover. The timing and magnitude of future climate change are similarly unclear. Despite these uncertainties, we believe that it is worthwhile to consider some of the potential broad-scale implications of climate warming in order to stimulate our thinking about major changes that watersheds may experience.

Several studies have addressed potential changes in the distribution of tree species in response to climate warming on the basis of $2xCO_2$ increases, using low resolution general circulation models (GCMs). Grid size of these

GCMs is coarse relative to the Pacific Northwest of the United States, so the GCMs do not reproduce the regional current climate or project future climate well, but may provide some idea of potential broad-scale effects of increasing atmospheric CO_2 (Jenne 1988). The greatest change in temperature predicted for this region by a group of widely published GCMs is +5°C (Wilson and Mitchell 1987). Studies employing vegetation modeling and analysis of lapse rates (Dale and Franklin 1989, Neilson et al. 1989, Franklin et al. 1991) suggest that this magnitude of warming would result in a 1,000 m upward shift in elevation of temperature-controlled features, such as boundaries of species habitat and snow hydrology zones. This would amount to latitudinal shifts in species ranges of hundreds of kilometers.

Climate change effects on watersheds are likely to be diverse. Here we comment on some examples of direct effects on watershed hydrology, vegetation and habitat distributions, and land use and disturbance considerations.

Hydrology Considerations

Watershed hydrology can be expected to change in direct response to climate change and as an indirect result of changes in vegetation structure and composition that affect hydrology (Dolph et al., this volume). Direct effects of climate on hydrology, for example, will result from shifts in the location of boundaries between rain-dominated, transient snow (elevation band where snow may accumulate and melt several times each winter), and seasonal snowpack zones (Harr 1981). Climate warming could cause these boundaries to rise in elevation, thereby altering the quantity and storage time of water stored as snow within a watershed and the magnitude and timing of peak and low streamflows. The magnitude of such effects will depend on the distribution of boundaries between snow hydrology zones in relation to watershed topography. Some watersheds, for example, may have a reduced percentage of area in the transient snow zone after climate warming, and would therefore be expected to have reduced peak flows, assuming no change in the timing and amount of precipitation; that is, only the form (snow versus liquid) would be altered.

Potential effects of warming on snow hydrology can be seen by comparing the 68 km^2 Lookout Creek and the 83 km^2 French Pete Creek watersheds in the western Cascade Range in Oregon. These two watersheds are similar in area and total relief but differ somewhat in elevation. In the present climate, 77% of the Lookout Creek watershed is in the transient snow zone—more than twice the percentage of the French Pete watershed (Table 10.1). To give some sense of the potential magnitude of this contrast, the estimated 10-year return period discharge per km^2 is 1.5 m^3/s on Lookout Creek but only 0.5 m^3/s on French Pete Creek.

Assuming the scenario of a 5°C warming and an associated 1,000 m increase in elevation of boundaries between hydrologic zones, French Pete watershed would still have about a third of its area in the transient snow

Table 10.1. Percentage of basin areas in various hydrologic zones under present climate compared with a 1,000 m increase in boundaries between zones.

	Lookout Creek		French Pete Creek	
Hydrologic Zone	Present	+1,000 m	Present	+1,000 m
Snowpack	23	0	65	0
Transient	77	6	35	33
Rain dominated	0	94	0	67

zone; however, it would be in the upper third of the watershed rather than the lower third, as it is in today's climate. The Lookout Creek watershed would be nearly exclusively in the rain-dominated zone, and neither watershed would have any seasonal snowpack under the altered climate scenario. Based only on consideration of shifting hydrologic zones, these projected changes would be likely to result in substantially reduced summer low flows in both basins (more so in French Pete) and possibly reduced peak flows in the Lookout watershed. Such changes in streamflow would affect the biology of stream and riparian systems. Geomorphic processes would also be affected by change in the type, amount, and pathways of water movement in relation to topography.

Another effect of climate warming would be increased evapotranspiration, which could have profound hydrologic effects, even with no change in precipitation. Franklin et al. (1991) estimate that a 5°C increase in temperature would result in a 64% increase in potential evapotranspiration at the sites of selected meteorological stations in the western Cascade Mountains of Oregon and the western Olympic Peninsula in Washington. Actual evapotranspiration will depend on available soil moisture and the amount of leaf area, which appears to be set by the landscape water balance (Woodward 1987, Neilson et al. 1989). The very high leaf areas of Douglas fir-western hemlock (*Tsuga heterophylla*) forests are likely to result in a rapid drying of soils and reduction of base streamflows in response to warming. This would lead to plant death and thus a reduction in leaf area.

Vegetation and Habitat Change

Climate change would also lead to migration of species and shifts in boundaries between biomes. Neilson et al. (1992) have been analyzing limitations to the geographic distribution of major biomes (e.g., forest, grassland, desert) for the United States by relating seasonal timing and magnitude of temperature, precipitation, and runoff patterns to life-cycle and physiologic requirements of plants. Their success in simulating biome distributions from simple rules suggests that if climate warms to the extent predicted by GCM analyses, and plant species respond to moisture and temperature conditions as predicted, major shifts in species distribution should be expected. Significant unknowns include dispersal rates across either natural or managed landscapes (Davis 1988), possible effects of soil incompatibility as a limit on dispersal rates (Perry et al. 1990), and CO_2 induced changes in plant water-use efficiency (Bazzaz 1990).

Shifts in biome boundaries that are triggered by climate change can be viewed as creating zones of biome stasis, retreat, and invasion (Holland et al. 1991, Neilson 1991). For example, if a forest biome moves northward in response to climate warming, the forest may invade tundra to the north, but retreat on the southern forest biome border, permitting northward invasion by shrublands. If climate change is not so extreme that the biome is completely displaced, the central part of the forest biome may remain in the forest vegetation type. Actual elevational and latitudinal shifts of biome boundaries will be complicated by many physical and biological effects on the stability of individual boundary segments. For example, where environmental gradients are steep, such as mountain fronts, major climate change may lead to only minor horizontal shifts in the location of boundaries of species ranges. The most extreme latitudinal shifts in plant distributions may take place where environmental gradients are not steep.

This concept of biome displacement, in all its naked oversimplification, has important implications for management of landscapes and watersheds, if climate change of the projected magnitude should occur (Holland et al. 1991). Watershed management considerations may differ substantially among the different zones of vegetation change, depending on the watershed-specific change in vegetation. For example, the most profound changes in vegetation structure and the physical processes mediated by vegetation are likely to occur in the forest retreat and invasion zones; smaller change may occur in the stasis zone. In terms of species conservation, the leading edge of biome shift probably deserves particular emphasis because it would be expected to develop improved habitat for advancing species in the future; whereas the trailing edge may experience degrading habitat conditions for retreating species. Of course, the advance zone of one species is the retreat zone of another.

In the face of dramatic climate warming, both advance and retreat zones may benefit from intensive management to facilitate species dispersal and establishment. Under such circumstances, it may be desirable to reconsider restrictions on management of wilderness areas and reserves such as HCAs. Benefits of such management may include better control of disturbance regimes (e.g., wildfire) and establishment of vegetation with beneficial effects on habitat and on water quantity and quality. This issue of the interaction of land use and climate change demands great attention. How will land use accelerate or decelerate the transition between vegetation types? How will land use interact with projected climate change to moderate or exacerbate watershed functions? What are the effects of what seem to be subtle system changes created through land use, even where land use retains vegetation type and dominant species, such as reduced carbon stores (Harmon et al.

1990) and broad-scale albedo increases (W. Cohen, U.S. Forest Service, Forestry Sciences Laboratory, Corvallis, Oregon, pers. comm.) where natural mature and old-growth forests are converted to intensive plantation forests in the Pacific Northwest?

Concept, Model, and Data Needs

We have discussed three broad categories of potentially important issues in watershed management: long-term consequences of alternative management systems for timber production; conservation of particular species (and, more broadly, biological diversity); and climate change. Present watershed management activities will produce future conditions very different from the current ones—conditions that are evident to us only through long-range projection of landscape structure and function.

These considerations suggest the following needs:

1. Tools are needed to predict changes in biological and physical systems resulting from land use, natural disturbances, climate change, and changes in atmospheric chemistry. Models and related field observations at appropriate temporal and spatial resolutions are needed to link analysis of watersheds, biomes, and GCM grid cells to predict changes in species distribution, hydrologic and nutrient cycling processes, disturbance regimes, and other system features. It is particularly important that models incorporate water balance considerations at each of these scales. Models and field studies are needed to examine species migration, including limiting factors such as dispersal (natural as well as intentional and unintentional human-induced), impediments to establishment (e.g., soil compatibility for invading species), and disturbance as impediments and facilitators of dispersal. Models and related field experiments are needed to examine the consequences of alternative management scenarios, such as forest or fish harvesting, over the next century, and to address issues of changing land use, climate, and societal expectations. This can be approached through adaptive management (Walters 1986), as is proposed for conservation of the northern spotted owl (Thomas et al. 1990); adaptive management is also appropriate for other landscape-level experiments using land use treatments.

2. Spatially explicit local and regional data bases are needed for modeling and monitoring ecosystem change to provide a basis for management prescriptions and policy determination.

3. Policy needs include mechanisms to manage change involving mixed-owner and multiresource objectives in today's political and physical climates. For example, species conservation issues have reached a point where regional biodiversity management plans would facilitate management and policy decisions.

Climate change may create a need for a regional management strategy that sifts down to the local level to deal with lands having different owners with different objectives. For example, climate change will increase movement of species across ownership and political boundaries. Some coordination between neighboring landholders may help minimize negative effects.

For years watershed specialists have stressed the importance of a drainage basin perspective in addressing stream and riparian issues. As this idea has gained wider acceptance, emerging issues of long-range planning, species conservation, and climate change are increasingly forcing land use managers to peer over the drainage divides into neighboring watersheds and landscapes. Watershed research and management are in a rather immature state relative to these rapidly evolving issues. Development of social mechanisms for resolving conflicts over natural resource systems, as well as technical and analytical tools for predicting effects of human activities, will need to keep pace with shifting societal expectations. The themes in this chapter suggest some of the rich and complex array of questions that will confront managers seeking to sustain ecosystems and quality of stream and watershed resources in the face of an uncertain future.

Acknowledgments. Development of concepts and information in this paper was supported in part by funding for the USDA Forest Service's New Perspectives program and the National Science Foundation's Long-Term Ecological Research program at the H.J. Andrews Experimental Forest, Blue River, Oregon. We thank A. Hansen, R.J. Naiman, J. Pastor, T. Spies, and an anonymous reviewer for helpful reviews and discussions.

References

Bazzaz, F. A. 1990. The response of natural ecosystems to the rising global CO_2 levels. Annual Review of Ecology and Systematics **21**:167–196.

Cissel, J. 1990. An approach for evaluating stand significance and designing forest landscapes. COPE (Coastal Oregon Productivity Enhancement) Report (Oregon State University) **3**(4):8–11.

Dale, V.H., and J.F. Franklin. 1989. Potential effects of climate change on stand development in the Pacific Northwest. Canadian Journal of Forest Research **19**:1581–1590.

Davis, M.B. 1988. Ecological systems and dynamics. Pages 69–106 *in* Committee on Global Change, editors. Toward an understanding of global change. National Academy Press, Washington, D.C., USA.

Franklin, J.F., and R.T.T. Forman. 1987. Creating landscape patterns by forest cutting: ecological consequences and principles. Landscape Ecology **1**:5–18.

Franklin, J.F., F.J. Swanson, M.E. Harmon, D.A. Perry, T.A. Spies, V.H. Dale, A. McKee, W.K. Ferrell, S.V. Gregory, J.D. Lattin, T.D. Schowalter, D. Larsen, and J.E. Means. 1991. Effects of global climate change on forests in northwestern North America. Northwest Environmental Journal **7**(2):233–254.

Hansen, A., D. Urban, and B. Marks. 1991. Avian community dynamics: the interplay of human landscape trajectories and species life histories. *In* F. di Castri and A. Hansen, editors. Landscape boundaries: consequences for biodiversity and ecological flows. Springer-Verlag, New York, New York, USA, *in press*.

Harmon, M.E., W.K. Ferrell, and J.F. Franklin. 1990. Effects on carbon storage of conversion of old-growth forests to young forests. Science **247**:699–702.

Harr, R.D. 1981. Some characteristics and consequences of snowmelt during rainfall in western Oregon. Journal of Hydrology **53**:277–304.

Hemstrom, M. 1990. Alternative timber harvest patterns for landscape diversity. COPE (Coastal Oregon Productivity Enhancement) Report (Oregon State University) **3**(1):8–11.

Holland, M.M., P.G. Risser, and R.J. Naiman, editors. 1991. The role of landscape boundaries in the management and restoration of changing environments. Chapman and Hall, New York, New York, USA.

Houghton, R.A., and G.W. Woodwell. 1989. Global climate change. Scientific American **260**:36–44.

Jenne, R.L. 1988. Data from climate models, the CO_2 warming. National Center for Atmospheric Research, Boulder, Colorado, USA.

Keeling, C.D., R.B. Bacastow, A.F. Carter, S.C. Piper, T.P. Whorf, M. Heimann, W.G. Mook, and H.J. Roeloffzen. 1989. A three dimensional model of the atmospheric CO_2 transport based on observed winds. I. Analysis of observational data. Geophysical Monographs **55**:165–236.

Li, H. 1989. Spatio-temporal pattern analysis of managed forest landscapes: a simulation approach. Dissertation. Oregon State University, Corvallis, Oregon, USA.

Magnuson, J.L. 1990. Long-term ecological research and the invisible present. BioScience **40**:495–501.

Neilson, R.P. 1991. Climatic constraints and issues of scale controlling regional biomes. Pages 31–51 *in* M.M. Holland, P.G. Risser, and R.J. Naiman, editors. The role of landscape boundaries in the management and restoration of changing environments. Chapman and Hall, New York, New York, USA.

Neilson, R.P., G.A. King, R.L. DeVelice, J. Lenihan, D. Marks, J. Dolph, B. Campbell, and G. Glick. 1989. Sensitivity of ecological landscapes and regions to global climate change. United States Environmental Protection Agency EPA/600/3–89/073, National Technical Information Services, Washington, D.C., USA.

Neilson, R.P., G.A. King, and G. Koerper. 1992. Toward a rule-based biome model. Landscape Ecology, *in press*.

Perry, D., J.G. Borchers, S.L. Borchers, and M.P. Amaranthus. 1990. Species migrations and ecosystem stability during climate change: the belowground connection. Conservation Biology **4**:266–274.

Thomas, J.W., E.D. Forsman, J.B. Lint, E.C. Meslow, B.R. Noon, and J. Verner. 1990. A conservation strategy for the northern spotted owl. United States Forest Service, Portland, Oregon, USA.

Walters, C. 1986. Adaptive management of renewable resources. Macmillan, New York, New York, USA.

Wilson, C.A., and J.F.B. Mitchell. 1987. A doubled CO_2 climate sensitivity experiment with a GCM including a simple ocean. Journal of Geophysical Research **92**:13315–13343.

Woodward, F.I. 1987. Climate and plant distribution. Cambridge University Press, London, England.

11

Using Simulation Models and Geographic Information Systems to Integrate Ecosystem and Landscape Ecology

JOHN PASTOR AND CAROL A. JOHNSTON

Abstract

The movement of water downslope in a watershed or through the soil-plant-atmosphere continuum integrates processes occurring at scales ranging from a few meters to many square kilometers. The complexity of the landscape may derive from these pathways of water flow. However, ecosystem data are usually collected in areas of centimeters to meters, represented by plots or transects. Ecosystem models predict productivity, nutrient flows, or habitat for a particular plot, but landscape processes occurring at larger scales, such as floods, fires, or animal activities, may constrain or override these smaller scale ecosystem processes. Extrapolation of data or model predictions at fine scales to the larger landscape therefore requires an assumption of spatial homogeneity or knowledge of resource heterogeneity. Geographic information systems (GISs) are new tools that enable us to assess the consequences of spatial heterogeneity. When coupled with simulation models and data bases, they allow us to bridge ecosystem and landscape scales. Three examples are presented of how simulation models can be coupled with GIS: (1) multivariate analyses of landscape structure, (2) parameterizing mathematical models of landscape dynamics, which can be used to predict landscape behavior, and (3) extrapolating simulations of ecosystem properties across the entire landscape. The need for new hardware and software for dynamic GIS simulations is also discussed.

Key words. Climate change, geographic information systems (GISs), landscape models, riparian ecosystems, wetlands.

Introduction

Ecologists are addressing questions crucial to the future habitability of the globe: How do terrestrial and aquatic ecosystems interact? What are the effects of climate change on the earth's biota? How do animals move within and between habitats, and what effect does this have on ecosystems? What are the cumulative effects of numerous disturbances on landscape function? How can the natural resources required by society be more efficiently located and managed without diminishing global productivity? (Lubchenco et al. 1991).

The spatial and temporal heterogeneity of ecological systems makes the solutions to these problems difficult: the interactions between biological entities, from individuals to ecosystems, depend on their spatial distribution relative to one another and to their physical environment. Spatial heterogeneity makes it difficult to extrapolate from data collected at points, small plots, or along transects to larger watershed, regional, or global scales. Furthermore, the fluxes of organisms and materials between adjacent ecosystems are key components of this heterogeneity, often contributing to it as well as being affected by it.

Watersheds, in particular the riparian zones between terrestrial and aquatic ecosystems, are key landscape components where the problems of spatial and temporal heterogeneity arise. As water flows from ridgetop to stream outlet through forests and across gravel bars, or from rain and snow into soil and back to the atmosphere through plant stomates, it and its chemical constituents encounter abrupt changes in the physical and biological environment. The interaction of water with these boundaries may create complex feedback loops that stabilize the boundaries or cause them to change. Changes in the spatial patterns of landscapes and processes at boundaries are documented by long-term observations and explained by manipulative experiments (Tilman 1987). But these data need to be subsumed into a common framework for a quantitative and predictive description of watersheds and their riparian zones. In order to solve ecological problems at complex landscape, regional, and global scales, the tools of geographic information systems, data manipulation, and simulation models must be integrated. Ecologists have long been familiar with the latter two, but it is the recent development of the geographic information system (GIS) that allows a quantitative assessment of the causes and consequences of spatial heterogeneity in ecological systems (Burrough 1986).

A GIS is a computer system that can analyze spatial distributions and integrate these data with other tools to generate new information not previously available. A GIS data base includes not only standard ecological data but the exact georeferenced coordinates on the surface of the earth where the data were obtained. A GIS assembles these georeferenced data on different landscape properties, derives new data that are syntheses of the original data, and analyzes the new data to predict gradients of habitat, net primary productivity, or nutrient flows across the landscape (Figure 11.1). A key feature of a GIS is its ability to perform Boolean operations, enabling researchers to statistically determine spatial coincidence of ecological processes and resources (Turner 1987*a*, *b*, *c*, 1990; Pastor and Broschart 1990). The same capability can be used with a time series of data from, for ex-

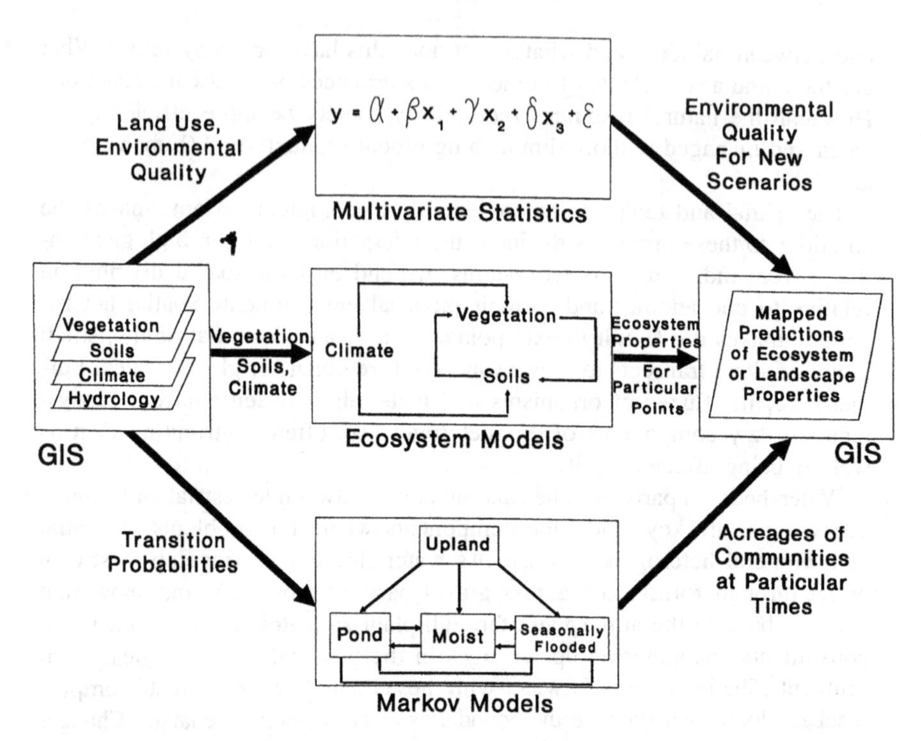

FIGURE 11.1. Alternative methods of interfacing simulation models with geographic information systems.

ample, historical air photos, to analyze temporal changes in the landscape (Johnston and Naiman 1990*a, b*). Such spatial patterns and their temporal changes can be analyzed statistically using various indices (Turner 1987*a, b, c*, 1990; Pastor and Broschart 1990) or using geostatistics (Journel and Huijbregts 1978). The data are stored in the GIS in either raster or vector format. In raster format, the landscape is subdivided into a checkerboard of grid cells, known as pixels. In vector format, the landscape is divided into irregular polygons whose borders are delimited by curvilinear boundaries rather than edges of square cells. Advantages and disadvantages of each are reviewed by Burrough (1986) and Johnson (1990).

Dynamic models of landscape and ecosystem processes have been reviewed recently (Shugart and West 1980, Loucks et al. 1981, Weinstein and Shugart 1983, Shugart 1984, Risser 1987, Baker 1989). Baker (1989) recognizes three types of landscape models: (1) whole landscape models that predict aggregate properties of the entire landscape, such as number of stands of a particular cover type in a township, (2) distributional landscape models that predict the distribution of land area among classes or states, and (3) spatially explicit landscape models that predict the spatial location and configuration of classes or states. Although these models predict landscape properties in increasing detail, they require more data storage and computational time to do so. Until the advent of geographic information systems, all these models suffered from computational and data management complexity, forcing the creators to simplify assumptions about spatial interactions among landscape components to either parameterize or run the models (Shugart et al. 1973, Baker 1989). However, the Boolean algebra of GISs explicit in their relational data base structures facilitates the large amount of bookkeeping required for analyses of landscape patterns by each of the general model classes described by Baker (1989).

Therefore, the power of GISs to predict new patterns or test hypotheses about the causes of these patterns is maximized when they are coupled with simulation models of ecosystem, landscape, or global processes (Pearlstine and Kitchens 1987, Gildea et al. 1987, Burke et al. 1990). Using a coupled GIS-simulation model, ecosystem properties across the landscape are predicted from the spatial distribution of controlling factors. This coupling can happen in three ways:

1. The correspondence or correlation of data from maps of different features can be analyzed statistically using multivariate techniques. For example, multiple regression models that predict the distribution of ecosystem properties from hypothesized and mapped controlling factors can be derived. This approach would produce a whole landscape model as defined by Baker (1989). As an example, we will show how a GIS can be used to predict changes in water quality caused by losses of different types of wetlands from different locations within a watershed.

2. A GIS can be used to extract data on rates of change of landscape features from historical data such as sequential air photos or satellite coverage. These data are then used to parameterize probabilistic models such as Markov chains. These are examples of what Baker (1989) calls distributional landscape models. Landscape properties not readily interpretable from the maps can then be inferred from the mathematical properties of these models. We will demonstrate this method using data on changes in wetlands in ponds created by beaver (*Castor canadensis*) in northern Minnesota, USA.

3. Models are driven for different combinations of controlling factors identified from a GIS data base, and model predictions for points or polygons are reentered into the GIS to update maps. This approach would produce spatially explicit landscape models (Baker 1989). For example, we will show how a model is used to predict changes in species growth at particular points with climate warming. The model predictions are then entered into the GIS to predict changes in the range and abundance of these species.

Whole Landscape Models: Multivariate Statistical Analyses of Landscape Patterns

The use of multivariate statistics to simplify complex relationships between environmental features into a few orthogonal, algebraic rules is well embedded in ecological research (Orlóci 1975, James and McCulloch 1990). This history and the well-characterized behavior of these techniques suggests that such an approach would prove useful in deriving models describing the relationship between a particular environmental feature and the spatial distribution of surrounding features (Baker 1989). Surprisingly, there have been only a few attempts at such an approach (Alig 1986, Johnston et al. 1988, Johnston 1989).

Assessing the cumulative effects of disturbances on environmental quality across the landscape is a complex problem that is well suited to such an approach. The cumulative alteration of the landscape by humans through industry, agriculture, and forestry is a chronic type of disturbance both spatially and temporally. Landscape alteration includes fragmentation of preexisting ecosystems (Harris 1984) as well as changing ecosystem processes across large portions of the landscape. Both of these changes affect the flows of materials and organisms and have implications at watershed, regional, and global scales. Among the most dramatic losses of ecosystems in recent times have been the loss of old-growth forests in tropical and temperate regions (see chapters by Franklin, Swanson et al., and Risser, this volume) and the loss of wetlands from riparian zones.

Wetlands continue to be lost at a rapid rate (Tiner 1984, Johnston 1989). There is abundant evidence that riparian wetlands retain sediments (Johnston et al. 1984), heavy metals (Giblin 1985), and fecal coliform bacteria (Godfrey et al. 1985), and convert dissolved nitrate to gaseous nitrous oxides through denitrification (Barlett et al. 1979). Riparian wetlands therefore have a major effect on water quality (Holland et al. 1990). Their cumulative loss could cause significant degradation of the environmental quality and functioning of a watershed. The cumulative impacts of chronic disturbances have been recognized as potentially serious for over a decade (Council on Environmental Quality 1978), but until recently methodologies for assessing the consequences of cumulative losses and alterations of ecosystems have been lacking (Gosselink and Lee 1987).

If a watershed is a mosaic of landscape elements linked by the downslope movement of water, then the quantity and quality of water at the output, an aggregate property of the whole landscape, should be predictable from the statistical distribution of the attributes of that mosaic (Preston and Bedford 1988) assessed at the proper scale (Allen et al. 1984). Mathematically, the problem is to predict a property at a given point from a statistical description of the environment surrounding and contributing material to that point. A geographic information system allows the statistical attributes of a watershed mosaic to be summarized and described. Using multivariate statistics, landscape attributes can then be coupled with water quality data for streams draining those watersheds. Thus the cumulative effects of changes in landscape mosaics on watershed properties can be evaluated by simplifying spatial relationships using multivariate statistics.

We will illustrate this technique with a study on the impact of cumulative losses of wetlands on water quality in central Minnesota, USA (Johnston et al. 1988, 1990). Fifteen major watersheds in the Minneapolis-St. Paul metropolitan area were chosen for study. For each watershed, stream monitoring data from the STORET data base (U.S. Environmental Protection Agency, Office of Water and Hazardous Materials), land use data from recent and historic air photos, and topographic and soil maps were available.

Various watershed attributes were measured from digitized maps of land use, topography, and soils using the ERDAS geographic information system. Note that statistical data, rather than maps, were the primary output from the GIS. Areal features such as land use and wetland types were summarized across the entire watershed or within buffer zones around wetlands and streams (created with the GIS). Linear features such as stream length for each stream order were also measured. Watershed shape was also described using elongation and compactness ratios (United States Geological Survey 1978). The diverse land uses adjacent to each stream were described by dividing the number of runs, or strings of adjacent cells with identical classifications, by the number of cells bisected by the stream (Cairns et al. 1968). In addition, the average wetland stream order position relative to the position of the water quality sampling point was estimated from the synthetic descriptor:

$$\frac{\sum_{j=1}^{j} (j - i)Ai}{\sum_{j=1}^{j} Ai} \tag{1}$$

where j = stream order of water sampling point and A_i is the total area of the wetlands in the *i*th order of streams. This index is a simple way of describing the distribution of wetland acreage relative to stream order and distance upstream from the sampling point.

Thirty-three watershed attributes summarized using the GIS were then reduced to eight principal components that explained 86% of the variance between watersheds. Because the variables were measured in different units, the principal components were extracted from the eigenvectors of the correlation matrix among all variables. The correlation matrix scales the spatial relationships between 0.0 and 1.0, thus removing distortions caused by expressing different variables in different units with different ranges (James and McCulloch 1990). In order of decreasing importance, these principal components represented wetland extent, wetland proximity to sampling point and watershed area, ratio of agricultural to urban land use, length of third-order streams and watershed diversity, forested riparian area, elongation of watershed headlands, soil erodibility, and extent of herbaceous marsh. The scores of each of the watersheds for each of the principal components were then related to physical, chemical, and microbiological parameters of water draining from each watershed using stepwise multiple regression.

The first principal component, which was related to wetland extent and explained 29% of the variance in landscape properties, was significantly correlated with specific conductivity, chloride, and lead concentrations in stream water. Principal component 2, which was related to wetland position, explained an additional 14% of landscape variance. It was significantly associated with decreased concentrations of inorganic suspended solids, fecal coliform, nitrates, specific conductivity, flow-weighted ammonium and total phosphorus concentrations, and decreased proportion of total phosphorus in dissolved form. Only specific conductivity was related to both of the principal components. Thus both spatial extent and spatial distribution of wetlands affect water quality, and each affects different parameters.

These findings have significant implications for regulatory agencies and land managers. Because wetland position, rather than wetland extent, was strongly correlated with many water quality parameters that are goals of environmental protection agencies, regulations that specify no net loss of wetlands, while necessary, may not in and of themselves accomplish desired goals if development results in a redistribution of wetlands even with no net loss. Land managers and regulatory decision makers need to consider the spatial distribution of wetlands as well as wetland abundance when seeking to regulate the loss of wetlands as a tool in environmental protection.

Distributional Landscape Models: Parameterizing Markov Chains Using Geographic Information Systems

Changes in the spatial distribution of land classes can be summarized with geographic information systems by overlaying maps of different dates and analyzing spatial coincidence of the same land class from one date to the next. However, it may be difficult to discern patterns in such tabulated data, particularly when the map is broken into many classes or when changes occur in many directions. Mathematically, the problem is one of discerning patterns of change in the entire landscape, rather than interfacing point and regional data as demonstrated above. Changes from one land class or state to another can be mathematically described as probabilities that a given pixel will remain in the same state or be converted to another state. Linear algebra can then be applied to discern patterns not readily perceived by inspection.

Mathematically, the expected change in landscape properties can be summarized by a series of transition probabilities from one state to another over a specified period. The maximum likelihood estimates of these probabilities for a specified period are

$$p_{i,j,\tau} = n_{i,j} \Big| \sum_{j=1}^{n} n_{i,j} \qquad (2)$$

where $p_{i,j,\tau}$ is the probability than state i has changed to state j during time interval τ for any given pixel and $n_{i,j}$ is the number of such transitions across all pixels, hectares, and so forth, in a landscape of n states (Anderson and Goodman 1957). These are obtained by overlaying maps from one date on another (using a GIS) and tabulating the number of all such transitions. However, the time interval τ between the maps is often something other than the desired time step of the transition probabilities. This frequently happens when using historic air photos taken at multiple-year intervals to construct the base maps even though the desired time step for transition probabilities is annual. The probabilities can then be normalized to annual probabilities assuming first-order kinetics:

$$p_{i,j} = 1 - e^{[\ln(1-p_{i,\tau})]}/\tau \quad \text{when} \quad i \neq j$$

$$\text{or} \tag{3}$$

$$p_{i,i} = 1 - \sum_{j=1}^{n} p_{i,j} \quad \text{when} \quad i = j$$

When assembled in a matrix and used to generate a temporal series, known as a Markov chain, these transition probabilities form a simulation model of changes in areas of different cover types over time:

$$n_{t+1} = An_t \tag{4}$$

where n is a column vector of the areal distribution of cover types in the landscape at time t or t+1 and A is the matrix of transition probabilities.

Markov models have been assembled to simulate succession of forest cover types (Shugart et al. 1973, Johnson and Sharpe 1976, Hall et al. 1991), land use classes (Turner 1987*c*), replacement probabilities of trees in a stand (Waggoner and Stephens 1970), and heathland succession (Jeffers 1988). Jeffers (1988) and Baker (1989) review numerous other examples of Markov models for various types of landscapes.

These earlier models were parameterized using long-term plot measurements or almost anecdotal data on successional times. Using a GIS to calculate these probabilities from historic air photos is less biased because it makes full use of the complex patterns seen in the landscape rather than data from only a few points.

To further discern major pathways of successional change, the techniques of linear algebra are useful. All matrices have a set of mathematical properties known as eigenvalues. Associated with each eigenvalue is a unique eigenvector. The eigenvalues and eigenvectors satisfy the equation:

$$A\mu = \lambda\mu \tag{5}$$

where A is the matrix of transition probabilities and is a particular eigenvalue associated with the eigenvector μ.There are as many eigenvalues in the solution to this equation as there are states in the system, and each element in each eigenvector corresponds to one of these states. Markov matrices have the nice property of having the dominant eigenvalue equal to 1. Thus, by definition from Eq. 5, the eigenvector associated with this eigenvalue will give the stable areal distributions of landscape elements (wetland classes, cover types, etc.) for this particular set of transition probabilities. In other words, this particular eigenvector tells us how the landscape will look under steady state conditions. We will illustrate this technique by analyzing successional trends in boreal wetlands created when beaver build dams across streams.

When beaver dam streams, they create complex patterns of wetlands across the landscape (Johnston and Naiman 1990*a*, *b*). The wetlands behind and in front of the dams range from flooded areas of open water with submersed macrophytes, to seasonally flooded areas typically dominated by *Iris* or tussock-forming sedges (*Carex*), to moist meadows dominated by the grass Canada bluejoint (*Calamagrostis canadensis)*, sedges (*Carex, Scirpus cyperinus*), and others. As the ponds are abandoned, the dams break and the ponds drain; at some future time the dam may be repaired and the pond reoccupied and reflooded. These changes in the hydrologic regime invoke changes in the distribution and abundance of different wetlands.

To illustrate, we have assembled transition probabilities of wetlands in beaver ponds in four watersheds in Voyageurs National Park, northern Minnesota, during four decades since the 1940s. Four wetland types were recognized and mapped from air photos: (1) valley bottoms not currently in ponds, (2) flooded areas, mainly open water, (3) seasonally flooded areas with emergent macrophytes in standing water persisting most of the year, and (4) moist meadows dominated by graminoids without standing water but with soil that remains saturated most of the year. We then calculated the dominant eigenvector for each of these sixteen transition matrices.

The transition probabilities derived for the beaver ponds are not constant, either for a given watershed through time or across watersheds for a specified period. Thus, even though the complexity of the different maps has been reduced to a set of tabulated probabilities of change, specific patterns may still be difficult to discern from simple inspection of these probabilities.

Now, if the transition probabilities are changing over time, we can use the eigenvector to tell us if there is some pattern in these changes or if they are random across the landscape. Changes in the stable areal distribution of one of these watersheds are shown in Figure 11.2. Three important patterns emerge: (1) at steady state, all valley bottoms are occupied by ponds; (2) seasonally flooded areas are always a small portion of the landscape that fluctuates seemingly at random; (3) the proportion of landscape in the moist class appears to rise and fall in inverse relation to that in the flooded state.

This third property is particularly interesting because flooded wetlands differ substantially from moist wetlands with regard to carbon and nutrient dynamics (Naiman and Melillo 1984, Naiman et al. 1986, 1988). Therefore, alternations of these states may cause fluctuations in water quality and carbon storage over time. To determine whether this pattern is common to the

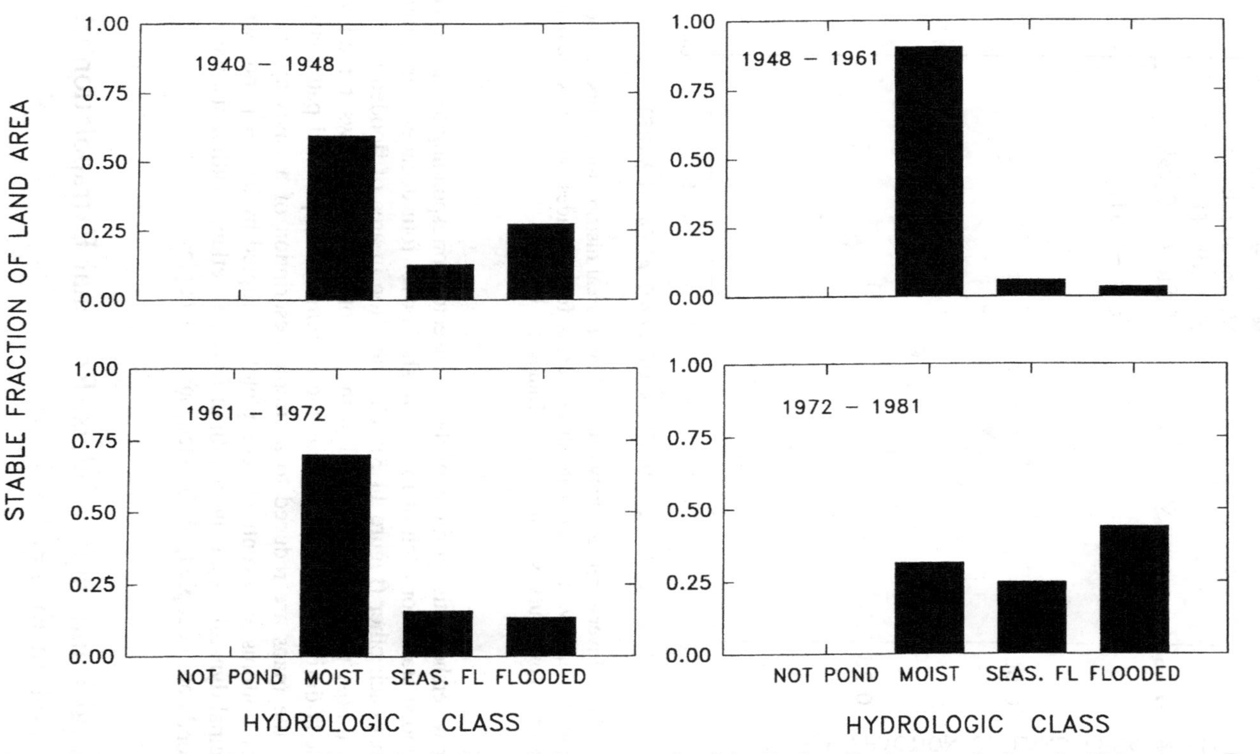

FIGURE 11.2. Stable areal distributions of hydrologic classes in beaver ponds of the Sucker Creek watershed on the Kabetogama Peninsula, Voyageurs National Park, Minnesota, USA. Note how the stable distributions change from decade to decade.

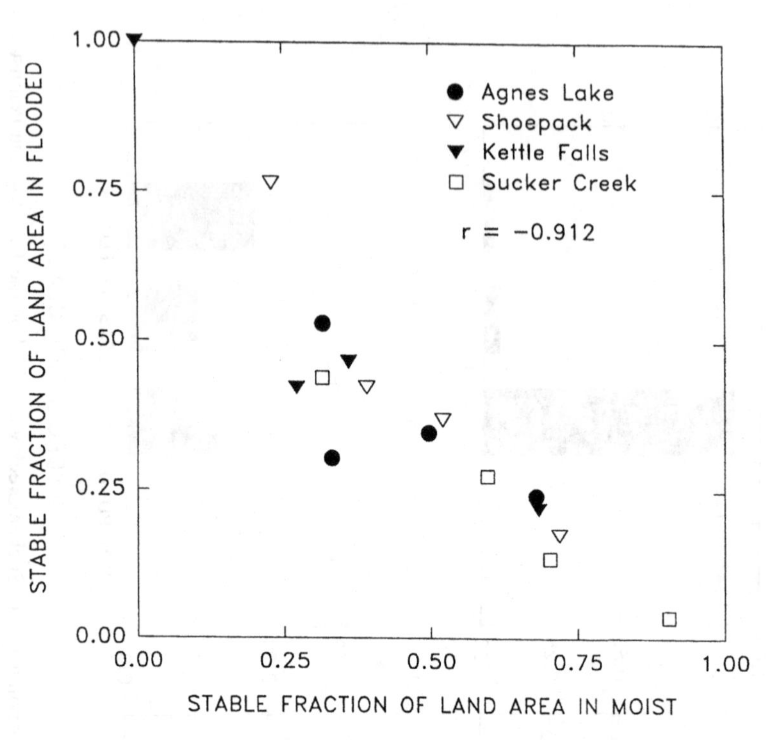

FIGURE 11.3. Inverse relation between the stable areal distribution of flooded ponds and moist meadows in four watersheds across four decades on the Kabetogama Peninsula, Voyageurs National Park, Minnesota, USA.

other watersheds, the portions of the eigenvector corresponding to the flooded and moist states for each of four watersheds over four decades were plotted against each other (Figure 11.3). The variation in area of flooded wetlands was almost perfectly correlated with that of moist meadows for all watersheds during all decades. Thus the extremely complicated patterns seen from the maps are reduced to a simple description of a landscape whose valley bottoms are eventually completely occupied by beaver ponds whose principal dynamic is an alternation of flooded wetlands with moist meadows as ponds are occupied, abandoned, and reoccupied.

Spatial Landscape Models: Regional Extrapolation of Ecosystem Properties

The current generation of ecosystem simulation models predicts ecosystem properties for a combination of climate, soils, and vegetation at a given point (Shugart 1984, Parton et al. 1987, Agren et al. 1991). The process of interpolating these properties between simulated points and integrating the spatial interpolation to yield changes in landscape properties, such as carbon balance and species distribution, is problematic. Mathematically, the problem is the opposite of the one encountered for the multiple regression, whole-landscape model. There, a statistical description of the landscape is used to predict a whole watershed property, such as water quality at the outlet; here, properties predicted for points are interpolated to yield a description of the surrounding landscape.

Conceptually, the simplest approach is to run an ecosystem model for every grid cell or polygon identified by a GIS. The ecosystem property predicted by the model, Z, for each cell or polygon, p, is then given by the general equation:

$$Z_p = f(R,O,G,C)_p - g(R,O,G,C)_{p,q} + h(R,O,G,C)_{q,p} \quad (6)$$

where R is the availabilities of resources, O is the suite of organisms, G represents geologic and topographic properties influencing resources and organisms, C represents climatic influences on the same, and q is each of the surrounding grid cells or polygons from which seeds, water, and so forth are imported to or exported from cell or polygon p. The functions f, g, and h describe ecosystem properties from the state variables R, O, G, and C (function f) and their export from p to surrounding cells q (function g) or import from surrounding cells q to p (function h). These functions can be simple multivariate regressions as described above, or they can be complex ecosystem models run for each point. Equation 6 thus describes the mass balance of material or organism movement across the landscape.

Global circulation models of the atmosphere are the most complex examples of such an approach (Schlesinger and Mitchell 1985), because the mass, momentum, and water content of air masses in 5 x 5 degree cells across the globe are simulated in at least one and sometimes two or even three dimensions. At a finer scale, this approach has been used to simulate the movement of water and sediment within a watershed (Sklar et al. 1985, Sklar and Costanza 1986, Young et al. 1989), watershed-vegetation-atmosphere interactions (Running et al. 1989), and the spread of fire in a forested landscape (Kessel 1979, Kessel et al. 1984).

Although such approaches are appealing because they provide a spatially complete and explicit description or simulation of a landscape, they are demanding of both data and computer resources. Accurate maps of data used to drive or test the models need to be developed for the entire landscape in question for each pixel or polygon to be simulated independently. In principal the functions f, g, and h can be replaced by more complex simulation models, such as the JABOWA/FORET forest models (Botkin et al. 1972, Shugart 1984), but for some problems this may require computers of a size and architecture not yet available, as will be discussed later.

Burke et al. (1990) present an alternative, simpler method for extrapolating ecosystem models across the landscape. In their approach, maps of driving features for CENTURY, a grassland ecosystem model developed by Parton et al. (1987), were compiled and digitized into a GIS. The maps were then overlaid, and unique combinations of driving variables, such as soil texture and climate, were identified. A composite map was then produced. Model runs were made for each of the combinations of driving variables, and model predictions were then entered into the GIS for those polygons representing the unique combinations of driving variables. Using this approach, they mapped productivity and other ecosystem properties across the Great Plains.

Another approach is to make model runs for selected points and mathematically interpolate predictions between the points. There are numerous statistically valid methods of interpolation between points in a network (Upton and Fingleton 1989). Many of these assume that the value of an ecosystem property for any interpolated point varies directly with the value at the nearest network point and inversely with distance from it. Some methods assume that properties across the network of points are not spatially correlated. This assumption is most likely to hold when the distance between the network points is much greater than the spatial scale of an ecosystem process likely to affect the simulations. For example, a network of points several hundreds of kilometers apart is spatially independent with respect to the dispersal of seeds several hundred meters from each point.

If the network of points is dense enough such that spatial independence between points for a given process cannot be reasonably assumed, then kriging is the interpolation method of choice (Journel and Huijbregts 1978, Krige 1981). Kriging incorporates the degree of spatial autocorrelation between points as well as the values and distances between these points. Most often used in mineral exploration for interpolating geochemical and mineralogical data between boreholes, kriging is also becoming an accepted tool for interpolating hydrological and ecological data (Burgess and Webster 1980*a*, *b*; Robertson 1987, Robertson et al. 1988, Delhomme 1978). Maps produced by kriging will differ from those produced by other forms of interpolation depending on the strength and degree of anisotrophy of spatial autocorrelation. In particular, kriging will smooth gradients across the landscape. Greater degrees of spatial autocorrelation will produce smoother gradients and more separated isopleths. In contrast, interpolation methods that assume spatial independence will emphasize gradients, and isopleths will be closer. Predictions of the two methods will converge where spatial autocorrelation is weak or absent.

We will illustrate this method for predictions of changes in species distribution with climate warming across eastern North America. Changes in climate force adjustment in feedbacks between vegetation and soil or vegetation and the flow of water across the landscape; these adjustments in turn amplify the responses of ecosystems to climatic warming (Pastor and Post 1988, Agren et al. 1991). Vegetation-soil feedbacks therefore introduce a high degree of spatial complexity in the regional responses of forests and grasslands to global warming (Pastor and Post 1988, Burke et al. 1990).

To examine spatial variability of forest response to climate change, the growth of mixed species forests was simulated for twenty points across eastern North America using the LINKAGES model (Pastor and Post 1986). Previous papers have presented maps of these points and results from other model runs (Solomon 1986, Pastor and Post 1988). The model simulates annual establishment, growth, and death of individual trees, and the decay of litter from them in a 1/12 ha forest plot. The model assumes that species migrate to new sites to the extent that temperature and soil water availability are optimal and that growth is limited by temperature, water, nitrogen, or light, whichever is most restrictive.

Forests at each point were simulated for 200 years using current monthly temperatures and precipitation. These climatic properties were then altered linearly for the next 100 years to reach a simulated climate corresponding to a 2 × current CO_2 concentration for each site (Mitchell 1983), followed by 200 more years under the new climate. Simulations were run for a sandy soil (10 cm available water per meter of soil depth) and a silty clay loam soil (18.3 cm available water per meter) for each point. These two soils represent extremes of soil water holding capacity. The simulations therefore indicate the bounds of possible changes in species ranges as well as other predicted ecosystem properties induced by a changing climate for each site.

Properties predicted at the network of points were interpolated to subcontinental scales by entering model simulations into the ERDAS geographic information system. An Albers equal area projection of North America east of the 100th meridian was rasterized into 20,000 m × 20,000 m pixels. Each pixel was assigned an interpolated value for the desired ecosystem property based on data for the simulated points weighted by the following:

$$W = (1-Q)/Q \quad (7)$$

where Q = D/S, and D is the calculated distance from the pixel center to the simulated point and S is the search radius from the desired pixel. The search radius was set at one-quarter of the maximum distance between the simulated points, or 1,224 km. The value of W ranges from 1 at the simulated point to 0 an infinite distance away from the simulated point.

Using the interpolation technique in Eq. 7, both current and projected future distributions were mapped for sugar maple (*Acer saccharum*) and black and white spruce (*Picea mariana*, *P. glauca*) on sands and silty clay loams (Figures 11.4 and 11.5). Current distributions predicted from the model and interpolated across eastern North America correspond to independent maps of the distribution of these species (Little 1971, 1977). The range of sugar maple on silty clay loams represents a slight departure from its actual range in the southeastern United States. Here, the range of sugar maple extends slightly beyond the southern Appalachians into the coastal plain. In fact, the actual range extends down the crest of the Appalachians into northern Geor-

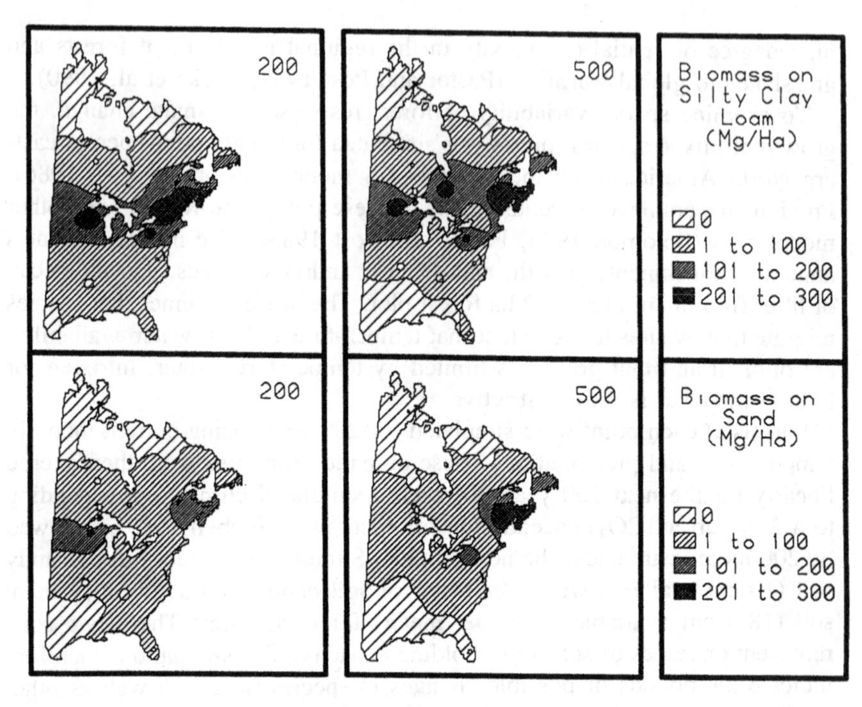

FIGURE 11.4. Simulated distribution of maple after 200 years under current climate followed by 100 years of climatic warming and 200 more years under a greenhouse climate (year 500). The simulations are shown for silty clay loams and sands east of the 100th meridian. Note the eastward migration of maple from year 200 to 500 on sands but the northward migration on silty clay loams.

gia. The simulated range departs from the actual range because the network of points in the southeastern United States was too sparse to accurately simulate the terminus in the Great Smoky Mountains. This discrepancy emphasizes the need to maintain a sampling density high enough so that interpolation does not distort data while at the same time making efficient use of computer time.

Nevertheless, the coupling of the simulation model with the GIS does indicate some general patterns in species range shifts with global warming. First, sugar maple has a different range shift than spruce. Second, the regional shift in species ranges depends on the local distribution of soil types. For example, the range of sugar maple is constricted eastward on sandy soils under a warmer and drier climate because of severe drought stress in the center of the continent. Silty clay loams mitigate against drought stress for sugar maple, and its range shifts northward on these soils because of intolerably high summer temperatures at the southern edge of its current range. In contrast, the ranges of spruce, which is more drought tolerant, shift north-

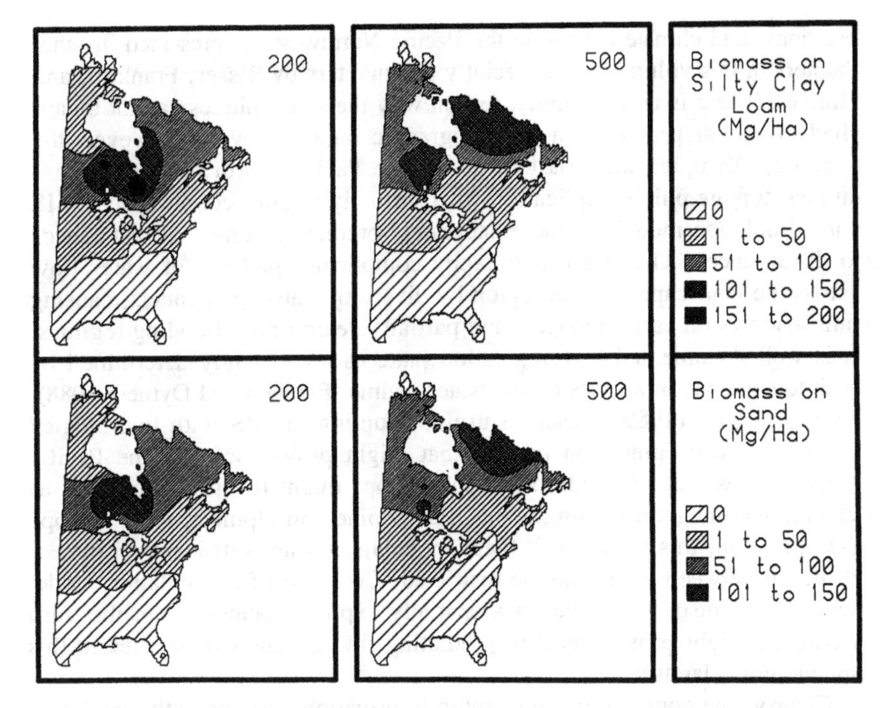

FIGURE 11.5. Simulated distribution of spruce after 200 years under current climate followed by 100 years of climatic warming and 200 more years under a greenhouse climate (year 500). The simulations are shown for silty clay loams and sands east of the 100th meridian. Note, in contrast to Figure 11.4, that spruce migrates northward on both soils from year 200 to year 500.

ward on both silty clay loams and sands because of summer warming at the southern edge of the ranges.

We can predict potential net changes in maple or spruce biomass over the entire subcontinent by using the GIS to integrate areas of biomass change for each species. Potential spruce biomass increased 23% on silty clay loams but declined 18% on sands under the warmer climate. In contrast, maple biomass decreased 13% on silty clay loams and 42% on sands. Thus both the spatial distribution and potential biomass of each species can be estimated by interpolating ecosystem predictions between given points using a GIS.

Where Do We Go From Here?

Although the examples presented were derived from eastern forests and wetlands, the general approach is useful for problems of the Pacific Northwest. Examples of the cumulative impact of wetland loss, ecosystem-animal interactions, and climate change in the Pacific Northwest are presented in other chapters in this volume (see especially the chapters by Risser, Franklin, and Elmore). There is no conceptual reason why these techniques cannot be applied to similar problems in this geographic region as well. However, the Cascade, Olympic, and Coast ranges of the Pacific Northwest present additional topographic complications that virtually require coupling of a GIS with simulation models to more accurately predict responses of watersheds to disturbances. The steep topography complicates patterns of water flow across the landscape and the region by diverting water to channels, causing rain shadows on eastern slopes, and partially determining flooding regimes. It is well documented that vegetation patterns are strongly determined by such topographic influences on the water regime (Franklin and Dyrness 1988). Running et al. (1989) present a useful coupling of GIS with hydrologic/biogeochemical simulation models that might prove useful in the Pacific Northwest as well. In addition, mountaintops might function as islands as climate warming causes migration of subalpine and alpine species upslope (see Franklin, this volume); if these mountaintops are sufficiently isolated, dispersal between them may be restricted. Coupling of a simulation model with a seed dispersal model in a spatially explicit manner, as in the third example, might prove useful in predicting future ranges of species in this mountainous landscape.

Clearly, the coupling of geographic information systems with simulation models is a powerful tool for describing and predicting the behavior of watersheds, regions, and the globe. Equally clear, however, is that these techniques are in their infancy. Both new hardware and software and new concepts are needed to advance these tools.

The examples given, as well as others cited, couple GIS and simulation models in a relatively static mode. That is, GIS output is used to parameterize or run models and then results are mathematically analyzed or remapped. However, GIS and models need to be coupled in a more dynamic manner to accurately predict the spread of fire, disease, or water across the landscape. This requires that models be reparameterized or model runs adjusted "on the fly." That is, if the spatial configuration of a map changes such that delivery of water, organisms, or materials to a point is significantly altered, then model runs for that point must be adjusted immediately. For example, if one pixel catches fire in a simulation, then all adjacent pixels should too, even if model simulations for those adjacent pixels show low inherent probability for fire initiation. Initiation and continuation of a process is a property of points simulated by models, but the spread of those properties is a spatial process simulated by a GIS.

Both new software and new hardware are required to couple models and GIS in a more dynamic sense. The UNIX operating system, with its stream-socket approach to passing information between models, holds some promise as a controller linking models and GIS. In fact, some interesting advances along these lines are already being made (W. Lauernroth, T. Kirchner, and W. Parton, Colorado State University, pers. comm.). Additional hardware is also needed to handle the huge volume of calculations required for each pixel or polygon. Massive parallel processors may be one solution to the problem. Each processor could be assigned to one particular pixel or polygon, making independent but occasionally interacting model runs. The entire bank of processors becomes one giant simulated map.

Perhaps even more important, new concepts are needed to unify ecosystem and landscape ecology in a common theoretical framework. Two questions are central to this problem, and may help guide the formulation of hypotheses and experiments to test them.

1. *What processes at the landscape, regional, or global level control ecosystem processes at local scales and what are their spatial distributions?* We have shown that water quality is controlled at the output by the extent and spatial distribution of wetlands in a watershed, that animal habitat alterations change the distribution of wetlands in the landscape, and that global changes in climate control the distribution of species and effects on properties at a point. Many other examples can be found throughout the chapters of this book. Whole, distributional, or spatially explicit landscape models each provide answers to this question, but the nature of the models and thus the answers are scale dependent. Thus the scales of processes explicit in hypotheses framed to answer this question need to be clearly stated. Only then can the proper model and GIS be chosen to provide an answer.

2. *How do ecosystem properties percolate across the landscape to affect entire watersheds?* The dispersal of animals and seeds from one point to adjacent points can affect the aggregate spatial and functional properties of entire landscapes, as shown by the analysis of landscape stabilities in boreal wetlands affected by beaver. How ecosystem properties percolate across the landscape depends partly on how processes interact across nested scales. For example, beaver build dams at a particular point based on particular local configurations of topography, hydrology, and access to food, but the formation of the large pond behind the dam affects a large piece of the landscape. At present, there are few models that explicitly incorporate processes occurring at nested scales, the ZELIG version of the FORET model being a notable exception (Smith and Urban 1988). Furthermore, most commercially available GISs now use single grid cell or vector resolution rather than multiple levels of resolution within the same data layer. Hypotheses of the effects of processes occurring at nested scales explicitly described by simulation models, as well as more sophisticated data base manipulations at nested scales, are therefore required.

Clearly, the physical scale addressed by these questions is much larger than those traditionally subsumed by experiments. However, it is the scale at which managers must normally operate. Experimentally testing the interactions between ecosystem properties and managed landscapes therefore requires a closer collaboration between researchers and managers. The coupling of geographic information systems with simulation models not only provides new tools for both researchers and managers but provides a framework for designing experiments to assess the consequences of new management techniques at ecosystem, watershed, and global scales. Knowledge is traditionally thought to flow from researchers to managers. However, the future development of simulation models and geographic information systems will require closer collaboration between these two groups.

Acknowledgments. We thank Geoff Poole and two anonymous reviewers for helpful comments on the manuscript. The ideas in this paper were sharpened by discussions of the senior author with colleagues at the symposium. We thank these colleagues and the organizers of the symposium for the opportunity to present these ideas. This research was supported by NSF grants BSR-8817665 and DIR-8805437–01, by the Department of Energy's Global Carbon Cycle Program under contract No. DE-AC05–84OR21400 with Martin Marieta Energy Systems, Inc., and by the U.S. Environmental Protection Agency agreements EPA CR-814083–010 and EPA CR-815482–010.

This is Publication Number 69 of the Center for Water and the Environment and Number 11 of the NRRI GIS Laboratory.

References

Agren, G.I., R.E. McMurtrie, W.J. Parton, J. Pastor, and H.H. Shugart. 1991. State-of-the-art of models of production-decomposition linkages in conifer and grassland ecosystems. Ecological Applications **1**:118–138.

Alig, R.J. 1986. Econometric analysis of the factors influencing forest acreage trends in the Southeast. Forest Science **32**:119–134.

Allen, T.F. H., R.V. O'Neill, and T.W. Hoekstra. 1984. Interlevel relations in ecological research and management: some working principles from hierarchy theory. United States Forest Service General Technical Report RM-110, Fort Collins, Colorado, USA.

Anderson, T.W., and L.A. Goodman. 1957. Statistical inference about Markov chains. Annals of Mathematical Statistics **28**:89–110.

Baker, W.L. 1989. A review of models of landscape change. Landscape Ecology **2**:111–133.

Barlett, M.S., L.C. Brown, N.B. Hanes, and N.H. Nickerson. 1979. Denitrification in freshwater wetland soil. Journal of Environmental Quality **8**:460–464.

Botkin, D.B., J.F. Janak, and J.R. Wallis. 1972. Some ecological consequences of a computer model of forest growth. Journal of Ecology **60**:849–872.

Burgess, T.M., and R. Webster. 1980*a*. Optimal interpolation and isarithmic mapping of soil properties. I. The semi-variogram and punctual kriging. Journal of Soil Science **31**:315–331.

Burgess, T.M., and R. Webster. 1980*b*. Optimal interpolation and isarithmic mapping of soil properties. II. Block kriging. Journal of Soil Science **31**:333–341.

Burke, I.C., D.S. Schimel, C.M. Yonker, W.J. Parton, L.A. Joyce, and W.K. Lauenroth. 1990. Regional modeling of grassland biogeochemistry using GIS. Landscape Ecology **4**:45–54.

Burrough, P.A. 1986. Principles of geographical information systems for land resources assessment. Clarendon Press, Oxford, England.

Cairns, J., Jr., D.W. Albaugh, F. Busey, and M.D. Chanay. 1968. The sequential comparison index: a simplified method for non-biologists to estimate relative differences in biological diversity in stream pollution studies. Journal of the Water Pollution Control Federation **40**:1607–1613.

Council on Environmental Quality. 1978. Recommendations for implementing the procedural provisions of the National Environmental Policy Act. 40 CFR, Sec. 1508.7.

Delhomme, J.P. 1978. Kriging in the hydrosciences. Advances in Water Research **1**:251–266.

Franklin, J.F., and C.T. Dyrness. 1988. Natural vegetation of Oregon and Washington. Oregon State University Press, Corvallis, Oregon, USA.

Giblin, A.E. 1985. Comparison of processing of elements by ecosystems. II. Metals. Pages 158–179 *in* P.J. Godfrey, E.R. Kaynor, S. Pelczarski, and J. Benforado, editors. Ecological considerations in wetlands treatment of municipal wastewater. Van Nostrand Reinhold, New York, New York, USA.

Gildea, M.P., B. Moore, C.J. Vorosmarty, J.M. Melillo, K. Nadelhoffer, and B.J. Peterson. 1987. Impacts of land use on nitrogen cycling in large systems: the Mississippi River basin. *In* The influence of land use pattern on landscape function: ecological theory and management implications. Second Annual Landscape Ecology Symposium, March 9–11, 1987, University of Virginia, Charlottesville, Virginia, USA.

Godfrey, P.J., E.R. Kaynor, S. Pelczarski, and J. Benforado. 1985. Ecological considerations in wetlands treatment of municipal wastewater. Van Nostrand Reinhold, New York, New York, USA.

Gosselink, J.G., and L.C. Lee. 1987. Cumulative impact assessment in bottomland hardwood forests. Center for Wetland Resources Report LSU-CEI-86–09, Louisiana State University, Baton Rouge, Louisiana, USA.

Hall, F.G., D.B. Botkin, D.E. Strebel, K.D. Woods, and S.J. Goetz. 1991. Large-scale patterns of forest succession as determined by remote sensing. Ecology **72**:628–640.

Harris, L.D. 1984. The fragmented forest. University of Chicago Press, Chicago, Illinois, USA.

Holland, M.M., D.F. Whigham, and B. Gopal. 1990. The characteristics of wetland ecotones. Pages 171–198 *in* R.J. Naiman and H. Décamps, editors. The ecology and management of aquatic-terrestrial ecotones. UNESCO, Paris, and Parthenon Publishing Group, Carnforth, United Kingdom.

James, F.C., and C.E. McCulloch. 1990. Multivariate analysis in ecology and systematics: panacea or Pandora's box? Annual Review of Ecology and Systematics **21**:129–166.

Jeffers, J.N.R. 1988. Practitioner's handbook on the modelling of dynamic change in ecosystems. John Wiley and Sons, New York, New York, USA.

Johnson, L.B. 1990. Analyzing spatial and temporal phenomena using geographical information systems. Landscape Ecology **4**:31–43.

Johnson, W.C., and D.M. Sharpe. 1976. An analysis of forest dynamics in the northern Georgia Piedmont. Forest Science **22**:307–322.

Johnston, C.A. 1989. Human impacts to Minnesota wetlands. Journal of the Minnesota Academy of Science **55**:120–124.

Johnston, C.A., G.D. Bubenzer, G.B. Lee, F.W. Madison, and J.R. McHenry. 1984. Nutrient trapping by sediment deposition in a seasonally flooded lakeside wetland. Journal of Environmental Quality **13**:283–290.

Johnston, C.A., N.E. Detenbeck, J.P. Bonde, and G.J. Niemi. 1988. Geographic information systems for cumulative impact assessment. Photogrammetric Engineering and Remote Sensing **54**:1609–1615.

Johnston, C.A., N.E. Detenbeck, and G.J. Niemi. 1990. The cumulative effect of wetlands on stream water quality and quantity: a landscape approach. Biogeochemistry **10**:105–141.

Johnston, C.A., and R.J. Naiman. 1990*a*. Aquatic patch creation in relation to beaver population trends. Ecology **71**:1617–1621.

Johnston, C.A., and R.J. Naiman. 1990*b*. The use of a geographic information system to analyze long-term landscape alteration by beaver. Landscape Ecology **4**:5–19.

Journel, A.G., and C.J. Huijbregts. 1978. Mining geostatistics. Academic Press, London, England.

Kessell, S.R. 1979. Gradient modeling: resource and fire management. Springer-Verlag, New York, New York, USA.

Kessell, S.R., R.B. Good, and A.J.M. Hopkins. 1984. Implementation of two new resource management information systems in Australia. Environmental Management **8**:251–270.

Krige, D.G. 1981. Lognormal-de Wijsian geostatistics for ore evaluation. South African Institute of Mining and Metallurgy Monograph Series, Geostatistics 1, Johannesburg, South Africa.

Little, E.L. 1971. Atlas of United States trees. Volume 1. Conifers and important hardwoods. United States Forest Service Miscellaneous Publication 1146.

Little, E.L. 1971. 1977. Atlas of United States trees. Volume 4. Minor eastern hardwoods. United States Forest Service Miscellaneous Publication 1342.

Loucks, O.L., A.R. Ek, W.C. Johnson, and R.A. Monserud. 1981. Growth, aging and succession. Pages 37–85 *in* D.E. Reichle, editor. Dynamic properties of forest ecosystems. Cambridge University Press, Cambridge, England.

Lubchenco, J., A.M. Olson, L.B. Brubaker, S.R. Carpenter, M.M. Holland, S.P. Hubbell, S.A. Levin, J.A. MacMahon, P.A. Matson, J.M. Melillo, H.A. Mooney, C.H. Peterson, H.R. Pulliam, L.A. Real, P.J. Regal, and P.G. Risser. 1991. The sustainable biosphere initiative: an ecological research agenda. Ecology **72**:371–412.

Mitchell, J.F.N. 1983. The seasonal response of a general circulation model to changes in CO_2 and sea temperatures. Quarterly Journal of the Royal Meteorological Society **109**:113–152.

Naiman, R.J., C.A. Johnston, and J.C. Kelley. 1988. Alteration of North American streams by beaver. BioScience **38**:753–763.

Naiman, R.J., and J.M. Melillo. 1984. Nitrogen budget of a subarctic stream altered by beaver (*Castor canadensis*). Oecologia (Berlin) **62**:150–155.

Naiman, R.J., J.M. Melillo, and J.E. Hobbie. 1986. Ecosystem alteration of boreal forest streams by beaver (*Castor canadensis*). Ecology **67**:1254–1269.

Orlóci, L. 1975. Multivariate analysis in vegetation research. Dr. W. Junk, The Hague, The Netherlands.

Parton, W.J., D.S. Schimel, C.V. Cole, and D.S. Ojima. 1987. Analysis of factors controlling soil organic matter levels in Great Plains grasslands. Soil Science Society of America Journal **51**:1173–1179.

Pastor, J., and M. Broschart. 1990. The spatial pattern of a northern conifer-hardwood landscape. Landscape Ecology **4**:55–68.

Pastor, J., R.H. Gardner, V.H. Dale, and W.M. Post. 1987. Successional changes in nitrogen availability as a potential factor contributing to spruce declines in boreal North America. Canadian Journal of Forest Research **17**:1394–1400.

Pastor, J., and W.M. Post. 1986. Influence of climate, soil moisture, and succession on forest carbon and nitrogen cycles. Biogeochemistry **2**:3–27.

Pastor, J., and W.M. Post. 1988. Response of northern forests to CO_2-induced climate change. Nature **334**:55–58.

Pearlstine, L.G., and W.M. Kitchens. 1987. A succession modelling approach to wetland impact assessment. National Wetlands Newsletter **9**:13–15.

Preston, E., and B. Bedford. 1988. Evaluating cumulative effects on wetland functions: a conceptual overview and generic framework. Journal of Environmental Management **12**:565–583.

Risser, P.G. 1987. Landscape ecology: state of the art. Pages 3–14 *in* M. G. Turner, editor. Landscape heterogeneity and disturbance. Springer-Verlag, New York, New York, USA.

Robertson, G.P. 1987. Geostatistics in ecology: interpolating with known variance. Ecology **68**:744–748.

Robertson, G.P., M.A. Huston, F.C. Evans, and J.M. Tiedje. 1988. Spatial variability in a successional plant community: patterns of nitrogen availability. Ecology **69**:1517–1524.

Running, S.W., R.R. Nemani, D.L. Peterson, L.E. Band, D.F. Potts, L.L. Pierce, and M.A. Spanner. 1989. Mapping regional forest evapotranspiration and photosynthesis by coupling satellite data with ecosystem simulation. Ecology **70**:1090–1101.

Schlesinger, M.E., and J.F.B. Mitchell. 1985. Model projections of the equilibrium climatic response to increased carbon dioxide. Pages 81–148 *in* M.C. MacCracken and F.M. Luther, editors. Projecting the climatic effects of increasing carbon dioxide. United States Department of Energy DOE/ER-0237, Washington, D.C., USA.

Shugart, H.H. 1984. A theory of forest dynamics. Springer-Verlag, New York, New York, USA.

Shugart, H.H., Jr., T.R. Crow, and J.M. Hett. 1973. Forest succession models: a rationale and methodology for modeling forest succession over large regions. Forest Science **19**:203–212.

Shugart, H.H., Jr., and D.C. West. 1980. Forest succession models. BioScience **30**:308–313.

Sklar, F.H., and R. Costanza. 1986. A spatial simulation of ecosystem succession in a Louisiana coastal landscape. Pages 467–472 *in* R. Crosbie and P. Luker, editors. Proceedings of the 1986 Summer Computer Simulation Conference, Society for Computer Simulation.

Sklar, F.H., R. Costanza, and J.W. Day, Jr. 1985. Dynamic spatial simulation modeling of coastal wetland habitat succession. Ecological Modelling **29**:261–281.

Solomon, A.M. 1986. Transient response modeling of forests to CO_2-induced climate change: simulation modeling experiments in eastern North America. Oecologia (Berlin) **68**:567–579.

Smith, T.M., and D.L. Urban. 1988. Scale and resolution of forest structural pattern. Vegetatio **74**:143–150.

Tilman, D. 1987. The importance of the mechanisms of interspecific competition. American Naturalist **129**:769–774.

Tiner, R.W., Jr. 1984. Wetlands of the United States: current status and recent trends. United States Fish and Wildlife Service. United States Government Printing Office, Washington, D.C., USA.

Turner, M.G., editor. 1987*a*. Landscape heterogeneity and disturbance. Springer-Verlag, New York, New York, USA.

Turner, M.G. 1987*b*. Land use changes and net primary production in the Georgia landscape: 1935 to 1982. Environmental Management **11**:237–247.

Turner, M.G. 1987*c*. Spatial simulation of landscape changes in Georgia: a comparison of three transition models. Landscape Ecology **1**:29–36.

Turner, M.G. 1990. Spatial and temporal analysis of landscape patterns. Landscape Ecology **4**:21–30.

United States Geological Survey. 1978. National handbook of recommended methods for water-data acquistion. USGS Office of Water Data Coordination, Reston, Virginia, USA.

Upton, G.J.G., and B. Fingleton. 1989. Spatial data analysis by example. John Wiley and Sons, New York, New York, USA.

Waggoner, P.E., and G.R. Stephens. 1970. Transition probabilities for a forest. Nature **225**:1160–1161.

Weinstein, D.A., and H.H. Shugart. 1983. Ecological modeling of landscape dynamics. Pages 29–55 *in* H.A. Mooney and M. Godron, editors. Disturbance and ecosystems: components of response. Springer-Verlag, New York, New York, USA.

Young, R.A., C.A. Onstad, D.D. Bosch, and W.P. Anderson. 1989. AGNPS: a nonpoint-source pollution model for evaluating agricultural watersheds. Journal of Soil and Water Conservation **44**:168–173.

12

Consideration of Watersheds in Long-Term Forest Planning Models: The Case of FORPLAN and Its Use on the National Forests

K. NORMAN JOHNSON

Abstract

Long-term planning models for the national forests have focused on determining sustainable levels of timber production over time. Recent forest planning modeling has sought economically efficient harvest schedules under broad constraints, while leaving it largely for plan implementation to discover whether the prescribed actions fully meet the variety of environmental objectives in the plans. As demonstrated in a case study of cumulative watershed effects on Oregon's Willamette National Forest, such an approach can easily result in a demand for higher harvest levels than are achievable. Over the next few years, requirements to demonstrate the compatibility of timber production with water and wildlife objectives will bring about major changes in forest planning methodology. Simulating cumulative effects will take precedence over demonstrating economic efficiency. Traditional harvest scheduling models like FORPLAN still will have a place in national forest planning, but they will no longer be the central focus of the effort.

Key words. Forest planning, cumulative effects, FORPLAN, linear programming, hydrologic recovery, trade-off analysis.

Introduction

The National Forest Management Act (NFMA) mandated that integrated forest plans be developed for the national forests. Since 1980, the FORest PLANning Model (FORPLAN) has been the primary analytical tool for this effort.

FORPLAN is a linear programming-based planning system for estimating sustainable levels of timber harvest over time while meeting other objectives. Its use in national forest planning has been controversial, partly because FORPLAN emphasizes natural resource relations through time more than across space. Over the last few years in Region Six of the United States Forest Service, the effects of timber harvest on watershed health and stability have become an increasingly important focus of national forest planning. Since FORPLAN is the primary analytical tool for this planning, much effort was expended in attempting to represent the relationship between timber harvest and watershed stability in that model. This article addresses three questions about forest planning analysis and FORPLAN's role in it: (1) Why has FORPLAN played such a central role in national forest planning? (2) How well has it dealt with the cumulative effects of timber harvesting on watersheds? (3) What does the future hold for forest planning models like FORPLAN as consideration of cumulative effects becomes increasingly important?

Why FORPLAN?

FORPLAN has gained a significant role in national forest planning because it deals with two major themes that have been at the center of this effort: (1) scheduling of timber harvest and (2) pursuit of economic efficiency.

Cutting Timber for the Public Good

More than anything else, the Forest Service builds roads and cuts timber. These two actions dwarf all other management activities, including range improvement, stream improvement, trail building, and siting of recreation facilities, whether they are measured by budget, personnel, or environmental impact. As a result, most national forest planning has addressed timber management scheduling (when, where, how, and how much timber to cut and grow) and road management scheduling (when, where, and how many roads to create and destroy).

Given the dominance of road building and timber harvesting, it is understandable that the NFMA would address their control. But the NFMA is very one-sided in its treatment: almost all attention is focused on timber harvest scheduling (Dana and Fairfax 1980). From the language of the Act, which uses terms like "allowable sale quantity" and "mean annual increment," to the major provisions dealing with harvest level, minimum rotation age, marginal lands, clearcut size, harvest method, protection of soil and water resources, and maintenance of plant and animal diversity, the Act aims primarily at providing direction for timber management.

It is not surprising, therefore, that the national forests would turn to a model that has its roots in the scheduling of timber harvest. As documented in a history of forest planning by Iverson and Alston (1986), FORPLAN descended from a long line of timber harvest scheduling models.

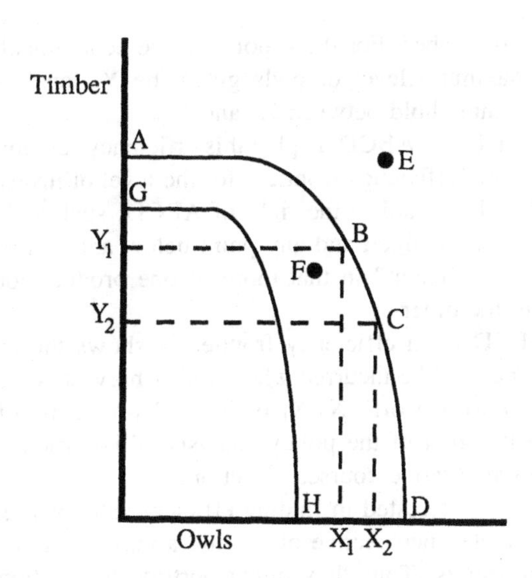

FIGURE 12.1. Two hypothetical owl-timber trade-off curves (ABCD and GH) under different assumptions about production possibilities.

These models share several common properties (Johnson and Tedder 1983, Davis and Johnson 1987): (1) The portion of the forest on which timber harvest will be permitted is represented as a set of timber stand groupings that combine areas of similar species, size classes or ages, and other characteristics. (2) Estimates of timber yield over time are made for the groupings as a function of these characteristics, management intensity, and management practice. (3) A planning horizon equal to two or three rotation lengths, broken into periods of 5 to 10 years each, is recognized. (4) Constraints are placed on rate (flow) of harvest and minimum rotation age. FORPLAN reflects its connection to these models with a land organization for harvest scheduling based on groupings of homogenous vegetative classes, choices for management of these groupings based on management emphasis (timber, wildlife, etc.) and management intensity, use of vegetation age and rotation length as organizing principles, a long time horizon (150–200 years), long planning periods within this horizon (10 years), and a variety of harvest flow constraints.

Pursuit of Economic Efficiency

For a level of investment, "economic efficiency" can be defined as the maximum level of one output given the level of another output, provided that both goods have a positive value (Davis and Johnson 1987). Looking at a trade-off between owls and timber, as an example (Figure 12.1), point B shows X_1 amount of owls for Y_1 amount of timber and point C shows X_2

of owls for Y_2 of timber. For these points to be economically efficient, X_1 must be the maximum level of owls given the Y_1 level of timber and a similar relation must hold between X_2 and Y_2.

If all points on curve ABCD display this efficiency relationship, the curve could be called an "efficiency frontier" for the level of investment involved. Then any point above and to the right of ABCD, such as E, is impossible to achieve with that budget; and any point below and to the left of it, such as point F, is "inefficient" in that more of one product could be obtained without loss in the other.

Because ABCD is an efficiency frontier, it shows the minimum loss in owls (X_2-X_1) that will be incurred when timber harvest is increased from Y_2 to Y_1. Thus we "trade off" X_2-X_1 owls for Y_1-Y_2 units of timber. These trade-offs are the grist of the policy-analysis mill—crucial information for choosing from alternative courses of action.

Economists are interested in finding efficient solutions, estimating trade-offs, and the related perspective of finding a maximum or minimum value subject to constraints. Thus they might portray the relationships in Figure 12.1 as a problem in maximizing the amount of timber harvest subject to different levels of owl protection and, in theory at least, delineate the trade-off curve outlined there.

FORPLAN reflects this focus on efficiency with its optimization (i.e., linear programming) approach to problem solution. In FORPLAN, some quantity such as timber volume or present net worth is maximized subject to constraints that reflect the resources available (the "investment level") and constraints that reflect policy objectives, such as a nondeclining yield of timber harvest volume over time and a maximum area of clearcutting per period (Johnson et al. 1986, Johnson 1987). Choices appear as alternative management practices and investment levels for homogenous land groupings.

It is more difficult to find justification in the NFMA for this focus on efficiency than for the focus on timber harvest scheduling. While the NFMA mentions that efficiency should be considered, and attained to the degree feasible, it does not emphasize efficiency as an overriding concern. Rather it stresses protection of the forest environment, including conservation of soil and water resources and maintenance of biodiversity (Wilkerson and Anderson 1987).

The focus on efficiency and optimization models in forest planning arose from three separate threads that became intertwined: (1) the interest of Forest Service economists in land management models that would display the efficiency frontiers shown in Figure 12.1— an interest that resulted in Daniel Navon's (1971) Timber Resource Allocation Method (RAM), which introduced harvest scheduling through linear programming to the national forests; (2) the fascination of the land management planning branch of the Forest Service with systems analysis and cost-effectiveness, an enthusiasm first expressed in development of the Resource Capability System (Betters 1978);

and (3) the interest in economic efficiency and trade-off analysis of the Committee of Scientists who assisted the Forest Service in writing the regulations to implement the NFMA (Teeguarden 1986).

With the rewrite of the regulations in 1982 under the Reagan administration, the fusion of forest planning and economic efficiency analysis was complete. In fact, the regulations now read, in large part, like the work of a doctoral student in forest economics. The "greatest good for the greatest number in the long run" has been recast as "net public benefit"; "benchmarks" have been added that require delineation of the maximum production of timber and other outputs possible on each national forest; and economic terminology has been used whenever possible.

This rewrite reflected the Reagan administration's preoccupation with economic development and commodity production as central objectives for the national forests. In the years that followed, suggestions to decrease the timber harvest level were met with great resistance by the administration's Office of Management and Budget, except for cases where timber sales did not make money. Resource uses that produced revenue were emphasized, and those that did not were barely tolerated.

In such an environment, it is not surprising that a model like FORPLAN would gain preeminence in forest planning. With its ability to find maximums and minimums given an objective and constraints, its sophistication in handling costs and revenues, and its detailed representation of the major commodity marketed by the Forest Service (i.e., timber), FORPLAN was a natural instrument for representing alternatives in a way that fit the political environment of the time.

How Well Has FORPLAN Done?

To explain the different approaches taken toward considering the cumulative effects on watersheds in forest planning, I will cite the recent history of forest planning on the Willamette National Forest (NF) in western Oregon. The Willamette NF has many properties that make it an instructive case study. Its proximity to Oregon's major population centers and travel routes makes it a highly visible national forest. It has harvested the most timber volume of any national forest in the nation. Its water resources are of importance for domestic water supplies, anadromous fisheries, and recreation use. Finally, the sequence of approaches taken by forest planners on the Willamette NF to represent the cumulative effect of timber harvest on watersheds illustrates both the past and future of forest planning. We will look at three successive attempts on the Willamette NF to represent cumulative effects of timber harvest on watersheds: (1) forestwide analysis, (2) analysis based on major watersheds, and (3) analysis based on subwatersheds.

Forestwide Analysis

As mentioned previously, FORPLAN descended from a long line of timber harvest scheduling models. These models were employed mainly to control the overall flow of timber harvest volume through time. Once the starting inventory, the growth rate of existing and future stands under different levels of investment, and the permitted forest practices were established, the models estimated a set of actions for each period that would provide the highest total harvest level that could be maintained through time.

An even flow of timber harvest traditionally has been viewed by public agencies as a protector of all resource values (Duerr and Duerr 1971). Under this thinking, a stable timber supply implies a stable level for other resources and outputs such as water and wildlife. Thus the cumulative effects of timber harvest actions on all forest resources are assumed to be a function of the flow of timber harvest achieved, which in turn implies that the geographic resolution for modeling the cumulative effects of a set of activities on timber supply (i.e., the entire forest) also suffices for other resources.

The U.S. Forest Service used FORPLAN to represent this traditional forestwide approach to setting harvest levels in the Draft Forest Plan for the Willamette NF (United States Forest Service 1987). Sustainable harvest levels were modeled using stand groupings organized for timber yield projection, a wide variety of timber management choices, a 150-year planning horizon, and constraints to ensure nondeclining yield of timber volume and minimum harvest ages near culmination of mean annual increment. Enhancements of the traditional approach included recognition of reduced harvest rates in visual zones and riparian areas to reflect the longer rotations there.

Before release of the Draft Plan, Willamette NF hydrologists assessed the plan's implications for the 33 major watersheds on the forest. That assessment suggested that almost one-third of these watersheds had significant risk of probable deterioration under the plan—that is, a high risk of cumulative downstream effects on water quality, physical stream characteristics, and fish habitat as a result of the activities contemplated. These deleterious effects resulted partly from harvest in riparian areas, partly from harvest on potentially unstable slopes, and partly from the rate of harvest applied across the forest (United States Forest Service 1987).

While these results suggested that it was incorrect to assume that an optimized timber harvest schedule would ensure protection of watersheds, the Forest Service published its Draft Plan anyway. Response to the proposal was swift and sharp. As the governor of Oregon said in his comments on the plan (State of Oregon 1988:13): "Thus one-third of the Forest's watershed area is vulnerable to management induced declines under your proposed plan—a disturbing result.... These results give us great concern. Consciously planning to put key watersheds under risk seems inconsistent with the mandate you have for protecting the resources under your control on a

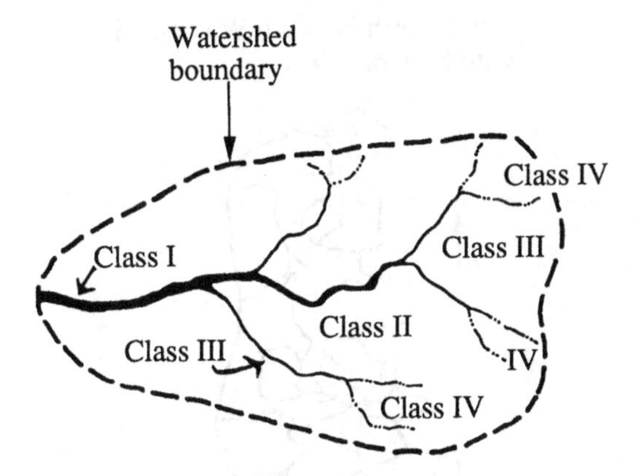

FIGURE 12.2. Stream hierarchy used on the Willamette National Forest. To qualify for Class I, streams must be the source of municipal or domestic water supply, be of large volume, or contain anadromous fish. Class II streams must have a resident trout population or enough volume to significantly affect a Class I stream. Remaining perennial streams qualify for Class III. Remaining (intermittent) streams belong in Class IV.

publicly owned Forest. You must make every effort to redistribute your proposed activities to lessen these risks.... In addition, you need to demonstrate these results over a number of decades into the future to insure that unacceptable cumulative effects are not building over time."

Analysis by Major Watershed

With the realization that forestwide modeling of timber and road-building activities would allow adverse cumulative watershed effects, Willamette NF personnel expanded the FORPLAN analysis to recognize major watersheds as the Final Forest Plan was prepared. Generally these watersheds reflected Class I and II streams (Figure 12.2). Thirty-three watersheds were recognized within FORPLAN, ranging in size from 4,000 to 50,000 hectares, with an average of 22,000 hectares (Figure 12.3). These watersheds generally encompass streams with significant year-round fish habitat and recreational use.

Including watershed definition in the FORPLAN model added another stratum to the definition of stand groups, thus increasing the total number of stand groupings recognized and creating a linear program too large to be solved easily. Therefore, forest planners dropped slope steepness (used primarily to designate logging method) and some other strata. Now, however, timber harvest and road building could be controlled over time by major watershed—a potential improvement over the previous formulation relative to moderating cumulative effects.

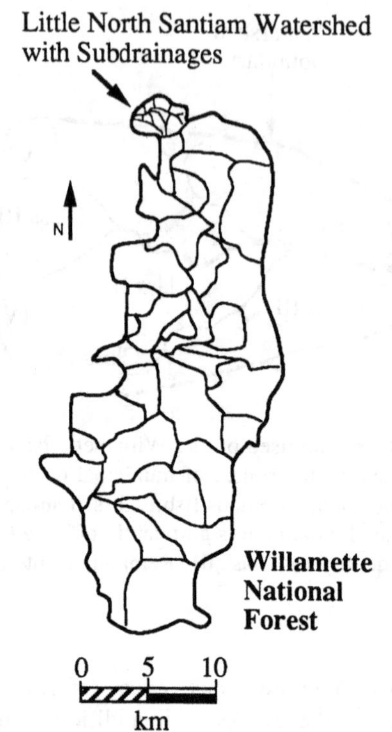

FIGURE 12.3. Outline of the Willamette National Forest, its major watersheds, and the subdrainages of the Little North Santiam watershed.

Willamette NF personnel described the cumulative effects of timber harvest on watersheds through assessing the relative risk of the watersheds failing to meet Oregon State standards for water quality or plan objectives for stream condition (Beilharz 1990). To assess this risk, they evaluated three components: (1) riparian conditions, (2) potential risk of sediment production, and (3) quantity and timing of runoff (Beilharz 1990).

Willamette NF personnel developed prescriptions for riparian areas that would maintain current stream temperatures. These prescriptions required a reduced rate of harvest and were used in the Draft Forest Plan. The Final Forest Plan calls for no scheduled harvest in riparian areas in order to reduce watershed risk and to contribute to diversity, wildlife habitat, and recreation.

Willamette NF personnel viewed potential sediment production as resulting from debris slides and surface erosion. They believed that sediment from both sources could be mitigated with site-specific measures. They assumed that risk of accelerated debris slides varies by amount and location of management activities on high risk soils. Much of this high risk ground occurs on certain soil types along ephemeral streams. The Draft Plan scheduled these areas for harvest, whereas the Final Plan withdraws them from it.

In assessing runoff patterns, Willamette NF personnel keyed on peak flows because of their potential for damage to stream channels, riparian areas, and forest facilities such as bridges and roads. They believed that larger-size peak flows cause significant damage, especially when combined with large amounts of sediment and landslides. Further, they viewed the risk of increased flows as mostly a function of the increased snow accumulation that occurs in clearcuts in the transient snow zone (below 1,500 m) combined with rain-on-snow events.

Willamette NF personnel developed a risk assessment tool called "hydrologic recovery" to estimate the likelihood of changes in snow accumulation and melt patterns under different forest conditions. The values taken by this tool, called the aggregate recovery percent (ARP), are a function of the diameter and crown closure of timber stands and measure the ability of the stands to intercept wind during rain-on-snow events.

A desired level of hydrologic recovery was identified for each large watershed (Class I or II streams) and each smaller subwatershed (Class II or III streams). These desired levels were based on the beneficial uses in each area and sensitivity of the area to damage from increased peak flows.

Hydrologic recovery assumptions and objective levels will be the focus of several Willamette NF Plan monitoring questions. Forest personnel acknowledged that assessing risk of increased peak flows could be improved by considering factors such as aspect and road location. However, analysis of the effects of the Forest Plan on peak flows and hydrologic recovery was limited by availability of information.

Representing hydrologic recovery relationships with a linear programming model like FORPLAN requires two conditions: (1) that each hectare can be given a recovery coefficient as a function of the stand condition and (2) that the total recovery level for each watershed of interest can be represented as a linear sum of the recovery condition of each hectare in the watershed. These conditions are met here. Thus hydrologic recovery values for a watershed can be represented in a linear program even through recovery after disturbance for a particular hectare is a nonlinear function of stand age (Johnson 1991).

When FORPLAN was required to achieve and maintain hydrologic recovery at acceptable levels through time in the 33 large watersheds shown in Figure 12.3, timber harvest was redistributed among the watersheds. But on reviewing the new tentative plan, field specialists still believed that the results would allow, and often require, certain smaller subdrainages to be "overcut" relative to hydrologic recovery objectives.

Controlling hydrologic recovery by watershed assumes that it is acceptable for one subdrainage within the watershed to compensate for deficiencies in the recovery of other subdrainages as long as overall watershed recovery remains above a certain level. After reviewing the results, watershed specialists on the Willamette NF determined that scheduling timber harvest by subdrainage would be needed to meet Forest Plan goals for water quality and aquatic habitat.

Analysis by Subwatershed

For purposes of management and analysis, the Willamette NF has been divided into approximately 450 subwatersheds, ranging in size from 200 to 8,000 hectares and averaging 1,000 hectares. Each major watershed shown in Figure 12.3 contains several of these subwatersheds. For example, the most northern major watershed recognized, the Little North Santiam, contains 11 subwatersheds. The subdrainage streams provide refuge habitat during high flows in the mainstem, and their stream channels can be more susceptible to channel scour.

During project planning, the potential for peak flows in a subdrainage is assessed by a hydrologist to determine a recommended ARP value that reflects the level of hydrologic recovery desired. This value takes into consideration land type and channel characteristics and appropriate mitigation to describe field conditions that influence streamflow patterns during rain-on-snow events (Beilharz 1990). To approximate these ARP values in forest planning, subdrainages were classified according to land type and beneficial uses. Information on channel characteristics was not available for all subdrainages, so it could not be used at this level of planning.

Portrayal of hydrologic recovery, and associated ARP values, in linear programming models like FORPLAN depends on the number of watersheds or subwatersheds represented not becoming unmanageably large. Recognizing watersheds and subwatersheds adds strata to the definition of timber stand groupings and adds constraints to the control of timber harvest. Size limitations in linear programming models generally do not allow for recognition of the 8,000–10,000 stand groupings and 2,000–3,000 recovery constraints that would be needed to model hydrologic recovery at the subdrainage level.

Willamette NF personnel therefore conducted a hydrologic recovery analysis at the subdrainage level for planned harvest over the next ten years in a side model separate from the FORPLAN analysis (Merzenich 1990). By comparing the current condition of each subdrainage with the condition needed to meet specified hydrologic recovery levels, they could estimate whether harvestable areas were available. When they combined this analysis with a harvest dispersion requirement at the subdrainage level, they concluded that harvest should be reduced 5–7% (25–30 million board feet per year) below the level estimated in the FORPLAN analysis that recognized major watersheds. That new harvest level became the basis for the Forest Plan recently published (United States Forest Service 1990*a*).

As discussed above, the Draft Plan suggested that one-third of the watersheds on the Willamette NF would have a high risk of significant deterioration under the harvest schedule proposed. The Final Plan suggests that all watersheds will have a low risk of significant deterioration. Control of hydrologic recovery by subdrainage, combined with withdrawal of riparian areas and potentially unstable areas adjacent to ephemeral streams from the timber base, resulted in a major reduction in projected cumulative impacts of timber harvest on watersheds. These controls and withdrawals also caused a decline in projected harvest levels, as discussed above.

Because the Forest Service has not examined hydrologic recovery at the subdrainage level beyond the first decade, it is not known whether the projected harvest will lead to difficulties in future decades. But with the high harvests of the last few years, Willamette NF personnel feel that hydrologic recovery will be more binding on timber harvest in the next ten years (1991–2000) than after the year 2000.

The analysis does suggest that the traditional control on timber harvest—finding the maximum even flow of timber volume over time from the entire forest—is inadequate as the control on cumulative effects. It also reenforces the argument that assessment of cumulative effects must be done at a level of geographic resolution below that of the whole forest.

We might argue that the hydrologic recovery modeling, as done on the Willamette NF, utilizes physical relationships that have been only partly tested and understood, and therefore we should not trust the results. I would argue that until we can represent those relationships in forest planning, as the Willamette NF personnel have done, we cannot begin to fully evaluate their significance. And a forest planning model that cannot represent the key perspective on cumulative effects, in the terms that the relevant specialists wish to have it done, fails a fundamental test of adequacy for interdisciplinary planning.

Forest Planning Models: What Does the Future Hold?

Assessment of cumulative watershed effects is but one of the many such assessments needed in forest planning. The recent Thomas Report on the northern spotted owl (*Strix occidentalis caurina*) calls for habitat conditions, where timber management will occur on the Willamette NF, similar to those needed for hydrologic recovery (Thomas et al. 1990). Visual management, big game management, harvest dispersion, and other objectives also require geographic-specific assessments of the cumulative effects of timber harvests. Forcing the analysis to a more aggregated level, as often happens due to size limitations in models like FORPLAN, makes it difficult to assess these effects realistically.

In addition, many of these relationships may not fit within the simple structures permitted in linear programming models. In the case of hydrologic recovery on the Willamette NF, we could represent the cumulative effects of timber harvest on a subdrainage as a linear sum of the activities that occurred. Size limitations precluded use of a linear programming model. If cumulative effects are a nonlinear sum of the activities that might occur, as can easily happen, then a linear programming formulation will not suffice (Davis and Johnson 1987).

Recently the Forest Service conducted an extensive review of forest planning with the assistance of many consultants (United States Forest Service 1990*b*). The recommendations about forest planning models in that review largely endorse the status quo. In their discussion about economic analysis in forest planning, the authors recommend that "FORPLAN Version II should continue to be used as the primary tool for economic and trade-off analysis in forest planning" (Hoekstra et al. 1990:32).

This endorsement largely misses the point that a trade-off assessment model is only as good as its ability to represent trade-offs. Information from national forests throughout the nation suggests that FORPLAN, and models like it, overstate the outputs that can be simultaneously produced from a forest, with the result that commodity targets are set unrealistically high in relation to other objectives in the forest plans. The Willamette NF experience described above gives but one example of this problem. Looking back at our owl-timber efficiency frontier (Figure 12.1), we might say that curve ABCD overstates what can be produced and that a curve such as GH might better represent the possibilities. And I believe that such a result is suggested by the Thomas Report (Thomas et al. 1990).

In the future, setting the level of timber harvest activity on the national forests should pass from aggregate models like FORPLAN to analysis procedures that more realistically assess cumulative effects and environmental thresholds. The detail needed in that assessment may force classical optimization techniques, like linear programming, to the sidelines for many of the key decisions, as happened on the Willamette NF. Models like FORPLAN will still have a role in pointing out efficiency considerations in setting harvest priorities, but they will not be the center of attention they once were.

Such a needed shift should increase the likelihood that the solutions will be environmentally feasible. Unfortunately, it may also increase the likelihood that the solutions will be inefficient (i.e., inside the efficiency frontier). This outcome would be directly opposed to our current results, in which solutions appear efficient but are often infeasible (i.e., outside the efficiency frontier).

Thus approaches that rely more on simulation than optimization may underestimate the commodities that can be provided on public lands, and therefore increase the pressure to reexamine land allocation decisions and objectives. Still, the NFMA and many other environmental laws clearly state that environmental protection must come before pursuit of economic efficiency. With such a mandate, a shift in the methodology of forest planning is long overdue.

Acknowledgments. In writing this paper, I benefited significantly from discussions with Sarah Crim and the planning staff of the Willamette National Forest.

References

Beilharz, M. 1990. Watershed analysis process, final resource and land management plan. Willamette National Forest, Eugene, Oregon, USA.

Betters, D. 1978. Analytical aids in land management planning. Department of Forest and Wood Science, Colorado State University, Fort Collins, Colorado, USA.

Dana, S.T., and S.K. Fairfax. 1980. Forest and range policy. Second edition. McGraw-Hill, New York, New York, USA.

Davis, L.S., and K.N. Johnson. 1987. Forest management. Third edition. McGraw-Hill, New York, New York, USA.

Duerr, W.A., and J.B. Duerr. 1971. The role of faith in forest resource management. Pages 45–55 *in* Man and the ecosystem. City Printers, Burlington, Vermont, USA.

Hoekstra, T., et al. 1990. Analytical tools and information: critique of land management planning. Volume 4. United States Forest Service, Policy Analysis Staff, Washington, D.C., USA.

Iverson, D., and R. Alston. 1986. The genesis of FORPLAN: a historical and analytical review of Forest Service planning models. United States Forest Service General Technical Report INT- 214, Intermountain Forest and Range Experiment Station, Ogden, Utah, USA.

Johnson, K.N. 1987. Reflections on the Development of FORPLAN. Pages 45–52 *in* T. Hoekstra, A. Dyer, and D. LeMaster, editors. FORPLAN: an evaluation of a Forest Planning Tool. United States Forest Service General Technical Report RM-140, Rocky Mountain Forest and Range Experiment Station, Fort Collins, Colorado, USA.

Johnson, K.N. 1991. Considering cumulative effects of timber harvest on wildlife and watersheds: implications for forest planning in the Douglas-fir region. Department of Forest Resources, Oregon State University, Corvallis, Oregon, USA, *in preparation*.

Johnson, K.N., T. Stuart, and S.A. Crim. 1986. FORPLAN Version 2: an overview. Land Management Planning Systems Section, United States Forest Service, Washington, D.C., USA.

Johnson, K.N., and P.L. Tedder. 1983. Linear programming vs. binary search in periodic harvest level calculation. Forest Science **29**:569–582.

Merzenich, J. 1990. Spatial disaggregation process-FORPLAN modeling. Willamette National Forest, Eugene, Oregon, USA.

Navon, D.I. 1971. Timber RAM: a long-range planning method for commercial timberlands under multiple-use management. United States Forest Service Research Paper PSW-70, Pacific Southwest Forest and Range Experiment Station, Berkeley, California, USA.

State of Oregon. 1988. Comments on the draft Willamette National Forest Plan. Governor's Office, State Capital, Salem, Oregon, USA.

Teeguarden, D. 1986. The Committee of Scientists perspective on the analytical requirements for forest planning. Pages 19–23 *in* T. Hoekstra, A. Dyer, and D.

LeMaster, editors. FORPLAN: an evaluation of a Forest Planning Tool. United States Forest Service General Technical Report RM-140, Rocky Mountain Forest and Range Experiment Station, Fort Collins, Colorado, USA.

Thomas, J.W., E.D. Forsman, J.B. Lint, E.C. Meslow, B.R. Noon, and J. Verner. 1990. A conservation strategy for the northern spotted owl: report of the Interagency Scientific Committee to address the conservation of the northern spotted owl. United States Forest Service, Bureau of Land Management, Fish and Wildlife Service, and National Park Service, Portland, Oregon. United States Government Printing Office 1990–791–171/20026, Washington, D.C., USA.

United States Forest Service. 1987. Draft environmental impact statement for the land and resource management plan for the Willamette National Forest. Eugene, Oregon, USA.

United States Forest Service. 1990*a*. Final environmental impact statement for the land and resource management plan for the Willamette National Forest. Eugene, Oregon, USA.

United States Forest Service. 1990*b*. Critique of land management planning. Policy Analysis Staff, Washington, D.C., USA. 13 volumes: FS-452 to FS-465.

Wilkerson, C., and H.M. Anderson. 1987. Land and resource planning in the national forests. Island Press, Washington, D.C., USA.

13

Integrating Management Tools, Ecological Knowledge, and Silviculture

CHADWICK DEARING OLIVER, DEAN R. BERG, DAVID R. LARSEN, AND KEVIN L. O'HARA

Abstract

Forest management is becoming highly technical in both the natural and management sciences. Natural forests are in a constant state of disturbance and regrowth, rather than in a stable, steady state as previously thought. The contemporary social attitude is to reduce the extremes of natural and man-caused "boom and bust" cycles that affect animal and plant populations. Forest managers must apply specific measurable criteria at the landscape, stand structure, and operational levels to reduce the extremes and achieve desired social goals. Active management is needed to maintain the targeted array of stand structures and landscape patterns by doing specific silvicultural operations at specific times. Several discrete steps are involved: (1) Identify the measurable criteria to be targeted. (2) Determine existing stand structures and landscape patterns. (3) Develop alternative silvicultural systems suitable for each stand. (4) Project the changing stand structures and landscape patterns resulting from the alternative systems. (5) Analyze the alternative systems and select the best one for each stand. (6) Implement the operations. (7) Monitor the results to ensure that objectives are achieved. Each step can be performed with varying degrees of detail, technical sophistication, and precision. The process is begun with incomplete knowledge, but adaptive management techniques can be used to make improvements along the way.

Key words. Stand dynamics, adaptive management, basin silviculture, forest structure.

Introduction

The many demands placed on forests appear conflicting but can often be met simultaneously and synergistically through appropriate management systems. A general forest management system is evolving within various organizations to deal with these demands. Most elements are not new or unique, and forest managers are already performing many of them. This chapter attempts to give organization and direction to the evolving system, which is described as a combination of activities, each of which can be implemented by using whatever technical sophistication is suitable to the occasion. Because any system to manage forest resources must be based on the natural processes that govern forest development, an understanding of ecological processes is necessary before a management system can be effective. This chapter therefore first describes the ecological context of forest management before considering the evolving management system.

Changes in Ecological Knowledge

Changing Perspective

Understanding of forest development patterns has increased dramatically over the past few decades (Oliver and Larson 1990). Forest ecosystems were once thought to be in a steady state, with small, rare perturbations only temporarily displacing the forest from its original, stable condition. This concept was compatible with the view that "climax" forests are the most stable and therefore the most desirable forests. Ecological values, it was believed, could be maintained by simply preserving forests from active management and exploitation. The concept found support in the romantic 18th century back-to-nature philosophy of Rousseau and in the early ideas of Thoreau. The steady state philosophy was taught in ecology courses in universities until recently. It is not surprising, therefore, that many environmentalists and foresters still assume that this concept of forest ecosystems is the prevailing one.

It is now scientifically accepted that most forests are constantly changing as a result of both natural and man-made disturbances and subsequent regrowth (Pickett and White 1985, Oliver and Larson 1990). The plants that grow following each disturbance form stands with structures different from previous ones. Stand structure is the physical and temporal distribution of trees in a stand. The distribution can be described by species; by vertical or horizontal spatial patterns; by size of trees or tree parts, including the crown volume, leaf area, stem, stem cross section, and others; by tree ages; or by combinations of the above (Oliver and Larson 1990).

The forest landscape—the aggregate of the stands—changes as the component stands change, and this results in changes in habitat suitability for terrestrial and aquatic animals, in resistance of stands to disturbances, in hydrologic properties of the soil, in the usefulness of timber for various wood products, in aesthetic appeal, and so forth. There is no single stand structure of optimum value to satisfy all the diverse human demands—even if it were possible to prevent the structures from changing as the trees grew or were destroyed by disturbances.

Natural disturbances have been quite large in much of the world, sometimes affecting hundreds of thousands of hectares (Oliver and Larson 1990). Large disturbances in the past created "boom and bust" cycles in both terrestrial and aquatic habitats. For example, a windstorm combined with a fire, or a volcanic tephra deposition, generally silted and warmed streams over a large area and created unsuitable habitats for most salmonid fish species for several years or decades after the event (Franklin et al. 1988, Bisson et al. 1987, Sullivan et al. 1987, Keller and Swanson 1979). On the other hand, such a disturbance would favor large populations of terrestrial animals that live in open areas (Raedeke 1988). As the forest regrew, streams would become cooler and less nutrient rich and more suitable to the salmonids. Many terrestrial herbaceous plants would be excluded as the forest canopy closed. Different aquatic and terrestrial animals would predominate. Some animals that survive and flourish only in open places would migrate to more recently disturbed areas; but on a very large disturbed area some individuals would not be able to migrate and would die, to be replaced by those better adapted to the change. Similar expansions and contractions of animal species would occur as the forest structure continued to change.

Today, the many demands on land make the impacts of large natural disturbances more extreme (Franklin, this volume). Forest areas are now smaller and more dissected by urban areas and highways. A large natural disturbance affects a greater proportion of the remaining forest; and cities, highways, and farms often block animals from migrating. Some areas have been set aside in national parks and wilderness areas to leave unchanged in the aftermath of natural events; however, it is uncertain what balance of ecological objectives can be achieved in these preserved areas.

In addition, forests have various necessary consumptive uses. If timber is not harvested in one place to make paper and other products, it will be harvested somewhere else in the world; or plastic, aluminum, steel, and brick will be used instead. Such substitution may have environmental disadvantages, since these products require far more fossil fuel to manufacture, and create more air pollution in the processing and use, than wood products do (Oliver et al. 1991).

Management Alternatives

The changing perspective of stability in natural ecosystems leads to two approaches to managing forest landscapes: (1) Large areas are set aside where human intervention is prevented, such as in natural parks and other controlled access areas in many parts of the world. (2) A landscape unit, such as a drainage basin, is managed to create and maintain on a moderate scale the variety of stand structures and habitats that are found naturally on a larger scale. Some structures should be in large enough areas to allow interior species (species that avoid edges and live in stands with certain structures) to survive; however, the objective is to allow species to survive disturbances that cover smaller areas. Stand sizes would be based on the minimum areas of uniform stands for interior species, operational logistics, and other constraints. The first approach is being used already. This article concentrates on the second.

Disturbances are created artifically in the second approach. If a natural disturbance did occur, only that part of the landscape in a susceptible structure would be destroyed; and stands of the other structures could be manipulated to recreate the destroyed structure as quickly as possible. The smaller impacts of natural and artificial disturbances should also provide more stable fish habitat, since stream sediments would be introduced periodically over time and space rather than all at once (Potts and Anderson 1990, Swanson et al. 1984, Bilby 1981, Bilby and Likens 1980). The amount and frequency of disturbances could be maintained within acceptable levels by managing the timing and extent of planned disturbances.

Management could be done with or without accompanying commodity extraction (e.g., timber harvest); however, the cost of the manipulations makes commodity extraction attractive to defray other management costs (Berg 1990).

The Management System

Deciding to manage a pattern of stand structures across a landscape is a long way from achieving that goal. Very real problems exist. The following questions address some of the issues: What are the objectives to be protected and managed? What are the targeted landscape patterns needed to reach these objectives? How can these patterns be achieved? What if some of the targeted patterns are incorrect? How can the management plan ensure achieving the objectives?

The remainder of this chapter will describe an evolving system for managing landscapes for a variety of goals. Many of the techniques are commonly used by forest managers. Others have been suggested but not yet widely implemented (Boyce 1985). Still others are quite new to forestry but were developed in North America in the 1930s and have been used extensively in Japanese industrial and service management (Deming 1982, Ishikawa 1986, Scherkenbach 1986, Taguchi 1986).

Defining Objectives and Measurable Criteria

The first task of resource management is to translate general social demands into progressively more specific objectives and, ultimately, into measurable criteria. Specific objectives are necessary: only when goals are expressed unambiguously, through measurable criteria, can management ensure that they will be achieved.

Part of the management process is refining the landowners' goals into specific objectives. The task can be complicated when owners confuse objectives with the means of achieving them. For example, they often assert that they want uneven-aged management when they really mean mixed sizes of trees, which can be achieved through even-aged management (Oliver 1980).

On private lands, the process of arriving at specific objectives is often carried out through private consultations between the resource manager and the landowner. The public's interests on private lands are protected by means of public laws. On public lands, objectives are reached through numerous methods—with varying degrees of rancor. Often confrontations, political rhetoric, and legal maneuvers lead to compromises. Since natural resources follow natural laws, political compromises and human laws cannot resolve natural resource issues unless they are in harmony with natural laws.

More orderly "gaming" techniques have been developed for resolving apparently conflicting resource objectives (Walters 1986). Using such techniques, advocates meet, agree on the scientific principles governing the resource's behavior (with the help of experts), and then project into the future the values and consequences of their different objectives. Participating advocates soon realize that every activity in a forest ecosystem sets in motion a series of long-term consequences. By emphasizing the relationship of goals to system structures (e.g., timber goals or wildlife goals to stand structures), the focus becomes the management of stand structures and landscape patterns to achieve the goals. The specific forest manipulations are then viewed as activities to achieve the objectives, rather than objectives in themselves. Most of the extreme ("claimed" but not really essential) objectives prove undesirable when actually projected; but the fundamental goals of all parties are similar and readily agreed upon (Walters 1986).

For natural resource objectives to be achievable, they must be compatible with natural processes, existing (or achievable) technology, and social norms. Regional managers are often the ones most aware of local scientific, logistical, and social constraints to management. Organizational structures that combine management planning and implementation and allow the most flexibility during implementation ensure that the goals will be realistic and increase the probability of successful management (Reich 1983).

Common Focus of Management

Translating specific objectives into measurable criteria is often done by natural resource experts. Specific objectives can be expressed in many forms—such as number of animals, degree of water turbidity, or monetary value of trees. Objectives become clearer when the measurable criteria have a common focus. A convenient common focus is the forest itself, expressed in terms of (1) landscape patterns, (2) stand structures, and (3) operations. Other specific objectives can be expressed as measurable criteria if there is a relation between them and these three aspects of common focus. If there is no relation, manipulating the forest will not affect these objectives and can be done independently of them.

Specific landscape patterns, stand structures, and forest operations are surrogates: their exact relation to the various objectives is not completely known and can only be approximated through measurable criteria and dealt with hypothetically. It is hypothesized that the economic, ecological, and social objectives will be met if specifically targeted patterns, structures, and operations are achieved.

The measurable criteria, as well as the objectives, will be refined and changed with time. Management is easier when there is an orderly process for incorporating the changed objectives and measurable criteria, as will be discussed below.

Landscape Patterns

Not all objectives can be expressed in terms of individual stand structures. At times, the arrangement of stands across the landscape—the landscape pattern—is important. For example, combinations of stands of all structures may be appropriate for streamside areas; large woody debris may be needed from older stands, and direct sunlight in the stand initiation stage (see Figure 13.1) increases primary production within the streams (Oliver and Hinckley 1987, Bisson and Sedell 1984, Heede 1985). Similarly, too much open structure on a landscape may allow excess runoff (Troendle and King 1987), soil slumping, or erosion (Cederholm and Reid 1987) and would not be good for aquatic species (Sullivan et al. 1987). Attempts to preclude all open structures may create a continuous, closed forest that could be destroyed by a single large disturbance (Christensen et al. 1989). The result is then too much open area at one time. In regard to water quality, relatively little area in the open structure of the stand initiation stage may be desired, since intact forest soils filter water; on the other hand, more open areas may be preferred if water quantity is a goal, since less evapotranspiration will occur. Continuous corridors between stands in stages other than the stand initiation stage are important for certain migrating animals (Hunter 1990).

Measurable criteria for certain objectives can be expressed as the proportion and distribution of stands of each structure across a landscape. The relation between the objectives and the landscape patterns can be approximated and then gradually refined through monitoring and analysis, as will be discussed.

It is logistically helpful to divide the landscape into units and manage for the landscape pattern within each unit. Landscape unit sizes will vary with objectives and ecological considerations. One approach is to have the landscape unit be larger than the home range of one of the widest ranging species (except in cases of such wide-ranging species as migratory waterfowl). In the Olympic Experimental State Forest (Washington State Department of Natural Resources 1990), a landscape unit is between 1,600 and 6,100 hect-

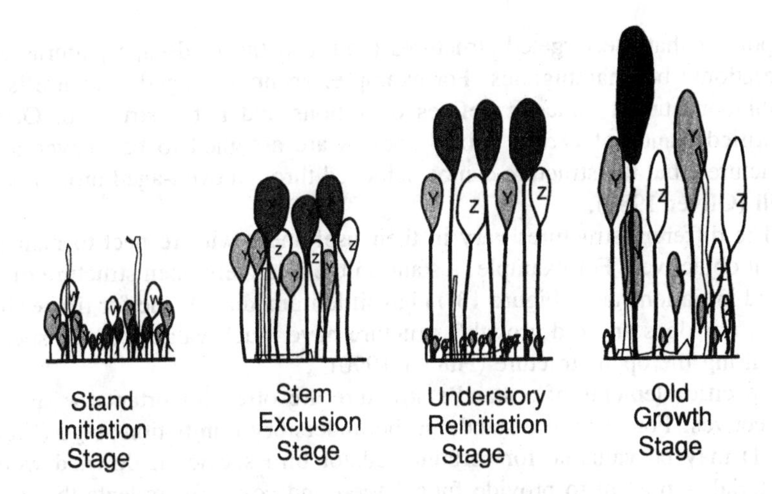

FIGURE 13.1. Stand structures change with time following disturbances, as is shown here for mixed species stands following stand-replacing disturbances (after Oliver 1981; letters represent different species). Many species and individuals invade immediately following the disturbance—during the stand initiation stage. Then individuals of a few species dominate and kill other individuals and exclude new plants from growing (stem exclusion stage). Eventually the older trees invading during the stand initiation stage are not vigorous enough to exclude young plants from invading the understory (understory reinitiation stage). Still much later, the younger stems can grow to the overstory (old growth stage). Other structures can develop in single species stands and stands following less severe disturbances (Oliver and Larson 1990). Different structures more completely meet different social demands, goals, and objectives; consequently, the different structures can be targeted as measurable criteria in management. Specific structural elements (e.g., snags, downed logs), as well as particular species, can also be identified as measurable criteria.

ares to accommodate elk (*Cervus canadensis* var. *roosevelti*). Natural boundaries for the landscape units can be drainage basin boundaries.

Managing in order to maintain certain structures across a landscape can present practical problems where a single drainage basin landscape unit is in several ownerships. Some landowners must make sacrifices to achieve the common landscape goal—a problem analogous to the one addressed in the federal Clean Air Act of 1990 (many industries in a large area will have to coordinate efforts to reduce pollution below a certain level, partly by selling excess portions of their pollution quotas).

Stand Structures

It is possible to translate many management objectives into specific measurable stand structure requirements. Wildlife objectives can be translated into habitat requirements, and snow catchment and evaporation rates can be related to stand structures, as can desirable riparian forest conditions. It is important that the targeted structures (and also the landscape patterns and operations) be unambiguous. For example, an uneven-aged structure is an ambiguous target, since it defines operations and not a structure. Often, stratified stands of certain mixed species are assumed to be uneven-aged structures, but the structure can be achieved through even-aged processes as well (Oliver 1980).

The different structures vary in their usefulness with respect to management objectives. For example, a stand in the relatively open structure of the stand initiation stage (Figure 13.1) has little immediate value for timber harvest, nor does an "old growth" structure have much value for bird species requiring the open structure (Hunter 1990).

Specific elements of a stand's structure are often important for specific objectives. For example, stands in the understory reinitiation stage (Figure 13.1) may be valuable for certain predator bird species if downed woody material is present to provide fungal food and cover for rodents that serve as prey (Maser and Trappe 1984).

Individual Forestry Operations

Individual operations include timber harvesting, thinning, planting, weed control, site preparation, road building, road use, and other forest management activities. They can be used to mimic natural disturbances and to alter existing stands so that they can grow to desirable structures—and thus create desirable landscape patterns.

Certain objectives can be achieved by adjusting specific management operations that are directly related to objectives. For example, road building near streams can be reduced or limited to noncritical seasons to avoid siltation of streams, which may interfere with salmon spawning. Unstable slopes will eventually collapse and silt streams with or without management, but timber harvesting and road building operations may accelerate the rate of collapse. It may be appropriate to designate unstable areas and restrict entry during a given period—such as a decade—to maintain a biologically acceptable rate of stream siltation.

Implementing the Management System

Management for each landscape area begins by planning the landscape pattern, then the individual stand structures to achieve the landscape patterns, and finally the individual operations (Figure 13.2). Implementing the plan begins at the operations level. Individual management operations (e.g., road building, thinning, harvesting, planting, creating snags) are used to achieve specific stand structures. These stand structures are distributed across a landscape to produce specific landscape patterns. Landscape patterns can be achieved over time only if the individual stands are capable of growing to certain structures; and stand structures are partly dependent on the sites,

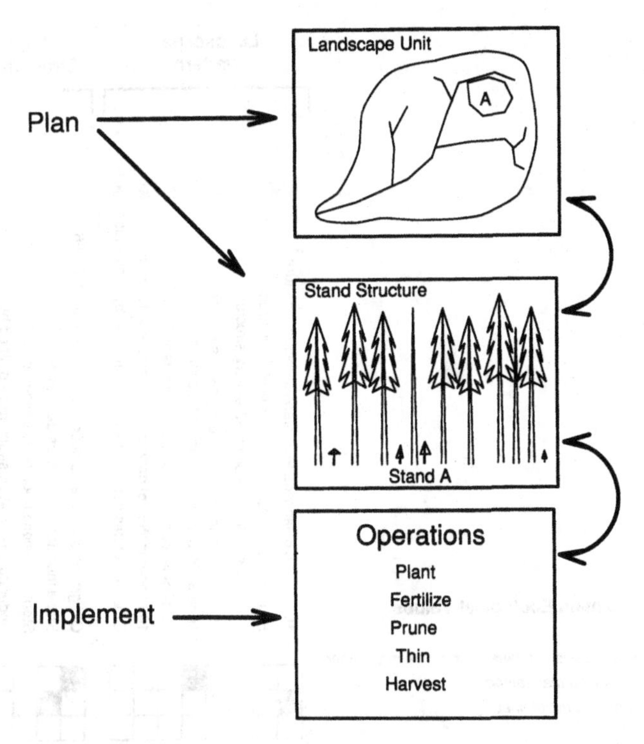

FIGURE 13.2. Planning is done first at the landscape level; then the stand structure changes are planned for each stand; and finally each operation needed to achieve the structure is planned. Implementation is in the opposite order: operations are carried out to achieve desired stand structures, which in turn achieve desired landscape patterns.

species, time, and feasibility of operations. Consequently, the planning will be iterative, with informational feedback among the three levels. A set of specific, measurable criteria at the operation, stand, and landscape levels has been developed for the Olympic Experimental State Forest (Washington State Department of Natural Resources 1990; see Figure 13.3 for an example). Several steps are involved in implementing management of landscapes. The steps can be carried out at various levels of detail, technical sophistication, and precision—and at various costs—depending on the landowners' needs.

Step 1. Identify the Measurable Criteria to Be Targeted

Knowledge of appropriate measurable criteria to achieve objectives may be incomplete. In fact, scientific knowledge is never complete; but it is improved with time, effort, additional data, and analyses. At one time, removal of woody debris was considered to be the best stream management practice

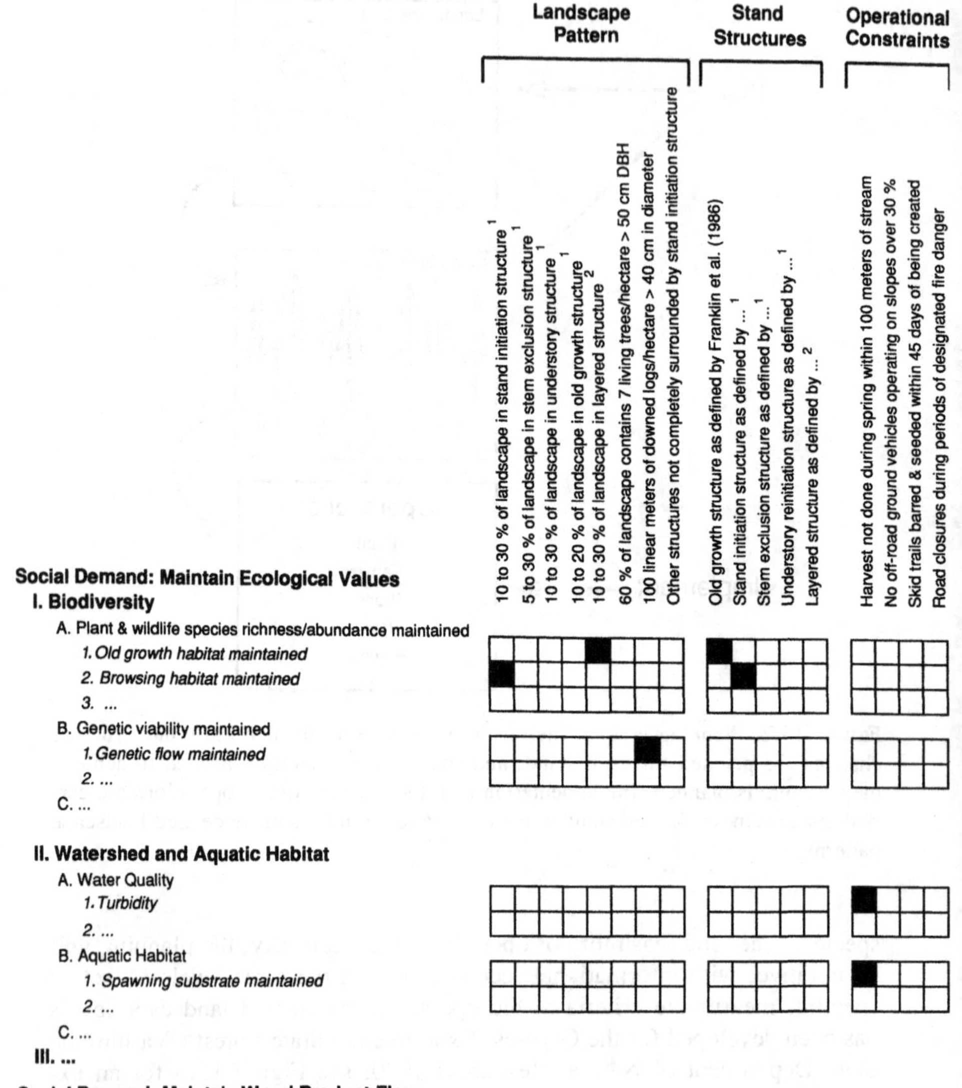

FIGURE 13.3. Part of a matrix (adapted from Washington State Department of Natural Resources 1990) showing the relation between social demands (left side), increasingly specific management goals (roman-numeral items), objectives (capital-letter items), and subobjectives (numbered items in italic) and target measurable criteria (top) at the landscape, stand, and operational levels to meet these demands. Specific measurable criteria to meet specific subobjectives are filled in (black boxes) on the matrix. More than one subobjective can be met with a single measurable criterion, and more than one measurable criterion may be needed to meet a single subobjective.
[1]Structures described in Oliver (1981).
[2]"Layered" structure described in Washington State Department of Natural Resources (1990).

and felling of snags was recommended (Washington State Department of Natural Resources 1976). Currently, retention of woody debris and snags is generally considered desirable (Bilby and Wasserman 1989, Harmon et al. 1986, Hunt 1988, Bryant 1983). Even-aged management and uneven-aged management have alternately been urged for Pacific Northwest forests (Kirkland and Brandstrom 1936, Munger 1950, Isaac 1956).

Management—even the decision to do nothing—does not wait for perfect knowledge. It is carried out with the best available scientific knowledge at a given time; and the targeted measurable criteria are generally the best estimates based on available knowledge (Lee, this volume). The manager must distinguish between the best scientific knowledge and partisan rhetoric in the natural resource management field. Scientists address a similar problem in assessing the merit of scientific articles and proposals by using other scientists for peer review. A scientific review process may be appropriate to determine the best knowledge relating goals to measurable criteria in forest management.

New scientific knowledge must be systematically incorporated into the system to guide management to those activities that efficiently lead to the landowners' goals. The management system can also be used to generate and update scientific knowledge in an "adaptive management" framework (Holling 1978, Baskerville 1985, Walters 1986, Lee and Lawrence 1986), as will be discussed.

Step 2. Determine Existing Stand Structures and Landscape Patterns

Forested landscapes rarely contain the most appropriate balance of structures for the given objectives. Past attempts to create "regulated forests" (stands of different ages on equal areas) seldom succeeded, because it takes a long time to create them, even when stand development is accelerated through silvicultural activities.

It will be necessary to work toward achieving the targeted landscapes with each forest operation. All operations should "move" the stand structures (hence landscape patterns) toward their targets. How soon the targeted landscape is achieved will depend on the initial and targeted landscape patterns, growth rates of the trees, funding, and incentives for landowners to manipulate their stands. As the targets are updated, the operations can be similarly adjusted.

An inventory to determine the existing stand structures and landscape patterns is fundamental to forest management. Forest inventories have traditionally been used to measure timber volume growth; but they are increasingly measuring more elements of stand structures. Inventories can be done from ground surveys, aerial photographs, satellite imagery, or a combination of these (Paine 1981), depending on the funds and technology available and the degree of precision needed. Maps, files, and geographic information systems (GIS) allow efficient storage and retrieval of the information (Baskerville and Moore 1988).

Even when (and if) the targeted stand structures and landscape patterns are achieved, it will be necessary to inventory and manipulate the stands as they grow, so that the pattern can be maintained across the dynamic landscape.

Step 3. Develop Alternative Silvicultural Systems Suitable for Each Stand

An operation must be effective in a given stand to achieve certain stand structure objectives. For example, because the objective of a thinning operation is to create future stand structures with large trees and wide spacings, thinning is ineffective if done after the stand has grown so dense that the trees remaining after thinning will be easily damaged or blown over by wind.

Several operations (including no activity) may be effective in a given situation. Each operation will cause the stand to develop into a different structure, and each structure will have different future times when certain operations will be effective. These times are referred to as "windows" (Oliver et al. 1986).

Growth combined with the operations performed on a stand can be viewed as a silvicultural system (Figure 13.4). Alternative choices will lead to different systems with different future stand structures, choices of operations, and costs. Each system will satisfy different goals to different extents and at different times.

The professional knowledge of silviculturists and engineers about forest and road operations is needed to determine effective systems. Experience, decision keys (Chew 1989, O'Hara et al. 1989, 1990), and various stand density charts (Reineke 1933, Gingrich 1967, Drew and Flewelling 1979, Long et al. 1988) help the professional forester determine which operations will be effective in specific situations. The stands are projected forward in time based on this knowledge, as well as on yield tables and growth models, with varying degrees of precision and accuracy (Davis and Johnson 1987). Most projection systems emphasize structures related to timber values and are more valuable for broad-area projections than stand-specific changes in structure. More flexible projection systems are being developed (Pukkala 1989, Koop 1989, Larsen 1991).

Step 4. Project the Changing Stand Structures and Landscape Patterns Resulting from the Alternative Systems

Alternative landscape patterns can be projected into the future as combinations of alternative silvicultural systems for the different stands (Figure 13.5). The process of projecting is aided by files, maps, and GIS data (Reutebuch and Hawkins 1987).

Step 5. Analyze the Alternative Silvicultural Systems and Select the Best One for Each Stand

Many landscape pattern scenarios are produced when all possible silvicultural systems for each stand are combined and projected with all possible systems for every other stand in the landscape unit. The most desirable land-

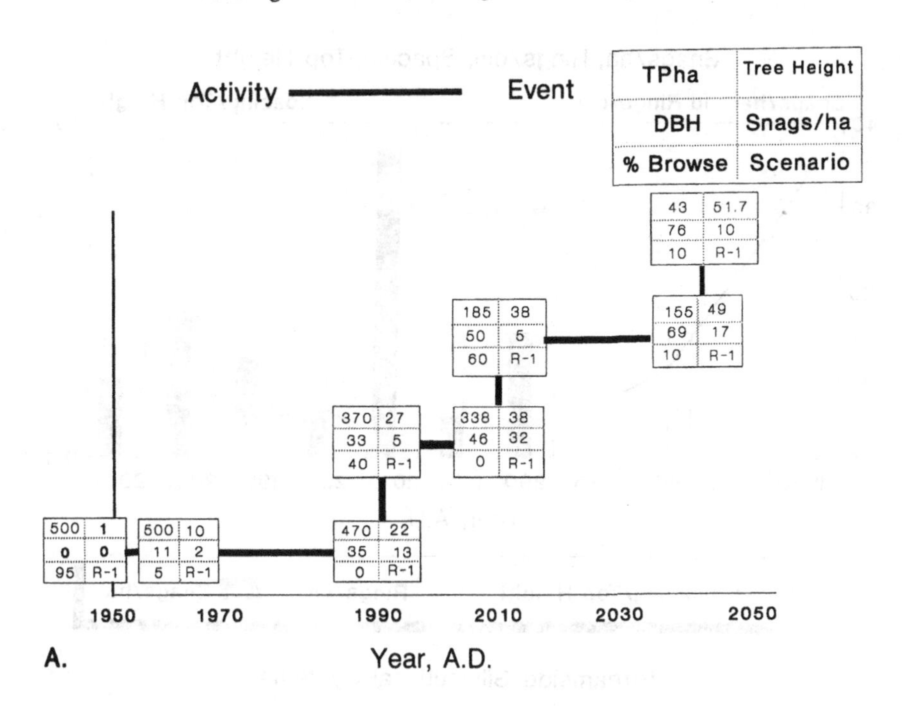

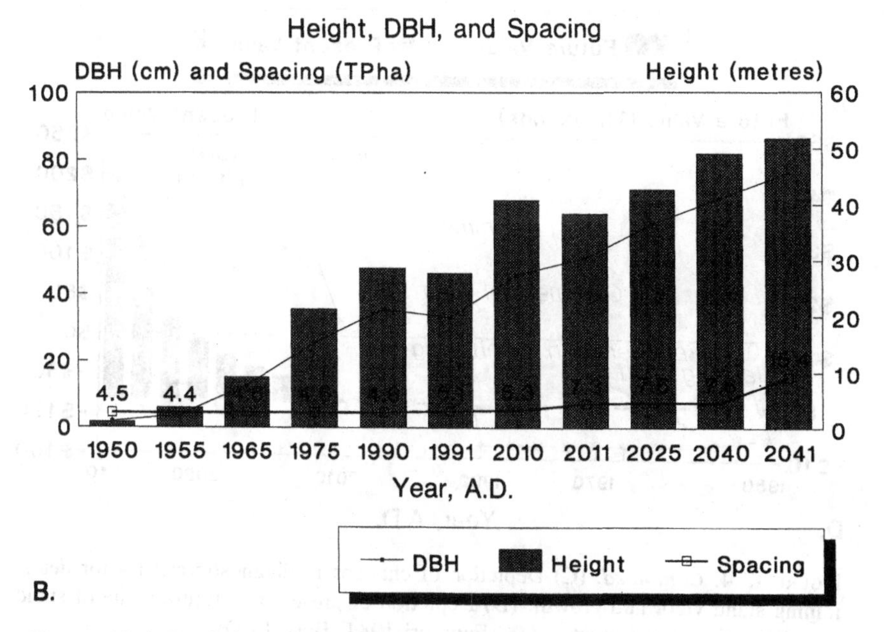

FIGURE 13.4. Depiction of one alternative silvicultural system for a specific stand, showing its history and projected future; depiction technique similar to Fujimori (1984) and Berg (1990). (A) Sample stand pathway chart, showing activities (lines) and events (boxes) similar to networking models (PERT; Levin and Kirkpatrick 1971). Events are points in time showing stand structures. Activities are disturbances or silvicultural operations that change a stand from one structure to another. (B) Depiction of changes in stand attributes.

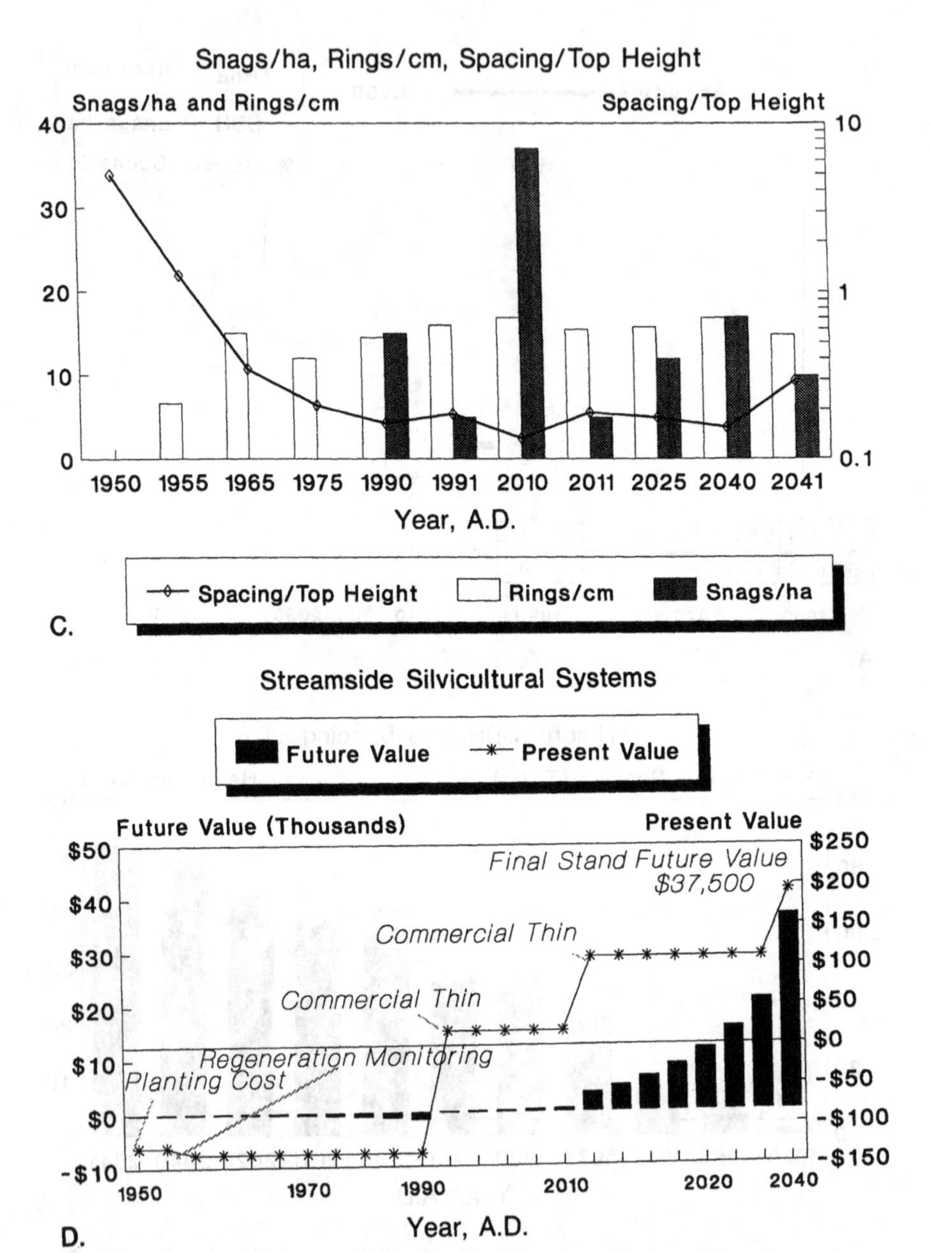

FIGURE 13.4. *Continued.* (C) Depiction of changes in diagnostic criteria for determining stand vigor and growth. (D) Depiction of present and future value of stand as an economic investment at 6% (Fujimori 1984, Berg 1990).

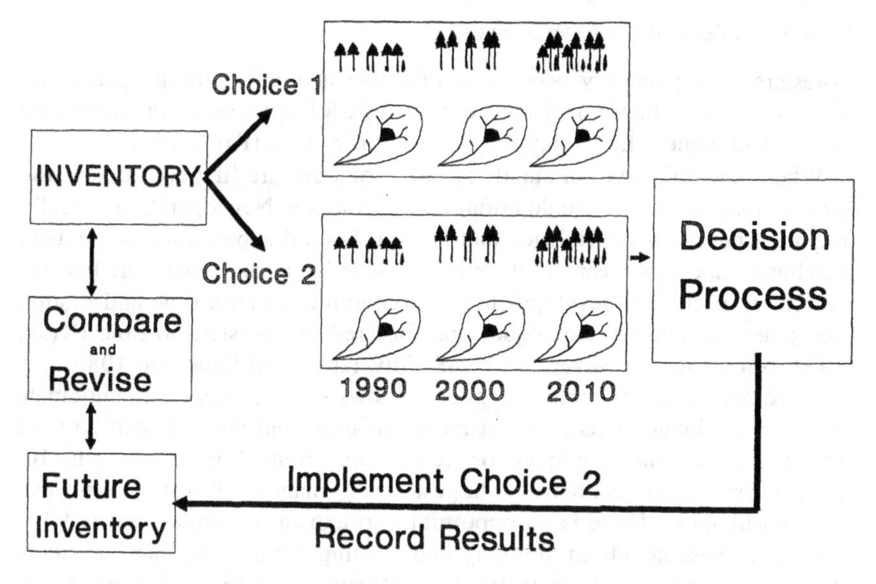

FIGURE 13.5. The various steps involved in managing forests can be viewed as a process of projecting options and making choices.

scape pattern (and change in pattern, since it is projected in time) needs to be chosen for its ability to achieve the targeted objectives at different times.

Various decision techniques can be used to determine the degree to which various scenarios will achieve the measurable criteria. Sometimes all targeted objectives cannot be completely met by any combination of silvicultural systems. Then a decision must be made about which systems to implement and which targets will not be reached. Careful early planning with the landowners can minimize conflicts by prioritizing objectives. Decision keys, linear programming, or other techniques can be used depending on the management infrastructure, the ability to quantify the objectives, and the complexity of the objectives (Davis and Johnson 1987). Other decision methods such as decision matrix techniques, analytic heirarchy processes, Delphi techniques, and multi-attribute utility theory can also help determine which alternatives to choose (Harker 1989).

On public (and increasingly on private) land the objective of maximizing economic return is often given low priority. The question often becomes: How much economic incentive is necessary to continue management? Different approaches may be needed to ensure continued management. For example, it may be appropriate to evaluate the economics of alternative scenarios on a landscape unit basis, rather than on a stand-by-stand basis. It may also be appropriate to write off all management activities as "expenses" of harvesting when computing taxes if a "cost" of harvesting is restoration of the landscape to its original condition.

Step 6. Implement the Operations

Foresters have generally been quite effective at implementing operations. Time lines, schedules, decision keys, and PERT (program evaluation and review technique) charts may aid specific operations (Krick 1962).

When new objectives or stands of new structures are first confronted, the forester may be required to do unfamiliar operations. New operations usually require a period of inefficiency and high costs until appropriate techniques, machines, and labor can be developed (Scherkenbach 1986). Off-line research on machinery development and ergonomics can use time-and-motion and other engineering techniques and thus reduce the start-up time (Krick 1962, Sundberg and Silversides 1988, Silversides and Sundberg 1989).

Development of silvicultural systems allows forest managers to anticipate future operational needs many years in advance, and the time can be used to develop efficient techniques for the new operations. For example, the 10- to 30-year age class contains the largest area of managed forest land in western Washington. There is an opportunity to expand economic and wildlife values in these stands by thinning and pruning them when they are about 10 to 30 years old (Oliver 1991). Cost-effective thinning and pruning systems need to be developed to realize this opportunity.

Implementing a management system requires that no planned operations be left out. Managers cannot eliminate parts of silvicultural systems they consider undesirable and expect to achieve the management goals. If parts of the systems are altered, the management system must be reformulated to incorporate the present and anticipated changes (Baskerville 1992).

Step 7. Monitor the Results to Ensure that Objectives Are Achieved

When a decision is made and a management direction is being implemented, it is always possible that the targeted operations, structures, and landscape patterns are not being met—or that they are not leading to the desired objectives. In the long term, management will achieve the stated goals only if (1) there is a biologically functional relationship between the objective and the stand structures, landscape patterns, or operations, and (2) managers use this relationship to control and regulate the availability to the objective. The functional relationships between the various objectives and the stands, landscape patterns, and operations need to be understood.

Where long-range plans for stand structures are made, deviations from the objectives are often not detected for many decades. Such possibilities for error are especially prevalent where managers are asked to develop untested silvicultural systems where functional relations have not been well studied. Forest managers do not have the time or resources to establish permanent research plots and wait (often decades) for results before proceeding with innovative management systems.

Early detection of deviations from the objective will allow early adjustment. Methods of early detection can be incorporated into a management system using techniques described as process control (Ishikawa 1986, Taguchi 1986), adaptive management (Walters 1986), and sequential analysis (Wald 1947). These techniques incorporate concepts of Bayesian analysis (Berger 1980, West and Harrison 1989).

The process of deciding which operations, silvicultural systems, and landscape patterns to implement involves projecting the impacts of the operations and changes in stand structures and landscape patterns. After implementation, the operations, stands, and landscapes need to be monitored to determine if the projected changes in constraints, structures, and patterns are achieved. Equally important is monitoring to determine if the landowners' objectives are being achieved as expected (Galloway 1990).

The values obtained by monitoring are compared with the projected values to determine deviations from the expected outcome. Statistical tools allow the manager to determine trends and significant deviations. Each projection can be considered a hypothesis, and each management activity an experiment (Lee and Lawrence 1986). The comparison of the projected (hypothesized) result with the achieved (monitored) result acts as a test of the experiment. Results of the monitoring can then be used both to improve the scientific information for future use and to confirm or correct the current management direction.

Recognizing all management practices as experiments formalizes the method through which many field resource managers gain knowledge—often referred to as experience. Such interweaving of research and management practices helps make research more immediately applicable and responsive to management. In some areas, variations of normal operations, stand structures, and landscape patterns can be done as experiments. Such variations could provide even more scientific insight into effective management scenarios and underlying ecosystem processes.

Deviations between the projected and monitored activities, structures, and landscape patterns could have several causes—incorrect projections, incorrect implementation of correct projections, or incorrect monitoring. Various techniques have been developed to help detect deviations and causes of deviations from the projected system. The techniques involve analyses such as control charts, Pareto diagrams, and cause-and-effect diagrams (Ishikawa 1986, Taguchi 1986). Improving the system is most effective in a cooperative environment that fosters a commitment to making a system work and rewards managers for their efforts (Deming 1982, Reich 1983, Scherkenbach 1986).

Comparisons with Present Management Systems

Much of the management system described above recognizes the activities already performed by resource managers. Many tools currently used can be incorporated into the system. The described system does differ from many forest management practices in important, subtle ways.

1. Each stand is treated as a unique system, based on its present structure, its growth potential, and its contribution to the landscape objectives. A uniform prescription is not applied to many stands or age classes or over large areas.

2. The system incorporates unanticipated changes in objectives, knowledge, and growth patterns using an adaptive management process. Management has always incorporated unanticipated changes, but since the incorporation was not planned, the process has been referred to as crisis management.

3. Management deals with many simultaneous but changing objectives on a given area over time. In the past, multiple use objectives have often resulted in dividing a forest into smaller areas allocated to individual, predominant uses.

4. The informational feedback (e.g., process control, adaptive management) under Step 6 above incorporates research into the management process, thus ensuring that research and management communicate efficiently.

5. Informational feedback and steady improvement can occur through process control and adaptive management techniques at all steps.

A constant problem with resource management is the cost of planning, monitoring, and implementing. The system described can be accomplished using various levels of detail, technical sophistication, and precision and at various costs, depending on the landowners' needs. Management is generally most effective when the components of management are clearly recognized, because these components can thus be incorporated into the management system where they are most appropriate.

References

Baskerville, G.L. 1985. Adaptive management, wood availability and habitat availability. Forest Chronicle **61**:171–175.

Baskerville, G.L. 1992. Forest analysis. *In* M.J. Kelty, B.C. Larson, and C.D. Oliver, editors. The ecology and silviculture of mixed-species forests: a festschrift for David M. Smith. Kluwer Academic Publishers, Boston, Massachusetts, USA.

Baskerville, G.L., and T. Moore. 1988. Forest information systems that really work. Forest Chronicle **64**:136–140.

Berg, D.R. 1990. Active management of streamside forests. Master's Thesis. College of Forest Resources, University of Washington, Seattle, Washington, USA.

Berger, J.O. 1980. Statistical decision theory and Bayesian analysis. Springer-Verlag, New York, New York, USA.

Bilby, R.E. 1981. Role of organic debris dams in regulating the export of dissolved and particulate matter from a forested watershed. Ecology **62**:1234–1243.

Bilby, R.E., and G.E. Likens. 1980. Importance of organic debris dams in the structure and function of stream ecosystems. Ecology **61**:1107–1113.

Bilby, R.E., and L.J. Wasserman. 1989. Forest practices and riparian management in Washington State: data based regulation development. Pages 87–94 *in* R.E. Gresswell, B.A. Barton, and J.L. Kerschner, editors. Practical approaches to riparian resource management. United States Bureau of Land Management, Billings, Montana, USA.

Bisson, P.A., R.E. Bilby, M.D. Bryant, C.A. Dolloff, G.B. Grette, R.A. House, M.L. Murphy, KV. Koski, and J.R. Sedell. 1987. Large woody debris in forested streams in the Pacific Northwest: past, present, and future. Pages 143–190 *in* E.O. Salo and T.W. Cundy, editors. Streamside management: forestry and fishery interactions. Contribution 57, Institute of Forest Resources, University of Washington, Seattle, Washington, USA.

Bisson, P.A., and J.R. Sedell. 1984. Salmonid populations in streams in clearcut vs. old-growth forests of western Washington. Pages 121–129 *in* W.R. Meehan, T.R. Merrell, Jr., and T.A. Hanley, editors. Fish and wildlife relationships in old-growth forests, American Institute of Fishery Research Biologists, Juneau, Alaska, USA.

Boyce, S.G. 1985. Forestry decisions. United States Forest Service General Technical Report SE-35, Southeastern Forest Experiment Station, Asheville, North Carolina, USA.

Bryant, M.D. 1983. The role and management of woody debris in west coast salmonid nursery streams. North American Journal of Fisheries Management **3**:322–330.

Cederholm, C.J., and L.M. Reid. 1987. Impact of forest management on coho salmon (*Oncorhynchus kisutch*) populations of the Clearwater River, Washington: a project summary. Pages 373–398 *in* E.O. Salo and T.W. Cundy, editors. Streamside management: forestry and fishery interactions. Contribution 57, Institute of Forest Resources, University of Washington, Seattle, Washington, USA.

Chew, J.D. 1989. An expert system for the diagnosis of stand treatment needs. Pages 288–297 *in* Artificial Intelligence and Growth Models for Forest Management Decisions Conference Proceedings. International Union of Forest Research Organizations, September 18–22, 1989, Vienna, Austria.

Christensen, N.L., J.K. Agee, P.F. Brussard, J. Hughes, D.H. Knight, G.W. Minshall, J.M. Peek, S.J. Pyne, F.J. Swanson, J.W. Thomas, S. Wells, S.E. Williams, and H.A. Wright. 1989. Interpreting the Yellowstone fires of 1988. BioScience **39**:678–685.

Davis, L.S., and K.N. Johnson. 1987. Forest management. Third edition. McGraw-Hill, New York, New York, USA.

Deming, W.E. 1982. Out of the crisis. Massachusetts Institute of Technology Center for Advanced Engineering Study, Cambridge, Massachusetts, USA.

Drew, T.J., and J.W. Flewelling. 1979. Stand density management: an alternative approach and its application to Douglas-fir plantations. Forest Science **25**:518–532.

Franklin, J.F., P. Frenzen, and F.J. Swanson. 1988. Re-creation of ecosystems at Mt. St. Helens: contrasts in artificial and natural approaches. Pages 1–37 *in* J. Cairns, editor. Rehabilitating damaged ecosystems. Volume 2. CRC Press, Boca Raton, Florida, USA.

Franklin, J.F., F. Hall, W. Laudenslayer, C. Maser, J. Nunan, J. Poppino, C.J. Ralph, and T. Spies. 1986. Interim definitions for old-growth Douglas-fir and mixed-conifer forests in the Pacific Northwest and California. United States Forest Service Research Note PNW-447, Pacific Northwest Forest and Range Experiment Station, Portland, Oregon, USA.

Fujimori, T. 1984. Pruning: basics and applications. Japan Forestry Technology Association, May 25, 1984. In Japanese.

Galloway, G.E. 1990. Adaptive management: a necessary process for the advancement of agroforestry (translated from Spanish). Pages 57–82 *in* Agroforestería 1: Prácticas agroforestales en los Andes. Proceedings of a Seminar in Cotopaxi, Ecuador, April 2–7, 1990. United Nations Food and Agriculture Organization.

Gingrich, S.F. 1967. Measuring and evaluating stocking and stand density in upland hardwood forests in the Central States. Forest Science **13**: 38–53.

Harker, P.T. 1989. The art and science of decision-making: the analytic heirarchy process. Pages 3–36 *in* B.L. Golden, E.A. Wasil, and P.T. Harker, editors. The analytic hierarchy process. Springer-Verlag, New York, New York, USA.

Harmon, M.E., J.F. Franklin, F.J. Swanson, P. Sollins, S.V. Gregory, J.D. Lattin, N.H. Anderson, S.P. Cline, N.G. Aumen, J.R. Sedell, G.W. Lienkaemper, K. Cromack, Jr., and K.W. Cummins. 1986. Ecology of coarse woody debris in temperate ecosystems. Advances in Ecological Research **15**:133–302.

Heede, B.H. 1985. Interactions between streamside vegetation and stream dynamics. Pages 54–58 *in* Proceedings of Riparian Ecosystems and Their Management: Reconciling Conflicting Uses, Tucson, Arizona. United States Forest Service General Technical Report RM-120, Rocky Mountain Forest and Range Experiment Station, Fort Collins, Colorado, USA.

Holling, C.S. 1978. Adaptive environmental assessment and management. John Wiley and Sons, New York, New York, USA.

Hunt, R.L. 1988. Management of riparian zones and stream channels to benefit fisheries. United States Forest Service General Technical Report NC-122, North Central Forest Experiment Station, St. Paul, Minnesota, USA.

Hunter, M.L., Jr. 1990. Wildlife, forests, and forestry. Prentice-Hall, Englewood Cliffs, New Jersey, USA.

Isaac, L.A. 1956. Place of partial cutting in old-growth stands of the Douglas-fir region. United States Forest Service Research Paper 16, Pacific Northwest Forest and Range Experiment Station, Portland, Oregon, USA.

Ishikawa, K. 1986. Guide to quality control. Second revised edition, edited for clarity. Asian Productivity Organization, Quality Resources, White Plains, New York, USA.

Keller, E.A., and F.J. Swanson. 1979. Effects of large organic material on channel form and fluvial processes. Earth Surface Processes **4**:361–380.

Kirkland, B.P., and A.J.F. Brandstrom. 1936. Selective timber management in the Douglas-fir region. Charles Lathrop Pack Forestry Foundation, Washington, D.C., USA.

Koop, H. 1989. Forest dynamics: SILVI-STAR: a comprehensive monitoring system. Springer-Verlag, New York, New York, USA.

Krick, E.V. 1962. Methods engineering. John Wiley and Sons, New York, New York, USA.

Larsen, D.R. 1991. Silvicultural planning with a Bayesian framework. *In* Proceedings of a Symposium for System Analysis in Forestry, March 3–7, 1991, Charleston, South Carolina, USA.

Lee, K.N., and J. Lawrence. 1986. Adaptive management: learning from the Columbia River Basin fish and wildlife program. Environmental Law **16:**431–460.

Levin, R.I., and C.A. Kirkpatrick. 1971. Quantitative approaches to management. Second edition. McGraw-Hill, New York, New York, USA.

Long, J.N., J.B. McCarter, and S.B. Jack. 1988. A modified density management diagram for coastal Douglas-fir. Western Journal of Applied Forestry **3**:88–89.

Maser, C., and J.M. Trappe. 1984. The seen and unseen world of the fallen tree. United States Forest Service General Technical Report PNW-164.

Munger, T.T. 1950. A look at selective cutting in Douglas-fir. Journal of Forestry **48**:97–99.

O'Hara, K.L., C.D. Oliver, S.G. Pickford, and J.J. Townsley. 1989. A prototype expert system for silvicultural decision-making on the Okanogan National Forest, Washington, U.S.A. Pages 299–307 *in* Artificial Intelligence and Growth Models for Forest Management Decisions Conference Proceedings, International Union of Forest Research Organizations, September 18–22, 1989, Vienna, Austria.

O'Hara, K.L., C.D. Oliver, S.G. Pickford, and J.J. Townsley. 1990. A prototype decision rule base for planning and anticipating effects of silvicultural activities over broad areas. AI Applications in Natural Resource Management **4**(1):25–34.

Oliver, C.D. 1980. Even-aged development of mixed species stands. Journal of Forestry **78**(4):201–203.

Oliver, C.D. 1981. Forest development in North America following major disturbances. Forest Ecology and Management **3**:153–168.

Oliver, C.D. 1991. Thinning and pruning 10-to-30 year old plantations in western Washington: investment, social, wood supply, and environmental consequences. Manuscript in review. College of Forest Resources, University of Washington, Seattle, Washington, USA.

Oliver, C.D., and T.M. Hinckley. 1987. Species, stand structures, and silvicultural manipulation patterns for the streamside zone. Pages 259–276 *in* E. O. Salo and T. W. Cundy, editors. Streamside management: forestry and fishery interactions. Contribution 57, Institute of Forest Resources, University of Washington, Seattle, Washington, USA.

Oliver, C.D., J.A. Kershaw, Jr., and T.M. Hinckley. 1991. Effect of harvest of old growth Douglas-fir and subsequent management on carbon dioxide levels in the atmosphere. Pages 187–192 *in* Are forests the answer? Proceedings of the 1990 Society of American Foresters National Convention. Society of American Foresters, Bethesda, Maryland, USA.

Oliver, C.D., and B.C. Larson. 1990. Forest stand dynamics. McGraw-Hill, New York, New York, USA.

Oliver, C.D., K.L. O'Hara, G. McFadden, and I. Nagame. 1986. Concepts of thinning regimes. Pages 246–257 *in* C.D. Oliver, D.P. Hanley, and J.A. Johnson, editors. Douglas-fir: stand management for the future. Contribution 55, Institute of Forest Resources, University of Washington, Seattle, Washington, USA.

Paine, D.P. 1981. Aerial photography and image interpretation for resource management. John Wiley and Sons, New York, New York, USA.

Pickett, S.T.A., and P.S. White, editors. 1985. The ecology of natural disturbance and patch dynamics. Academic Press, Orlando, Florida, USA.

Potts, D.F., and B.K.M. Anderson. 1990. Organic debris and the management of small stream channels. Western Journal of Applied Forestry **5**:25–28.

Pukkala, T. 1989. Methods to describe the competition process in a tree stand. Scandinavian Journal of Forest Research **4**:187–202.

Raedeke, K.J., editor. 1988. Streamside management: riparian wildlife and forestry interactions. Contribution 59, Institute of Forest Resources, University of Washington, Seattle, Washington, USA.

Reich, R.B. 1983. The next American frontier. Time-Life Books, New York, New York, USA.

Reineke, L.H. 1933. Perfecting a stand-density index for even-aged forests. Journal of Agricultural Research **46**:627–638.

Reutebuch, S.A., and G.E. Hawkins. 1987. Using a computer-aided planning package to assess the impact of environmental restrictions on harvesting systems. Pages 16–25 *in* Proceedings of the Eighth Annual Council on Forest Engineering, August 18–22, 1985, Lake Tahoe, California, USA.

Scherkenbach, W.W. 1986. The Deming route to quality and productivity: road maps and roadblocks. Mercury Press, Rockville, Maryland, USA.

Silversides, C.R., and U. Sundberg. 1989. Operational efficiency in forestry. Volume 2: Practice. Kluwer Academic Publishers, Boston, Massachusetts, USA.

Sullivan, K., T.E. Lisle, C.A. Dolloff, G.E. Grant, and L.M. Reid. 1987. Stream channels: the link between forests and fishes. Pages 39–97 *in* E.O. Salo and T.W. Cundy, editors. Streamside management: forestry and fishery interactions. Contribution 57, Institute of Forest Resources, University of Washington, Seattle, Washington, USA.

Sundberg, U., and C.R. Silversides. 1988. Operational efficiency in forestry. Volume 1: Analysis. Kluwer Academic Publishers, Boston, Massachusetts, USA.

Swanson, F.J., M.D. Bryant, G.W. Lienkaemper, and J.R. Sedell. 1984. Organic debris in small streams, Prince of Wales Island, southeast Alaska. United States Forest Service General Technical Report PNW-166.

Taguchi, G. 1986. Introduction to quality engineering: designing quality into products and processes. Asian Productivity Organization. UNIPUB, Kraus International Publications, White Plains, New York, New York, USA.

Troendle, C.A., and R.M. King. 1987. The effect of partial and clearcutting on streamflow at Deadhorse Creek, Colorado. Journal of Hydrology **90**:145–157.

Wald, A. 1947. Sequential analysis. Dover, New York, New York, USA.

Walters, C. 1986. Adaptive management of renewable resources. Macmillan, New York, New York, USA.

Washington State Department of Natural Resources. 1976. Washington forest practice rules and regulations. Washington Forest Practice Board, July 16, 1976.

Washington State Department of Natural Resources. 1990. Olympic Experimental State Forest draft management plan, July 1990. Washington State Department of Natural Resources, Forks, Washington, USA.

West, M., and J. Harrison. 1989. Bayesian forecasting and dynamic models. Springer-Verlag, New York, New York, USA.

Part 3
Innovative Approaches for Mitigation and Restoration of Watersheds

14

The Science and Politics of BMPs in Forestry: California Experiences

RAYMOND M. RICE

Abstract

Best management practices (BMPs) are the result of political compromises based on what is scientifically known, technically feasible, economically reasonable, and socially acceptable. Consequently, forestry BMPs vary from state to state on the Pacific coast, and their detail and rigor seem to be related to the degree of urbanization and the economic importance of the forest products industry in each state. California, being the most urbanized and having a forest products industry that is a relatively minor part of its economy, has the most restrictive forestry BMPs. Similar rules are likely to be adopted in the other states as they become more urbanized. Studies spanning changes in forest practice regulations suggest that California's increasingly strict forest practice rules have reduced erosion and maintained water quality. Researchers have also found that most erosion comes from a tiny fraction (0.5–1.8%) of the terrain. Therefore, discriminant functions have been developed to identify sites at risk of causing large amounts of erosion if logged or roaded. Methodologies, such as the Bayesian approach, can assist managers in choosing an acceptable risk threshold that optimally balances competing demands for forest-related resources. If forest managers can become accustomed to rigorously evaluating competing values and site conditions, greater improvements in erosion control may be obtained without reducing harvests.

Key words. Erosion, forest roads, timber harvesting, discriminant functions, risk assessment.

Introduction

Best management practices (BMPs) are defined in federal regulations as what is practicable in view of "technological, economic, and institutional considerations" (Council on Environmental Quality 1971). Therefore, BMPs are political compromises taking into account what is scientifically known, technically feasible, economically reasonable, and socially acceptable. BMPs can be procedural or prescriptive. In California, for example, U.S. Forest Service BMPs are largely procedural, describing the steps to be taken in determining how a site will be managed. In contrast, BMPs on private land are almost exclusively prescriptions of practices to be employed in response to site conditions. Prescriptive BMPs usually include a practice and some way of determining when and where the practice should be applied.

In this paper it will be argued that because of the political component of BMPs, California is a bellwether of future changes in forest practice regulations in other Pacific coast states. Data will be presented suggesting that California's prescriptive BMPs have resulted in reduced erosion and sedimentation. Recent erosion studies in California will be reviewed, but the main focus will be the ways of estimating erosion hazard. Finally, an objective method of estimating and managing erosion risk will be presented. The method has been tested and found effective in estimating risk, but problems impede its use as a management tool. These will be explored.

Forestry Best Management Practices

Most of the measures to protect water quality in current forestry BMPs owe their origin to the Federal Water Pollution Control Act Amendments of 1972 (PL 92–500), which mandated the control of nonpoint sources of water pollution. Silviculture was one of the nonpoint sources specifically mentioned in the Act. Section 208 also required the states to develop areawide management plans to reduce water quality degradation. Forest practice regulations are part of each state's efforts to satisfy the requirements of Section 208. It was recognized early that forestry regulations would have to be mainly prescriptive rather than reactive. Three considerations make the prescriptive approach appropriate: (1) Because most forestry-related pollutants are natural substances, such as sediment, their origin may be difficult to determine. (2) The practice that results in pollution may be difficult or impossible to correct once the pollution has occurred. (3) The level of pollution is the result of the interaction of a practice and the subsequent weather. If rules were based on the measured level of pollution, an appropriate practice followed by an extreme storm might become a violation, while a careless disregard for the environment might go unnoticed if followed by benign weather. It is unknown whether current forest practice rules will result in achieving PL 92–500's specific water quality targets.

The wide variability in the forest practice rules of the Pacific coast states is exemplified by requirements for approval of timber harvest plans. In California, operations must be described by a Registered Professional Forester, approved by the Department of Forestry, and conducted by a Licensed Timber Operator (State of California, n.d.). In Washington, forest practices are divided into four classes: two classes require approval, one can proceed five days after notifying the Department of Natural Resources (if the Department fails to object), and one requires no notification (Washington State Forest Practices Board 1988). In Oregon, notification is the rule and only operations involving certain practices require approval of written plans (State of Oregon 1987). In Alaska, the rules have yet to be written, but indications are that they will be still more lenient (Alaska Department of Natural Resources 1989). This progression suggests that the detail and rigor of regulations are linked to the urbanization of each state and the relative economic importance of its forest products industry. California has the most detailed and rigorous rules, followed by Washington, Oregon, and Alaska. As political entities, BMPs respond to public perceptions of what is acceptable, as well as increased scientific understanding of relevant processes. It appears that demographics play a large role in determining BMPs in each state. If this inference is correct, California's northern neighbors can see in its Forest Practice Rules what lies ahead as their states become more urbanized and industrially diversified.

The Effectiveness of BMPs

The evolution of forest practice regulations in California provides a few clues to the effectiveness of BMPs. What follows are only clues because the results of different locales and different experimental methods will be compared. Between 1945 and 1973, forest practice rules dealt primarily with forest regeneration and fire protection. However, in 1959 and 1960 some rules concerning erosion control and stream protection were adopted. The effectiveness of these rules may be inferred by contrasting the results of two watershed experiments. In 1959, prior to adoption of the rules, a tractor-yarded partial cut removed 9,900 m^3 of timber (12% of the merchantable volume) from the 10.4 km^2 Castle Creek watershed in the Sierra Nevada (Rice and Wallis 1962). During the first two postlogging years, suspended sediment discharge increased fivefold. In 1971–72, a tractor-yarded partial cut removed 84,600 m^3 (38% of the merchantable volume) from the 4.1 km^2 South Fork Caspar Creek watershed in northwestern California (Rice et al. 1979). During those two years, suspended sediment discharge also increased about fivefold. This relative increase was held constant even though the Caspar Creek logging removed about twenty times more volume per hectare than the Castle Creek logging. Furthermore, Caspar Creek presented a greater erosion hazard than Castle Creek. It was somewhat steeper (average slope 34% versus 17% in Castle Creek), had soils developed from marine sediments that were considerably more erodible than those developed from the igneous rocks of Castle Creek, and had postlogging unit area peak discharges that were four times higher.

Table 14.1. Changes in erosion rates due to mass movements and gullies caused by forest roads and logging measured in studies conducted in 1975–76 and 1985–86.

	Erosion Rate (m^3/ha)		
Area	1975–76	1985–86	Ratio
Roads	132.6	27.6	0.21
Harvest areas	17.5	11.0	0.63

Two studies of erosion provide a more valid measure of the effectiveness of BMPs. In 1973 the Z'Berg-Nedjedly Forest Practice Act began an era of increasingly restrictive forest practice regulations (Arvola 1976, Martin 1989). In 1975 and 1976 a study of logging-related erosion measured 57 plots on private land averaging 4.5 ha (Rice and Datzman 1981). Plots were selected from strata based on slope, annual precipitation, geologic parent material, yarding method, and time since logging. Each plot was 201 m wide and included a landing and the area yarded to it. The average age of the plots was 4.6 years, most of them having been logged prior to the new rules. In 1976, erosion was measured on 344 1.6-km segments of Forest Service logging roads (McCashion and Rice 1983). McCashion and Rice estimated the average age of these roads to be 11.5 years, but expressed little confidence in the accuracy of their age information. Road plots were selected from strata based on slope, annual precipitation, geologic parent material, road standard, and time since construction or reconstruction of the road segment. These results can be contrasted with the data collected during a 1985–86 study that measured 0.81 ha plots randomly located on private roads and harvest areas where operations had been completed between November 1978 and October 1979 (Lewis and Rice 1990). This comparison (Table 14.1) reveals that road erosion in the 1985–86 study has dropped to one-fifth of its value in the 1975–76 study but that harvest area erosion was still 63% of its former value. Since roads are responsible for most of the erosion associated with forest operations, the aggregate reduction in erosion over the decade was about two-thirds.

These comparisons argue that changes in forest practices have resulted in lower erosion rates and, presumedly, improved water quality in California. Although the rules changed considerably during that time, the addition of 48 Forest Practice Inspectors by the California Department of Forestry may be of at least equal importance. Correct application of a BMP is the responsibility of the timber operator. Human nature being what it is, compliance varies with the operator's motivation and understanding of regulations. In the course of measuring 426 plots in northwestern California, Durgin et al. (1988) observed that compliance with regulations tended to diminish with distance from the point of entry to the harvest area. This suggests that the effectiveness of BMPs will depend, in part, on the level of review and enforcement. Whatever the reason, there were environmental dividends from California's investment in better forest practices.

Erosion Hazard Ratings

Any effort to apply BMPs should be governed by an estimate of the erosion hazard. This reasonable assumption has led to some rather unreasonable schemes. Perhaps foremost were attempts to adapt the Universal Soil Loss Equation (USLE; Wischmeier and Smith 1965) to a forested environment. The USLE was a good procedure in its time and place, but its time was 1960 and its place was on agricultural lands east of the Rocky Mountains. It was an inappropriate starting point for an index of erosion hazard resulting from forest management. It was developed in a totally different environment from the Pacific coast forests, and it estimated erosion due to processes that were of little importance in mountainous forests. In mountains, slope is much more important than it is on agricultural lands because of the dominance of mass erosion processes. For the same reason, long-duration rainfall amounts and subsurface water replace short-term rainfall intensity and overland flow in determining erosion.

An erosion hazard rating (EHR) was made part of the Forest Practice Rules for the Coast Forest Practice District in California in 1973. It was developed by four scientists (Henry Anderson, Bill Colwell, Paul Zinke, and the author) in a few days. The basic structure of the EHR came from Anderson's (1974) regression analysis, modified by the group's collective professional judgment. It is doubtful that any forestry EHRs have a less questionable parentage. When the EHR was tested empirically (Datzman 1978), it had a coefficient of determination (r^2) with measured erosion of 0.01. In an empiric test of another EHR, Llerena et al. (1987) found: "The rankings by the proposed rating system showed poor agreement with those based on actual measurements." Datzman's (1978) findings led to the 1980 revision of the EHR. The new procedures dealt with surface erosion and mass erosion separately. Beyond that, the new methods were a step backward. The surface EHR was patterned after one used by the Forest Service and has never been tested empirically. The Board of Registration of Geologists and Geophysicists objected that the mass EHR would require foresters to practice geology without a license. Consequently, the mass erosion procedure is only found hidden in definitions of the terms "slide areas," "unstable areas," and "unstable soils." These terms are found mainly in rules related to roads and landings. In harvest areas, slope and the surface EHR regulate practices. Presumedly, harvest-related mass wasting is dealt with only indirectly through the consideration of slope.

Quite apart from the lack of a sound scientific basis, most EHRs err by taking too broad an approach. The 1980 EHR was applied to areas no smaller than about 4 ha. Erosion in the forest almost always occurs in a tiny fraction of the operating area. Inspection of Datzman's (1978) data revealed that 68% of all erosion measured occurred on just 4 of 102 plots. In a companion study of road-related erosion, only 0.6% of the road length had events displacing more than 15 m^3 of eroded material (Rice and Lewis 1986). Durgin et al. (1988) reported: "Almost all the measured erosion was produced on 12% of the study area—and nearly all erosion was concentrated in a few geographic areas." Others have also noted that most erosion from forest operations occurs on a few critical sites (Dodge et al. 1976, Peters and Litwin 1983). Peters and Litwin concluded that the key to reducing adverse environmental effects lay in developing a way to identify high risk sites. What was needed was an EHR that estimated the risk that serious erosion would occur, not how serious the erosion might be.

Managing Erosion Risk

The Critical Sites Erosion Study

The Peters and Litwin (1983) report led to the Critical Sites Erosion Study (CSES), a study of the occurrence of critical sites (erosion $>189\ m^3/ha$) in harvest areas and on forest roads (Lewis and Rice 1989). The sampled population came from the areas covered by Timber Harvest Plans completed between November 1978 and October 1979. The 1978–79 period was chosen for the study because it was a year of heavy cutting and because enough time had elapsed for the occurrence of logging- or road-related mass wasting. Earlier studies had shown mass wasting to be the most important erosional process (Dodge et al. 1976, Rice and Datzman 1981, Peters and Litwin 1983). Due to the cooperation of most landowners, the study came close to obtaining a truly random sample of the target population. Access was granted to lands covered by 415 of the 638 Timber Harvest Plans (THPs) in northwestern California. They included all the THPs on industrial ownerships and covered 75% of the area in the total 638 THPs.

The sampled units were 0.81 ha plots. All erosion features displacing more than 10 m^3 were measured and tallied. The plots were classed as critical (erosion $>189\ m^3/ha$) or noncritical. Harvest area noncritical plots were randomly located with the probability of selection in proportion to the area covered by each THP (Lewis and Rice 1989). Road noncritical plots were randomly located on each THP with the probability of selection in proportion to length of roads in the THP area. Data from all harvest area critical sites were included in the harvest area analysis, and a randomly chosen two-thirds of the road-related critical sites were used in the road analysis.

Each plot was characterized by 172 variables to ensure that it was fully described. Only 31 variables were used in the statistical analysis of the harvest plot data and 25 variables with the road plot data. The fieldwork was carried out between May 1985 and December 1986. Each plot was visited

Table 14.2. Distribution of erosion features larger than 10 m^3 on critical plots (Lewis and Rice 1990).

Erosion Type*	Road Plots Percentage by:		Harvest Plots Percentage by:	
	Number	Volume	Number	Volume
Debris flow	17.0	18.4	35.3	45.4
Debris slide	43.4	31.5	47.1	41.7
Earthflow	2.8	21.0	2.0	0.6
Slump	12.3	4.1	2.0	0.8
Translational/Rotational	6.6	18.2	3.9	7.2
Deep-seated translational	7.5	3.4	3.9	1.9
Rotational	0.9	0.6	2.0	0.8
Total Mass Movements	90.6	97.2	96.1	98.5
Gully	9.4	2.8	0.0	0.0
Streambank	0.0	0.0	3.9	1.5
Total Other Types	9.4	2.8	3.9	1.5

*Using the nomenclature of Bedrossian (1983).

by an interdisciplinary team composed of a forester, a geologist, and a soil scientist. In addition to making measurements, the team attempted to discover why each critical event had occurred. They measured 57 management variables and only 13 site variables, but concluded that natural site conditions were most important (Durgin et al. 1988).

The results of the investigation confirmed previous findings. It was estimated that erosion features displacing more than 10 m^3 of soil occurred on only 12% of the plots (Durgin et al. 1988). Critical plots contained 65.4% of the erosion but occupied only about 2% of the road length and 0.5% of the harvested area (Lewis and Rice 1989). When the area of erosion features is considered rather than plot area, only 0.2% of the 1978–79 THP area was scarred by erosion features displacing more than 10 m^3. Mass wasting was also found to be the cause of almost all of the erosion (Table 14.2). The study confirmed the dominance of road-related erosion over harvest area erosion, which has been noted in studies since at least 1954 (Anderson 1954). Roads yielded 70% of the total erosion volume. The erosion rate on roads was 21.5 times that in harvest areas, a ratio close to the 17 reported by McCashion and Rice (1983).

Discriminant Analysis

Linear discriminant analysis (Fisher 1936) is a statistical procedure well suited for distinguishing unstable sites from stable ones. Its use to assess landslide potential on road alignments has been proposed by Duncan et al. (1987). Several studies have used discriminant analysis for problems similar to the CSES (Furbish and Rice 1983, Rice et al. 1985, Rice and Lewis 1986). In each of these studies the accuracy of the discriminant function was tested with data not used in its development. The test accuracies varied from 75 to 81%. Consequently, in the CSES all the data were used to develop the discriminant functions and none were held back for testing (Lewis and Rice 1990). The accuracies of the equations were estimated using bootstrapping techniques (Efron and Gong 1983). The discriminant function for identifying critical and noncritical sites on forest roads was:

$$\begin{aligned} DS = & -0.0281 - 0.1142*SLOPE \\ & + 22.91*HCURVE + 1.0075*HUE, \end{aligned} \tag{1}$$

and for logged areas it was:

$$\begin{aligned} DS = & \ 5.032 - 0.1633*SLOPE + 20.69*HCURVE \\ & - 1.215*WEAKROCK, \end{aligned} \tag{2}$$

where:

DS	is the discriminant score;
SLOPE	is the terrain slope in degrees;
HCURVE	is the horizontal curvature of the road centerline in Eq. 1 and of the terrain in Eq. 2 (Horizontal curvature is the radius of a circle passing through the measurement site and two other points on the same contour at distances of about 18 m. It was coded negative in swales and positive on ridges, being zero on planar slopes.);
HUE	is the Munsell hue of moist subsoil (Y = yellow, YR = yellow-red) coded: 1 if the hue is 5Y, 2 if 2.5Y, 3 if 10YR, 4 if 7.5YR, 5 if 5YR;
WEAKROCK	is coded +1 if a bedrock specimen crumbles or deforms under hammer blows and -1 if the specimen fractures. This variable is a simplification of a more refined scale of rock strengths proposed by Williamson (1984).

The variables in these equations are good surrogates for the factors affecting slope stability. Slope indexes the partitioning of the force of gravity into a normal component (promoting stability) and a tangential component (promoting failure). Horizontal curvature indexes the convergence of subsurface water and zones of accumulation of colluvium (both conditions promoting failure). And HUE most likely indexes subsurface water, because most of the yellower soils had a bluish cast due to reduced iron. WEAKROCK separates stable and unstable geologic materials.

The estimated accuracies of the equations, corrected for bias using bootstrapping, are 78% for roads (Eq. 1) and 69% for harvest areas (Eq. 2). The accuracy of Eq. 2 was lower than that in the earlier harvest area studies (Furbish and Rice 1983, Rice et al. 1985), but those studies were developed and tested in limited environments, whereas Eq. 2 was developed from data spanning a variety of conditions. Therefore, Eq. 2 may be more general and more stable if applied outside the range of its developmental data.

The overall risk of a critical site in the population is called a prior probability. It must be known or estimated before a discriminant score can be used to estimate the risk at a site. The predicted risk at a site is known as a posterior probability. Prior probabilities are usually defined as the ratio of the area being identified (critical sites) to the total area (all 1978–79 harvest areas or roads). Lewis and Rice (1990) estimated the prior probability of a harvest area critical site to be 0.0050 and a road critical site to be 0.0177. The posterior probability of a critical site is:

$$PP = 1/\{1 + [(1 - PC)/PC]\exp(DS)\} \tag{3}$$

where PP is a posterior probability and PC is the prior probability of a critical site. Using Eq. 1 or 2, the posterior probabilities are the risks that road construction or timber harvesting will result in more than 189 m^3/ha of erosion. All estimates of posterior probability are not equally precise (Figures 14.1a and 1b). The wide error bands around sites with high posterior probabilities may partially explain the documented tendency of experts to overestimate risk (McGreer and McNutt 1981). There are many stable sites that appear identical to unstable ones, within the precision of our measurements.

Acceptable Risk

In order to use an erosion hazard rating to manage risk, a threshold of acceptable risk must be set. Often this is done intuitively, but more objective methods have been proposed. Rice and Pillsbury (1982) developed a method to estimate a threshold for use in an area, such as a whole harvest unit. It requires the collection of data from the whole unit in question so that local hazard, as well as the prior probability, can affect the choice of an acceptable risk threshold. Lewis and Rice (1990) proposed an equation that uses only the prior probabilities and measurements of the site being evaluated. Both methods, however, require that all four possible outcomes of a prediction (Figure 14.2) be explicitly evaluated. Condition A is the correct identification of a stable site. It carries the net value (V) of changes in all resources affected by the activity. Condition B is the incorrect designation of a stable site as unstable. The value (v) of this result has the value of Condition A minus the cost of any mitigation undertaken as the result of the misclassification. Condition C is the failure to identify an unstable site. It carries the value of Condition A minus the cost (D) of the resulting damages. Condition D is the correct identification of an unstable site. Its value is that of Condition C minus any residual excess damage (d) to resources that may occur, even with mitigation. According to the Bayesian rule (Green 1978), the cutpoint (TC) in the discriminant function which minimizes the expected cost is:

$$TC = \ln PC/(1 - PC)[(D - d) - (V - v)]/(V - v) \tag{4}$$

Evaluating the four conditions of Figure 14.2 is difficult, but doing so

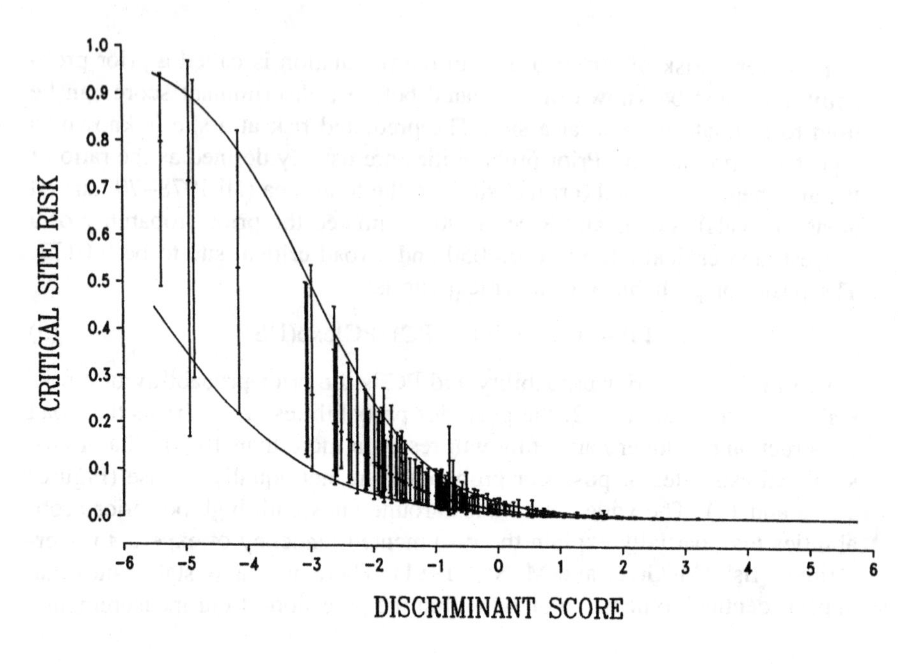

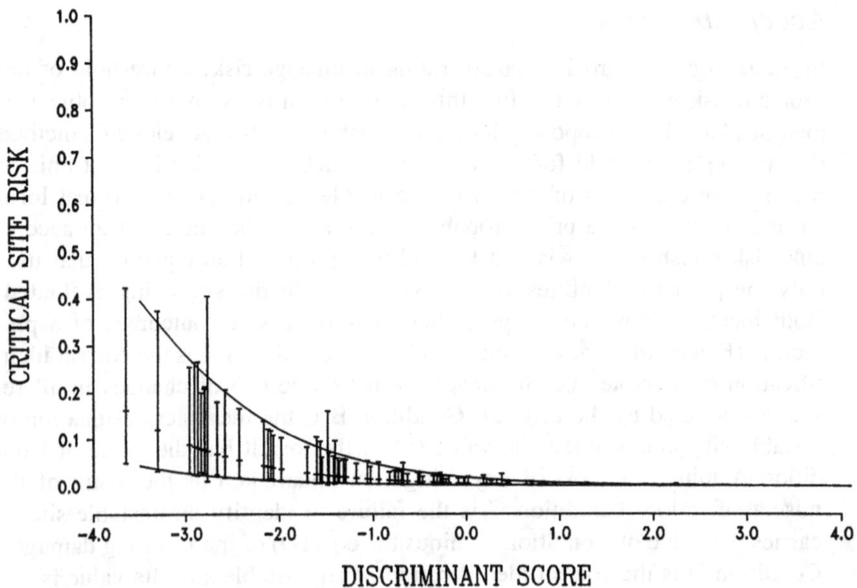

FIGURE 14.1. Variability associated with critical site predictions based on Eq. 1 for roads (top) and Eq. 2 for harvest areas (bottom). The bars show the bootstrap standard deviation at each discriminant score. The curves were fitted to the standard deviations using logistic regression.

Actual Condition	Predicted Condition Stable	Predicted Condition Unstable
Stable	A	B
Unstable	C	D

FIGURE 14.2. The matrix of possible results when predicting the stability of a site.

makes it possible to estimate the acceptable risk threshold that will maximize return from competing resources, based on a manager's value system. It also provides a framework for displaying and discussing alternatives.

The individual manager's value system will affect both the benefits and the costs of operations, whether using the system proposed here or sticking with intuition. A small survey was conducted to gain some insight into the effect of managers' value systems on clearcutting in steep inner gorges and the resulting landslide erosion. Part of the data came from responses to a questionnaire mailed to 54 Forest Supervisors and District Rangers in northwestern California and western Oregon. The same questionnaire was passed out at a meeting of the Jedediah Smith Chapter of the Society of American Foresters (SAF). Replies were obtained from 48 Forest Service personnel and 12 employed in private industry. Statistical treatment of these data was not appropriate because the respondents were self-selected and not independent observers. Although such a sample does not warrant rigorous extrapolation to the population of forest managers, the results may be instructive. Any insights must be tempered by recognition of the limitations of the sampling and the small number of industry responses.

A reprint of an article describing the proposed acceptable risk procedure (Rice et al. 1985) was mailed with the questionnaire. At the SAF meeting, the questionnaire was distributed after a lecture on using discriminant analysis to manage landslide risks associated with clearcutting. The respondents were asked to rate each of the four decision outcomes (Figure 14.2) on a scale ranging from -1,000 (the worst outcome) to +1,000 (the best outcome). They were asked to incorporate in their rating all factors that would customarily be weighed when considering how to harvest timber from a potentially unstable streamside area—not only the economic and environmental considerations, but the social, political, and personal career effects that might result.

The results were much as might have been expected (Table 14.3). Both public and private respondents gave high ratings for harvesting timber on stable land and mitigating high risk sites. The industrial foresters were more

Table 14.3. Average responses and management styles of Forest Service and privately employed personnel concerning logging high risk terrain.

Concern*	Private	Forest Service
A – harvesting a stable site	+992	+728
B – unnecessary mitigation	−733	−367
C – causing a landslide	−517	−831
D – mitigating a landslide	+617	+791
Environmental concerns |D-C| = E	1,133	1,622
Utilization concerns |A-B| = U	1,725	1,094
Management style U/E	1.96	0.79†

* A, B, C, and D are rated on a scale ranging from −1,000 (the least desirable result) to +1,000 (the most favorable result).

† One respondent with a management style of 2,000 was omitted. If he had been included, the value would be 42.4. The inclusion of this respondent in other means did not greatly change them.

concerned about being able to harvest timber, while their Forest Service counterparts expressed about equal concern for timber harvest and landslide prevention. Private foresters' appraisals of the loss from failing to cut timber on stable terrain was nearly twice that of Forest Service people, and they also attached a smaller penalty to causing a landslide.

The responses summarized in Table 14.3 were used to create an index of managerial style. The range between the reward for preventing a landslide and the penalty attached to causing one was taken as a measure of environmental concern. Timber utilization concerns were indexed by range between the penalty for carrying out unnecessary mitigation on stable land and the reward associated with harvesting timber on stable land. Private foresters' utilization concerns were 58% greater than those of public land managers. That difference was reversed for environmental concerns. Forest Service managers concerns for the environment were 43% greater than those of their private counterparts.

An index of managerial style was created by dividing timber utilization concerns by environmental concerns. Public land managers favored the environment, with a score of 0.79. Private foresters exhibited a wider range of concerns, but displayed a decided utilization bias with a score of 1.96.

Only Forest Service data were used for more detailed analysis of the effects of managerial style. The private data set was too small. The Forest Service managers were divided into six relatively homogenous groups at styles of 0.4, 0.7, 0.9, 1.1, and 2.0. Six terrains of varying landslide risk were hypothesized. The most stable had a prior probability of 1.45%, equivalent to the landslide risk in the inner gorges of the Six Rivers National Forest, California (Furbish 1981). The other five terrains had prior probabilities of 5%, 10%, 15%, 20%, and 25%—all extremely hazardous. Managers' acceptable risk thresholds were estimated using the method of Rice and Pillsbury (1982). Each manager's value system was tested in ten simulations. In each simulation a terrain was created by a random drawing of data points, defined in terms of variables in Furbish and Rice's (1983) equation.

The simulations showed managerial style to have much more influence on the threshold probability than does the prevailing risk of the area, as indexed by its prior probability. Landslide risk had its greatest influence on managers with middle managerial styles. Acceptable risk threshold probabilities are, however, only means to ends. What effect did they have on land management? For these simulations, it was assumed that mitigation consisted solely of not harvesting hazardous sites. Surprisingly, neither managerial style nor prior probability had much effect on timber utilization. The greatest difference attributable to management style was less than 2%. Landslide risk was more influential, but only ranged from 89% utilization by the most environmentally concerned managers on the most hazardous terrain to 99.5% utilization by the most timber production-oriented managers on the most stable terrain. Erosion was a different story. As would be expected, erosion was closely tied to landslide risk for all managerial styles. Style, itself, mainly separated the most environmentally concerned managers from the others.

The average of the responses of the private foresters was contrasted with the average of the Forest Service respondents for the six terrains to gain some indication of the possible differences that private foresters' managerial styles might yield. The Forest Service managers had risk thresholds much below their private counterparts (Figure 14.3). That disparity, however, translated into almost imperceptible differences in timber utilization. On the other hand, compared with the average Forest Service style, private foresters' decisions had produced a fairly constant 14% more erosion. The excess erosion associated with the private management style increased from 12 m^3/ha on the least hazardous terrain to 287 m^3/ha on the most hazardous terrain. If these simulations reflect reality, the private foresters' style would be justified if the environmental costs of the 14% increase in erosion were offset by operational economies or other environmental benefits.

The differences seen in the simulations may be an artifact of the questionnaire. The Forest Service responses suggest that actual behavior may not reflect the managers' stated value systems. For example, timber harvests are severely constrained on inner gorge areas amounting to about 12% of the Six Rivers National Forest. This constraint would be justified if the prior probability of a landslide in that terrain was approximately 25%. It is actually only 1.45% (Furbish 1981). It may be that private foresters similarly are unaware of their implicit value systems and believe that their timber utilization orientation is greater than it really is. If the maximum benefit is to be gained from the erosion risk management method just presented, forest managers must become accustomed to setting acceptable risk thresholds explicitly and quantitatively. Until they are able to do so, their actions may not reflect their intentions.

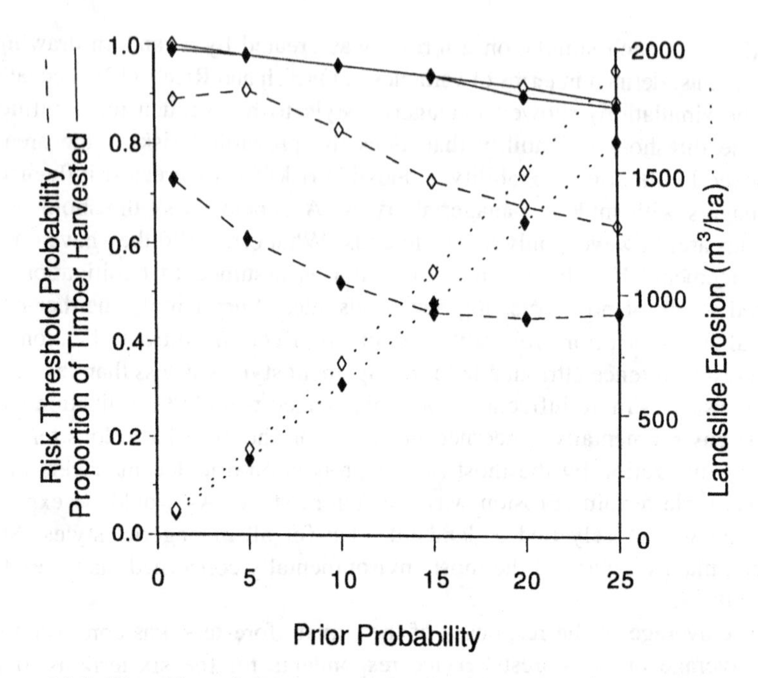

FIGURE 14.3. The effect of the managerial styles of public and private forest managers on the choice of an acceptable risk threshold to use when harvesting timber in an inner gorge, and the resulting erosion and timber utilization. The interaction between six prior probabilities (0.0145, 0.05, 0.10, 0.15, 0.20, 0.25) and the average managerial style of each of the groups. ◆ = Forest Service. ◇ = Private.

Summary and Conclusions

Best management practices, being partly the result of political processes, respond to the moods of the electorate. As states become more urbanized, forest practice rules protecting water quality tend to become more restrictive. The change in erosion and sedimentation accompanying the evolution of California's forest practice rules suggests that the tightening of rules has reduced erosion and improved water quality. Research has found that most erosion in forests comes from a small fraction of the terrain and that site conditions are the most important determinate of erosion risk. This suggests that identifying high risk sites is the key to effective erosion control. Erosion hazard ratings, however, tend to be poorly grounded in science and too broad in scope. One result, perhaps, is that some public land managers are unnecessarily restricting harvests on stable terrain. The discriminant functions presented here permit accurate estimation of erosion risk at a site and the choice of an optimal risk threshold. If forest land managers were to apply risk management tactics based on scientifically developed discriminant analyses, the impediments to forestry might be more in line with the resulting water quality benefits.

References

Alaska Department of Natural Resources. 1989. Alaska forest practices act review. Land and Resources Section, Anchorage, Alaska, USA.

Anderson, H.W. 1954. Suspended sediment discharge as related to streamflow, topography, soil, and land use. Transactions of the American Geophysical Union **35**:268–281.

Arvola, T.F. 1976. Regulation of logging in California, 1945–1975. California Department of Forestry and Fire Protection, Sacramento, California, USA.

Bedrossian, T.L. 1983. Watershed mapping in northern California. California Geology **36**:140–147.

California, State of. N.d. California Code of Regulations, Title 14. Natural Resources, Division 1.5, Department of Forestry.

Council on Environmental Quality. 1971. Guidelines for implementation of PL 92–500, 40 CFR 1508.7.

Datzman, P.A. 1978. The erosion hazard rating system of the Coast Forest District: how valid is it as a predictor of erosion and can a better prediction equation be developed? M.S. Thesis. Humboldt State University, Arcata, California, USA.

Dodge, M., L.T. Burcham, S. Goldhaber, B. McCully, and G. Springer. 1976. An investigation of soil characteristics and erosion rates on California forest lands. California Division of Forestry, Sacramento, California, USA.

Duncan, S.H., J.W. Ward, and R.J. Anderson. 1987. A method for assessing landslide potential as an aid to forest road placement. Northwest Science **61**:152–159.

Durgin, P.B., R.R. Johnston, and A.M. Parsons. 1988. CSES critical sites erosion study. Volume 1. Causes of erosion on private timberlands in northern California. California Department of Forestry and Fire Protection, Sacramento, California, USA.

Efron, B., and G. Gong. 1983. A leisurely look at the bootstrap, the jackknife and cross validation. American Statistician **37**:36–48.

Fisher, R.A. 1936. The use of multiple measurements in taxinomical problems. Annals of Eugenics **7**:179–188.

Furbish, D.J. 1981. Debris slides related to logging of streamside hillslopes in northwestern California. M.S. Thesis. Humboldt State University, Arcata, California, USA.

Furbish, D.J., and R.M. Rice. 1983. Predicting landslides related to clearcut logging in northwestern California, USA. Mountain Research and Development **3**:253–259.

Green, P.W. 1978. Analyzing multivariate data. Dryden, Hillsdale, Illinois, USA.

Lewis, J., and R.M. Rice. 1989. CSES critical sites erosion study. Volume 2. Site conditions related to erosion on private timberlands in northern California. California Department of Forestry and Fire Protection, Sacramento, California, USA.

Lewis, J., and R.M. Rice. 1990. Estimating erosion risk on forest lands using improved methods of discriminant analysis. Water Resources Research **26**:1721–1733.

Llerena, C.A., H. Zhang, and R.L. Rothwell. 1987. Test of an erodibility rating system for the foothills of central Alberta, Canada. Pages 155–161 *in* Forest hydrology and watershed management. International Association of Hydrological Sciences Publication 167, Wallingford, Oxfordshire, United Kingdom.

Martin, E.F. 1989. "A tale of two certificates": the California forest practice program, 1976 through 1988. California Department of Forestry and Fire Protection, Sacramento, California, USA.

McCashion, J.D., and R.M. Rice. 1983. Erosion on logging roads in northwestern California: how much is avoidable? Journal of Forestry **81**:23–26.

McGreer, D., and J. McNutt. 1981. Mass failure hazard index. Pages 17–23 *in* R.O. Blusser, technical director. Research on the effects of mass wasting of forest lands on water quality and the impact of sedimentation on aquatic organisms. Stream Improvement Technical Bulletin 344. National Council for Air and Stream Improvement, New York, New York, USA.

Oregon, State of. 1987. Forest practice rules, northwest Oregon region.

Peters, J.H., and Y. Litwin. 1983. Factors influencing soil erosion on timber harvested lands in California: report to the California Department of Forestry and Fire Protection. Western Ecological Services Company, Novato, California, USA.

Rice, R.M., and P.A. Datzman. 1981. Erosion associated with cable and tractor logging in northwestern California. Pages 362–374 *in* Erosion and sediment transport in Pacific Rim steeplands. International Association of Hydrological Sciences Publication 132, Wallingford, Oxfordshire, United Kingdom.

Rice, R.M., and J. Lewis. 1986. Identifying unstable sites on logging roads. Pages 239–249 *in* Proceedings of the International Union of Forestry Research Organizations 18th World Congress, Division 1, Volume 1, September 7–21, 1986, Ljubljana, Yugoslavia. International Union of Forestry Research Organizations Secretariat, Vienna, Austria.

Rice, R.M., and N.H. Pillsbury. 1982. Predicting landslides in clearcut patches. Pages 303–311 *in* Recent developments in the explanation and prediction of erosion and sediment yield. International Association of Hydrological Sciences Publication 137, Wallingford, Oxfordshire, United Kingdom.

Rice, R.M., N.H. Pillsbury, and K.W. Schmidt. 1985. A risk analysis approach for using discriminant functions to manage logging-related landslides on granitic terrain. Forest Science **31**:772–784.

Rice, R.M., F.B. Tilley, and P.A. Datzman. 1979. A watershed's response to logging and roads: the South Fork of Caspar Creek, 1967–1976. United States Forest Service Research Paper PSW-146, Pacific Southwest Forest and Range Experiment Station, Berkeley, California, USA.

Rice, R.M., and J.R. Wallis. 1962. How a logging operation can affect streamflow. Forest Industries, November 1962, pages 38–40.

Washington State Forest Practices Board. 1988. Washington forest practices rules and regulations, Title 222 Washington Administrative Code. Olympia, Washington, USA.

Williamson, D.A. 1984. Unified rock classification system. Bulletin of the Association of Engineering Geologists **21**:345–354.

Wischmeier, W.H., and D.D. Smith. 1965. Predicting rainfall-erosion losses from cropland east of the Rocky Mountains: guide for selecting practices for soil and water conservation. Agricultural Handbook 282. United States Department of Agriculture, Washington, D.C., USA.

15

Best Management Practices and Cumulative Effects from Sedimentation in the South Fork Salmon River: An Idaho Case Study

WALTER F. MEGAHAN, JOHN P. POTYONDY, AND KATHLEEN A. SEYEDBAGHERI

Abstract

Poor land use, including intensive unregulated logging from 1940 through the mid-1960s, contributed to massive cumulative effects from sedimentation in Idaho's South Fork Salmon River (SFSR) by 1965. Severe damage to valuable salmon and steelhead habitat resulted. The BOISED sediment yield prediction model was used to evaluate the effects of historical and alternative land management on Dollar Creek, a representative 46.1 km^2 tributary watershed in the SFSR basin. Present-day management practices, properly implemented, have the potential of reducing sediment yields by about 45 to 95% compared with yields caused by the historical land use in Dollar Creek. Cumulative effects analysis is a useful tool for evaluating management alternatives. Some increases in sedimentation are unavoidable even using the most cautious logging and roading methods. However, much of the sedimentation in the SFSR and other drainages could have been avoided if logging and road construction had followed current best management practices.

Key words. Cumulative effects, sedimentation, best management practices, erosion control, mitigation.

Introduction

The National Environmental Policy Act of 1969 (PL 91–190) required federal agencies of the United States to prepare environmental impact statements (EISs) for projects that might significantly affect environmental quality. In 1978 the Council on Environmental Quality stated that an EIS must consider direct, indirect, and cumulative impacts. A cumulative impact was defined as "the impact on the environment which results from the incremental impact of the action when added to other past, present, and reasonably foreseeable future actions. . . . Cumulative impacts can result from individually minor but collectively significant actions taking place over a period of time."

The South Fork of the Salmon River (SFSR) in Idaho is a quintessential example of the cumulative effects of forest practices resulting in sedimentation in a western United States river. The purpose of this article is to provide an overview of the history of land use and sedimentation in the SFSR and to illustrate, using a cumulative effects assessment procedure, how use of current best management practices (BMPs) might have reduced the impacts.

Description of the South Fork Salmon River

The SFSR drainage, part of the Columbia River Basin, is an area of 3,350 km^2 in central Idaho (Figure 15.1). The watershed is almost entirely within the Idaho batholith. Approximately one-third of this land lies within the Boise National Forest, and about two-thirds is within the Payette National Forest (United States Forest Service 1985).

Elevations within the drainage range from 820 to 2,830 m, with about half of the area between 1,500 and 2,300 m. The region is characterized by steep slopes with overstory vegetation dominated by ponderosa pine (*Pinus ponderosa*) and Douglas-fir (*Pseudotsuga menziesii*) at the lower elevations, and lodgepole pine (*Pinus contorta* var. *latifolia*), grand fir (*Abies grandis*), Engelmann spruce (*Picea engelmannii*), and subalpine fir (*Abies lasiocarpa*) at the higher elevations (United States Forest Service 1985).

Summers are typically hot and dry, with warm-season precipitation occurring primarily during high intensity thunderstorms. Winters are characterized by heavy snows and cold temperatures. Long-duration, low intensity storms are common in autumn, winter, and spring and may generate considerable snowmelt in addition to storm rainfall. On average, about 65% of the annual precipitation falls as snow.

The granitic bedrock of the batholith produces shallow, coarse-textured soils that are essentially without cohesion. When disturbed, such soils exhibit high erosion rates from both surface and mass erosion processes. The combination of steep slopes, erodible soils, and high climatic stresses can cause dramatic increases in sediment production. For example, Megahan and Kidd (1972) reported on a six-year study designed to compare natural sediment production with that from forest roads and from areas disturbed by cable logging. Cable logging had minimal impacts, causing an increase in unit area sediment yields of only 0.6 times compared with undisturbed slopes. In contrast, forest roads accelerated unit area sediment yields by a factor of

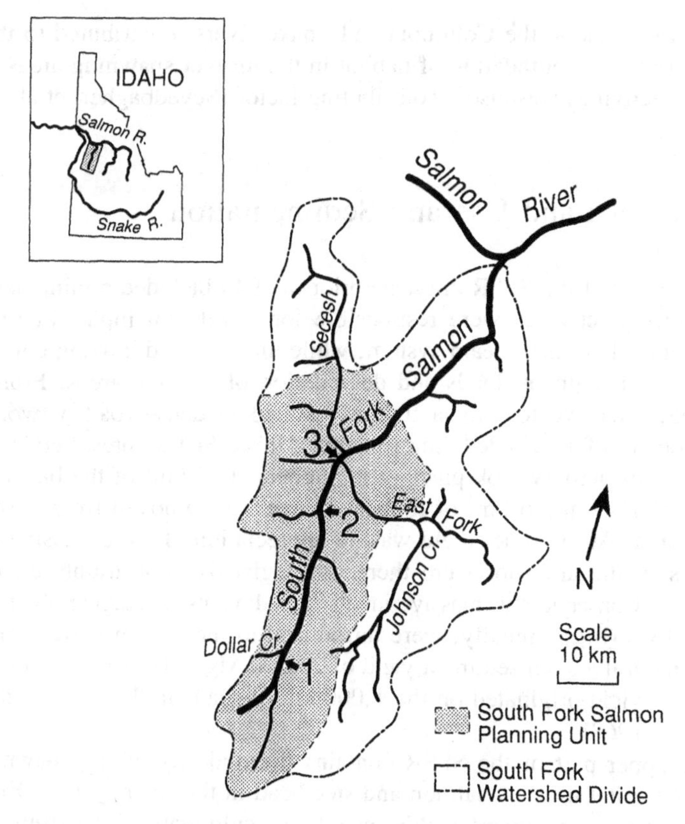

FIGURE 15.1. Location and plan map of the South Fork Salmon River watershed, showing the location of the Stolle (1), Poverty Flat (2), and Glory (3) spawning areas.

220 times from sheet and gully erosion. Mass erosion on the roads (primarily road-fill failures) caused an additional increase of 550 times in unit area sediment yields compared with undisturbed slopes.

The SFSR basin has a wealth of resources, including minerals, recreation, timber, water, forage, wildlife, and fish. The SFSR is considered particularly crucial as a source of spawning and rearing habitat for anadromous fish populations, including salmon and steelhead (*Oncorhynchus* spp.). According to Platts and Partridge (1983), the SFSR historically supported Idaho's largest population of summer chinook salmon (*Oncorhynchus tshawytscha*). The number of returning adults declined from an estimated high of 10,000 fish in the mid-1950s to an estimated low of 250 in 1979. The reduction in returning adult salmon led to the closure of a popular and economically important anadromous sport fishery on the SFSR in 1965. Downstream influences of commercial and sport fishing, as well as eight mainstream hydroelectric dams on the Columbia and Snake rivers, contributed to this decline. However, degradation of habitat in the upriver spawning areas due to land use activities was also a contributing factor (Seyedbagheri et al. 1987).

History of Land Use and Sedimentation

Development of the SFSR drainage prior to 1940 included mining and grazing. Mining activities were responsible for significant inputs of sediment and chemicals to the stream system, while uncontrolled grazing contributed to increased sediment loads and degradation of riparian areas. From 1945 to 1965, intensive logging activities resulted in dense road networks and other sources of accelerated sedimentation (United States Forest Service 1985). Most of this activity took place in the upper 1,020 km^2 of the basin, where a total of 1.15 million m^3 of timber volume was removed from about 15% of the area. At the time, there was no appreciation of the excessive erosion hazards in the area nor were there any legislative constraints on logging activities. Consequently, massive inputs of sediments, averaging about 73,000 metric tons (Mg) annually, were introduced to the stream system in excess of the natural annual sediment yield of 23,200 Mg. About 85% of the excess sediment yield originated on the 1,000 km of roads in the area (Arnold and Lundeen 1968).

The upper part of the SFSR contains the majority of the spawning and rearing habitat used by salmon and steelhead in the river system. Effects of the accelerated sediment yields over time, culminating in extreme runoff events in 1964 and 1965, created sediment greatly in excess of the transport capabilities of the river. The result was widespread channel aggradation, estimated at about 2.0 million Mg within the main river and its tributaries (Arnold and Lundeen 1968). Sand (0.1 to 2.0 mm diameter) was the dominant particle size of deposited sediments. The aggraded sediments contributed to a reduction in anadromous fish production to its lowest level, 20% of potential (United States Forest Service 1985). Aggradation occurred in fish spawning and rearing areas often to depths of a meter or more. For example, the "Swimming Hole" in the SFSR, used as a resting area by migrating fish, was completely filled with sediment (Figure 15.2a).

In 1965, a moratorium on logging and road construction was imposed by the Forest Service in the South Fork Planning Unit in the headwaters of the basin. This unit is the largest (1,480 km^2) of five major subdivisions of the SFSR drainage basin and includes all of the 1,020 km^2 included in the Arnold and Lundeen (1968) study referenced above plus some additional land in the East Fork of the South Fork and the Secesh River drainages (Fig. 15.1). A watershed rehabilitation program that focused on road erosion control was implemented in conjunction with the moratorium (Megahan et al. 1980).

FIGURE 15.2. Sediment storage in the "Swimming Hole," one of the important resting areas for migrating salmon and steelhead in the South Fork Salmon River. The hole was completely filled with fine sediments in 1965 (top). By 1982 (bottom), the accumulated sediments had been flushed from the pool, exposing the original boulder bottom.

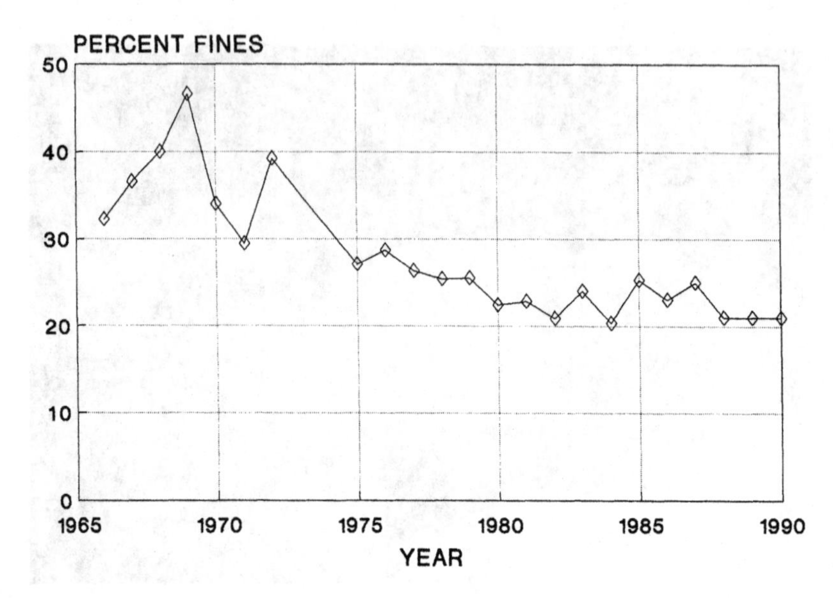

FIGURE 15.3. Time trend in the average percentage of fine sediments (<4.75 mm diameter) in the substrate of the Stolle, Glory Hole, and Poverty Flat spawning areas.

The logging moratorium and the watershed rehabilitation program reduced sediment inputs to the river by eliminating new sources of sediment and reducing sediment supplies from previously disturbed areas. This allowed the river to flush the previously aggraded sediments (Figure 15.2b). Monitoring results in the years following the moratorium declaration showed a fairly rapid recovery of the fish habitat, as indicated by changes in streambed elevation and in average particle size on and beneath the surface of the streambed (Megahan et al. 1980).

Based on assessments of the monitored habitat indicators, in 1977 it was decided that the river had recovered enough (Figure 15.3) to allow the implementation of a new ten-year Land Management Plan on the South Fork Planning Unit. The plan recognized the anadromous salmonids as the most valuable resource of the unit. All land use activities were contingent on continued salmonid habitat improvement. Several timber harvest practices designed to minimize sedimentation (many of which are now BMPs) were required. Efforts were made to minimize disturbance to soil and streams, reduce erosion on disturbed areas, provide sediment buffer zones, and schedule disturbances to disperse the resultant sediment to streams in space and time (Cole and Megahan 1980).

In addition to the land use plan, a comprehensive monitoring program was implemented in 1978 to evaluate the effects of management activities through studies conducted at sites, in tributary streams, and in the SFSR main channel. Timely feedback to land managers regarding existing or potential problems was an integral part of the monitoring process. In addition, the South Fork Salmon River Monitoring Committee, consisting of soil, water, and aquatic specialists from various concerned agencies and organizations, was established to review monitoring results and make recommendations (Cole and Megahan 1980).

The early monitoring efforts showed few major impacts from new land use activities and evidence of continued habitat improvement. However, between 1981 and 1985, increases in fine sediments were observed on and beneath the surface in spawning areas (Platts et al. 1989). Subsequent data on the percentage of fine sediments (<4.75 mm diameter) below the surface of spawning areas showed additional increases in 1986 and 1987 followed by a return to lower levels in 1988 through 1990 (Figure 15.3). The data suggest that the river is fluctuating about an apparent equilibrium with respect to sediment supply and transport. Major factors contributing to sediment supply along the river include one or more of the following: increased sediment inputs from natural floods in two major tributaries, the breach of a river oxbow, and the effects of past and current road erosion in the basin. Because of these numerous sources of sediment input, it is difficult to determine conclusively whether or how the post-1977 logging activities are related to the recent fluctuation in the percentage of fine sediment in spawning areas. However, the land management plan for the South Fork Planning Unit required that soil-disturbing activities cease if the river failed to continue to improve, regardless of the cause. Accordingly, no timber harvest was permitted in the South Fork Planning Unit beginning in 1984.

The National Forest Management Act of 1976 (PL 94–588) called for the development of integrated forest plans for national forest lands. Completion of the forest plans for the Payette National Forest in 1988 and for the Boise National Forest in 1990 supersedes implementation of the land management plan for the old South Fork Planning Unit. The new forest plans permit only limited land disturbance in the SFSR basin until river conditions improve.

Development of Best Management Practices in Idaho

The sedimentation damage to the SFSR occurred long before the development of BMPs in Idaho. It was not until 1974 that Idaho adopted a comprehensive Forest Practices Act (Title 38, Chapter 13, Idaho Code) modeled after the Oregon Forest Practices Act. The stated purpose of the Forest Practices Act (FPA) was to provide authority for rules "to assure the continuous growing and harvesting of trees and to protect and maintain the forest soil, air, and water resources, and wildlife and aquatic habitat." Regulated forest practices include timber harvest, road construction and maintenance, reforestation, use of chemicals or fertilizers, and slash management. The first set of rules and regulations was issued in 1976 (Idaho Administrative Procedures Act 20.15).

In 1979, the Idaho Forest Practices Water Quality Management Plan (Idaho Department of Health and Welfare 1979) was completed consistent with the intent of Section 208 of the federal Clean Water Act Amendments (PL 92–500, PL 95–217). This plan identified the rules and regulations associated with the Idaho FPA (with recommended modifications) as approved BMPs for the state of Idaho.

An interdisciplinary task force was set up in 1983 to look at nonpoint source pollution due to forest practices, and the adequacy of the existing BMPs and the regulatory processes for protecting water quality. The 1985 report of the task force (Idaho Department of Health and Welfare 1985) documented varying degrees of compliance and made recommendations for changes in the FPA and in the associated rules and regulations.

One of the provisions of the Idaho antidegradation policy, adopted in 1988 (Idaho Executive Order 88.23), is the establishment of a coordinated monitoring program with inputs to the policy revision process. Part of the monitoring program is designed to assess the effectiveness of BMPs. This includes monitoring of BMP implementation, pollutant source and transport, and beneficial uses. The monitoring program focus for forestry BMPs is the impact of sediment on cold water biota and salmonid spawning (Clark 1990). This is consistent with the historical emphasis of Idaho forestry BMPs on control of erosion and the prevention of stream channel sedimentation.

A 1989 interdisciplinary team water quality audit report (Idaho Department of Health and Welfare 1989) showed that, when applied, BMPs were 99% effective in preventing obvious excess sediment from entering streams. When they were not applied, excess sediment was delivered to waters in 70% of the cases. Precise quantification and an assessment of cumulative effects were not part of the audit. However, it was noted that 80% of the streams evaluated during the audit had intermediate or high levels of sedimentation from past activities.

Cumulative Effects Assessments in Idaho

The Forest Service has developed procedures to predict the cumulative effects of forest practices. The Northern Region did the initial development work in northern Idaho in the early 1970s. Procedures are designed to evaluate increases in average water yield resulting from alternative levels of timber harvest. Water yield increases are estimated on the basis of the percentage of timber volume removed and the amount of hydrologic recovery based on time after cutting (United States Forest Service 1974).

Subsequent work in Idaho by both the Northern and the Intermountain regions of the Forest Service led to the development of the R1-R4 Guide (Cline et al. 1981) to predict average annual sediment yields from forested watersheds under alternative patterns of land use. The procedures are designed for use throughout the state but are considered to be best adapted to the granitic watersheds in southern and central Idaho. Estimates of annual sediment yields are made for undisturbed watersheds and for watersheds disturbed by timber harvest, road construction, and forest fire. On-site erosion estimates are based on the type and age of disturbance and are adjusted for any applied mitigation measures. Downstream sediment yields are estimated using a sediment delivery ratio varying by watershed size.

How Might the SFSR Damage Have Been Avoided?

What occurred in the SFSR is an excellent case study of river response to poor watershed management. Ignorance of the consequences of soil disturbance and the lack of legislative constraints contributed to the problem. The cumulative effects were dramatic and indisputable. But what might have happened given present-day knowledge, BMP direction, and cumulative effects assessment procedures? We evaluated this by applying the sediment yield cumulative effects modeling procedure to a reconstruction of the original disturbance scenario in the SFSR compared with present-day management methods.

The Boise National Forest version of the R1-R4 sediment yield prediction procedure called BOISED (Potyondy et al. 1991) was chosen for this. The BOISED model is based on a large body of locally derived data for establishing average quantities and time trends of both surface and mass erosion. Additional data were available to evaluate the effectiveness of alternative BMPs for reducing surface erosion. Local data were also used to define the downslope delivery of sediments derived from surface and mass erosion. Aside from a delivery ratio approach that provides an estimate of the total volume of sediment that reaches the outlet of a channel, there is no provision for channel sediment routing. It is our opinion that BOISED does a reasonable job of estimating the total quantity of sediment yielded by alternative forest management practices but may not provide an accurate display of time trends. Given this caveat, BOISED is still the best available tool for evaluating the cumulative effects of sedimentation from alternative land use practices in the SFSR.

Because we did not have time to apply BOISED to the entire SFSR basin, we selected a tributary watershed, Dollar Creek, for illustration (Figure 15.1). Dollar Creek is a 46.1 km^2 watershed located near the upper end of the SFSR drainage. Watershed characteristics and management activities are representative of major tributaries of the SFSR drainage.

We reconstructed the logging and road construction history for Dollar Creek from records available for the Cascade Ranger District of the Boise National Forest. A plan map of the disturbances is shown in Figure 15.4. A total area of 3.3 km^2 was cut in two timber sales, in 1958 and 1965. The first

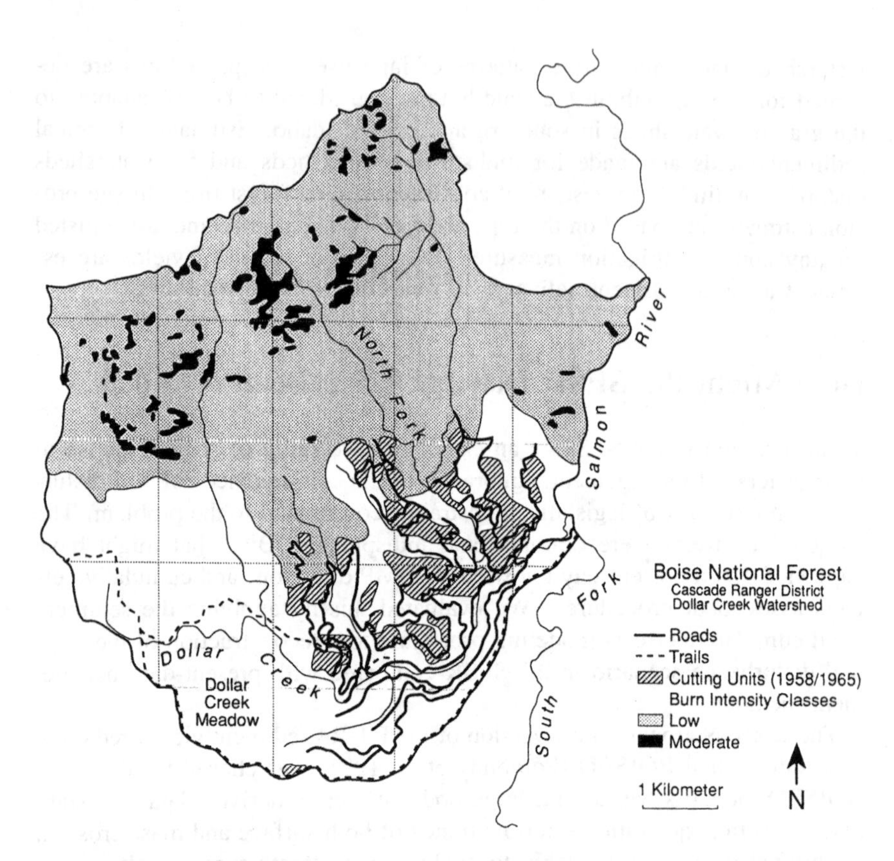

FIGURE 15.4. Plan map of the Dollar Creek drainage showing historical logging activities within the basin.

sale covered 2.1 km^2 and included about 90% selection logging and 10% clearcut. The second covered 1.2 km^2 and took place on much of the same area as the 1958 sale, using mostly clearcutting. In both sales, about 70% of the area was cut by jammer logging, a ground cable system that requires a dense road network. The remaining area was tractor logged. Approximately 17.7 km of roads were built in 1956 and 19.3 km additional in 1957 before the first logging in 1958. For the 1965 logging, the entire road system was reconstructed and about 9.7 km of roads were added. By 1965, a total of 46.7 km of road had been built. Additional soil disturbance occurred in the watershed from a wildfire that burned across the upper part of the Dollar Creek watershed in July and August 1989, burning 58% of the drainage. Fire severity was relatively low, with 94% of the burning at low intensity and 6% at moderate intensity.

In addition to estimates of sediment yields for the undisturbed watershed conditions in Dollar Creek, four logging/roading scenarios were modeled

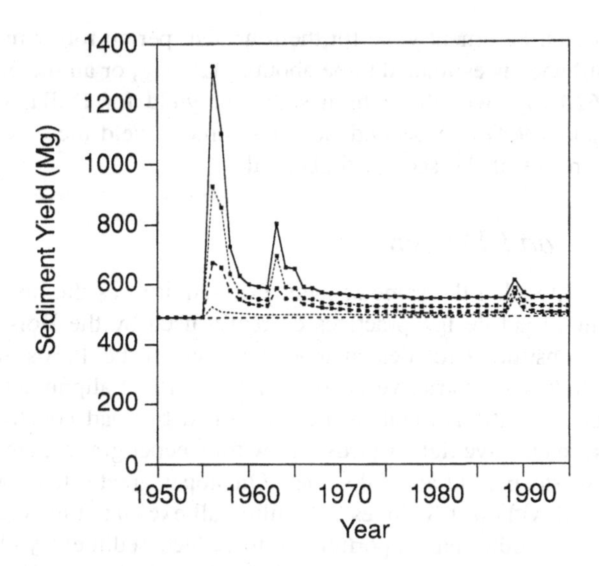

FIGURE 15.5. Modeled time trend in sediment yields from Dollar Creek for alternative patterns of land use. –■– Historical record. ·····■····· Improved design. ---■--- Maximum mitigation. .... Helicopter. ---- Undisturbed.

to compare the cumulative effects of past activities to what might occur using present-day BMPs. These scenarios portray: (1) a reconstruction based on the historical record of management practices, (2) improved road design, (3) maximum mitigation, and (4) helicopter logging. In all cases except helicopter logging, the same mix of silvicultural methods (clearcutting and selection cutting) and logging systems (jammer and tractor logging) was used, and the same lengths and locations of logging roads were assumed. Effects of the 1989 wildfire are also modeled for each of the four scenarios.

Undisturbed Watershed

The average annual sediment yield for Dollar Creek, assuming undisturbed watershed conditions, amounts to 450 Mg at the confluence of Dollar Creek with the main stream of the SFSR. This figure provides a benchmark for evaluating the sedimentation caused by the alternative land use activities for the period 1950 to 1995 (Figure 15.5).

Historical Record

This estimate reflects sediment yield resulting from development that occurred in the Dollar Creek watershed as discussed above (Figure 15.5). Maximum sediment yield increase occurred in 1956 following the first series of road construction activities and decreased to a relatively constant, low level within four years. Sediment yield for the four-year period following the first logging disturbance is estimated to be about 3,420 Mg, or an increased yield of about 1,620 Mg over the natural sediment yield for Dollar Creek (an increase of about 90%). A second, lower sediment yield increase occurred in 1965 as a result of the second timber sale.

Improved Road Design

This scenario modeled the same sequence of activities as the historical record except that road design practices currently used by the Boise National Forest were substituted for design practices used in the 1950s and 1960s. Changes include less restrictive horizontal and vertical alignment standards and a reduction in the amount of area disturbed by road construction. To achieve this, roads have tighter turns, allow for steeper grades, are narrower, and may have steeper cut and fill slopes. On slopes greater than 60%, roads are constructed without fill slopes by hauling all excavated materials to safe disposal areas. Additional opportunities to reduce sediment yield through improved road location were not modeled. Nevertheless, total accelerated sediment yields for the four-year period following disturbance amount to 910 Mg, or about 51% over natural (Figure 15.5). This represents a reduction in the accelerated sediment yield of about 45% as a result of the improved road design, compared with the historical management.

Maximum Mitigation

This scenario modeled the same sequence of activities as the historical record, including the improved road design, and added maximum erosion mitigation standards and guidelines from the Boise National Forest Land and Resources Management Plan. In some instances, these practices exceed the BMPs required by Idaho's Forest Practices Act rules and regulations. Included are dry seeding of all cut and fill slopes, scattering straw mulch on cuts and fills along with a binding agent such as asphalt on cut slopes if needed, placing filter windrows on fill slopes at drainage crossings, using erosion control netting around culvert inlets and outlets, and graveling the road surface, including ditches. Application of this level of erosion mitigation reduces accelerated sediment yields from Dollar Creek to an estimated 430 Mg for the four-year period following disturbance (about 24% over natural). This treatment provides a reduction in the accelerated erosion caused by the historical land use of about 73% (Figure 15.5).

Helicopter Logging

This scenario modeled timber harvest in Dollar Creek assuming present-day helicopter logging technology. Both selection and clearcutting silvicultural systems were used, but all the logging was assumed to occur in a single year, 1958. Using current helicopter flight distances of up to 1.6 km, a 4.8 km road system was required to access all the cutting units, compared with the 46.7 km of road constructed to harvest the timber with tractor and cable logging systems. This reduced road length is the primary reason the accelerated sediment yield amounts to just 60 Mg for the four-year postlogging period, for an increase over natural of only 3% (Figure 15.5). In this case, the sediment yield is reduced by approximately 95% compared with the historical scenario.

Wildfire Sediment Yield

The increase in sediment yield shown for 1989 is the modeled sediment yield attributed to wildfire. Just before the fire, actual sediment yield for Dollar Creek (the historical record) was about 505 Mg per year. The fire increased the peak sediment yield to 560 Mg per year, for an increase of about 12% (Figure 15.5).

Conclusion

The results of this modeling exercise demonstrate that existing cumulative effects analysis procedures provide a means of evaluating alternative forest management practices. We estimate that the historical sediment yields could have been reduced considerably by present-day BMPs, by amounts ranging from about 45 to 95%. Numerous other scenarios could be evaluated; selection of the most suitable would have to be based on an evaluation of the benefits versus the costs of each alternative. All land use alternatives caused some increase in sediment yields, ranging from 3 to 51% over natural levels. Thus there is some sedimentation cost of timber harvest in these granitic areas that must be accepted if such activities are to take place.

Present-day BMPs make it possible to reduce sediment yields. Cumulative effects analysis procedures allow forest managers to compare the sediment yields caused by alternative forest disturbances. Given the application of current knowledge and environmental constraints over the period of logging development, much of the sedimentation damage that occurred in the SFSR might have been avoided.

References

Arnold, J.F., and L.L. Lundeen. 1968. South Fork Salmon River special survey, soils and hydrology. United States Forest Service, Boise National Forest, Boise, Idaho, USA.

Clark, W.H. 1990. Coordinated nonpoint source water quality monitoring program for Idaho. Idaho Department of Health and Welfare, Division of Environmental Quality, Boise, Idaho, USA.

Cline, R., G. Cole, W. Megahan, R. Patten, and J. Potyondy. 1981. Guide for predicting sediment yields from forested watersheds. United States Forest Service, Northern Region, Missoula, Montana, and Intermountain Region, Ogden, Utah, USA.

Cole, G.G., and W.F. Megahan. 1980. South Fork Salmon River: future management. Pages 396–405 *in* American Society of Civil Engineers, editor. Proceedings, Symposium on Watershed Management, July 21–23, 1980, Boise, Idaho. Volume 1. American Society of Civil Engineers, New York, New York, USA.

Council on Environmental Quality. 1978. CEQ's NEPA regulations, 40 CFR, Sections 1,500–1,508. Washington, D.C., USA.

Idaho Department of Health and Welfare. 1979. Forest practices water quality management plan. Idaho Department of Health and Welfare, Division of Environment, Boise, Idaho, USA.

Idaho Department of Welfare. 1985. Silvicultural nonpoint source task force final report. Idaho Department of Health and Welfare, Division of Environment, Boise, Idaho, USA.

Idaho Department of Welfare. 1989. Final report: Forest practices water quality audit 1988. Idaho Department of Health and Welfare, Division of Environmental Quality, Boise, Idaho, USA.

Megahan, W.F., and W.J. Kidd, Jr. 1972. Effects of logging and logging roads on erosion and sediment deposition from steep terrain. Journal of Forestry **70**:136–141.

Megahan, W.F., W.S. Platts, and B. Kulesza. 1980. Riverbed improves over time: South Fork Salmon. Pages 380–395 *in* American Society of Civil Engineers, editor. Proceedings, Symposium on Watershed Management, July 21–23, 1980, Boise, Idaho. Volume 1. American Society of Civil Engineers, New York, New York, USA.

Platts, W.S., and F.E. Partridge. 1983. Inventory of salmon, steelhead trout, and bull trout: South Fork Salmon River, Idaho. United States Forest Service Research Note INT-324, Intermountain Forest and Range Experiment Station, Ogden, Utah, USA.

Platts, W.S., R.J. Torquemada, M.L. McHenry, and C.K. Graham. 1989. Changes in salmon spawning and rearing habitat from increased delivery of fine sediment to the South Fork Salmon River, Idaho. Transactions of the American Fisheries Society **118**:274–283.

Potyondy, J.P., G.L. Cole, and W.F. Megahan. 1991. A procedure to estimate sediment yields from forested watersheds. Pages 46–54, Section 12, *in* Interagency Committee on Water Data, Subcommittee on Sedimentation. Proceedings of the Fifth Interagency Sedimentation Conference, Las Vegas, Nevada, March 18–21, 1991. Volume 2. Document 1991 0–288–410. United States Government Printing Office, Washington, D.C., USA.

Seyedbagheri, K.A., M.L. McHenry, and W.S. Platts. 1987. An annotated bibliography of the hydrology and fishery studies of the South Fork Salmon River. United States Forest Service General Technical Report INT-235, Intermountain Research Station, Ogden, Utah, USA.

United States Forest Service. 1974. Forest hydrology, hydrologic effects of vegetation manipulation. Part 2. Document 1974–799–640/152. United States Government Printing Office, Washington, D.C., USA.

United States Forest Service. 1985. South Fork of the Salmon River appendix for the Payette and Boise National Forests proposed land and resource management plans. United States Forest Service, Intermountain Region, Ogden, Utah, USA.

16

Management for Water Quality on Rangelands Through Best Management Practices: The Idaho Approach

KENDALL L. JOHNSON

Abstract

Nonpoint sources of pollution from rangelands are becoming a steadily larger public issue. Recent water quality concerns, notably the 1987 Clean Water Act, have impelled states to address nonpoint control, chiefly through best management practices (BMPs) to reach water quality goals. Rangeland BMPs are based on study and experience gained from observing the effects of grazing management and range improvements on the hydrological processes of watershed uplands and riparian zones. Grazing practices providing moderate grazing intensity within a system that includes seasonal rotation and periodic rest will tend to maximize infiltration rates and decrease sediment production from uplands through production of increased biomass. The same approach will tend to reduce effects at the four main riparian points of stress (riparian vegetation, channel morphology, water column, and streambanks). Virtually every range improvement technique, properly conducted, is a potential BMP for abatement of nonpoint source pollution. Nearly all rangeland BMPs must be based on a professional understanding of ecological principles.

Idaho is leading state efforts to improve nonpoint source abatement of pollution. The state has adopted a rather complex but fairly effective two-pronged approach consisting of stream protection and pollution control. Under the latter effort, Idaho has developed an antidegradation policy and water quality standards. BMPs are the chief management instrument, combined with long-term monitoring and feedback processes to make incremental changes. The agricultural (including grazing) water quality program to reduce nonpoint source pollution relies on voluntary, cost-share-encouraged cooperation through soil conservation districts. Given the scale of the problem, this is probably the only feasible approach, and is likely to produce tangible results in the years ahead.

Key words. Rangeland, water quality, best management practices, watershed management, range management, grazing.

Introduction

Rangelands are defined by their relative lack of water, either seasonally or annually, and by highly variable soils, frequent salinity, sparse vegetative cover, and sharp climatic extremes. All intensive and most extensive uses of such lands, if improperly managed, may generate negative hydrologic effects in the form of nonpoint (geographically diffuse) source pollution, principally through erosion and subsequent transfer and deposition of eroded materials. Grazing by domestic livestock is easily the most visible and probably the most generalized example of a rangeland use that creates pollution, although some forms of outdoor recreation, especially those involving off-road vehicles, are beginning to challenge grazing (Wilshire and Nakata 1976). Nonpoint source pollution from rangeland is steadily becoming a larger public issue. Recent legislation based on water quality concerns, in particular the federal Water Quality Act of 1987 (Clean Water Act), authorizes attempts to reduce such pollution. Best management practices (BMPs) are the chief means employed to reach Clean Water Act goals. BMPs are defined as the most effective and practicable means of preventing or reducing nonpoint sources of pollution. As such, they should be responsive to natural conditions, and represent reasonable and well-considered trade-offs between technical feasibility and social, economic, and political realities.

The objectives of this chapter are (1) to briefly review the technical status of land use practices employed on rangelands (both uplands and bottomlands) as a modifier of site hydrology, (2) to identify possible BMPs from among the practices, and (3) to trace their application to rangeland in the state of Idaho, as a case study of attempts to reduce nonpoint sources of pollution.

Rangeland Management As Watershed Management

The hydrologic conditions of rangeland sites reflect complex interactions of climate, soil, and vegetation, including amount and intensity of precipitation; soil depth, texture, structure, bulk density, and compaction; ground cover of living and dead vegetation; and grazing intensity (Branson et al. 1981). Grazing effects on the hydrologic cycle are expressed primarily in the surface-related processes of interception, infiltration, runoff, and percolation to recharge both soil moisture and groundwater (Gifford and Hawkins 1978). The effects of grazing are strongly influenced by spatial orientation: the hydrologic effects of grazing on watershed uplands are different in kind and intensity from those associated with riparian zones, simply as a function of proximity to bodies of surface water. But it is at the point where water meets ground that good range management becomes good watershed management in both land types.

Grazing of Watershed Uplands

The limitations that relative aridity imposes on land use practices and the recovery potential of uplands affect all management programs. Restoration of lands damaged through improper land use is both limited in scale and extended in time. Thus uplands, particularly compared with better-watered lands, are best managed through prevention of degradation than restoration of potential. For example, reduction of erosion and sediment flux through effective land management practices is preferable to rehabilitation measures. Therefore, the primary goal of watershed management on uplands must be to favorably influence the fundamental hydrologic processes of infiltration, runoff, and erosion through better vegetation management.

Numerous alterations of soil and vegetation characteristics as a result of abusive grazing have been demonstrated on a variety of sites and plant communities (Blackburn 1984). These effects may produce lowered infiltration capacities, increased overland flow, and increased erosion, with the relative intensity of change determined by the natural variation in physical site factors and the artificial variation in grazing management.

Much of the natural variation in site factors has not been accounted for in studies of the hydrologic effects of livestock grazing, and documentation of the intensity and duration of grazing has been poor (Blackburn 1984). Nonetheless, there is wide recognition that soil and vegetation strongly influence rangeland hydrology (Elmore, this volume). Efforts to explore natural variation of within-site influences include Devaurs and Gifford (1984), who used a rainfall simulator to provide data on spatial variability from small plots located within large-plot boundaries. Sampling fenced, unfenced, and rototilled conditions on each of three range sites, they found that considerable variability in measured infiltration and physical soil properties occurred on relatively uniform sagebrush-grass communities in southwestern Idaho. The study supported those of Achouri and Gifford (1984), Sharma et al. (1980), Tricker (1981), and Viera et al. (1981), which had also found great spatial variability of hydrologic properties within short distances. It appears there is a need to characterize the variability associated with individual measurements in land management approaches.

Wilcox et al. (1988) and Wilcox and Wood (1989) investigated factors influencing infiltration and interrill erosion from semi-arid slopes (0–70%) in New Mexico. As expected, the soil and vegetation factors shown by many other studies to influence infiltration and erosion are also important on steep slopes. Both infiltrability and interrill erosion are positively related to slope gradient, possibly because interflow increases with higher slope angles, and sediment concentration is greater.

The understanding that heavy, continuous grazing has adverse hydrologic effects (increased runoff and erosion) is widespread in the literature, with studies reported in every decade of this century (e.g., Rich 1911, Sampson and Weyl 1918, Forsling 1928, Stewart and Forsling 1931, Cottam and Evans 1945, Love 1958, Ellison 1960, Smeins 1975, Wood et al. 1986, Takar et al. 1990). Many of these and other studies indicate that light or moderate continuous grazing generally differ little in their effects on hydrologic relations, and some studies find no difference from ungrazed conditions (Blackburn 1984).

In a review of grazing effects on rangeland hydrology, Gifford and Hawkins (1978) observed that only the infiltration function within the hydrologic cycle has been studied sufficiently to allow an initial analysis of grazing-runoff relations. They supported that observation by summarizing 22 studies located within the western rangeland of North America, mostly within the Great Plains. Conclusions drawn from the analysis were: (1) grazing affected infiltration at any intensity (ungrazed rates were statistically different beyond the 90% level); (2) light and moderate grazing were statistically identical; and (3) heavy grazing was statistically different from light and moderate grazing. As a rough approximation, they suggested that light and moderate grazing on relatively porous soils could be expected to reduce f_c (final or constant infiltration rate) to about three-fourths of the ungrazed condition, and heavy grazing to about two-thirds of the light to moderate condition.

A review of the literature since the Gifford and Hawkins (1978) publication reveals relatively few additional papers directly addressing the infiltration process as affected by grazing on rangeland. The hydrologic effects of sheep grazing on steep slopes (30–70%) were studied by Wilcox and Wood (1988) in New Mexico. Light grazing (10 ha/AU) reduced infiltrability 12–17% lower than on ungrazed slopes. (AU = animal unit, generally defined as one mature cow of approximately 1,000 lb or 455 kg, either dry or with calf up to six months of age, or their equivalent.) Sediment production for grazed slopes was higher initially but not at final infiltration rates. Both results are comparable to those reported for moderate slopes.

The influence of livestock trampling on soil hydrologic characteristics under intensive rotation grazing was investigated by Warren et al. (1986*c*). They found that infiltration decreased and sediment increased significantly on a silty clay soil devoid of vegetation subject to intense periodic trampling. The effects—expressed in soil bulk density, aggregate size distribution and stability, and surface microtopography—increased with increasing stocking rate.

Busby and Gifford (1981) found no consistent effect on infiltration rates of forage removal and soil compaction on sandy loam soils of chained and unchained piñon-juniper (*Pinus edulis–Juniperus osteosperma*) sites in Utah. (Chaining is a land-treatment process involving use of a large ship anchor chain drawn between two large crawler tractors to pull down or uproot brush or small trees. The soil thus disturbed covers grass seed previously broadcast on the site.) Interrill erosion was not significantly affected by forage removal treatments, but no consistent relation was found with soil compaction.

The effects of range fertilization and subsequent grazing on infiltration rates and sediment production were assessed by Wood et al. (1986). Fertilized areas were stocked at double the rate of nonfertilized areas (1.2 ha/AU as opposed to 2.5 ha/AU); infiltration rates and sediment production varied between years, depending on precipitation, but not within a year. It was thought that fertilizer-induced increases in plant production mitigated the effects of higher stocking rates.

Takar et al. (1990) found in a study on Somalian rangeland that heavy communal grazing on sand and clay soils under two cover types (shrub understory versus interspaces) produced significantly greater infiltration rates and interrill erosion on the sand site, regardless of cover type or season.

During the 1980s, substantial interest was evident in the effects of specialized grazing systems on hydrologic processes (Blackburn et al. 1980). All the research data now available were derived from studies conducted in the Great Plains, almost entirely in the southern plains of Texas and New Mexico. Studies were located on the Edwards Plateau (McGinty et al. 1979; McCalla et al. 1984*a, b*; Thurow et al. 1986, 1988; Warren et al. 1986*a, b*), the Rolling Plains (Wood and Blackburn 1981*a, b*; Pluhar et al. 1987), and Fort Stanton (Gamougoun et al. 1984; Weltz and Wood 1986*a, b*). One study from the central plains (Abdel-Magid et al. 1987) is available. The results of the 14 studies are summarized in Table 16.1.

Considering the several studies grouped by location, Thurow et al. (1986) on the Edwards Plateau affirmed that total organic cover and bulk density of the soil surface were the variables most strongly correlated with infiltration rate, while total aboveground biomass and midgrass bunch growth form were most strongly correlated with sediment production. The primary role of cover or biomass is to decrease kinetic energy of raindrops, reducing the disaggregating effect of rain on the soil surface, and in turn keeping soil pore structure open and serving as a barrier to overland flow. Thus oak mottes (primarily *Quercus virginiana*), compared with bunchgrasses (mainly *Bouteloua curtipendula*, *Stipa leucotricha*, and *Aristida* spp.) and sodgrasses (mainly *Hilaria belangeri*), maintained highest infiltration rates and lowest sediment production under all grazing treatments. (A motte is a grove of trees growing on prairie land. In this study, the oak trees were cut by hand and carefully removed to facilitate equipment access; accumulated oak leaf litter was left on the site.) The heavy continuous and intensive rotation pastures had lower total organic cover and lower total aboveground biomass than moderate continuous and livestock exclusion pastures, and thus had significantly lower infiltration rates and higher sediment production.

These fundamental relations were affirmed under all other Edwards Plateau research (Table 16.1). In addition, it was found that during drought the infiltration rate and interrill erosion deteriorated on heavy continuous and heavily stocked intensive rotation pastures in a stair-step pattern of decreasing condition, with a resultant inability to recover to predrought levels. Infiltration rates were seasonally cyclic as well in both treatments, but no seasonal trend occurred in moderate continuous pasture (Thurow et al. 1988), apparently because of greater cover stability. Infiltration rate declined and

Table 16.1. Summary of grazing system effects on infiltration rates and sediment production.

Site	Soils	Sampling Locations	Grazing System/ Pasture	Grazing Intensity (ha/AU)	Infiltration Rate (f_c) (cm/hr)			Sediment Production (Sp) (kg/ha)			Notes
EDWARDS PLATEAU, TEXAS											
McGinty et al. (1979)[a]	Sandy clay loam	Shallow soil	Continuous	5.0	4.4			211			f_c and Sp averaged by pasture
		Intermediate	Deferred rotation (Merrill 4-pasture)	5.0	10.4			134			
		Deep	Exclusion	—	10.4			160			
Warren et al. (1986*a*, *b*)[b]	Silty clay	Midgrass interspace Shortgrass interspace	Intensive rotation (3-pastures)	4.8	14.3			1,200			f_c and Sp averaged by pastures
					15.1			1,035			
					15.4			1,435			
					A	B		A	B		
McCalla et al. (1984*a*, *b*)[a]	Silty clay	Midgrass interspace (Col. A) Shortgrass interspace (Col. B)	Heavy continuous	0.3–12*	10.5	8.0		3,200	2,600		Data reported for both sampling locations at highest point, midgrass (March 1980)
			Moderate continuous	8.1	19.0	14.0		100	500		
			Short duration	3.2–4.9†	19.5	16.5		100	500		
			Exclusion	—	17.0	9.0		600	1,600		
					A	B	C	A	B	C	
Thurow et al. (1986)[a]	Silty clay	Oak mottes (Col. A)	Heavy continuous	4.6	20.0	—	6.5	74	—	5,600	Data reported for all three sampling locations
		Bunchgrass (Col. B)	Moderate continuous	8.1	20.0	19.5	16.0	2	180	1,400	
		Sodgrass (Col. C)	Intensive rotation	4.6	19.0	13.0	10.0	13	1,600	2,400	
			Exclusion	—	20.0	16.0	14.5	4	220	1,000	

Thurow et al. (1988)[a]	Silty clay	Ten plots randomly assigned to midgrass and shortgrass	Heavy continuous	4.6	6.0			700			Mean annual Sp values from natural rainfall events
			Moderate continuous	8.1	7.6			250			
			Moderate high-intensity, low frequency	4.6	12.8			120/563			
ROLLING PLAINS, TEXAS											
					A	B	C	A	B	C	
Wood and Blackburn (1981*a*, *b*)[a]	Clay	Shrub canopy (Col. A)	Heavy continuous	4.6	15.9	8.1	6.6	22.3	114.6	192.6	Data reported for all three sampling locations
		Midgrass interspace (Col. B)	Moderate continuous	6.5	14.7	11.4	5.1	22.6	27.9	143.6	
			Deferred rotation	6.2	17.0	13.9	7.9	9.7	14.4	62.9	
		Shortgrass interspace (Col. C)	Rested		14.9	13.1	7.2	13.4	9.5	56.1	
			High-intensity, low frequency	6.5	12.8	8.2	6.7	19.4	39.2	77.7	
			Rested	—	13.1	9.6	6.1	17.7	27.8	35.6	
			Exclusion	—	17.2	16.5	—	2.3	4.4	—	
			Exclusion	—	15.7	13.9	—	7.5	17.1	—	
					A	B		A	B		
Pluhar et al. (1987)[b]	Clay loam	Midgrass (Col. A)	Moderate continuous	5.8	8.9	8.5		35	30		Data reported for both sampling locations
		Shortgrass (Col. B)	Deferred rotation	5.8	8.1	6.8		71	54		
			Intensive rotation-14	3.6	6.4	5.5		105	105		
			Intensive rotation-42	3.6	8.5	7.9		75	53		
			Exclusion	—	8.8	—		23	—		
NEW MEXICO											
Gamougoun et al. (1984)[a]	Mixed loam	Random plots	Heavy continuous	—		6.0		130			f_c and Sp averaged by pasture, reported for 1980
			Moderate continuous	2.3		5.8		250			

Table 16.1. Continued.

Site	Soils	Sampling Locations	Grazing System/ Pasture	Grazing Intensity (ha/AU)		Infiltration Rate (f_c) (cm/hr)		Sediment Production (Sp) (kg/ha)		Notes
Gamougoun Cont.			Rotation, grazed	—		4.7		470		
			Exclusion	—		7.3		20		
				A	B	A	B	A	B	
Weltz and Wood (1986*a*, *b*)[c]	Mixed loam (Fort Stanton)	Random plots (Cols. A, B)	Heavy continuous	13.5	—	2.6	—	335	—	f_c and Sp reported for both research locations
	Fine loam (Fort Sumner)		Moderate continuous	18.0	26.6	4.9	5.5	300	73	
			Short duration, rested	14.0	13.3	3.9	7.0	218	26	
			Short duration, grazed	14.0	13.3	2.3	3.8	565	260	
			Exclusion	—	—	7.4	6.3	65	21	
WYOMING										
				A	B	A‡	B‡			
Abdel-Magid et al. (1987)[d]	Sandy loam	Plots at transect (Cols. A, B)	Continuous	3.0	2.25	9.7	7.2	Not measured		Grazing system x year interaction on f_c for 1983 and 1984
			Deferred rotation	3.0	2.25	8.1	9.0			
			Short duration	3.0	2.25	9.0	10.4			

Note. Measurement equipment employed:

[a]Blackburn et al. (1974)

[b]Meyer and Harmon (1979)

[c]Bertrand and Parr (1961)

[d]Haise et al. (1956)

*Stocking rate varied with changes in forage production and breeding season.

†Stocking rate changed due to destocking during 1980 drought.

‡Estimates soil compaction rather than rainfall infiltration.

sediment production increased following the short-term intense grazing periods inherent in intensive rotational management, especially during periods of drought or winter dormancy (Warren et al. 1986*a*). Rest, rather than intensive livestock activity, appears to be the key to soil hydrologic stability (Warren et al. 1986*b*).

Evaluation of heavy and moderate continuous grazing, rested and grazed deferred-rotation, rested and grazed high intensity-low frequency, and grazing exclusion on the Rolling Plains of Texas showed responses consistent with those above (Wood and Blackburn 1981*a*, *b*). All hydrologic effects were closely related to the effect of the several grazing treatments on vegetation standing crop and cover (Pluhar et al. 1987).

In New Mexico, heavy rotation grazing had lower infiltration rates than continuous treatments, while sediment production was similar in all treatments except when livestock were concentrated in heavy rotation. Overall, rotation grazing responded no better than continuous grazing when both were stocked heavily, and grazing exclusion did not confer significant hydrologic advantages over moderate stocking (Gamougoun et al. 1984).

Weltz and Wood (1986*a*, *b*) showed that grazing exclusion produced higher infiltration rates than any grazing treatment. Moderate continuous grazing had higher infiltration rates than short-duration and heavy continuous grazing, which responded similarly. The terminal infiltration rates immediately after short-duration grazing were about half those of the same pasture after resting.

Sediment production was least from grazing exclusion, and was higher but statistically the same for all grazing treatments, except at one location where higher sediment production on short-duration grazing was attributed to reduced biomass and more bare ground.

Summarily, the composite results of these studies appear to support the conclusions of Gifford and Hawkins (1978): (1) grazing influences infiltration and sediment production; (2) heavy grazing nearly always results in lower infiltration rates and higher sediment production; (3) moderate grazing affects infiltration and sediment production, but is frequently not statistically different from grazing exclusion; (4) all effects are related to reduced biomass and increased bare ground; (5) periods of rest after grazing appear to be the key to soil hydrologic stability; and (6) there is considerable spatial variability of hydrologic properties within short distances.

It is likely that BMPs for upland grazing will be based on these relations. The grazing practice should provide moderate grazing intensity within a system that includes seasonal rotation and periodic rest. The relative aridity of most upland ranges means that improvements in soil and vegetation characteristics as a result of management will tend to be slow.

Grazing of Riparian Areas

Riparian areas are lands directly influenced by the presence of flowing water—creeks, streams, rivers, ponds, lakes, and other bodies of surface or subsurface water. Although such lands occupy only a small portion of the total rangeland area, frequently regarded as less than 1% (Chaney et al. 1990), the physical and vegetative characteristics conferred by the presence of water make riparian areas not only different but disproportionately important in the overall ecological relations of the landscape. Thomas et al. (1979) described three characteristics common to western rangeland riparian areas in addition to small relative size: (1) well-defined habitat zones within a larger and much drier area, (2) much higher productivity in both plant and animal communities than the drier area, and (3) a critical source of plant and animal diversity within rangelands. These factors underlie the management of many ecologically and economically important riparian uses.

Grazing by domestic livestock is, of course, the most extensive use of riparian areas geographically. It may be said that riparian areas have made possible the general use by livestock of relatively arid uplands throughout the western rangeland region. Livestock, especially cattle, are drawn to riparian areas within an otherwise arid landscape for the same reasons other animals are drawn there. Besides readily available water, they find shade and protective cover, as well as a much wider diversity, greater quantity, and higher quality of forage. In addition, the inherent qualities of riparian areas are attractive for a wide variety of human activities. Consequently, riparian areas are the most modified land form in the West (Chaney et al. 1990), and the resulting environmental and economic costs have engaged increasing attention (Elmore, this volume).

The numerous publications dealing with riparian concerns, ranging from scientific investigation through management suggestions to personal opinion, indicate the degree of attention given to the land type. It is well beyond the purpose of this article to describe that voluminous literature, but it is clear that a large portion of it is devoted to the effects of livestock grazing on riparian zones, especially on public lands. Among the grazing publications are comprehensive reviews of literature, including Skovlin (1984) and Kauffman and Krueger (1984). The subject of livestock use and management in riparian areas also has formed an important part of many symposia, workshops, and special programs (e.g., Johnson and McCormick 1978, Nelson and Peek 1981, Johnson et al. 1985, Gresswell et al. 1989). And riparian areas are receiving extensive management attention (United States Bureau of Land Management 1990, Elmore and Beschta 1987) .

The effects of livestock grazing on riparian habitats may be divided into four main areas of concern (Platts 1989): (1) numerous alterations of streamside soils and vegetation, including soil compaction and changes in vegetation (decreased vigor, lower biomass, changed species composition) which reduce infiltration rates and other riparian functions; (2) changing channel morphology, including either wider and more shallow streams or entrenched streams, depending on substrate, which may affect water flow characteristics; (3) altered characteristics of the water column itself, including warmer temperatures, greater nutrient loading, higher bacterial counts, and increased suspended sediment, all directly affecting water quality; and (4) degradation of streambanks through caving and sloughing, resulting in increased sediment pollution.

In recent investigations of streamside soils and vegetation, Bohn and Buckhouse (1985) evaluated responses of riparian soils to grazing management. Small riparian pastures were moderately stocked at 3.2 ha/AUM (animal unit month), to achieve 70% utilization, and managed under either four-pasture rest rotation, deferred rotation, continuous (season-long), or exclusion over five years. Rest-rotation grazing and no grazing favored higher infiltration rates and lower sediment production, while deferred rotation and continuous grazing had neutral or negative effects. These results were regarded as the process of recovery linked with a period of rest also found in other studies (Gifford 1981).

Kauffman et al. (1983*a*) studied livestock impacts on riparian plant community composition, structure, and productivity on fall-grazed pasture stocked at 1.3 to 1.7 ha/AUM. After three years, four plant communities of ten sampled (including two meadow types and a shrub type), showed significant differences in standing phytomass compared with excluded areas. These pastures were utilized more heavily than the others, but few differences were found otherwise. Studies have shown that the impact of cattle on riparian meadows tends to follow a seasonal trend dependent on both their behavior and the degree of utilization of streamside vegetation (Gillen et al. 1985, Marlow and Pogacnik 1986).

Clifton (1989) reported on the effects of vegetation and land use on channel morphology, noting that spatial and temporal variability in mountain streams may be attributed to local prevailing conditions, including composition and structure of vegetation, physiography, and land use. Comparisons of grazed and ungrazed reaches and a forested reach with large organic debris showed greatest width in the forested reach and maximum depth in the ungrazed reach. Spatial variability resulted from vegetative conditions, temporal variability from changes in grazing management.

Efforts to further define grazing effects on the water column include Green and Kauffman (1989), who studied the biological interaction between riparian vegetation, groundwater, and microorganisms involved in nutrient uptake. They hypothesized that decreasing the residence time of water in the riparian zone (through erosional downcutting as a result of poor grazing management) reduces the potential for biological interaction, resulting in sharply altered patterns of material loading and, more generally, declines in resource values associated with water quality. Maintaining management systems that allow biological processes to approach a natural equilibrium will best ensure a stable output of resource values. An example of grazing management designed to improve or restore riparian functions including biological interactions was provided by Tohill and Dollerschell (1990).

Bacterial contamination by grazing animals has received less recent study, although the variation of bacterial populations in stream water flowing from watersheds subjected to livestock grazing has been examined. In one study

(Skinner et al. 1984*a*), bacterial counts were not significantly different between sampling sites on individual streams. Seasonal variations occurred within a stream and between streams, but only the former could be related directly to runoff. Bacterial counts serving as indicators of fecal pollution were low.

In a second study (Skinner et al. 1984*b*), bacterial indicators of pollution varied in their ability to detect change between grazing treatments (continuous and deferred rotation) as well as between streams. Fecal coliform and streptococci indicated nonpoint source pollution, but sample variation that could not be fully accounted for by differences in grazing treatment may be partly explained by beaver (*Castor canadensis*) damming of streamflow. These studies support the observations of many researchers (e.g., Doran and Linn 1979) that bacterial counts decline rapidly in streams after removal of livestock, and that variations in bacterial indicators are affected by a number of factors in addition to grazing, including rainfall runoff (Jawson et al. 1982) and bottom sediments (Stephenson and Rychert 1982). Hussey et al. (1986) found little change in the bacteriological water quality in streams of mountain watersheds in Wyoming after ten years of grazing management, recreational activities, and wildlife use.

Impacts of cattle grazing on streambanks were studied by Kauffman et al. (1983*b*). Streambank erosion, disturbance, and undercutting were compared between grazing and exclusion treatments, vegetation types (herbaceous, shrub, and tree), and meander positions. Stocking rate was approximately 1.3 to 1.7 ha/AUM. No significant differences were found between vegetation types or meander locations, but significantly greater erosion and disturbances occurred in grazed areas, somewhat at variance with results found elsewhere with similar utilization (Buckhouse et al. 1981). This discrepancy probably illustrates the differences among streams in susceptibility to disturbance.

Summarily, the composite results of these and many other studies of the effects of livestock grazing on riparian habitats appear to support the conclusion of Platts (1989) that riparian-sensitive grazing strategies should include one or more of the following features: (1) including a riparian pasture within a grazing system to allow the riverine-riparian habitat to be managed separately from the uplands; (2) fencing streamside corridors to allow stream-riparian habitats to rehabilitate (a last resort); (3) changing the kind of livestock (from cattle to sheep on certain ranges) for better grazing compatibility with rangeland types and other uses; (4) adding more rest to the grazing cycle; (5) reducing intensity of use on streamside forage; (6) controlling the timing (often season) of forage use so grazing occurs during periods most compatible with riverine-riparian processes; and (7) managing grazing programs as specified and required in properly prepared allotment management plans.

Development of BMPs based on these features will reduce the effects of livestock grazing on all four major points of stress (riparian vegetation, channel morphology, water column, and streambanks) of riparian habitats. Given the presence of abundant water, improvement of riparian zones as a result of management should be rapid.

Range Improvements

Treatments to improve range condition and productivity, commonly called range improvements, are often necessary in coordinated management programs. They can be best considered as special aids in achieving the objectives of range management (Vallentine 1989). As such, they must be tools to reach a desired level of productivity based on technical, economic, cultural, political, and social factors (Herbel 1983).

In an extensive review and appraisal of range improvements, Vallentine (1989) suggested that the benefits obtainable from range improvements include: (1) increased quantity of forage, (2) increased quality of forage, (3) increased animal production, (4) greater facility in handling and caring for range animals, (5) control of poisonous plants, (6) reduced fire hazard, (7) increased water yields, (8) control of erosion by stabilizing erosive soils, and (9) reduced conflicts between multiple uses of range resources.The full expected benefits from each practice accrue only when accompanied or followed by good grazing management. At the same time, well-planned and conducted programs of range improvement confer benefits on other range uses, such as wildlife habitat, especially on public land (Herbel 1984).

An extensive literature on range improvements is available, with a substantial portion reviewed by Vallentine (1989). The effects of range improvement on hydrological characteristics have received considerable study, including the influence of prescribed burning in piñon-juniper woodland (Roundy et al. 1978), use of herbicides in sagebrush rangeland (Balliette et al. 1986), mechanical treatment of southwestern rangeland (Tromble 1980), sagebrush clearing and reseeding in the Great Basin (Brown et al. 1985), and snow trapping in the Great Plains (Ries and Power 1981).

It is beyond the scope of this chapter to attempt a detailed review of the large improvement literature, other than to note that hydrologic response to range improvements varies with the physical and biological site conditions, the degree of planning and application of the treatment, and the posttreatment climatic conditions and grazing management (Vallentine 1989). Differences in risks and benefits between extensive practices (e.g., manipulation of animals and prescribed fire) and intensive practices (e.g., control of unwanted plants and revegetation) extend to hydrological effects as well. When properly conducted, intensive practices often result in much higher increases in vegetative productivity (Herbel 1983). The resultant increase in biomass may be expected to increase infiltration rates and decrease sediment production correspondingly.

When properly planned, carefully applied, and correctly managed, range improvements play an important role in pollution control programs. In this sense, every improvement technique is a potential BMP for abatement of nonpoint source pollution.

Range Management and Water Quality

It is apparent from the foregoing that water management goals on rangeland will depend on the degree of reference to upland watersheds or riparian areas. For uplands, management will be focused on programs to attain maximum on-site use of precipitation. Healthy, diverse site vegetation will produce greater total aboveground biomass, which in turn will provide greater protection against rain splash, maintenance of soil physical properties, higher infiltration rates, and lower interrill erosion. If improvements in these characteristics are attained, range management can reduce or eliminate upland sources of pollution to rangeland water bodies.

For riparian areas, management will emphasize programs to reduce or eliminate sources of pollution directly affecting surface water. These include maintenance or improvement of streamside vegetation to facilitate riparian functions, protection and stabilization of streambanks to maintain or improve channel morphology, and reduction of negative effects on the water column itself. Attainment of these objectives will contribute directly to better water quality.

The Nature of Rangeland BMPs

Although development of best management practices for livestock grazing is in one sense a young program, in another and fundamentally more important sense it is an old approach. From its advent in western ecosystems until very recently, livestock grazing has focused on upland resources and needs. Not so long ago, many riparian areas and other sources of livestock water were regarded as "sacrifice areas," or at best ignored in development of management programs designed for upland conditions. Since establishment of the U.S. Forest Service in 1906 and the U.S. Grazing Service in 1934 (later the Bureau of Land Management in 1946), management of livestock grazing, especially in the last two decades, has been a search for best management practices. More particularly, grazing management on many private and nearly all public lands is now cast within the concept of an allotment management plan for grazing. Generically, the allotment management plan is based on resource concerns previously identified for the geographic area, and prescribes a management program for the entire unit. The specifications of management—the site-specific practices to implement the plan—may be regarded as BMPs focused on grazing management systems and range improvements.

Due to the relative lack of site-specific information, rangeland BMPs, like management practices generally, must be based on a professional understanding of ecological principles. They fall into three general categories (Clawson and George 1990).

1. *Process*. These are procedures-oriented practices through which site-specific methods can be developed to achieve water quality standards. For example, a prescribed change in grazing management to achieve improvement of riparian functions may reach its primary goal while developing one or more specific practices at the same time.

2. *Prescription*. These are practices through which measurable management objectives on specific sites can be attained. For example, planting shrubs to stabilize a critical section of streambank may be an effective practice under appropriate conditions. The physical complexity of rangeland sites, however, may lead to prescriptive BMPs too numerous for effective management.

3. *Performance*. These practices are goal oriented but difficult to monitor and enforce. BMPs stated too generally immediately encounter both spatial and temporal difficulties of implementation.

Because BMPs are usually a combination of practices tailored to meet established standards, a management suite of BMPs generally contains one or more of each of the three types, combined into a general program of recommended on-site practices. Most of the programs will be process or prescription BMPs, usually either systems of grazing management or different kinds of range improvements. Their application, along with monitoring of results and subsequent amendments, constitutes a kind of successive-approximation approach toward established standards. As such, BMPs themselves provide useful additional information over time. Improvements in vegetation characteristics through best management practices are often directly reflected in improved hydrologic processes. The converse is equally true; lands managed toward a lower range condition will as often reflect lower hydrologic condition.

Water Quality Programs: The Idaho Approach

Statutory Overview

In concert with the other states, Idaho derives much of its regulatory framework regarding water quality from the state's federal involvement in the national water pollution control efforts under the Federal Water Pollution Control Act of 1972, as amended by the Water Quality Act of 1987 (the Clean Water Act, 33 U.S.C.; 1251 et seq.). Pursuant to these statutes, the state seeks to prevent impairment of the beneficial uses of water as a matter of public trust and protection of public health and welfare. Methods include the prevention of pollution by enforcing water quality standards (including best management practices).

It is important to note that Idaho's water policy is directed toward maintaining the beneficial uses of water—as opposed to clean water per se. The list of beneficial uses includes the traditional consumptive values—domestic, municipal, commercial, and irrigation—and newer instream values for fish and wildlife habitat (including spawning and rearing of salmonid populations), recreation, aesthetics, transportation, and water quality itself.

Because of this fundamental orientation, the state's strategy is directed at water quality to support beneficial uses through a coordinated two-pronged approach (Turner and O'Laughlin 1991): (1) resource protection of streams and channels by limiting development, and (2) pollution control by preventing contamination or remedying existing water degradation.

The resource protection thrust is implemented primarily through laws regulating minimum streamflows and stream channel alterations, and regulation of specific activities such as well drilling, dredging, and surface mining. The Idaho Water Resource Board may protect streams as "natural" (no dams or stream alteration), "recreational" (special consideration of dams or stream alteration), or "interim protected" (pending protection).

The pollution control thrust, more relevant to the present writing, is founded primarily in the Idaho Environmental Protection and Health Act of 1972, which directs the establishment of an antidegradation policy and water quality standards for prevention and abatement of pollution. The Idaho Water Pollution Abatement Act of 1970 establishes activities directed toward nonpoint source pollution control.

Pollution Control

The Antidegradation Policy

As required by Section 303 of the Clean Water Act, Idaho has developed an antidegradation policy to prevent pollution by (1) protecting water quality sufficient to maintain all existing uses, (2) allowing lowered quality for important new uses if existing uses are maintained, and (3) maintaining and protecting "high quality" waters. These stipulations have been promulgated in a State Antidegradation Agreement (Governor's Executive Order 88–23) hammered out during several years of difficult negotiations between competing interests (Forge 1989). A full copy of the final agreement and the executive order are presented as appendices A and B in Dunn (1990).

The main features of the agreement center on the establishment of six biennial Basin Area Meetings which, in addition to considering technical, social, and economic trends affecting water quality, accept nominations for "stream segments of concern" within each basin. Upon acceptance by the Water Quality Advisory Working Committee (composed of 18 gubernatorial appointees representing agency and resource interests), these segments receive additional emphasis in abating nonpoint sources of pollution. Nearly 200 stream segments of concern from the six basin areas of the state were designated by the committee (Dunn 1990).

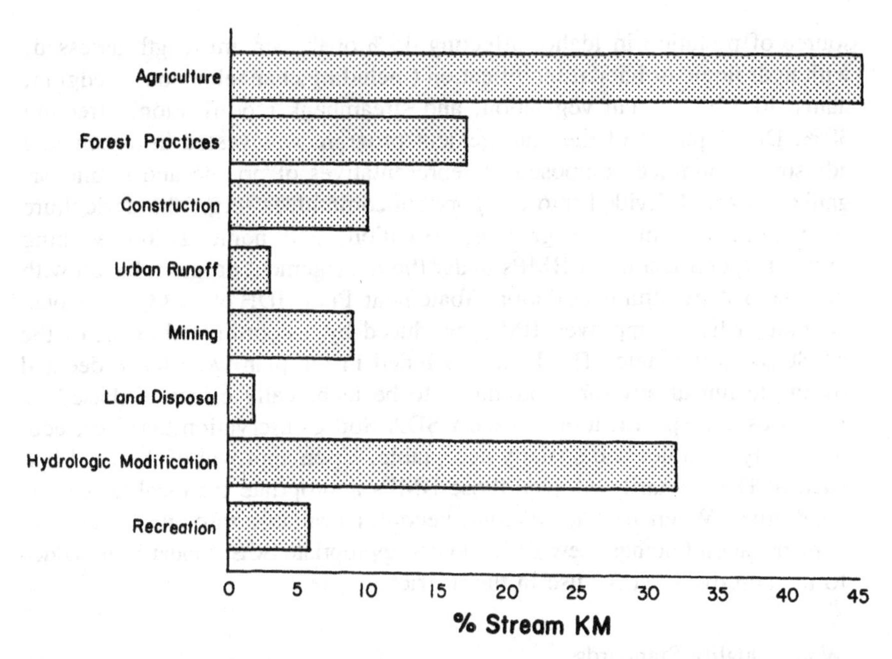

FIGURE 16.1. Categories of nonpoint source pollution affecting beneficial uses of Idaho waters, in percentage of stream kilometers assessed (modified from Idaho Department of Health and Welfare 1989).

High quality or outstanding resource waters recommended for designation by the legislature may be protected from all nonpoint sources of pollution that may lower (a statistically significant change relative to a beneficial use) water quality. Nominations for outstanding resource water designation have been made, but so far none have been designated by the legislature.

The agreement also establishes a water quality monitoring program, to be developed by a monitoring working committee, and provides for review and evaluation every three years, beginning in 1991. The Division of Environmental Quality (DEQ) within the Idaho Department of Health and Welfare (IDHW) is designated as lead agency in implementing the general provisions of the antidegradation policy, including development of water quality monitoring programs. The Idaho Department of Lands (IDL) is responsible for implementing policy requirements in mining and forestry practices, while the Soil Conservation Commission is charged with coordinating the agricultural (including grazing) provisions of the policy through the Soil Conservation Districts (SCDs).

Idaho developed a Nonpoint Source Management Program (Bauer 1989) that was approved by the Environmental Protection Agency in 1989. The management program is directed at sources of pollution identified in the initial assessment (Figure 16.1). Agriculture is by far the largest nonpoint source of pollution in Idaho, affecting 45% of the stream length assessed, with hydrologic or habitat modification (including channelization, dredging, dams, loss of riparian vegetation, and streambank modification) affecting 30%. Development of the management program was vested in a technical advisory committee composed of representatives of private and public organizations and divided into category subcommittees (e.g., the agriculture subcommittee containing grazing, irrigation, and nonirrigation working groups). Identification of BMPs under the management program began with the Idaho Agricultural Pollution Abatement Plan (IDHW 1983). The plan contains a list of approved BMPs produced by a coordinated effort of the 51 SCDs in the state. The BMPs included in the plan were those deemed by the technical advisory committee to be technically effective (based on standards and specifications of the USDA Soil Conservation Service), economically feasible, and socially acceptable. Upon approval of the final list, each SCD voluntarily adopted those BMPs appropriate for local needs and conditions. When new technology becomes available through research or demonstration findings, new BMPs under appropriate SCS standards are added to the approved list for use in the districts.

Water Quality Standards

Nonpoint sources of pollution are managed under Idaho water quality legislation by applying best management practices to all land uses generating pollution. Federal land management agencies are required by the Clean Water Act to comply with state water quality standards and to be consistent with the nonpoint source management program of the state. BMPs are mandated for all forestry and mining practices, under the supervision of the Idaho Department of Lands. BMPs for agriculture, including grazing, are voluntary and vested in development of water quality plans covering lands within the boundaries of individual SCDs. The Idaho Soil Conservation Commission has responsibility to work with the districts in developing and coordinating plans and to administer cost-share programs for BMP implementation.

A water quality project developed by a SCD typically will follow the same pattern employed in application of all conservation practices within the district. Practices (BMPs) for a number of individual farms or ranches designed to retain soil on the field are negotiated between cooperating owners and the district. The BMPs will most likely include both hydrological and agronomic treatments chosen to fit the physical conditions obtaining on each site. For grazing operations, as noted earlier, the BMPs are likely to include a grazing system designed to improve infiltration rates and decrease sediment production on both uplands and bottomlands, together with appropriate range improvements designed to maintain or improve soil structure and vegetative production. Typically installation and operating costs will be assumed by both the cooperator and the district, using money from the Water Pollution Control Account granted by IDHW or conservation loan funds.

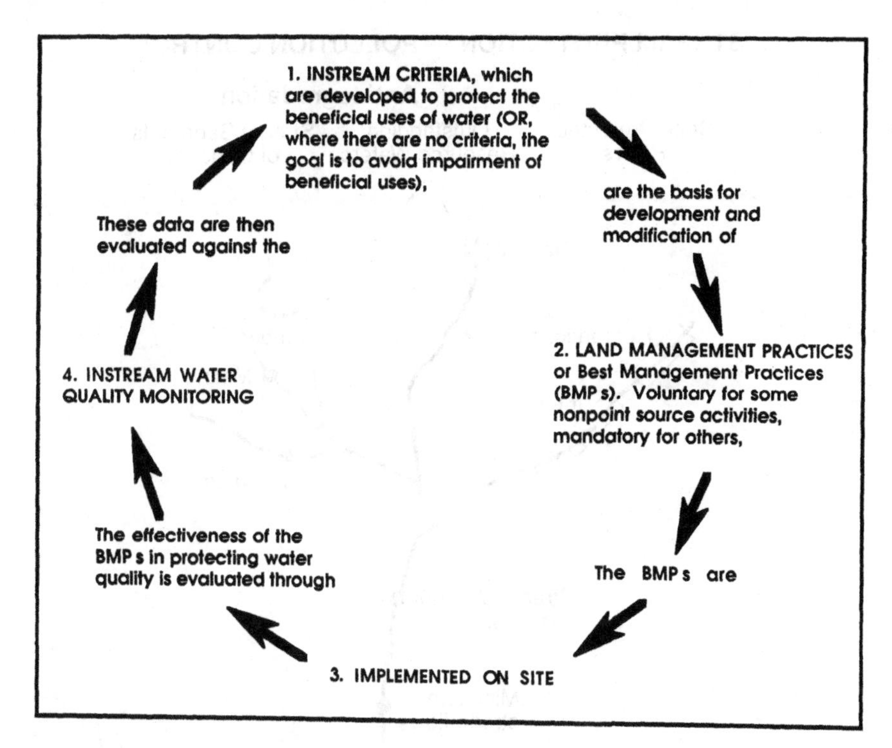

FIGURE 16.2. The feedback loop process for nonpoint source control (modified from Clark 1990).

Approximately 30 water quality projects are in some phase of application within the 51 SCDs in Idaho (Turner and O'Laughlin 1991). All are directed toward abating agricultural effects on existing beneficial uses of water within stream segments of concern identified under the state's antidegradation policy.

Evaluation of the water quality project (and of the BMPs within it) is based on a feedback process (Figure 16.2) incorporated in the Idaho Water Quality Standards affecting nonpoint source pollution control. Its purpose is to upgrade BMPs in their relative effectiveness in reducing or controlling nonpoint source pollution, especially sediments and nutrients (Clark 1990).

Currently, formal evaluation (monitoring) is required only at the project level and not for the individual components of the project. Long-term evaluation of individual BMPs in meeting project goals will most likely prove helpful and effective, especially if conducted within the voluntary context of the SCD program. Determining effectiveness of BMPs requires expertise in resource management as well as water quality assessment for sedimentary, chemical, and bacterial loading. In turn, this would tend to require a comprehensive monitoring program, still in the process of being developed.

Turner and O'Laughlin (1991) observe that long-term professional monitoring of BMPs will probably be more useful to the feedback process than

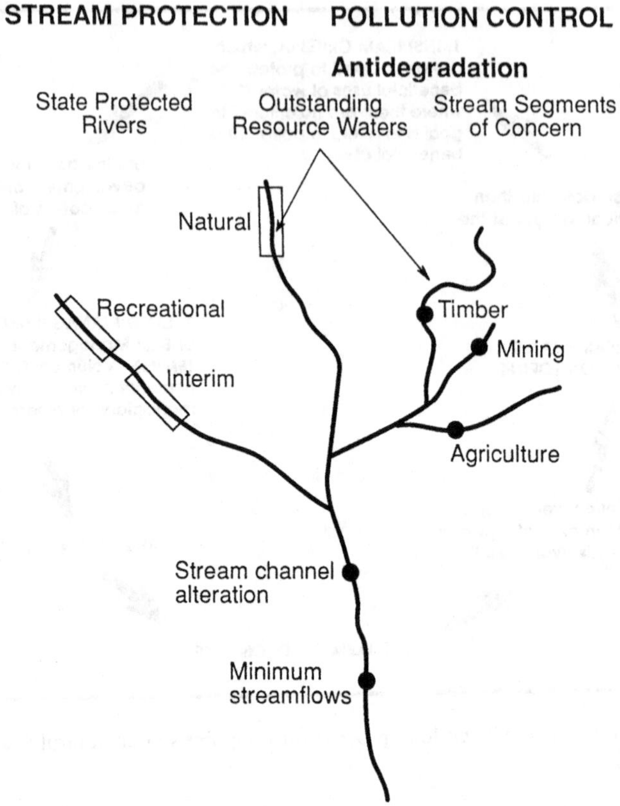

STREAM PROTECTION POLLUTION CONTROL

Antidegradation

State Protected Rivers
Outstanding Resource Waters
Stream Segments of Concern
Natural
Recreational
Interim
Timber
Mining
Agriculture
Stream channel alteration
Minimum streamflows

FIGURE 16.3. The complementary effects of stream protection and pollution control on a generalized watershed (modified from Turner and O'Laughlin 1991).

short-term technical analysis of instream criteria. Fairly broad-based estimates of effectiveness gained through field surveys would probably serve to improve BMPs through at least the first or second successive approximations.

The Agricultural Water Quality Program: An Appraisal

The Idaho water quality program appears to provide a complex but reasonably effective approach to management of nonpoint source pollution. The two tiers of the strategy to protect beneficial uses—stream protection and pollution control—complement each other if actively employed. Stream protection works to protect the stream in its channel while pollution control is directed at the water itself. A generalized application of the two approaches is shown as Figure 16.3.

As applied to agricultural lands, it appears that pollution control is the most important feature of the strategy, given the likelihood that protected rivers (either natural, recreational, or interim) will be located in timbered headwater areas. A similar qualification may attach to designation of outstanding resource waters under the antidegradation policy. But the process leading to designation of stream segments of concern has and will continue to focus attention on agriculturally based nonpoint sources of pollution. On such designated reaches as well as on agriculturally affected streams generally, it is the express policy of the State Board of Health and Welfare to address pollution emanating from agricultural lands by financially assisting soil conservation districts to develop abatement programs. Given the mission and organization of the districts, this policy necessarily means that pollution control programs must be cast within voluntary participation.

The voluntary nature of the Agricultural Water Quality Program seems to many to be a weakness and perhaps a fatal flaw (Ford 1989). But given the scale of agricultural effects (see Figure 16.1), and the thousands of individual agriculturists involved, a voluntary program may well be the only feasible approach. In addition, understanding of the physical disposition of pollutants in soil and water systems is incomplete, which means in turn that there is a lack of certainty that policy mandates will effectively resolve pollution problems (Reichelderfer 1991). In practical terms, this means that Idaho will focus its limited resources on mandated BMPs for forestry and mining, leaving the much larger and more diffuse agricultural nonpoint sources to the workings of a long-established conservation system. Ford (1989) quotes an EPA staff member as saying: "Any nonpoint control scheme, much more than point source control, rides on choices. Idaho is choosing where to focus very limited time and money."

The adequacy of the voluntary program in reducing nonpoint source pollution will reflect the technical effectiveness of the BMPs employed and, even more important, the degree of program acceptance by agriculturists. Ultimately, the acceptance level will derive from several personal and institutional factors (Turner and O'Laughlin 1991), expressed in the degree to which the programs individually and collectively maintain or improve downstream water quality. Perhaps the most applicable analogy is the Dust Bowl-induced experiment in applying soil conservation methodology to the nation's cropping systems through voluntary, cost-share-encouraged cooperation. While that program has not met expectations entirely (soil erosion is still a considerable problem on the nation's farms), it has achieved a great deal. It is reasonable to expect something similar from this program: it will not do everything, but it will do much to abate agricultural nonpoint source pollution.

But one must remember that the Idaho water quality program, while reasonably clear in its intent, is still largely a future program in execution. There has been only one designation of stream segments of concern, and no outstanding resource waters have been named. Local working committees are still designing sets of BMPs to employ on timbered stream segments of concern. There are just over half as many agricultural water quality projects as there are soil conservation districts in Idaho. The long-term monitoring program to make the feedback process functional is currently under development, but is still some distance from being operational. There are few data and fewer results in hand—yet. But Idaho will be among the first states to try to define BMPs to reach water quality standards, not just for a few stream segments of concern, or a few watersheds, but for the whole state: Idaho is on the leading edge of nonpoint pollution abatement in the West. There will be mistakes made, deficiencies uncovered, impossibilities raised. But it is fair to say that there will also be tangible gains derived from the simple fact that the state is doing something. Given the enormous social, economic, and political difficulties attending an extraordinarily complex physical situation, that is a considerable achievement, and one that may well point the way for other states.

References

Abdel-Magid, A.H., G.E. Schuman, and R.H. Hart. 1987. Soil bulk density and water infiltration as affected by grazing systems. Journal of Range Management **40**:307–309.

Achouri, M., and G.F. Gifford. 1984. Spatial and seasonal variability of field measured infiltration rates on a rangeland site in Utah. Journal of Range Management **37**:451–455.

Balliette, J.F., K.C. McDaniel, and M.K. Wood. 1986. Infiltration and sediment production following chemical control of sagebrush in New Mexico. Journal of Range Management **39**:160–165.

Bauer, S.B. 1989. Idaho Nonpoint Source Management Program 1989. Idaho Department of Health and Welfare, Boise, Idaho, USA.

Bertrand, A.R., and J.F. Parr. 1961. Design and operation of the Purdue sprinkling infiltrometer. Purdue University Research Bulletin 723, Lafayette, Indiana, USA.

Blackburn, W.H. 1984. Impacts of grazing intensity and specialized grazing systems on watershed characteristics and responses. Pages 927–983 *in* Developing strategies for rangeland management. National Research Council/National Academy of Sciences. Westview Press, Boulder, Colorado, USA.

Blackburn, W.H., R.W. Knight, M.K. Wood, and L.B. Merrill. 1980. Watershed parameters as influenced by grazing. Pages 552–572 *in* Symposium on Watershed Management 1980. American Society of Civil Engineers, New York, New York, USA.

Blackburn, W.H., R.O. Meeuwig, and C.M. Skau. 1974. A mobile infiltrometer for use on rangeland. Journal of Range Management **27**:322–323.

Bohn, C.C., and J.C. Buckhouse. 1985. Some responses of riparian soils to grazing management in northeastern Oregon. Journal of Range Management **38**:378–381.

Branson, F.A., G.F. Gifford, K.G. Renard, and R.F. Hadley. 1981. Rangeland hydrology. Second edition. Society for Range Management Range Science Series 1, Denver, Colorado, USA.

Brown, J.D., R.A. Evans, and J.A. Young. 1985. Effects of sagebrush control methods and seeding on runoff and erosion. Journal of Range Management **38**:195–199.

Buckhouse, J.C., J.M. Skovlin, and R.W. Knight. 1981. Streambank erosion and ungulate grazing relationships. Journal of Range Management **34**:339–340.

Busby, F.E., and G.F. Gifford. 1981. Effects of livestock grazing on infiltration and erosion rates measured on chained and unchained pinyon-juniper sites in southeastern Utah. Journal of Range Management **34**:400–405.

Chaney, E., W. Elmore, and W.S. Platts. 1990. Livestock grazing on western riparian areas. U.S. Environmental Protection Agency, Washington, D.C., USA.

Clark, W.H. 1990. Coordinated nonpoint source water quality monitoring program for Idaho. Idaho Department of Health and Welfare, Boise, Idaho, USA.

Clawson, W.J., and M.R. George. 1990. Best management practices (BMPs) – control of nonpoint source pollution (NPS) for rangeland. California Cooperative Extension Range Resources Notes **14** (2):3–6.

Clifton, C. 1989. Effects of vegetation and land use on channel morphology. Pages 121–129 *in* R.E. Gresswell, B.A. Barton, and J.L. Kerschner, editors. Practical approaches to riparian resource management: an educational workshop. United States Bureau of Land Management, Billings, Montana, USA.

Cottam, W.P., and F.F. Evans. 1945. A comparative study of the vegetation of the grazed and ungrazed canyons of the Wasatch Range, Utah. Ecology **26**:171–181.

Devaurs, M., and G.F. Gifford. 1984. Variability of infiltration within large runoff plots on rangelands. Journal of Range Management **37**:523–528.

Doran, J.W., and D.M. Linn. 1979. Bacteriological quality of runoff water from pastureland. Applied and Environmental Microbiology **37**:985–991.

Dunn, A.K. 1990. Water Quality Advisory Working Committee designated stream segments of concern. Idaho Department of Health and Welfare, Boise, Idaho, USA.

Ellison, L. 1960. Influence of grazing on plant succession of rangelands. Botanical Review **26**:1–78.

Elmore, W., and R.L. Beschta. 1987. Riparian areas: perceptions in management. Rangelands **9**:260–265.

Ford, P. 1989. Idaho points the way to stream quality. High Country News **21**(23):15–17.

Forge, G.W. 1989. Idaho's antidegradation program: a model. Pages 49–53 *in* Water quality standards for the 21st century: proceedings of a national conference. United States Environmental Protection Agency, Washington, D.C., USA.

Forsling, C. L. 1928. The soil protection problem. Journal of Forestry **26**:944–997.

Gamougoun, N.D., R.P. Smith, M.K. Wood, and R.D. Pieper. 1984. Soil, vegetation, and hydrologic responses to grazing management at Fort Stanton, New Mexico. Journal of Range Management **37**:538–541.

Gifford, G.F. 1981. Watershed responses to grazing management. Pages 147–159 *in* D.M. Baumgartner, editor. Interior west watershed management. Cooperative Extension, Washington State University, Pullman, Washington, USA.

Gifford, G.F., and R.H. Hawkins. 1978. Hydrologic impact of grazing on infiltration: a critical review. Water Resources Research **14**:305–313.

Gillen, R.L., W.C. Krueger, and R.F. Miller. 1985. Cattle use of riparian meadows in the Blue Mountains of northeastern Oregon. Journal of Range Management **38**:205–208.

Green, D.M., and J.B. Kauffman. 1989. Nutrient cycling at the land-water interface: the importance of the riparian zone. Pages 61–68 *in* R.E. Gresswell, B.A. Barton, and J.L. Kerschner, editors. Practical approaches to riparian resource management: an educational workshop. United States Bureau of Land Management, Billings, Montana, USA.

Gresswell, R.E., B.A. Barton, and J.L. Kerschner, editors. 1989. Practical approaches to riparian resource management: an educational workshop. United States Bureau of Land Management, Billings, Montana, USA.

Haise, H.F., W.W. Donnan, J.T. Phelan, L.F. Lawhon, and D.G. Shockley. 1956. The use of cylindrical infiltrometers to determine the intake characteristics of irrigated soils. United States Agricultural Research Service ARS 41–47.

Herbel, C.H. 1983. Principles of intensive range improvements. Journal of Range Management **36**:140–144.

Herbel, C.H. 1984. Manipulative range improvements: summary and recommendations. Pages 1167–1178 *in* Developing strategies for rangeland management. National Research Council/National Academy of Sciences. Westview Press, Boulder, Colorado, USA.

Hussey, M.R., Q.D. Skinner, and J.C. Adams. 1986. Changes in bacterial populations in Wyoming mountain streams after 10 years. Journal of Range Management **39**:369–370.

Idaho Department of Health and Welfare. 1983. Idaho agricultural pollution abatement plan. Boise, Idaho, USA.

Idaho Department of Health and Welfare. 1989. Idaho water quality status report and nonpoint source assessment. Boise, Idaho, USA.

Jawson, M.D., L.F. Elliott, K.E. Saxton, and D.H. Fortier. 1982. The effect of cattle grazing on indicator bacteria in runoff from a Pacific Northwest watershed. Journal of Environmental Quality **11**:621–627.

Johnson, R.R., and J.F. McCormick, technical coordinators. 1978. Strategies for protection and management of flood plain wetlands and other riparian ecosystems. United States Forest Service General Technical Report WO-12.

Johnson, R.R., C.D. Ziebell, D.R. Patton, D.F. Ffolliott, and R.H. Hamre, technical coordinators. 1985. Riparian ecosystems and their management: reconciling conflicting uses. United States Forest Service General Technical Report RM-120.

Kauffman, J.B., and W.C. Krueger. 1984. Livestock impacts on riparian ecosystems and streamside management implications . . . a review. Journal of Range Management **37**:430–438.

Kauffman, J.B., W.C. Krueger, and M. Vavra. 1983*a*. Impacts of cattle on streambanks in northeastern Oregon. Journal of Range Management **36**:683–685.

Kauffman, J.B., W.C. Krueger, and M. Vavra. 1983*b*. Effects of late season cattle grazing on riparian plant communities. Journal of Range Management **36**:685–691.

Love, L.D. 1958. Rangeland watershed management. Proceedings, Society of American Foresters **1958**:198–200.

Marlow, C.B., and T.M. Pogacnik. 1986. Cattle feeding and resting patterns in a foothills riparian zone. Journal of Range Management **39**:212–217.

McCalla, G.R., II, W.H. Blackburn, and L.B. Merrill. 1984*a*. Effects of livestock grazing on infiltration rates, Edwards Plateau of Texas. Journal of Range Management **37**:265–269.

McCalla, G.R., II, W.H. Blackburn, and L.B. Merrill. 1984*b*. Effects of livestock grazing on sediment production, Edwards Plateau of Texas. Journal of Range Management **37**:291–294.

McGinty, W.A., F.E. Smeins, and L.B. Merrill. 1979. Influence of soil, vegetation and grazing management on infiltration rate and sediment production of Edwards Plateau rangeland. Journal of Range Management **32**:33–37.

Meyer, L.D., and W.C. Harmon. 1979. Multiple-intensity rainfall simulator for erosion research on row sideslopes. Transactions of the ASAE **22**:100–103.

Nelson, L. Jr., and J.M. Peek, co-chairs. 1981. Proceedings of the wildlife-livestock relationships symposium. Forest, Wildlife and Range Experiment Station, University of Idaho, Moscow, Idaho, USA.

Platts, W.S. 1989. Compatability of livestock grazing strategies with fisheries. Pages 103–110 *in* R.E. Gresswell, B.A. Barton, and J.L. Kerschner, editors. Practical approaches to riparian resource managment: an educational workshop. United States Bureau of Land Management, Billings, Montana, USA.

Pluhar, J.J., R.W. Knight, and R.K. Heitschmidt. 1987. Infiltration rates and sediment production as influenced by grazing systems in the Texas Rolling Plains. Journal of Range Management **40**:240–244.

Reichelderfer, K. 1991. Agriculture and water quality: Is a little knowledge good or dangerous? Renewable Resources Journal **9**:7–11.

Rich, J.L. 1911. Recent stream trenching in the semi-arid portion of southwestern New Mexico, a result of removal of vegetation cover. American Journal of Science **32**:237–245.

Ries, R.E., and J.F. Power. 1981. Increased soil water storage and herbage production from snow catch in North Dakota. Journal of Range Management **34**:485–488.

Roundy, B.A., W.H. Blackburn, and R.E. Eckert, Jr. 1978. Influence of prescribed burning on infiltration and sediment production in the pinyon-juniper woodland, Nevada. Journal of Range Management **31**:250–253.

Sampson, A.W., and L.H. Weyl. 1918. Range preservation and its relation to erosion control on western grazing lands. United States Department of Agriculture Bulletin 675.

Sharma, M.L., G.A. Gander, and C.G. Hunt. 1980. Spatial variability of infiltration in a watershed. Journal of Hydrology **45**:101–122.

Skinner, Q.D., J.C. Adams, A.A. Beette, and G.P. Roehrkasse. 1984*a*. Change in bacterial populations downstream in Wyoming mountain drainage basin. Journal of Range Management **37**:269–274.

Skinner, Q.D., J.E. Speck, Jr., M. Smith, and J.C. Adams. 1984*b*. Stream water quality as influenced by beaver within grazing systems in Wyoming. Journal of Range Management **37**:142–146.

Skovlin, J.M. 1984. Impacts of grazing on wetlands and riparian habitat: a review of our knowledge. Pages 1001–1103 *in* Developing strategies for rangeland management. National Research Council/National Academy of Sciences. Westview Press, Boulder, Colorado, USA.

Smeins, F.E. 1975. Effects of livestock grazing on runoff and erosion. Pages 267–274 *in* Watershed management. American Society of Civil Engineers, New York, New York, USA.

Stephenson, G.R., and R.C. Rychert. 1982. Bottom sediment: a reservoir of *Escherichia coli* in rangeland streams. Journal of Range Management **35**:119–123.

Stewart, G., and C.L. Forsling. 1931. Surface runoff and erosion in relation to soil and plant cover on high grazing lands of central Utah. Journal of American Society of Agronomy **23**:815–832.

Takar, A.A., J.P. Dobrowolski, and T.L. Thurow. 1990. Influences of grazing, vegetation life-form, and soil type on infiltration rates and interrill erosion on a Somalian rangeland. Journal of Range Mangement **43**:486–490.

Thomas, J.W., C. Moser, and J.E. Rodieck. 1979. Wildlife habitats in managed rangelands: the Great Basin of southeastern Oregon. Riparian zones. United States Forest Service General Technical Report PNW-80.

Thurow, T.L., W.H. Blackburn, and C.A. Taylor, Jr. 1986. Hydrologic characteristics of vegetation types as affected by livestock grazing systems, Edwards Plateau, Texas. Journal of Range Mangement **39**:505–509.

Thurow, T.L., W.H. Blackburn, and C.A. Taylor, Jr. 1988. Infiltration and interrill erosion responses to selected livestock grazing strategies, Edwards Plateau, Texas. Journal of Range Management **41**:296–302.

Tohill, A., and J. Dollerschell. 1990. "Livestock" the key to resource improvement on public lands. Rangelands **12**:329–336.

Tricker, A.S. 1981. Spatial and temporal patterns of infiltration. Journal of Hydrology **49**:261–277.

Tromble, J.M. 1980. Infiltration rates on root plowed rangeland. Journal of Range Management **33**:423–425.

Turner, A.C., and J. O'Laughlin. 1991. State agency roles in Idaho water quality policy. Idaho Forest, Wildlife and Range Policy Analysis Group Report No. 5, University of Idaho, Moscow, Idaho, USA.

United States Bureau of Land Management. 1990. Riparian-wetland initiative for the 1990's. Washington, D.C., USA.

Vallentine, J.F. 1989. Range development and improvements. Third edition. Academic Press, San Diego, California, USA.

Viera, S.R., D.R. Nielsen, and J.W. Bigger. 1981. Spatial variability of field-measured infiltration rate. Soil Science Society of America Journal **45**:1040–1048.

Warren, S.D., W.H. Blackburn, and C.A. Taylor, Jr. 1986*a*. Effects of season and stage of rotation cycle on hydrologic condition of rangeland under intensive rotation grazing. Journal of Range Management **39**:486–491.

Warren, S.D., W.H. Blackburn, and C.A. Taylor, Jr. 1986*b*. Soil hydrologic response to number of pastures and stocking density under intensive rotation grazing. Journal of Range Management **39**:500–504.

Warren, S.D., T.L. Thurow, W.H. Blackburn, and N.E. Garza. 1986*c*. The influence of livestock trampling under intensive rotation grazing on soil hydrologic characteristics. Journal of Range Management **39**:491–495.

Weltz, M., and M.K. Wood. 1986*a*. Short-duration grazing in central New Mexico: effects on infiltration rates. Journal of Range Management **39**:365–368.

Weltz, M., and M.K. Wood. 1986*b*. Short-duration grazing in central New Mexico: Effects on sediment production. Journal of Soil and Water Conservation **41**:262–266.

Wilcox, B.P., and M.K. Wood. 1988. Hydrologic impacts of sheep grazing on steep slopes in semiarid rangelands. Journal of Range Management **41**:303–306.

Wilcox, B.P., and M.K. Wood. 1989. Factors influencing interrill erosion from semiarid slopes in New Mexico. Journal of Range Management **42**:66–70.

Wilcox, B.P., M.K. Wood, and J.M. Tromble. 1988. Factors influencing infiltrability of semiarid slopes. Journal of Range Management **41**:197–206.

Wilshire, H.G., and J.K. Nakata. 1976. Off-road vehicle effects on California's Mojave Desert. California Geology **29**:123–130.

Wood, M.K., and W.H. Blackburn. 1981*a*. Grazing systems: their influence on infiltration rates in the Rolling Plains of Texas. Journal of Range Management **34**:331–335.

Wood, M.K., and W.H. Blackburn. 1981*b*. Sediment production as influenced by livestock grazing in the Texas Rolling Plains. Journal of Range Management **34**:228–231.

Wood, M.K., G.B. Donart, and M. Weltz. 1986. Comparative infiltration rates and sediment production on fertilized and grazed blue grama rangeland. Journal of Range Management **39**:371–374.

17

Riparian Responses to Grazing Practices

WAYNE ELMORE

Abstract

Fur trappers and settlers during the early 1800s reported extensive stands of willows and wide, wet meadows along stream systems throughout the western rangelands. By the early 1900s, many of these stream systems were severely damaged or eliminated because of improper livestock use. Although grazing practices initiated in the mid-1930s have dramatically improved uplands, riparian conditions have continued to decline in most areas. Many grazing management strategies are being used to restore riparian systems, but there is little understanding of basic stream processes or riparian vegetation requirements. Resource managers must thoroughly understand the relationship between the natural stress in individual stream systems and the management stress of the various grazing systems before prescribing solutions. Riparian exclosures throughout the West have proven that livestock grazing is not necessary to improve stream riparian systems. However, recent experience has shown that with proper grazing, livestock can be present while stream systems are improving.

Key words. Riparian, grazing, cattle, streams, livestock.

Introduction

"Riparian" is defined by Webster as "relating to or living or located on the bank of a natural watercourse (as a stream or river, lake)...." Other definitions range from broad to very specific, but basically riparian areas are the green zones or lands associated with or affected by water sources (Naiman 1990, Gregory et al. 1991). They are the willow (*Salix*), sedge (*Carex*), rush (*Juncus*), and grass communities lining streams, rivers, lakes, and springs. Recorded observations by early Oregon explorers and residents provide a glimpse of how riparian areas may have looked. For example, in 1826, Peter

Skene Ogden, after traveling through the Crooked River Basin of eastern Oregon, wrote "all the different forks were well wooded (willows and aspin) and all have been examined" (Ogden 1824–26). Most of that scene is gone. The Indian word Ochoco, for which our central Oregon mountains are named, means "streams lined with willows" (McArthur 1982), yet today willow is uncommon. Senior ranchers in central Oregon tell stories about the problems once encountered gathering cattle in the "thick willow stands" on Big Summit Prairie (pers. comm.). The "thick willow stands" have been reduced to scattered clumps. The historic evidence in general indicates that most riparian zones have changed dramatically within the last hundred years or so, and that the chief cause has been improper livestock grazing.

The Riparian System

In recent years, the specific management of riparian areas has typically been the primary responsibility and interest of wildlife and fisheries biologists. Management changes have been judged mainly by their ability to improve habitat for big game, songbirds, and fish. But riparian areas are more than just habitat for wildlife. They are physical systems functioning to filter water, store nutrients, stabilize banks, and assist in the recharge of underground aquifers (Naiman and Décamps 1990). Like other uses of riparian areas, fisheries and wildlife habitat are products of those functions, but should not be considered the only reasons for managing riparian systems. In fact, they often represent the lowest economic value to be derived from proper riparian management.

To fully evaluate the benefits of riparian management and to incorporate alternative management strategies into land use plans, an understanding is needed of the basic functions of riparian areas (Swanson et al. 1982). Stream riparian functions are those processes influenced by the interaction of soil, water, and vegetation. They include physical filtering of sediment, bank stability, and water storage and recharge of subsurface aquifers, as discussed below.

1. *Physical filtering of sediment.* Riparian vegetation provides hydraulic roughness along channel margins and floodplains, slowing water velocity and "combing" out sediments and debris. This water clarification process also helps to build channel banks. Channels that were once wide and shallow typically become narrow and deep. Vegetation (grasses, sedges, rushes, etc.) is pushed down during high flows and forms a protective cover over the banks. Because of the roughness this vegetation provides, deposition of sediments is also encouraged. Where deposition occurs over long periods, extensive wet meadows or floodplains develop (Elmore and Beschta 1987).

2. *Bank stability.* Riparian vegetation can withstand high velocities of water and still maintain the positive factors of the bank-building processes. The grasses, forbs, sedges, rushes, shrubs, and trees produce a variety of fibrous

and woody roots that bind and hold soils in place. The combination of fibrous and woody rooted species has a mutually reinforcing effect: the woody roots provide physical protection against the hydraulic forces of high flows and allow the fibrous roots to bind the finer particles. This diversity of plant species is much more effective in promoting bank stability than is any single species alone.

3. *Water storage and recharge of subsurface aquifers*. Riparian systems with healthy vegetation slow the flow of water and allow it to spread and soak into the banks like a sponge, thus decreasing peak flows, maintaining local water tables, and extending baseflows through summer months (Wissmar and Swanson 1990). Yet streamside aquifers in many areas of the West have gone dry during the last century of intensive land use. For many degraded riparian systems (particularly those that have undergone channel incision and downcutting), high flows are contained in the channel and cannot access the floodplains, where water can slow and spread, recharging streamside aquifers. It is widely accepted that a water table can be lowered through the use of channels or drainage ditches. However, it is not as readily recognized that the process can be reversed: recovery of riparian vegetation and deposition of sediment in formerly degraded channels can increase water storage. When banks rebuild through sediment filtration, and channel degradation is reversed, the area for water absorption is increased and the recharge of streamside aquifers improves.

It is important to recognize that uplands must not be excluded from consideration of riparian areas, because they are an integral part of the system. For example, overland and subsurface flows also influence stream sediment loads, water cycles, and recharge of aquifers (Dunne and Leopold 1978). Improving upland vegetation through proper livestock use can increase infiltration rates, reduce overland flows, and add to the water stored by stream systems.

Observations of riparian systems in eastern Oregon that have shown substantial ecological improvement indicate that significant hydrologic changes are under way. These include increases in the baseflow (the relatively low discharge to which the stream returns after storms or snowmelt periods), reduction in the buildup of ice, more moderate water temperature regimes, and physical filtering of sediments by ice and vegetation. Almost all the negative features observed in degraded stream systems were improved when those streams were returned to good ecological condition. The information that is made available to the managers and users of natural resources must identify such hydrologic changes.

Management Evaluation

Every stream has a certain level of natural stress or sensitivity, depending on soil type, stream gradient, weather, geology, and other physical features.

Every management strategy (road construction, logging, grazing) also exerts a certain amount of stress on streams and their associated riparian areas. The ability of a given stream to handle management stress depends on its inherent level of natural stress. Some streams with high levels of natural stress (such as those with bentonitic soils and high erosion potential) can stand little management stress, while streams of low natural stress (such as those with flat gradients and sandy loam soils) can withstand a much higher degree of management stress. The natural stress of a stream must be considered when designing management strategies for recovery.

Management stress must not be confused with simple quantitative levels or numbers of use, whether that means number of livestock, recreationists, or logging trucks. This is particularly true of livestock, because reduction in numbers of livestock grazing an area has been employed as a treatment for streams in poor condition, yet often no recovery has occurred. For example, Bear Creek in the Bureau of Land Management (BLM) Prineville District in Oregon once supported 73 AUMs (animal use, or unit, months) of grazing under season-long management. This system caused more management stress than over 300 AUMs managed under a late winter and early spring grazing system. As a result of decreased management stress, the stream and its riparian areas are improving significantly, even with a fourfold increase in grazing use (Figure 17.1A and B). There are many other examples of the positive effects of proper riparian grazing in Oregon and elsewhere in the Great Basin (Platts and Raleigh 1984, Platts and Nelson 1985). If the natural potential of each stream reach, its riparian functions, and the expected stress of management are properly integrated, the characteristics of a healthy riparian system will soon be evident, as shown in Figure 17.2 (Elmore and Beschta 1987).

Restoring Riparian Condition

Understanding riparian system functions and the role that vegetation plays in these functions is essential to their management. Changes in grazing management can provide important opportunities to improve the ecological condition of stream systems while meeting the requirements of riparian vegetation.

Many riparian areas in poor ecological condition have become, in effect, upland exclosures. The attraction of livestock to streams and streamside areas during summer grazing periods often means that 90 to 95% of the adjacent upland areas receive little or no use (Krueger and Bonham 1986). Riparian areas in poor ecological condition often display all the negative results of improper grazing while none of the positive results of grazing in the upland areas are attained.

Exclusion of livestock is often proposed in order to initiate recovery of riparian areas along many stream systems. Although effective, exclusion is usually not necessary because riparian recovery and livestock grazing can

FIGURE 17.1. Twelve years (1976–88) of vegetation and channel responses before (top) and after (bottom) five years of rest followed by seven years of a late winter (February) and early spring (March) grazing system. Prior to 1976, grazing was season long. Grazing use increased from 72 AUMs in 1976 to over 300 AUMs in 1988. (Reduced numbers of juniper in the background are the result of efforts to improve upland ecological conditions.)

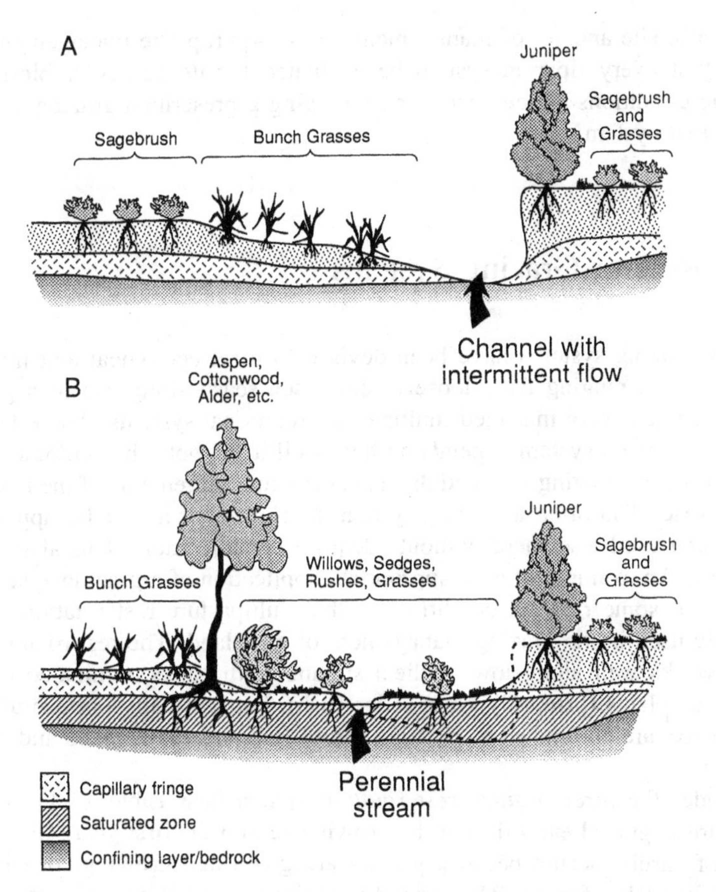

FIGURE 17.2. General characteristics and functions of riparian areas. (A) *Degraded riparian area*: little vegetation to protect and stabilize banks, little shading; lowered saturated zone, reduced subsurface storage of water; little or no summer streamflow; warm water in summer and icing in winter; poor habitat for fish and other aquatic organisms in summer or winter; low forage production and quality; low diversity of wildlife habitat. (B) *Recovered riparian area*: vegetation and roots protect and stabilize banks, improve shading; elevated saturated zone, increased subsurface storage of water; increased summer streamflow; cooler water in summer, reduced ice effects in winter; improved habitat for fish and other aquatic organisms; high forage production and quality; high diversity of wildlife habitat.

be managed together. Further, loss of natural resource benefits accruing to livestock grazing and reduced flexibility to manage for multiple uses make the practice undesirable. There are also substantial social and economic costs associated with fence construction and maintenance, adverse effects on wildlife habitat, and restriction of other uses. Rather than eliminating grazing, the goal should be prescription grazing, a management scheme designed for a specific site and set of management goals. Appropriate management dictates that every riparian system be evaluated for its particular biotic and abiotic conditions as the basis for developing a prescribed grazing strategy to fit that system.

The Role of Grazing Systems

Many grazing systems have been devised to meet ecological and management goals, ranging from loosely supervised season-long use of a general area to intensively managed multipasture rotational systems. The effectiveness of a given system depends on how well it fits both the ecological conditions of the grazing area and the management requirements of the livestock enterprise. Too often a grazing system developed for a specific application has been used elsewhere without adequate consideration of local site conditions. An example of an inappropriate application of a grazing system not suited to some riparian conditions is the multipasture rest rotation system widely used in the grazing management of grassland. The rest rotation system works well along low gradient streams with banks dominated by herbaceous plants (primarily grass-sedge-rush), but is usually not useful when the banks are dominated by riparian shrubs (Elmore 1988, Platts and Nelson 1985).

Under the *three-pasture rest rotation system* (see Table 17.1), a given pasture is grazed early during the growing season the first year. Little grazing of shrubs occurs because grasses are green and growing, providing a more palatable forage. The second year the pasture is grazed after upland grasses are seed-ripe (usually mid-July). The desired utilization rate for upland grasses in the system is 60%; but because the plants are dry and relatively unpalatable in late summer, by the time that level is attained 80 to 90% of riparian herbaceous vegetation has been used. Livestock observed in Oregon will begin using current annual growth of willows during the late summer months (mid-July through September), when use of herbaceous species reaches about 45%. Use of shrubs will increase into second-year twigs when herbaceous utilization reaches 65%, and into third-year wood at 85 to 90%. In the third year, the pasture is totally rested from livestock grazing.

This system provides total annual rest for each pasture on a regular basis and promotes plant vigor, seed production, seedling establishment, root production, and litter accumulation. The system was designed to meet the physiological needs of herbaceous but not woody plants, and is therefore inappropriate to shrub-dominated riparian areas, especially in the primary stages of shrub establishment. In the example above, willows may be losing three years of growth and only getting two years back per grazing system cycle. Although the physiological needs of herbaceous species are being met, those

Table 17.1. Typical grazing systems (annual unless otherwise indicated).

Grazing system	Year	Spring	Summer	Fall	Winter
Winter Grazing (Dormant Season)					/////////////
Early Growing Season Grazing		///////////////			
Deferred Grazing (Late Season)			//////////////////////		
Three-Pasture Rest Rotation	Year 1	//////////////////////////			
	Year 2		//////////////////////		
	Year 3				
Deferred Rotation Grazing (Three Pasture System)	Year 1	//////////////			
	Year 2	//////////			
	Year 3		//////////////////////		
Early Rotation Grazing	Year 1	/////////////			
	Year 2	/////////////			
Rotation Grazing	Year 1	//////////////////////////			
	Year 2		//////////////////////		
Season-Long Grazing		//////////////////////	//////////////////////	//////////////////////	
Spring and Fall Grazing		////////////////		/////////////	
Spring and Summer Grazing		////////////////////	////////////////////		

Note. GRAZE: ////////// REST: ——

of the willows are not, and the shrubs slowly decline (Figures 17.3A and B, 17.4A and B).

Appropriate amendments of the grazing system to meet this problem include restricting utilization of riparian herbaceous species to 50% or less during the seed-ripe treatment year (Clary and Webster 1989). Another is to make the riparian area a separate pasture managed according to its special physiological needs. A third is to add more pastures to achieve additional rest. Regrowth after summer use should also be considered. Some sedges and grasses will continue to grow until the end of October if moisture is available (Meyers 1989). If 30–45 days of regrowth occur, frequently this will restore enough bank cover to trap sediments and protect banks. The central point is that the effects of proposed management on riparian vege-

FIGURE 17.3. Ten years (1976–86) of continued improvement of herbaceous cover on a low gradient stream system under three-pasture rest rotation grazing. Channel shape and water table influences are evident before (top) and after (bottom).

tation must be incorporated into the grazing system employed in a particular area.

Other commonly used grazing systems (see Table 17.1) contain both advantages and disadvantages.

Winter grazing (dormant season) provides total growing period rest every year, since grazing occurs only during plant dormancy. This promotes plant

FIGURE 17.4. Ten years (1976–86) of continued channel degradation in a high gradient, high energy stream system under three-pasture rest rotation grazing. Willows, which show a downward trend from 1976 (top) to 1986 (bottom), are needed to help stabilize this laterally unstable channel.

Table 17.2. Generalized relationships between grazing system, stream system characteristics, and riparian vegatation response (adapted from Buckhouse and Elmore 1991).

Grazing System	Steep Low Sediment Load		Steep High Sediment Load		Moderate Low Sediment Load		Moderate High Sediment Load		Flat Low Sediment Load		Flat High Sediment Load	
No Grazing	Shrubs	+	Shrubs	+	Shrubs	+	Shrubs	+	Shrubs	+	Shrubs	+
	Herbs	+	Herbs	+	Herbs	+	Herbs	+	Herbs	+	Herbs	+
	Banks	0	Banks	0 to +	Banks	0	Banks	+	Banks	+	Banks	+
Winter or Dormant Season	Shrubs	+	Shrubs	+	Shrubs	+	Shrubs	+	Shrubs	+	Shrubs	+
	Herbs	+	Herbs	+	Herbs	+	Herbs	+	Herbs	+	Herbs	+
	Banks	0	Banks	0 to +	Banks	+	Banks	+	Banks	+	Banks	+
Early Growing Season	Shrubs	+	Shrubs	+	Shrubs	+	Shrubs	+	Shrubs	+	Shrubs	+
	Herbs	+	Herbs	+	Herbs	+	Herbs	+	Herbs	+	Herbs	+
	Banks	0	Banks	0 to +	Banks	+	Banks	+	Banks	+	Banks	+
Deferred or Late Season	Shrubs	−	Shrubs	−	Shrubs	−	Shrubs	−	Shrubs	−	Shrubs	−
	Herbs	+	Herbs	+	Herbs	+	Herbs	+	Herbs	+	Herbs	+
	Banks	0 to −	Banks	0 to −	Banks	0 to +	Banks	+	Banks	+	Banks	+
Three-Pasture Rest Rotation	Shrubs	−	Shrubs	−	Shrubs	−	Shrubs	−	Shrubs	−	Shrubs	−
	Herbs	+	Herbs	+	Herbs	+	Herbs	+	Herbs	+	Herbs	+
	Banks	0 to −	Banks	0 to −	Banks	0 to +	Banks	+	Banks	+	Banks	+

	Shrubs	–	Shrubs	–	Shrubs	–	Shrubs	–	Shrubs	–	Shrubs	–
Deferred	Herbs	+	Herbs	+	Herbs	+	Herbs	+	Herbs	+	Herbs	+
Rotation	Banks	0 to –	Banks	0 to –	Banks	0 to +	Banks	+	Banks	+	Banks	+
	Shrubs	+	Shrubs	+	Shrubs	+	Shrubs	+	Shrubs	+	Shrubs	+
Early	Herbs	+	Herbs	+	Herbs	+	Herbs	+	Herbs	+	Herbs	+
Rotation	Banks	0 to –	Banks	0 to +	Banks	+ to 0	Banks	+	Banks	+	Banks	+
	Shrubs	–	Shrubs	–	Shrubs	–	Shrubs	–	Shrubs	–	Shrubs	–
Rotation	Herbs	+	Herbs	+	Herbs	+	Herbs	+	Herbs	+	Herbs	+
	Banks	0 to –	Banks	0 to +	Banks	0 to +	Banks	+	Banks	+	Banks	+
	Shrubs	–	Shrubs	–	Shrubs	–	Shrubs	–	Shrubs	–	Shrubs	–
Season-	Herbs	–	Herbs	–	Herbs	–	Herbs	–	Herbs	–	Herbs	–
Long	Banks	0 to –	Banks	0 to –	Banks	–	Banks	–	Banks	–	Banks	–
	Shrubs	–	Shrubs	–	Shrubs	–	Shrubs	–	Shrubs	–	Shrubs	–
Spring and	Herbs	–	Herbs	–	Herbs	–	Herbs	–	Herbs	–	Herbs	–
Fall	Banks	0 to –	Banks	0 to –	Banks	–	Banks	–	Banks	– to 0	Banks	0 to +
	Shrubs	–	Shrubs	–	Shrubs	–	Shrubs	–	Shrubs	–	Shrubs	–
Spring and	Herbs	–	Herbs	–	Herbs	–	Herbs	–	Herbs	–	Herbs	–
Summer	Banks	0 to –	Banks	0 to –	Banks	–	Banks	– to 0	Banks	– to 0	Banks	0 to +

Note. – = decrease; + = increase; 0 = no change. Stream gradient: 0 to 2% = flat; 2 to 4% = moderate; >4% = steep.

vigor, seed and root production, and seedling establishment. Dormant woody riparian species are used to some degree, and therefore live twig growth would be removed. However, winter grazing can benefit riparian vegetation if use of riparian areas is low (due to an availability of livestock water elsewhere) or if drainages are colder, thus discouraging livestock use of riparian zones.

Early growing season grazing provides rest during much of the growing period, since use occurs before flower stalks emerge from the basal bud, and thereby promotes seed and root production in most years. Riparian vegetation benefits since regrowth always occurs and use on woody plants is kept to a minimum. Thus this kind of grazing can be beneficial to riparian system recovery but can also be detrimental to upland grasses if grazing always occurs during the critical part (shoot elongation) of their growing season.

Deferred grazing (after seed-ripe) every year can benefit sedge and rush communities, particularly in wide, low gradient valley systems, because it provides total growing period rest for each pasture every year and promotes seed and root production as well as seedling establishment. This system is usually detrimental to woody riparian vegetation, because it can quickly remove shrubs from the riparian zone as described above.

Deferred rotation grazing (three pasture system) provides total annual growing season rest for each pasture on a regular basis and promotes plant vigor, seed production, seedling establishment, root production, and litter accumulation. Woody riparian vegetation is usually not improved with this system.

Early rotation grazing provides rest for a portion of the growing period for each pasture and promotes plant vigor. Seed and root production are not necessarily enhanced. This system benefits riparian vegetation by allowing regrowth each year and by minimizing livestock use of woody plants.

Rotation grazing provides total growing season rest every other year for herbaceous plants to promote root growth and vigor. However, this system is usually detrimental to woody riparian vegetation but can be acceptable on low gradient, wide valley herbaceous grass sites.

Season-long grazing (spring, summer, and into late fall) is similar to spring and summer grazing except that grazing extends into plant dormancy. Since rest is never provided, the plants do not replace food reserves in roots; seed may or may not be produced. Concentration of livestock in riparian areas results in heavy use of woody and herbaceous riparian species.

Spring and fall grazing provides no rest, since grazing occurs in the spring during the critical growing period and again in the fall, after seed-ripe. Some rest may occur depending on when livestock are removed in the spring, but this system does not promote plant vigor, seed or root production, or litter accumulation. The system is detrimental to riparian vegetation due to heavy use of woody riparian species in the fall. Riparian areas can improve if fall use occurs when temperatures are cool, fall green up has occurred, or utilization is closely monitored.

Spring and summer grazing occurs during the critical part of the growing period and into the summer. The system provides no rest during the growing period for plant vigor or reproduction. Use occurs from early spring into July or August and usually results in heavy use of woody riparian species.

It becomes apparent that utilization of riparian vegetation in itself is usually not a major concern unless it affects riparian function. This situation occurs commonly under deferred grazing on sites where regrowth is limited and under rest rotation grazing where shrubs are needed for bank stability and sediment filtering. Therefore, the management challenge is to design a grazing strategy that fits the physical conditions of a given stream reach and allows recovery or maintenance of proper riparian function. Table 17.2 presents the generalized relationships between grazing systems, stream characteristics, and vegetation response.

Conclusion

We are at a critical time in the management of riparian areas and associated uplands. The requirements of riparian vegetation must be evaluated differently than in the past, and changes in grazing management instituted to improve the ecological condition of stream systems. Furthermore, the watershed, not just the stream system, must be the focal point, because the runoff from the uplands has a major impact on what happens in the stream system. If improvements in water quality and hydrologic regime are needed, then alteration of those practices affecting the whole watershed must be a significant component of riparian programs.

We should be aiming at prescription grazing, a management method designed for a specific site and set of management goals. Appropriate management dictates that every grazing area be evaluated for its unique combination of biotic and abiotic factors and then managed under a prescription rather than a blanket grazing policy.

Regrowth after summer use should be considered (Meyers 1989, Clary and Webster 1989). Some sedges and grasses will continue to grow until the end of October if moisture is available. Regrowth of 30–45 days will frequently restore enough bank cover to trap sediments and protect banks.

Drought periods appear to be important for channel narrowing. Observations in central Oregon (J. Heilmeyer, Range Technician, Bureau of Land Management, Prineville, Oregon, pers. comm., 1991) on streams with proper management have narrowed substantially with lower stream flows. Rhizomatous plants seeking lowered water tables are establishing on previous channel bottoms, allowing for stream riparian function at lower flows. This may well prove to be an important step in total stream channel recovery, and management should consider this during drought periods.

The benefits of these changes far outweigh the costs of present practices and policies. Members of the livestock industry can provide leadership in understanding and solving complex riparian questions, but it is unclear whether they will accept the responsibility. Both private and public lands must be considered, because riparian areas do not recognize differences in ownership. If management is not changed, either the benefits of resource production or the flexibility to manage for multiple uses will be lost. The American public is obviously concerned about the quality and quantity of usable water, and will continue to demand more from the management of natural resources. We must start now to meet those demands.

References

Buckhouse, J.C., and W. Elmore. 1991. Grazing practice relationships: predicting riparian vegetation response from stream systems. *In* T. E. Bedell, editor. Watershed management guide for the interior Northwest. Oregon State University Publication EM 8436.

Clary, W.P., and B.F. Webster. 1989. Managing grazing of riparian areas in the Intermountain Region. United States Forest Service General Technical Report INT-263, Ogden, Utah, USA.

Dunne, T., and L.B. Leopold. 1978. Water in environmental planning. W.H. Freeman, San Francisco, California, USA.

Elmore, W. 1988. Rangeland riparian systems. Pages 93–95 *in* California Riparian Systems Conference Proceedings. United States Forest Service General Technical Report PSW-110, Berkeley, California, USA.

Elmore, W., and R.S. Beschta. 1987. Riparian areas: perceptions in management. Rangelands **9**:260–265.

Gregory, S.V., F.J. Swanson, and W.A. McKee. 1991. An ecosystem perspective of riparian zones. BioScience, *in press*.

Krueger, W.C., and C.D. Bonham. 1986. Responses of low cattle densities on large areas of forested pastures. Pages 67–73 *in* Statistical Analyses and Modeling of Grazing Systems Symposium Proceedings, February 11, 1986, Kissimmee, Florida. Society for Range Management, Denver, Colorado, USA.

McArthur, L.A. 1982. Oregon geographic names. Western Imprints; The Press of the Oregon Historical Society.

Meyers, L.H. 1989. Grazing and riparian management in southwestern Montana. Pages 117–120 *in* R.E. Gresswell, B.A. Barton, and J.L. Kerschner, editors. Practical approaches to riparian resource management: an educational workshop. United States Bureau of Land Management, Billings, Montana, USA.

Naiman, R.J. 1990. Influence of forests on streams. Pages 151–153 *in* 1991 McGraw-Hill yearbook of science and technology. McGraw-Hill, New York, New York, USA.

Naiman, R.J., and H.Décamps, editors. 1990. The ecology and management of aquatic-terrestrial ecotones. UNESCO, Paris, and Parthenon Publishing Group, Carnforth, United Kingdom.

Ogden, P.S. Ogden's snake country journals, 1824–26. Hudson Bay Records Society (1950).

Platts, W.S., and R.L. Nelson. 1985. Impacts of rest-rotation grazing on streambanks in forested watershed in Idaho. North American Journal of Fisheries Management **5**:547–556.

Platts, W.S., and R.F. Raleigh. 1984. Impacts of grazing on wetlands and riparian habitat. Pages 1105–1117 *in* Developing strategies for rangeland management. National Research Council/National Academy of Sciences. Westview Press, Boulder, Colorado, USA.

Swanson, F.J., S.V. Gregory, J.R. Sedell, and A.G. Campbell. 1982. Land-water interactions: the riparian zone. Pages 267–291 *in* R.L. Edmonds, editor. Analysis of coniferous forest ecosystems in the western United States. Hutchinson Ross, Stroudsburg, Pennsylvania, USA.

Wissmar, R.C., and F.J. Swanson. 1990. Landscape disturbances and lotic ecotones. Pages 65–89 *in* R.J. Naiman and H.Décamps, editors. The ecology and management of aquatic-terrestrial ecotones. UNESCO, Paris, and Parthenon Publishing Group, Carnforth, United Kingdom.

18

The Changing Spokane River Watershed: Actions to Improve and Maintain Water Quality

RAYMOND A. SOLTERO, LYNN R. SINGLETON, AND
CLAY R. PATMONT

Abstract

Maintenance of water quality today requires legislative control as well as scientific information. Economic or other considerations can no longer be used to justify the degradation of aquatic systems by wastewater disposal. The Spokane River, along with its various tributaries, is presented as an example of water resource planning and management resulting in improved water quality. The Spokane River and its major reservoir, Long Lake, have had numerous water quality problems and associated controversies. Long Lake has experienced massive algal blooms and ever-expanding stands of macrophytes. The main source of nutrients (particularly phosphorus) was the city of Spokane's primary sewage treatment facility. Advanced wastewater treatment initiated by Spokane in 1977 markedly improved the trophic condition of the reservoir. A phosphorus waste-load allocation for all point and nonpoint source dischargers in the drainage (Washington and Idaho, USA) was initiated in the early 1980s to protect and sustain the improved water quality of Long Lake. The Washington State Department of Ecology later established a legally enforceable standard such that the total maximum daily phosphorus load to the reservoir was to be no higher than 259 kg to effect mid-mesotrophic conditions. Recently, regulators, dischargers, and scientists have joined forces in the management of phosphorus loads in the Spokane River drainage basin. These managers are demonstrating that the most effective way to ensure good water quality is through local control of degradation processes.

Key words. Algae, eutrophication, phosphorus, reservoir, river, trophic status, water quality management.

Introduction

Most water resource managers are quick to point out that more attention must be given to water resources to ensure adequate water supplies for an ever-increasing population. Previously emphasized criteria had their place in bringing about improved water quality, but today they narrow our understanding of the impacts that effluents have on the biological community. Karr stresses that the biological focus has been lost in describing water quality problems because of the simpler chemical and physical approaches; in the future, direct assessments of "biological integrity" and "ecological health" of water bodies will be required to stem the continuing decline of our nation's waters (Karr 1991).

The Spokane River, like many other waterways of the Pacific Northwest, has invited human settlement, beginning with the first Native Americans who established their camps along its shores. Much of the settlement of the community of Spokane Falls, which became Spokane in 1891, began between 1870 and 1890 (Fahey 1986). In the early years, Spokane was a clapboard milltown linked to the river primarily for food and power.

Washington Water Power Company (WWP) was formed in 1889 to consolidate river rights and construct power stations. The suitability of the relatively short Spokane River for hydroelectric sites is evident from its topography: the elevation declines about 320 m from its headwaters to the mouth. One of the first dams completed was at Post Falls, Idaho, in 1906. WWP recognized that it would have to continue to construct electric facilities to keep up with growth. Three more sites were acquired: Nine Mile (1908), Little Falls (1910), and Long Lake (1913). The construction of Little Falls Dam resulted in the loss of a substantial salmon and steelhead fishery (McDonald 1978). These dams and others significantly changed the character of the river.

As Spokane's population grew, the first sewer line was laid as a combined storm water and sanitary system to the river. Sewage was a problem in 1888 and the city council adopted the Waring system to treat the wastewater. This system was nothing more than a series of gravel shelves over which sewage ran before entering the river. As the sewer system expanded, the number of discharge points along the river increased. By the 1920s, the volume of waste was too great to effectively dilute, and raw sewage could be seen in the river, especially following a heavy rainfall. Typhoid fever became an annual problem, compelling the city to forbid use of private wells (probably contaminated by privies and cesspools) wherever city supplies were available. The municipal water source was changed to deep wells east of the city in 1908 when Coeur d'Alene, Idaho, started dumping sewage into the river (Fahey 1988).

The Washington State Department of Health in 1909 and again in 1929 ordered Spokane to cease and prevent any further dumping of sewage into the river. Obviously the city had taken little action, and apparently the state did not pursue the matter. Riverine sludge deposits, low dissolved oxygen levels, and pathogenic hazards were linked to the raw sewage discharges along the river in the 1930s (Pearse et al. 1933). Bond issues in 1933 and 1937 to install intercepter and transmission lines to a wastewater treatment facility were defeated because the citizenry could see no need for them.

Continuing into the 1950s, hundreds of thousands of cubic meters of untreated industrial and municipal waste were discharged to the river daily (Esvelt and Saxton/Bovay 1972). It was not until 1952, following a "terrific barrage of propaganda," that Spokane passed the bond issue (McDonald 1978). The city completed construction of the sewer system and its primary wastewater treatment plant in 1958. Although the plant was doing the job for which it was designed, during wet weather or long periods of snowmelt, hydraulically overloaded collection pipes and raw wastewater overflowed to the river at 45 different locations up to 140 times a year (Esvelt and Saxton/Bovay 1972).

The Spokane River story is not unlike other case studies (Beeton 1969, Edmondson 1970, 1972) where population growth resulted in untreated and treated wastewaters being dumped into surface waters. Eventually, waste disposal exceeded the waters' assimilative capacity and created water quality problems. The primary objective of this chapter is to point out to water resource managers the importance of having an array of approaches for managing problems. The detection of degradation and its cause(s) is fundamental. For the public to be willing to spend millions of dollars on water quality correction, credible and optimistic predictions must be made.

Initial Reaction to the Problem

The Spokane River, a young stream flowing in an alluvium-filled trough, was probably formed by glaciation and subsequent flooding 15 to 20 thousand years ago (Conners 1976). Its headwaters begin at the outlet of Coeur d'Alene Lake, Idaho, and empty into the Columbia River (Lake Roosevelt, Washington) approximately 180 km downstream. The drainage basin of the river and its tributaries is in excess of 17,200 km^2 and conveys an annual average of $7.5 \times 10^9 m^3$ of water. Two major tributaries join the river at km 116.7 (Latah, or "Hangman," Creek) and at river km 90.6 (Little Spokane River).

The largest reservoir on the Spokane River was formed by the completion of Long Lake Dam, approximately 24 km northwest and downstream from Spokane (Figure 18.1). Water can be discharged from two outlets: the spillways (centerline elevation 464.4 m above mean sea level; msl), and the power penstocks (centerline elevation 456.9 m above msl). At normal operating level (468 m above msl) the reservoir is about 35 km long and has a storage capacity of $304.9 \times 10^6 m^3$. Other morphometric data for the reservoir are presented in Table 18.1.

In 1959 the U.S. Geological Survey established a water quality station on the Spokane River approximately 1 km downstream of Long Lake Dam. Oxygen data from this site often revealed concentrations less than 5 mg/L between July and October. The Washington Water Pollution Control Commission (now the Washington State Department of Ecology) found oxygen

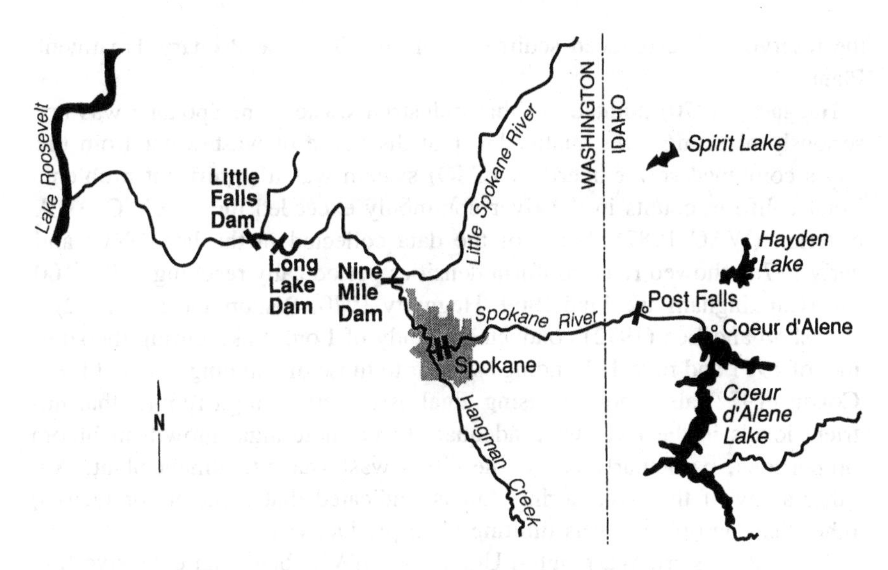

FIGURE 18.1. Map of the Spokane River system, USA.

levels less than 1 mg/L from the 9 m depth (power penstock elevation) to the bottom behind the dam on September 1, 1964 (E. Asselstine, Washington Water Pollution Control Commission, unpublished data). During the week of September 12, 1966, Cunningham and Pine (1969) found the reservoir thermally stratified and oxygen depleted at depths greater than 9 m. Their data showed an inverse relationship between percentage of volatile solids in the sediments and overlying oxygen concentrations. Anaerobiosis was attributed to volatile solids stabilization occurring in the sediments, with the assumption that phytoplankton made up the majority of the solids. They suggested that the primary sources of nutrients supporting algal growth in

Table 18.1. Morphometric data for Long Lake, Washington, USA, at maximum capacity (elevation 468.3 m above mean sea level).

Maximum length	35.4 km
Maximum effective length	5.8 km
Maximum width	1.1 km
Maximum effective width	1.1 km
Mean width	571.8 km
Maximum depth	54.9 km
Mean depth	14.6 km
Area	208.4×10^5 m^2
Volume	304.9×10^6 m^3
Shoreline length	74.3 km
Shoreline development	4.6 km
Bottom gradient	0.15%

the reservoir were reduced sediments and the Spokane Primary Treatment Plant.

Haggarty (1970) pointed out that industrial waste from Spokane was not seriously affecting river quality but that discharge of wastewater from the city's combined sewer overflow (CSO) system was a significant problem. Total coliform counts in the river commonly exceeded the state's Class A criterion (WAC 1987). Some of the data collected in the late 1960s and early 1970s showed river coliform densities periodically reaching $>10^4/100$ mL (Cunningham and Pine 1969, Haggarty 1970, Bishop and Lee 1972).

Bishop and Lee (1972) conducted a study of Long Lake during the summer of 1971 and revealed findings similar to those of Cunningham and Pine. Condit (1972) also showed, using algal assay during high runoff, that nutrient levels in the river were adequate to promote algal growth in bloom proportions, particularly below the city's wastewater treatment plant. Another assay of the river during August indicated that a factor (or factors) other than phosphorus was limiting algal productivity.

In 1972, Eastern Washington University (EWU) began an extensive limnological investigation of Long Lake and its major tributaries (Soltero et al., 1973, 1974, 1975, 1976, 1978). Seven sampling stations were located on the Spokane River and its tributaries and five on Long Lake. The length of the sampling season was usually seven months (June through December). These investigations demonstrated that the major source of nutrient influent to Long Lake was the Spokane Primary Treatment Plant (before advanced wastewater treatment). An internal density current was evident during these studies which altered vertical and longitudinal distribution of physical and chemical parameters when the reservoir was thermally stratified. This interflow of the river at the depth of the power penstocks isolated a wedge of water on the bottom which soon became anoxic. During stratification, oxygen depletion was most evident from July through September, when as much as 60% of the reservoir's volume was void of oxygen (anoxic). It was concluded that phytoplankton production was sufficient to substantiate algal decomposition as the cause of hypolimnetic anoxia. Strong correlations existed between maximal algal assay yields (dry weight of *Selenastrum capricornutum*) and the limiting nutrient for Long Lake euphotic zone samples taken in 1974 and 1975 after chelation of toxic heavy metals (Greene et al. 1975). Algal assay yields correlated well with indigenous reservoir phytoplankton standing crop and chlorophyll *a* concentrations. Assay yields and N:P ratios indicated that reservoir samples were primarily phosphorus limited, but co-limitation of nitrogen and phosphorus was also frequently determined. Elevated phosphorus concentrations were particularly evident in the hypolimnion during the summers of 1973 and 1977, with levels >0.3 mg/L P. These maxima were primarily attributed to reduced sediment release during these extraordinarily high water retention years. The reservoir was classified as eutrophic for all years of study based on levels of primary productivity, chlorophyll *a* concentrations, hypolimnetic oxygen depletion, and nitrogen and phosphorus concentrations.

Focused Solution to Long Lake Eutrophication

Water quality improvements were needed to address Long Lake's degraded state. The federal Clean Water Act (CWA) of 1972 provided the means to effect control in the Spokane River drainage basin. It compiled and expanded the intent of several previously passed water quality protection programs (Kovalic 1987). The CWA was founded on the following premises.

- There are no rights to pollute navigable waters. A discharge permit is needed to do so.
- Discharge permits limit the pollutants present and the concentrations. Permit condition violators are subject to fines and imprisonment.
- Permit conditions may require the best controls technology can produce.
- Limits or control measures more stringent than the minimum federal requirements are to be based on the quality of the receiving water. Higher standards are appropriate only for protecting the receiving water quality.

The CWA precipitated changes in the Spokane River Basin as it set up funding programs and wastewater treatment criteria that shaped water quality control measures. The Washington State Department of Ecology also established funding programs that could be matched with federal grants. Environmental changes via new treatment facilities were therefore relatively inexpensive to local communities and in general were viewed favorably.

The CWA has been amended several times, but the changes made in 1987 altered the way environmental control and protection are accomplished and funded. Although water quality protection was always the Act's foundation, historically it has been implemented in two distinct ways. Initially, technology (essentially secondary treatment) was to be in place, and if that level of treatment was inadequate, enhancements were to be made to protect water quality. This occurred at Spokane's plant. When technology was in place, the CWA was modified such that discharge permits would be written under established water quality standards that predicted effluent loads and treatment needed to maintain acceptable quality and set National Pollution Discharge Elimination System (NPDES) permit limits; and therefore treatment would be required accordingly.

Collectively, the EWU information, results of other Spokane River and Long Lake investigations, and the local community supported the position of the U.S. Environmental Protection Agency (EPA) and the Washington State Department of Ecology that phosphorus removal at Spokane's treatment plant would be acceptable and significantly improve reservoir quality. Spokane was directed by the Department of Ecology in November 1972 to provide advanced treatment for the city's wastewater. Advanced wastewater treatment (AWT) was defined as >85% biochemical oxygen demand and phosphorus removal, and >90% suspended solids removal. The EWU investigations continued to document the water quality of the river/reservoir system for data continuity just prior to and following (Soltero et al. 1979, 1980, 1981, 1982, 1983, 1984, 1985, 1986) operation of the new AWT plant.

Despite general agreement, the decision to require treatment beyond secondary treatment was somewhat controversial. Critics doubted that AWT would improve the trophic condition of the reservoir, because of internal phosphorus cycling. Thomas and Soltero (1978) extracted reservoir sediment cores 8 km above the dam in July 1974 to address these concerns. The overall sedimentation rate was estimated at 26 mm/yr between 1958 and 1973. Two cores were zoned into annual layers according to profile concentrations of organic matter, total nitrogen, total phosphorous, and diatom frustule density. Core analyses suggested that increased nutrient loading and resulting diatom production were related to the volume of sewage influent to the reservoir. Illite clays were found to be a major structural component of the sediments and played an important role in the distribution of phosphorus. During spring runoff, heavy loads of clay settle from the water column over a layer containing high phosphorus concentrations of primarily biological origin laid down over a clay layer deposited the previous year. This mechanism effectively prevents internal nutrient recycling, particularly of phosphorus, from the bottom sediments.

In 1974, the city began construction of the new AWT plant ($55 million), and questions were still being asked about how quickly the reservoir would recover under reduced phosphorus loadings. In order to further address these concerns, the Department of Ecology contracted Battelle Pacific Northwest Laboratories (Gasperino and Soltero 1977) to perform a water quality modeling study of Long Lake in 1975. The first phase of the project involved acquisition and preparation of the data collected by EWU during the first three years of study (1972–74) to calibrate the model. The EWU reservoir data collected during 1975 were used to verify the calibrated model. One of the test cases was, in part, to simulate chlorophyll *a*, phosphorus, and dissolved oxygen concentration profiles in the upper, middle, and lower ends of the reservoir under conditions of low river flow and 85% phosphorus removal at Spokane's new plant. Simulations revealed a substantial decrease in phosphorus and chlorophyll *a* levels in the reservoir and near elimination of anoxia within one year. It was predicted that eutrophic Long Lake would experience a rapid recovery with the establishment of AWT by the city.

During the upgrade of the city's treatment plant, a bypass of 4.2×10^5 m^3 of raw sewage was released to the river in October 1975 to facilitate rerouting of the sewage flow from the old to the new plant headworks. Twenty-two Long Lake residents filed a $5.2 million lawsuit contending that the raw sewage created a public health hazard, reduced property values, caused mental anguish, and resulted in toxic blue-green blooms in the late summer and fall of 1976 and 1977. Although the city of Spokane and the state of Washington were found liable, the Superior Court of Spokane County agreed that the toxic blooms were not caused by the October 1975 bypass (ruling by the Honorable Harold D. Clark, July 20, 1979, Docket No. 229268).

Phosphorus and nitrogen loading to the reservoir was altered little during and following the sewage bypass. The additional amount of phosphorus contributed because of the bypass was approximately 0.05% of the annual load influent to the reservoir. No significant differences in river or reservoir concentrations for nitrogen and phosphorus were determined during the bypass from those previously observed. The bloom potential of Long Lake has been high for a number of years, and blue-green pulses should have occurred previous to 1976 (Soltero and Nichols 1981).

The blue-green algal standing crop potential in Long Lake was not achievable before 1976, even with excessive nutrient loading, because of heavy metal inhibition (Soltero and Nichols 1981). Algal assay research has shown that *Anabaena* growth can be suppressed in the presence of heavy metals. Shiroyama et al. (1976) found that as Spokane River flows subside, metal concentrations decline, rendering Long Lake waters more suitable for *Anabaena flos-aquae* growth. Heavy metals, particularly zinc, have entered the reservoir at algicidal or algistatic levels, thereby influencing quantitative and qualitative composition of the phytoplankton community. Trend analysis has shown that total mean zinc concentrations in the Spokane River at Post Falls, Idaho, decreased 50% between 1973 and 1978 (Yake 1979). Zinc content of the 1973 and 1974 zones of the reservoir sediment cores was about half that determined for the 1968 to 1972 zones (Thomas and Soltero 1978). Greene et al. (1978) reported that a positive relationship existed between Spokane River flow and zinc concentrations influent to Long Lake. They also pointed out that zinc levels in the latter part of the growing season can fall to, or below, the lower range of algal growth inhibition, which is 30–100 μg/L. All of this information correlates well with EPA's abatement efforts in the Kellogg, Idaho, mining districts.

The new AWT facility was put into operation August 22, 1977, and phosphorus removal began December 15. Total and soluble reactive phosphorus loads in the sewage outfall decreased approximately 90%. This reduction significantly reduced phosphorus loads to the reservoir, with a concomitant decline in reservoir concentrations (Table 18.2). Prior to AWT, the sewage effluent was the major source of phosphorus to the reservoir. However, the Spokane River above the treatment plant became the primary source of phosphorus following AWT.

The degree of hypolimnetic anoxia observed for all post-AWT years was less than that determined for the low flow years 1973 and 1977, but similar to that observed in the high flow years of 1974 and 1975. Mean chlorophyll *a* concentrations for the growing season significantly declined with AWT and were directly related to decreased phosphorus loading to the reservoir

Table 18.2. Total phosphorus loading, euphotic zone chlorophyll *a* concentrations, and phytoplankton biovolumes for all years of study during the growing season (June-October), Long Lake, Washington, USA.

Year	TP Load (kg/day)	Chl *a* (μg/L)	Biovolume (mm^3/L)
	Pre-AWT		
1972*	1,246	18.7	8.20
1973	951	27.8	15.57
1974	1,167	17.0	7.09
1975	1,046	18.4	8.27
1977	689†	20.4	9.74
Mean	1,020	20.5	9.77
	Post-AWT		
1978	252	15.0	47.63 (4.07)‡
1979	210	15.2	5.39
1980	972(206)§	9.4	2.81
1981	243	11.6	2.77
1982	263	9.4	2.02
1983	215	10.2	2.84
1984	290	8.7	3.78
1985	193	7.9	3.15
Mean	238‖	10.9	3.25¶

*Based on samples collected at Seven Mile Bridge (RM 62.0); Nine Mile Dam (RM 58.1) was not sampled during 1972.

†Based on all data for the growing season, but load value was lower because of the initiation of secondary wastewater treatment in August.

‡Parenthetic figure excludes *Microcystis aeruginosa* biovolumes at stations 3 and 4 on September 12.

§Parenthetic figure excludes TP data collected on June 2 and 16 because of the large and variable loads following the eruption of Mount St. Helens, depositing a large quantity of ash across the watershed.

‖Mean calculation includes the parenthetic figure for TP load in 1980.

¶Mean calculation includes the parenthetic figure for biovolume in 1978.

(Table 18.2). Decreases in reservoir phosphorus and chlorophyll *a* concentrations, and hypolimnetic anoxia, indicated the reversion of eutrophic Long Lake to a more mesotrophic state following AWT.

Overall, a decline in reservoir phytoplankton biovolumes (approximately 2.5 fold with AWT) occurred in all classes of algae with the exception of blue-greens. Phytoplankton diversity became greater with AWT, as species increased to 84 following AWT and averaged 52 prior to the upgrade.

The city, in 1981, was granted an amended sewage discharge permit allowing for seasonal phosphorus removal. A critical growing season was defined for the reservoir as the period from June through October, and seasonal phosphorus removal began in 1982 with alum addition commencing April 1 and ending November 1. Secondary treatment was operational for the remainder of the year. A methodology for estimating an initiation date for phosphorus removal after April 1 was established based on predicted spring runoff and phosphorus residence times in the reservoir (Mires and Soltero 1983). An earlier termination date for alum addition was also developed by utilizing a time response model for increased nutrient loading to the reservoir (Mires et al. 1983). Both approaches defined a maximum allowable in-reservoir phosphorus concentration of 25 μg/L and estimated how long it would take Long Lake to reach this upper limit. These models demonstrated that the initiation and termination dates for chemical phosphorus removal could vary, and were approved by the Department of Ecology for implementation by the city in 1984. Field monitoring of the year-round (Soltero et al. 1979, 1980, 1981) and seasonal (Soltero et al. 1982, 1983, 1984, 1985, 1986) phosphorus removal schemes showed that seasonal alum addition was as effective in reducing phytoplankton standing crop and chlorophyll *a* concentrations in the reservoir as year-round AWT.

Most of the sewers in Spokane were of the combined storm and sanitary type with discharge directly to the river before 1958. An NPDES waste discharge permit was issued to Spokane by the Department of Ecology specifying "cleanup" of all CSO (combined sewer overflow) by June 30, 1977. Spokane's CSO abatement project (EPA 1979) has three goals: (1) to eliminate the discharge of untreated sanitary waste to the river, thereby reducing the potential public health hazard, (2) to remedy local drainage problems that cause property damage and create health hazards, and (3) to increase the hydraulic capacity of the AWT plant and maximize its potential as a regional facility.

The most significant impact that CSOs have had on the river's water quality has been on coliform counts (City of Spokane 1977, EPA 1979). A year-long fecal coliform survey was carried out over a 100 km stretch of the river from the Washington-Idaho border to Long Lake Dam (Merrill 1986). This study established 33 river and 14 reservoir sampling stations. The river met its Class A criterion for fecal coliform during dry weather, but values often increased a thousandfold following major storm events or snowmelt. Long Lake also experienced contamination in the upper end following storm events, probably indicating CSO had entered the reservoir. A more recent study (Soltero et al. 1990) was conducted to determine what impact four selected CSO outfalls near the city center had on river quality. A CSO event was likely to occur when precipitation exceeded 0.06 cm/hr. Data comparison of dry and wet weather flows demonstrated that the monitored outfalls seriously degraded water quality of the river. Hangman Creek (Figure 18.1) was also shown to contribute significantly to poor downstream quality during wet weather.

The city's approach to CSO abatement is phased. The first phase was completed in January 1990 with the construction of separate storm sewers in two large drainages, eliminating 84% of the estimated CSO that annually reached the river. This phase was at a cost of approximately $40 million. The second phase will probably cost in excess of $60 million, completing the construction of the storm sewers in the remainder of the city, including

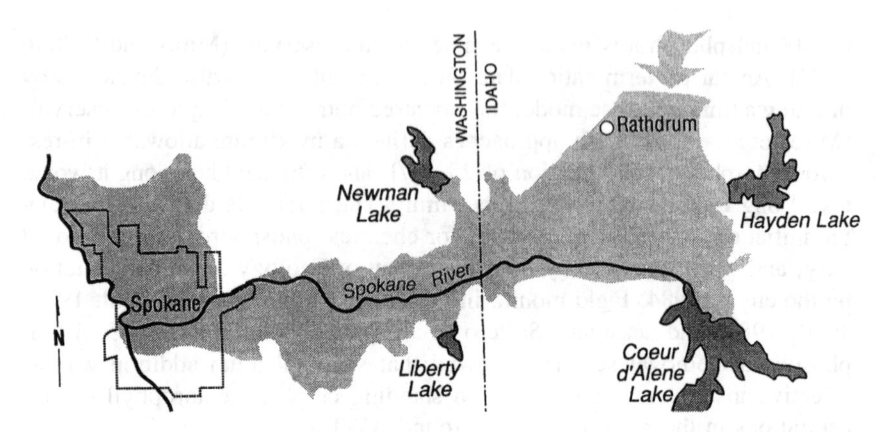

FIGURE 18.2. Map delineating the boundaries of the Spokane Valley-Rathdrum Prairie Aquifer.

the central business district. Completion time for this phase is estimated at eight to ten years.

Restatement of the Problem

Substantial population growth has occurred in the Spokane River drainage basin. With growth comes the issue of wastewater management. Do we treat on site with septic tank drainfield systems or should collection and treatment facilities be developed? These issues have arisen in the Spokane area and adjacent areas in Idaho because the primary drinking water source is the Spokane Valley-Rathdrum Prairie Aquifer (Figure 18.2). The aquifer was designated as a sole or principal source aquifer by EPA under provisions set forth in 1974 by the Safe Drinking Water Act (PL 93–523). As such, the aquifer is afforded greater protection from contamination sources (e.g., septic systems). The highly porous soils overlying this shallow aquifer make septic tank drainfield systems less desirable. As a result, wastewater collection, treatment, and disposal in the Spokane River have received the most interest to date. The Comprehensive Wastewater Management Plan (CWMP 1981) for the greater Spokane area identified the Spokane AWT plant as a regional facility. Construction of the collection system is under way. Funding options for environmental construction projects also changed. The grant-based program was replaced by a loan-based one. The local population now pays for their own environmental control projects. This change has heightened local interest in treatment issues.

Communities upstream of Spokane have grown and are either treating their wastewaters and discharging them to the river or are interested in doing so. The continued and expanded use of the Spokane River for wastewater disposal was challenged in 1979 when the community of Liberty Lake, Washington, was granted an NPDES permit. James A. Schasre and the Lake Spokane (Long Lake) Environmental Association filed a suit in Washington Superior Court. The resulting stipulated agreement required the Department of Ecology to study phosphorus loads to the Spokane River and determine the total maximum daily phosphorus load (TMDPL) from all sources that would be allowable and protective of Long Lake's improved water quality.

There are several ways to regulate and administer water quality in Long Lake. The Department of Ecology chose to promulgate a special condition for total phosphorus concentrations in the Spokane River in its water quality regulations (WAC 1987). Total phosphorus (TP) is regulated for several reasons.

- It is directly responsible for the identified water quality problems of low dissolved oxygen, excessive phytoplankton populations, and poor water quality in Long Lake during the summer period.
- It could be related directly to upstream sources.
- The average concentration of 25 μg/L in the euphotic zone equated to mesotrophic water quality conditions.

The goal of the water quality standard is to keep Long Lake at a mesotrophic condition during the high use summer period. Therefore, on the average and with median river flow, the seasonal June through October euphotic zone concentrations of TP will be at or below 25 μg/L. As a result, half of the year will have lower seasonal average euphotic zone TP concentrations and half will have higher ones, but on the average a mesotrophic condition will prevail. Median river flow and an average euphotic zone concentration of 25 μg/L TP equates to a daily phosphorus load of 259 kg to the reservoir.

The relationships between influent phosphorus concentrations and key water quality variables in Long Lake have been assessed using simple regression and modeling methods (Patmont et al. 1987). Two representative water quality parameters, phytoplankton biovolume and hypolimnetic dissolved oxygen (DO), are depicted in Figures 18.3 and 18.4. Based on these relationships and others, an influent TP concentration of 25 μg/L was an approximate threshold value leading to the development of mid-mesotrophic conditions within the reservoir. Under the median river flow condition, an influent TP of 25 μg/L would be expected to lead to an average chlorophyll *a* (chl *a*) concentration of 11 ± 2 μg/L, a peak chl *a* of 22 ± 5 μg/L, a median biovolume of 2.6 ± 1 mm^3/L, a median Secchi disc transparency of 3.3 ± 0.4 m, and a minimum hypolimnetic DO of 3.3 ± 1.3 mg/L. Overall, these water quality values are indicative of mid-mesotrophic conditions, consistent with other investigations conducted throughout the northern temperate zone (OECD 1982).

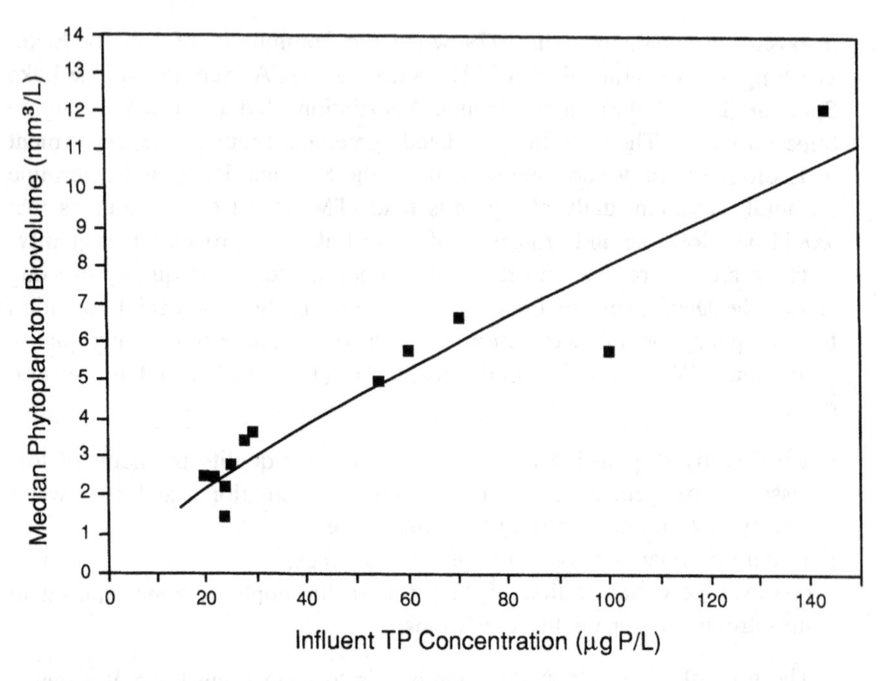

FIGURE 18.3. Long Lake, Washington, USA, phytoplankton standing crop as a function of influent total phosphorus concentration.

Additional Investigations

With the establishment of a legally enforceable TMDPL that is reasonably protective of reservoir water quality, the task remained to allocate or apportion this load among the numerous point and nonpoint sources (e.g., stormwater releases). However, the allocation process was complicated by a preliminary study of nutrient transport through the Spokane River system, which suggested that substantial uptake or attenuation of phosphorus occurred, particularly for discharges entering the river as much as 64 km upstream from the reservoir (Yearsley 1982).

In consideration of the potential financial benefits (i.e., reduced treatment) associated with incorporating attenuation into the overall TP allocation, the Department of Ecology funded a study of nutrient transport occurring from the start of the Spokane River at the Lake Coeur d'Alene, Idaho, outlet to its eventual discharge into Long Lake. The study was conducted during the low flow season of 1984 to determine if significant losses of TP occurred within the river system (Patmont et al. 1985). The study included sampling of 33 locations for nine days between July and September 1984. A detailed assessment of mass balances obtained during the attenuation study revealed that more than 40% of the influent TP load to the reservoir was lost during

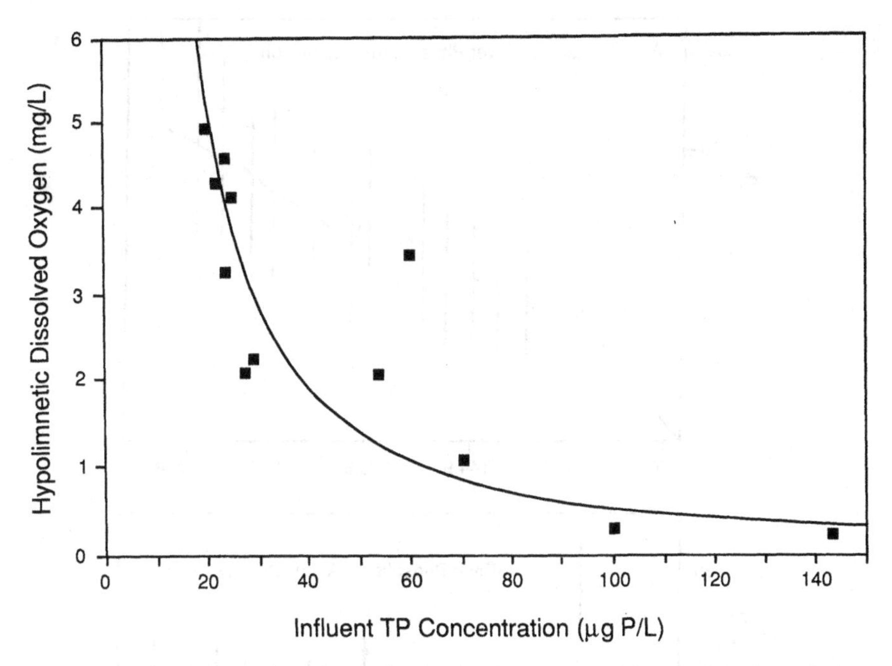

FIGURE 18.4. Minimum hypolimnetic dissolved oxygen concentration in Long Lake, Washington, USA, as a function of influent total phosphorus concentration.

river transport. The cumulative mass attenuation is depicted in Figure 18.5. Much of the observed TP loss probably occurred via in-river seasonal processes such as biological uptake by attached macrophytes and periphyton. Chemical adsorption onto river sediment may have also contributed to in-stream removals. River seepage into the adjacent aquifer system was probably an additional loss mechanism, particularly within impounded reaches (Patmont et al. 1985).

The magnitude of the attenuation process was found to be empirically related to both phosphorus and nitrogen concentrations within the river. Upper reaches of the river system (upstream of high nitrate aquifer inputs) appeared to be strongly nitrogen limited, as evidenced by a low ambient N:P ratio (less than 3:1 by weight). Phosphorus attenuation (expressed as a first-order rate derived from mass balances) in this upper river area was significantly lower than in more downstream reaches characterized by a higher N:P ratio (greater than 30:1). The degree of TP attenuation, therefore, was found to vary with respect to the location of an input along the length of the river.

In addition to their role in determining the TP attenuation that occurs between the point of discharge and its release into Long Lake and resultant reservoir trophic conditions, the in-river periphyton community's response to changes in nutrient loading was also of concern. In upper reaches of the

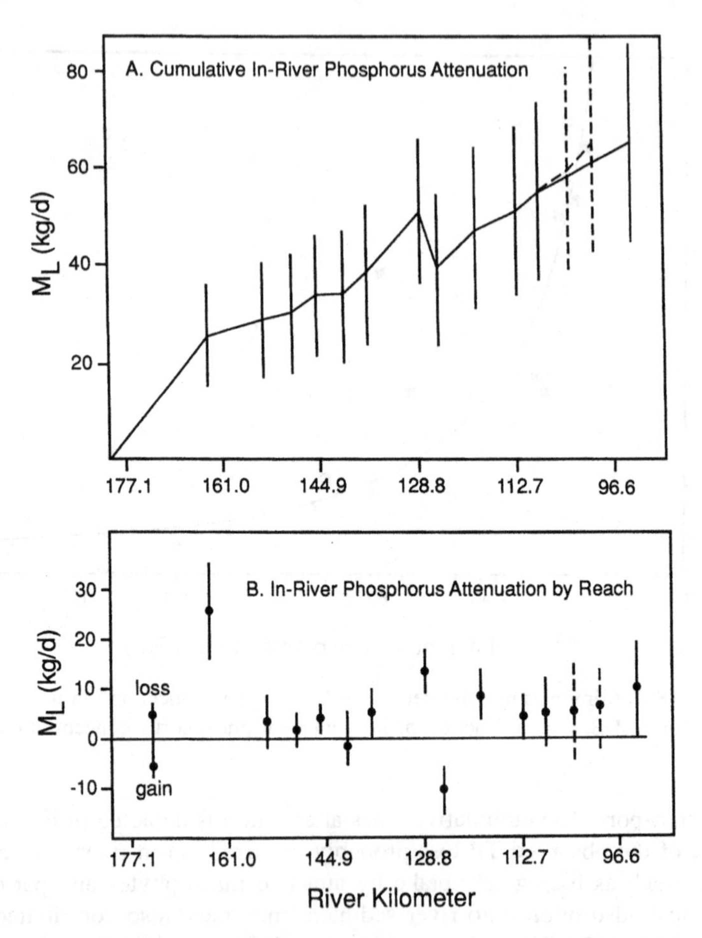

FIGURE 18.5. Mean total phosphorus attenuation (M_L) in the Spokane River aggregated across all dates (solid lines denote ± one standard error; dashed lines denote reaches that may have had sampling bias).

river characterized by nitrogen limitation, periphyton accumulations on the river bottom were relatively low (generally less than 50 mg chl a/m^2) and appeared more apt to be determined by lack of grazing and/or scouring losses than by phosphorus supplies. For this reason, and also because very large reductions in the river TP load would be required to minimize periphyton development, control of in-river periphyton accumulation was not determined to be a principal water quality objective.

Management Scenario

Implementation of the TMDPL was next. The regulators—Washington State Department of Ecology, Idaho Department of Health and Welfare, and EPA (who has legal responsibility to issue NPDES permits in Idaho)—agreed that the intentions were for local involvement, acceptance, and actual implementation. The regulators agreed that the dischargers should have one year to reach an agreement on allocation of the available phosphorus loads. If no resolution was made, an 85% TP removal requirement would be placed in the NPDES permits of all facilities. A Technical Advisory Committee (TAC) of regulators and dischargers (present and near future) was soon formed to address phosphorus management issues in the basin. In 1989 the committee developed and signed a memorandum of agreement, "The Spokane River Phosphorus Management Plan," which delineates which discharger(s) will implement 85% phosphorus removal when the maximum loading of 259 kg/d to the reservoir is exceeded. The plan represents a compromise meeting the basic needs of all parties; but, most important, it has local control. The TAC also moved toward another basin phosphorus control measure via a phosphate detergent ban in the areas with discharges to the river.

The management plan of the TAC represents the culmination of several years of technical and scientific effort and a water quality standard developed in a public forum. Nevertheless, it must stand the test of time. Two years have passed since the Long Lake water quality standard was developed and waste-load allocation was implemented. The standard, the water quality modeling used to evaluate compliance, the management plan, and the lack of continued annual water quality studies have all been criticized as not being adequate. Criticism to date has not focused on any one issue but seems related to disagreement or misconception about the founding principles and assumptions behind the TMDPL. Principally, lake management occurs on an annual basis at a minimum, and Long Lake is being managed similarly. Most of the concern is over day-to-day and embayment-to-embayment variations. Some believe that the entire TMDPL effort and all work to date is flawed and should be redone because these variations happen. There also appears to be a perception or expectation that water quality is controllable on a time scale of a few days.

The details of concern are not as important as the need they represent—the need for local involvement. Local control appears to be a most effective way to implement watershed management. All inhabitants of the watershed must realize that collectively they are the problem as well as the solution. And agencies must acknowledge the ongoing need for information and education once basinwide controls are in place. Public involvement and education are time consuming and costly, but ultimately represent time and effort well spent.

Conclusions

The Spokane River system has experienced serious water quality problems. Much of the research and agency actions taken to date have been the result of moving from one issue to the next. This management by reaction is best characterized as being propelled by the availability of funds and prevailing politics. Nevertheless, water quality progress was achieved by employing state-of-the-art technology and practices. Most likely the day will arrive when source controls are fully implemented and growth will cause the Long Lake standard (maximum loading of 259 kg/d TP) to be violated. At that time, the public and the members of the TAC (or a similar group) will face the tough and costly issues once again. Likely questions will be: Are there technological improvements in treatment that will allow the necessary TP reductions? Will point sources be removed from the Spokane River and the wastewater applied to land for irrigation? Can nonpoint source controls be adequately expanded? Should existing water quality standards be relaxed in consideration of the overriding public good (a public-driven option under state and federal law)? Or should standards become more stringent because of the demand for better quality water? All these questions bear directly on water quality impacts and benefits that are best resolved locally.

Today Spokane River constituencies recognize that long-range planning is required to sustain and improve river water quality. However, such efforts must accommodate future changes that will occur in the watershed. Unforeseen water quality problems will continue to arise, requiring adaptive management. Regulators, dischargers, and the public at large will be faced with situations requiring new technologies and methodologies to further protect water resources.

Acknowledgments. Funding support for the Spokane River/Long Lake, Washington, research has been provided in part by the Washington State Department of Ecology, City of Spokane, Washington Water Resources Center, and the Environmental Protection Agency. Important data and information were provided by numerous other agencies. Many individuals have also made significant contributions to this research and are gratefully acknowledged.

References

Beeton, A.M. 1969. Changes in the environment and biota of the Great Lakes. Pages 150–187 *in* Eutrophication: causes, consequences, correctives. National Academy of Sciences, Washington, D.C., USA.

Bishop, R.A., and R.A. Lee. 1972. Spokane River water quality study. Washington State Department of Ecology Technical Report 72–001, Olympia, Washington, USA.

City of Spokane. 1977. Facilities planning report for sewer overflow abatement. Spokane, Washington, USA.

Condit, R.J. 1972. Phosphorus and algal growth in the Spokane River. Northwest Science **46**:190–193.

Conners, J. 1976. Quaternary history of northern Idaho and adjacent areas. Dissertation. University of Idaho, Moscow, Idaho, USA.

Cunningham, R.K., and R.E. Pine. 1969. Preliminary investigation of the low dissolved oxygen concentrations that exist in Long Lake located near Spokane, Washington. Washington Water Pollution Control Commission Technical Report 69–1, Olympia, Washington, USA.

CWMP. 1981. Spokane County Comprehensive Wastewater Management Plan. Economic and Engineering Services, Inc., Olympia, Washington, USA.

Edmondson, W.T. 1970. Phosphorus, nitrogen, and algae in Lake Washington after diversion of sewage. Science **169**: 690–691.

Edmondson, W.T. 1972. Nutrients and phytoplankton in Lake Washington. Pages 172–193 *in* G.E. Likens, editor. Nutrients and eutrophication. American Society of Limnology and Oceanography, Lawrence, Kansas, USA.

EPA (United States Environmental Protection Agency). 1979. City of Spokane, Washington combined sewer overflow abatement project. Environmental Impact Statement EPA 910/9–78–053, Region X, Seattle, Washington, USA.

Esvelt and Saxton/Bovay Engineers, Inc. 1972. Action plan for better wastewater control, advanced waste treatment, high river quality, and better environment. Prepared for the City of Spokane, Washington, USA.

Fahey, J. 1986. The inland empire: unfolding years, 1879–1929. University of Washington Press, Seattle, Washington, USA.

Fahey, J. 1988. The Spokane River: its miles and its history. Spokane Centennial Trail Committee, Spokane, Washington, USA.

Gasperino, A.F., and R.A. Soltero. 1977. Phosphorus reduction and its effect on the recovery of Long Lake Reservoir. Battelle Pacific Northwest Laboratories, Richland, Washington, USA.

Greene, J.C., W.E. Miller, T. Shiroyama, R.A. Soltero, and K. Putnam. 1978. Use of laboratory cultures for *Selenastrum*, *Anabaena*, and the indigenous isolate *Sphaerocycstis* to predict the effects of nutrient and zinc interactions upon phytoplankton growth in Long Lake, Washington. Internationale Vereinigung für theoretische und angewandte Limnologie, Verhandlungen **21**:372–384.

Greene, J.C., R.A. Soltero, W.E. Miller, A.F. Gasperino, and T. Shiroyama. 1975. The relationship of laboratory algal assays to measurements of indigenous phytoplankton in Long Lake, Washington. Pages 93–126 *in* E.J. Middlebrooks et al., editors. Biostimulation and Nutrient Assessment Workshop. Ann Arbor Science, Ann Arbor, Michigan, USA.

Haggarty, T.G. 1970. Water pollution in the Spokane River. Washington Water Pollution Control Commission Technical Report 69–1, Olympia, Washington, USA.

Karr, J.R. 1991. Biological integrity: a long-neglected aspect of water resource management. Ecological Applications **1**:66–84.

Kovalic, J.M. 1987. The Cleanwater Act of 1987. Second edition. The Water Pollution Control Federation, Alexandria, Virginia, USA.

McDonald, D.L. 1978. Life along the Spokane. Statesman-Examiner, Inc., Colville, Washington, USA.

Merrill, K.R. 1986. Contamination of the Spokane River as indicated by the presence of fecal bacteria. M.S. Thesis. Eastern Washington University, Cheney, Washington, USA.

Mires, J.M., and R.A. Soltero. 1983. Methodology for assessing the initiation date of chemical phosphorus removal at Spokane's Advanced Wastewater Treatment Plant according to predicted spring runoff. City of Spokane, Washington Contract

OPR 81–1054. Completion Report. Eastern Washington University, Cheney, Washington, USA.
Mires, J.M., R.A. Soltero, and D.G. Nichols. 1983. Methodology for assessing the fall termination date of chemical phosphorus removal at Spokane's Advanced Wastewater Treatment Plant. City of Spokane, Washington. Completion Report. Eastern Washington University, Cheney, Washington, USA.
Organization for Economic Cooperation and Development (OECD). 1982. Eutrophication of waters: monitoring, assessment and control. Final Report. OECD Cooperative Programme on Monitoring of Inland Waters (Eutrophication Control), Environment Directorate, OECD, Paris, France.
Patmont, C.R., G.J. Pelletier, and M.E. Harper. 1985. Phosphorus attenuation in the Spokane River. Washington State Department of Ecology Contract C84–076, Harper-Owes, Seattle, Washington, USA.
Patmont, C.R., G.J. Pelletier, L.R. Singleton, R.A. Soltero, W.T. Trail, and E.B. Welch. 1987. The Spokane River Basin: allowable phosphorus loading. Washington State Department of Ecology Contract C0087074, Harper-Owes, Seattle, Washington, USA.
Pearse, Greeley and Hansen Hydraulic and Sanitary Engineers. 1933. Report on sewage disposal, Spokane, Washington. Chicago, Illinois, USA.
Shiroyama, T., W.E. Miller, J.C. Greene, and C. Shigihara. 1976. Growth response of *Anabaena flos-aquae* (Lyngb.) de Breb. in waters collected from Long Lake Reservoir, Washington. Pages 267–275 *in* R.D. Andrews III et al., editors. Terrestrial and aquatic ecological studies of the Northwest. Eastern Washington State College Press, Cheney, Washington, USA.
Soltero, R.A., A.F. Gasperino, and W.G. Graham. 1973. An investigation of the cause and effect of eutrophication in Long Lake, Washington. Office of Water Resources Research Project 143–34–10E-3996–5501. Completion Report. Eastern Washington State College, Cheney, Washington, USA.
Soltero, R.A., A.F. Gasperino, and W.G. Graham. 1974. Further investigation as to the cause and effect of eutrophication in Long Lake, Washington. Washington State Department of Ecology Project 74–025A, Olympia. Completion Report. Eastern Washington State College, Cheney, Washington, USA.
Soltero, R.A., A.F. Gasperino, P.H. Williams, and S.R. Thomas. 1975. Response of the Spokane River periphyton community to primary sewage effluent and continued investigation of Long Lake. Washington State Department of Ecology Project 74–144, Olympia. Completion Report. Eastern Washington State College, Cheney, Washington, USA.
Soltero, R.A., D.M. Kruger, A.F. Gasperino, J.P. Griffin, S.R. Thomas, and P.H. Williams. 1976. Continued investigation of eutrophication in Long Lake, Washington: verification data for the Long Lake Model. Washington State Department of Ecology Project WF-6–75–081, Olympia. Completion Report. Eastern Washington State College, Cheney, Washington, USA.
Soltero, R.A., and D.G. Nichols. 1981. The recent blue-green algal blooms of Long Lake, Washington. Pages 143–159 *in* W.W. Carmichael, editor. The water environment: algal toxins and health. Plenum, New York, New York, USA.
Soltero, R.A., D.G. Nichols, G.A. Pebles, and L.R. Singleton. 1978. Limnological investigation of eutrophic Long Lake and its tributaries just prior to advanced wastewater treatment with phosphorus removal by Spokane, Washington. Wash-

ington State Department of Ecology Project 77–108, Olympia. Completion Report. Eastern Washington University, Cheney, Washington, USA.
Soltero, R.A., D.G. Nichols, G.P. Burr, and L.R. Singleton. 1979. The effect of continuous advanced wastewater treatment by the City of Spokane on the trophic status of Long Lake, Washington. Washington State Department of Ecology Project 77–108, Olympia. Completion Report. Eastern Washington University, Cheney, Washington, USA.
Soltero, R.A., D.G. Nichols, and J.M. Mires. 1980. The effect of continuous advanced wastewater treatment by the City of Spokane on the trophic status of Long Lake, Washington during 1979. Washington State Department of Ecology Project 80–019, Olympia. Completion Report. Eastern Washington University, Cheney, Washington, USA.
Soltero, R.A., D.G. Nichols, and J.M. Mires. 1981. The effect of continuous advanced wastewater treatment by the City of Spokane on the trophic status of Long Lake, Washington during 1980. Washington State Department of Ecology Contract WG81–001, Olympia. Completion Report. Eastern Washington University, Cheney, Washington, USA.
Soltero, R.A., D.G. Nichols, and M.R. Cather. 1982. The effect of seasonal phosphorus removal by the City of Spokane's advanced wastewater treatment plant on the water quality of Long Lake, Washington. Washington State Department of Ecology Contract WF81–001, Olympia. Completion Report. Eastern Washington University, Cheney, Washington, USA.
Soltero, R.A., D.G. Nichols, and M.R. Cather. 1983. The effect of seasonal alum addition (chemical phosphorus removal) by the City of Spokane's Advanced Wastewater Treatment Plant on the water quality of Long Lake, Washington, 1982. City of Spokane, Washington Contract 414–430–000–534.40–3105. Completion Report. Eastern Washington University, Cheney, Washington, USA.
Soltero, R.A., D.G. Nichols, M.R. Cather, and K.O. McKee. 1984. The effect of seasonal alum addition (chemical phosphorus removal) by the City of Spokane's Advanced Wastewater Treatment Plant on the water quality of Long Lake, Washington, 1983. City of Spokane, Washington Contract 414–430–000–534.40–3105. Completion Report. Eastern Washington University, Cheney, Washington, USA.
Soltero, R.A., M.R. Cather, K.O. McKee, and D.G. Nichols. 1985. Variable initiation and termination of alum addition at Spokane's Advanced Wastewater Treatment Facility and the effect on the water quality of Long Lake, Washington, 1984. City of Spokane, Washington Contract 414–430–000–534.40–3105. Completion Report. Eastern Washington University, Cheney, Washington, USA.
Soltero, R.A., M.R. Cather, K.O. McKee, K.R. Merrill, and D.G. Nichols. 1986. Variable initiation and termination of alum addition at Spokane's Advanced Wastewater Treatment Facility and the effect on the water quality of Long Lake, Washington, 1985. City of Spokane, Washington Contract 414–430–000–501.34–3105. Completion Report. Eastern Washington University, Cheney, Washington, USA.
Soltero, R.A., L.M. Humphreys, and L.M. Sexton. 1990. Impact of combined sewer/stormwater overflows on the Spokane River, Washington. Bovay Northwest Inc. Project 1817–010. Completion Report. Eastern Washington University, Cheney, Washington, USA.
Thomas, S.R., and R.A. Soltero. 1978. Recent sedimentary history of a eutrophic reservoir: Long Lake, Washington. Journal of the Fisheries Research Board of Canada **34**:669–676.

WAC (Washington Administrative Code). 1987. Water quality standards for waters of the State of Washington. Chapter 173–201-WAC. Washington State Department of Ecology, Olympia, Washington, USA.

Yake, W.E. 1979. Water quality trend analysis: the Spokane River Basin. Washington State Department of Ecology, Technical Report DOE-PR-6, Olympia, Washington, USA.

Yearsley, J.R. 1982. An examination of the nutrient and heavy metals budget in the Spokane River between Post Falls and Hangman Creek. United States Environmental Protection Agency, Region X, Seattle, Washington, USA.

19

Unexamined Scholarship: The Land Grant Universities in the Inland West

EDWIN H. MARSTON

Abstract

Performance is evaluated of land grant universites in the eight-state inland West (Idaho, Montana, Wyoming, Nevada, Utah, Colorado, Arizona, and New Mexico) with regard to natural resource policy. The context is a western United States that is increasingly urbanized and is shifting away from natural resource extraction and processing toward a service economy. These and other factors have created a new public attitude toward natural resource development. The indications are that land grant universities are not fulfilling their charter obligations to their states and their region with regard to natural resource issues. While this area of the country has attracted some independent institutes and researchers, the land grants' exclusive loyalty to the traditional approaches to natural resources has contributed to a large policy vacuum in the West.

Key words. Land grant universities, natural resource policy, environmental policy, western United States.

Introduction

One well might ask what land grant universities in the inland, arid West have to do with watersheds, the subjects of this symposium. In my experience, not enough. In fact, the land grant universities in the eight-state region this chapter confines itself to (Idaho, Montana, Wyoming, Nevada, Utah, Colorado, Arizona, and New Mexico) appear distant not only from questions dealing with watersheds, but from all the larger natural resource and environmental issues the region has been struggling with for two decades. My object, then, is to answer the question: To what extent have the West's land grant universities aided the ongoing transformation of the region's approaches to the land and its natural resources?

I have suspected a gap between the West's changing attitudes to natural resources and the stance of the land grant universities for several years. For over a century, we Westerners have been cavalier and expedient in our treatment of natural resources. We have elevated "quick and dirty" to the level of a universal operating principle. Only when all else fails do we attempt to read the "instruction manual." The academic world in my part of the West seems almost invisible in the public forums where policies toward water, landscape, wildlife, energy resources, and other assets are debated and decided.

This chapter asks if land grants are really uninvolved in these broad natural resource issues, and, if so, why? These questions are important, but my main ambition is to raise for discussion the responsibilities of the land grant universities toward the West's natural resources. So far as I know, this is a first attempt to ask these questions.

The Region

The eight inland western states are experiencing a period of deep change. The foundations on which the region once rested—construction of dams to irrigate land and generate electricity, grazing of livestock, intensive logging of national forests, building of coal-fired power plants, and other natural-resource-based economies—are in decline or face strong opposition (*High Country News* Staff 1987, 1989). For example, the defeat of the proposed Two Forks Dam in Colorado, despite a $35 million environmental impact statement, shows that water projects can be built, if at all, only with great difficulty. In the once equally hallowed area of livestock grazing, a 20-year reform effort by mainstream environmentalists has faltered, and has been succeeded by an attempt to ban all livestock from the public land. The livestock industry appears to be taking this current push—one that is easier to conceive and implement politically than gradual change—much more seriously than it did the reform effort.

While the old-growth logging controversy of the Pacific Northwest is outside the eight-state, 2,240,350 km^2 region discussed here, it provides an especially dramatic example of the challenge to traditional natural resource extraction. In the inland West, a quieter controversy over deficit, or below-cost, logging on the national forests has led to a White House initiative to change the long-standing policy of subsidizing logging through the U.S. Forest Service road budget and other mechanisms (O'Laughlin 1986, Hof and Field 1987).

These and other natural resource and environmental conflicts occur against the background of a changing demographic and economic base, as the population and economy of the West become more like the nation's as a whole. The trend is visible in a recent study of the Four Corners states of Colorado, Utah, New Mexico, and Arizona. The study shows that these states are overwhelmingly urban, with 75% of the population living in metropolitan areas (Goerold 1990). Their economies show continuing decline in the already small agricultural and natural resource sectors and continuing growth in services. The per capita income of residents of the four states continues to grow, moving ever closer to the national average. The change in economic base has been accompanied by a change in attitudes toward natural resource extraction, as residents of the inland West become more like their fellow U.S. citizens when it comes to logging, dam building, and the like.

This, then, is the context: natural resource extraction under serious, homegrown, within-the-region attack for the first time since the settlement of the area; and the inland West as an increasingly urban, service-oriented, and higher income region. Against this background, I wish to consider the West's land grant universities.

The land grant universities were established by the Morrill Act of 1862 to strengthen rural America through research and teaching. The Morrill Act provided financing, through grants of land, to each state to support colleges in which the "leading objects shall be . . . to teach such branches of learning as are related to agriculture and the mechanic arts" (Eddy 1956:33). In the Midwest, which pushed for their creation, the land grants' research and teaching have concentrated on row crop agriculture, dairies, and orchards. In the region considered here, the emphasis is on forestry, grazing, mining, water development, and other activities largely centered on the region's vast public lands.

The land grants' charters put them, in theory, at the very center of the reevaluation of natural resource activities now occurring in the inland West. To a lesser extent, the West's state universities, which operate under broader educational and research charters, should also be deeply involved. The importance of land grant and state universities in the inland West cannot be overemphasized: they dominate the region to an extraordinary degree because private institutions are almost nonexistent. The private institutions include a very few small liberal arts colleges (e.g., Saint John's in Santa Fe, New Mexico, and Prescott College in Prescott, Arizona); two independent liberal arts colleges (Colorado College in Colorado Springs and the University of Denver); several small church colleges (e.g., Regis College in Denver and Westminster College in Salt Lake City); and one very large church school, Brigham Young University (Cass and Birnbaum 1989). As a result, in the inland West almost all research and teaching comes out of the land grant and state universities.

A Search of the Literature

The fact that the land grant faculty and their research have been invisible to the publisher of a public lands and environmental newspaper (*High Country News*, based in Paonia, Colorado) does not necessarily mean they are uninvolved in natural resource policy. I began work on this chapter with a search for papers, talks delivered at conferences, and reports dealing with the approaches taken by land grant universities to natural resources. I was not looking for papers about forestry, or mining, or range. I was looking for articles, talks, and reports that examined land grants from a natural resource policy perspective. It is a measure of the failure of the search that the most relevant document I turned up was James Hightower's book *Hard Tomatoes, Hard Times* (1973), which sparked a long-lived controversy in the 1970s. Based on work by the Agribusiness Accountability Project, the book concluded that land grant research and teaching have pursued efficiency and mechanized production of food at the expense of small farmers, rural communities, and high quality food. The accountability project blames declining rural populations, disappearing small towns, and rural poverty in part on the policies of land grant universities.

This thesis was incorporated into a lawsuit brought against the University of California in 1979 by plaintiffs who charged that the university's work discriminated against farm workers and small-scale farming, and in favor of large growers and manufacturers of farm machinery and agricultural chemicals (California Agrarian Action Committee et al. v. The Regents of California, 516427–5, Superior Court of the State of California, Oakland). The lawsuit elicited responses from the land grant institutions. The attack on research that promotes large farms brought a general rebuttal in *Science* (Tweeten 1983) entitled "The Economics of Small Farms." In a later issue of *Science,* the lawsuit itself was evaluated and judged, by two land grant-based authors, to be severely lacking in substance (Martin and Olmstead 1985). The court found in favor of the plaintiffs and against the land grant system.

The issues raised by Hightower were interesting but not directly relevant to natural resource and public land issues. However, the Hightower book, articles inspired by it, and the lawsuit were essentially all I found in a search of the Stanford University libraries and their various data bases. I expected most from the ERIC data base (Educational Resources Information Center) because it is a catch-all for policy materials. It indexes both journal articles and many unpublished research reports, program descriptions, lectures, curriculum and teaching guides, and instructional materials. I used the CD-ROM form of ERIC to search back 20 years. I also searched back to 1964 in the Public Affairs Information Service Bulletin (PAIS), which specializes in public policy articles. The searches were made by subject and key word, including Land Grant Universities, Land Grant Universities and Policy, and Land Grant Universities and Natural Resources. Almost certainly I missed some papers, talks, and conferences. But just as certainly there is no substantial body of literature on this subject. Confidence in the search was heightened by the substantial number of references that turned up about the land grant institutions established for black Americans, and their policies toward rural poverty and related subjects in the southeastern United States.

Based on this search, it appears little is being written by the land grants themselves or by outside observers on their relationship to natural resource policy in the West. While the land grants are undoubtedly reacting, and perhaps even adapting, to the West's changing attitude and behavior toward natural resource development, that reaction and adaptation appears to be implicit, unstated, and unexamined, insofar as can be determined from the literature.

Discussion

Why is the literature not bulging with materials telling of the land grants' efforts to reinterpret natural resource issues and determine how the new socioeconomic climate affects their research and education programs? I have no concise answer. But the collective failure of the land grant universities to publicly and thoroughly reexamine their relationship to natural resources makes them an anomaly in the West among natural resource institutions. Other federal bodies concerned with natural resources—such as the Bureau of Reclamation, the Army Corps of Engineers, the Forest Service (Juday 1990)—have put a great deal of effort into reexamining their missions and behavior with regard to natural resources. The Forest Service's New Perspectives Program and Change on the Range initiative, and the Bureau of Reclamation, with its move away from dam building, have been prominent examples of institutional change.

In some ways, the land grant universities and the natural resource agencies resemble each other. Their interests transcend state lines; they are all involved with natural resources; they are rural; they are very large organizations; and their roots are in the 19th century, giving them a great deal of inertia. But a Bureau of Reclamation or Forest Service has a great advantage over the land grant universities: the federal agencies are each unified under a single director. And in the cases of the Bureau and the Forest Service, the impetus to change has come at least partly from the top.

By comparison, the land grant universities in the eight states have no centralized leadership. Morover, collectively they are very large and are spread over a vast area, creating a barrier to an internal or external examination of how well they are fulfilling their missions. In the eight states considered here, there are eight land grants. In 1986, they had 120,000 students, 8,000 faculty, and a collective research budget of $350 milion (National Association of State Universities and Land-Grant Colleges 1987). (See Table 19.1.) Given this size, and the lack of a central organization, it is not surprising that the land grant universities have been unable to come together to examine themselves, as the Forest Service and other natural resource agencies have done.

This failure, for whatever reason, may be responsible for a second failure: the inability of these institutions to fill a very important, and nearly empty,

Table 19.1. Land grant universities in the inland West.

Name	Enrollment	Full-time Faculty	Research Funding (millions)
University of Arizona	34,461	1,328	$117.5
Colorado State University	18,084	1,077	$70.3
University of Idaho	8,584	770	$21.5
Montana State University	10,233	889	$13
University of Nevada-Reno	10,245	1,800	$12.8
New Mexico State University	17,241	923	$54.7
Utah State University	11,690	705	$44.6
University of Wyoming	10,109	485	$18
Total	120,647	7,977	$352.4

Note. The data above are from the years 1985 to 1987, depending on the school's reporting cycles. Source: National Association of State Universities and Land-Grant Colleges (1987).

niche in the West. The battle over natural resource policy in the region has been political and economic. Largely absent from this debate have been the voices of science and technology, telling what can and cannot be done in a physical sense. More generally, the debate has lacked the kind of disinterested expertise on natural resource policy one would expect to come out of academe. Also lacking has been the neutral ground academe could provide for the natural resource debate. In fact, the land grant universities in the inland West have failed to play the kind of role that the University of Washington and its Center for Streamside Studies have played in putting on this watershed conference, with its examination of technical and scientific issues in an economic and social context.

There is, of course, good reason why the land grant universities have not filled this niche, and it lies in the nature of the inland West—a nature that makes it difficult for independent, neutral expertise to thrive. The roots of the problem lie in the region's history. Until the early 1970s, the region was a closed shop, intellectually speaking (*High Country News* Staff 1989). The West in this century has been marked by ancestor worship—an unthinking acceptance of the economic activities and values established by the first settlers. This has made it difficult to question environmental impacts associated with cattle ranching, logging, water development, and mining. Until the 1970s, it was mainly non-Westerners who challenged those activities in order to protect such well-known areas as the Grand Canyon and Dinosaur National Monument. Otherwise, the region was left to its own devices.

The result was a remarkable degree of internal cohesion. Even today, certain natural resource activities may not be questioned. Long after big dams have acquired the status of pariahs nationally (Reisner 1986), almost no western federal representative or senator dares vote against a dam, including such environmentally sensitive men as Senator Tim Wirth of Colorado and former Arizona Congressman Morris Udall. The land grant universities have been both supports and captives of this regional cohesion.

Even ignoring that they get much of their funding from the agricultural committees of the Congress, state legislatures, livestock associations, and other organizations (Hightower 1973), the land grant universities would have found it hard, through the 1970s, to provide disinterested expertise on natural resource issues.

Finally, it may be that land grant universities have made sporadic efforts at change, and that they have not succeeded, or even gained much notice. If the campaign associated with Hightower's book is any guide, change does not come easily to land grant institutions. That campaign consisted of a major research project, congressional attention, and a successful lawsuit. Nevertheless, it has not had a substantial effect on the land grants' emphasis on efficiency, mechanization, and chemical farming. Indeed, the last 20 years have seen continued movement in the high tech, high capital direction, with genetic engineering likely to extend those trends. The course the land grants chose to follow is irrelevant here. The point is that despite the Hightower book, the lawsuit, the support of Robert Bergland, who was President Carter's secretary of agriculture, and pressure from a variety of U.S. senators and representatives, the land grant universities have not budged from their approach to agriculture.

However, the Hightower campaign may not be a good guide to the future of reform of the western land grant universities. The "hard tomatoes" critique was Jeffersonian: it attempted to push an agricultural economy that was becoming ever more Hamiltonian—ever more mechanized, large-scale, and capital intensive—back toward a Jeffersonian ideal of small farmers and natural processes.

The West's land grant issues are very different from those framed in the Hightower debate. The central challenge to the traditional commodity-producing West, and therefore to the land grant universities that are an integral part of it, is environmental rather than economic and social. While environmental issues have economic and social aspects, until now the environmental movement has been economically and socially conservative. It is my judgment that, with the possible exception of Earth First! and Greenpeace, the environmental movement avoids radical analyses and is far more vulnerable to charges of elitism than of Marxism. Environmental challenges to the West's approaches to natural resources start with calls for biological sustainability and diversity rather than with calls for social and economic justice. Wendell Berry (1977), among others, sees the two as related, but on the surface the environmental challenge is relatively free of economic and social criticism.

Analysis

By its very nature, this paper is contentious. The object of my analysis is to minimize ideological criticisms and to concentrate on pragmatic concerns. The broad question posed here is: How well have the land grants aided and adapted to the West's changing attitudes toward the land and natural resources? The question is based on the assumption that the West's attitudes toward natural resources are undergoing fundamental change. Even ten years ago, this assumption would have been challenged by many Westerners. Today, however, there are numerous examples to indicate that natural resource policy is rapidly changing. One striking example concerns the defeat of the proposed Two Forks Dam, which was to be located near Denver, Colorado. Charles F. Wilkinson (Moses Lasky professor of law at the University of Colorado) has concluded from that defeat, and other events, that the doctrine of prior appropriation—the very bedrock of natural resource policy in the West—is now very weak. In Wilkinson's view, the doctrine has simply become one of many legal determinants when it comes to water development. The others include the Endangered Species Act, the Clean Water Act, and the National Environmental Policy Act (Wilkinson 1991).

With that as background, I wish to pose and try to answer three questions.

Question 1: Have the faculties of the land grant institutions done research that has helped the West understand and define itself in this period of transformation?

Answer 1: This is a sweeping question about the work of 8,000 faculty, and I will give the only possible answer: yes and no. The "yes" refers to some of the research that concerns what I would call the detailed functioning of natural resource systems—forests, streams, atmospheric flows, and the like. But this article is not concerned with that type of research. It is concerned with the broader, more inclusive, synthetic, nonreductionist work that is necessary to the making of public policy decisions. And here the answer must be no. In terms of overview and synthesis, very little has come out of the land grant universities that has the capacity to help the West understand and define itself during this natural resource reformation. But important work about western issues has been done, and I wish to use two examples to illustrate the kind of research the land grant universities have failed to do.

This question is the heart of the paper. It is an attempt to set a standard by which the land grants can be judged. Attempts to evaluate academic work are very difficult. The thousands of land grant university faculty are engaged in a great deal of research, much of it having nothing to do with natural resources. With few exceptions, the work dealing with natural resources consists of narrow, detailed studies of range management practices, weed control measures, computer models of forest planning systems, management of multi-aged timber stands, economic analyses of logging practices, and the like. Presumably, the quality and relevance of this work varies according to a bell-shaped curve.

What appears to be missing is the higher order research and thought on the West's attempts to adapt its traditional natural resource economy and way of life to the search for sustainability. Work on sustainability would not replace reductionist or technical work on grazing, logging, watersheds, and the like; it would build on these results and create a framework to guide them. That has not happened. Instead, we have the land grants grinding away at nuts-and-bolts research projects without placing that activity in a context relevant to the emerging West.

To ground these generalizations, I wish to discuss two examples of the research that is missing from the land grant universities' *curriculum vitae*. The first piece of work was done by the *Sacramento Bee* in 1985 and published in a special series in September 1985. Titled "Selenium: A Conspiracy of Silence," it described widespread selenium contamination resulting from U.S. Bureau of Reclamation projects in 15 western states. The story the *Bee* revealed is usually associated with the contamination of the Kesterson Wildlife Refuge in California. But Kesterson was simply the first such example found. The newspaper's work showed that Kesterson was not a fluke; it was one instance of a wide-ranging problem associated with reclamation projects in the West.

I call this exposé "research" rather than journalism because the reporters, Thomas Harris and James Morris, gathered water samples, submitted them to testing laboratories for analysis, and interpreted their findings (Harris 1989). They were guided by a work plan put together by federal scientists, and then suppressed by top officials in the Bureau of Reclamation. The research plan then fell into the hands of the two journalists. The newspaper's efforts resulted in congressional hearings, repetition of the tests by other scientists, and substantial changes in federal reclamation policy and practice. At present, a $100 million research program, sparked by the *Sacramento Bee* series, is under way. It is important to note that contamination from selenium and other substances was widespread, affecting areas in 15 states. It required a level of scientific investigation two newspaper reporters could do by drawing on extensive selenium research, much of it detailed work done at land grant universities, going back 50 years. Yet the land grant faculty in those states—people who are their area's specialists in water quality, irrigated agriculture, soil types, and so on—never turned up the problem. Or, if they turned it up, they did not bring it to the attention of the public. This raises the selenium contamination issue from an isolated "scoop" to an indication that the land grants have serious deficiencies when it comes to research on broad natural resource questions.

The second piece of work was done by Frank Popper and Deborah Epstein Popper. He chairs the Urban Studies Department at Rutgers University, New Jersey; she is a member of the Geography Department at Rutgers. Their analysis of the 1980 U.S. census results revealed that much of the Great Plains—the region west of the 100th meridian and east of the Rocky Mountains from Canada to Texas—has reverted to pre-1890 population levels of less than two people per square mile, or 0.77 person per square kilometer.

This level is significant for both historic and practical reasons. Historian Frederick Jackson Turner, using data from the 1890 census, declared the western frontier closed in a famous paper he gave in 1893 ("The Significance of the Frontier in American History"). He declared the frontier closed partly because the population density in the West had risen above two people per square mile; therefore, he wrote, the West no longer had an identifiable frontier beyond which lay "empty" land waiting to be settled (Turner 1894). Today historians argue over the significance of Turner's thesis (see below), but it is, at the least, interesting that the Poppers find that in 1980 over 100 counties in the U.S. West had fewer than two people per square mile, or 0.77 person per square kilometer (Popper 1986).

This decline, which they were first to perceive and dramatize for a large audience, has enormous importance for federal, regional, state, business, and individual decisions made about the Great Plains. On the federal and regional levels, decisions about energy policy, farm policy, proposed irrigation projects such as Garrison and Oahe, and use of water out of the Ogallala Aquifer should all be influenced by population trends and population levels.

The Poppers have not been shy about drawing natural resource policy implications from their demographic research. They have attracted national media attention by proposing that the Great Plains again become a "buffalo commons," in which land would be replanted to short-grass prairie and repopulated by the wildlife that once roamed the region. The Poppers argue that five separate settlement attempts of the Plains by the United States have failed over the past century, with the failures marked by such cataclysms as the dust bowl of the 1930s and the energy bust of the 1980s. Now, the Poppers say, it is time to let the region return to nature.

The Poppers' ideas have attracted national attention. They have also attracted intense, hostile attention in the Plains states. So far as I know, no one has rebutted their demographic analysis. But the Poppers have been attacked repeatedly for being from New Jersey. Typically, Plains-based critics of the husband-wife team will ask why two researchers from an eastern state with enormous problems of its own are studying the Plains.

If that question is slightly rephrased, it fits in with the theme of this chapter: How is it that two "outsiders" from an eastern land grant university were the first to discover the depopulating of a region that has ample research institutions of its own, in the form of Great Plains-based land grant universities? It is the same kind of question that the *Sacramento Bee's* work raises: How is it that two newspaper reporters were able to make a major scientific discovery in a 15-state region with significant scientific resources, in the form of land grant universities and charters that require them to examine natural resource and agricultural issues?

This section began by asking how useful the land grants have been in helping the West understand and define itself in this period of transformation. Rather than attempt to answer the question in general, I cite these two pieces of research as examples of synthesizing work that can help us understand the period the West is passing through. The work done by the *Sacramento Bee* reveals a very difficult problem that attends long-term irrigation of certain arid lands. The work by the Poppers concerns population declines that they attribute to settlement strategies—mainly based on natural resource development—that failed to take into account the environmental realities of the West. This is the kind of material that individuals and policy makers need in order to make informed decisions about natural resources. I know of no examples of similar helpful, generalizing work that has come out of the land grant universities in the states we are concerned with here.

If the West's land grant universities are not raising and answering important natural resource policy questions, where are they putting their energy? Two examples help answer this question. The tabletop fusion debacle occurred when the University of Utah chose to promote a claim by two researchers that they had discovered a way to create nuclear fusion in a glass beaker. The claim, thanks in part to the backing it received from the university and from the Utah legislature, set off a year-long scientific circus before it was decided that there was no substance to the claim. The incident can be seen as an isolated fluke. But it can also be seen as part of a larger pattern, in which the West's research institutions, unable or unwilling to grapple with the natural resource problems that affect their home region, choose instead to chase big science.

This incident does not stand alone. An earlier, widespread example of this flight from western reality is provided by the West's unsuccessful pursuit of the superconducting super collider (SSC). While not specifically about land grants—most of the chasing was done by state universities—this example illustrates the general theme. In 1986, I noticed that state after state in the inland West was appropriating hundreds of thousands of tax dollars, and raising hundreds of thousands of additional dollars from the private sector, to pursue the SSC even though it was clear from the start that only Colorado had a chance of winning the machine. Investigation showed that all eight states considered here—Arizona, Utah, Montana, Wyoming, Idaho, New Mexico, Colorado, and Nevada—were pursing the project (Hinchman and Marston 1987).

The region, it appeared, could organize itself and raise relatively large amounts of money to pursue high tech research projects, but its public research and education institutions could not discover that reclamation projects were damaging scores of wildlife refuges and other bodies of water, and that a large chunk of the West was reverting to pre-1890 population levels. At least one conclusion to be drawn from the fusion incident in Utah and the more general pursuit of the SSC is that western institutions of research and teaching are in search of a vision and a mission having nothing to do with the West's natural resource issues. This apparent flight from their charter responsibilities is occurring even though the West's natural resource issues are compelling to a national and even international constituency.

Question 2: Have the West's land grant universities established themselves as places to which people turn for information on questions of natural resource policy and practice?

Answer 2: This question is more general than the first question, dealing as it does with the *perception* of authority. It, too, is open to a variety of answers. Public land users such as ranchers or timber companies are likely to turn to a land use college with questions. But it appears to me that the land grants have failed to establish themselves as authorities for the general public and the media in their areas of responsibility. For example, no land grant institution comes to mind as a center of excellence in the study of public land management, western water use and reclamation, Native American issues, range management, and energy resources. There are, of course, institutions that can tell you with great accuracy where the West's coal reserves lie, whether a given range would best support cattle or sheep, and what areas of the West have been heavily drilled for oil and gas. But, again, this chapter is not concerned with such questions. It is concerned with the synthesis of reductionist research in order to help individuals and institutions make natural resource policy decisions.

The failure of faculties at the West's public institutions to take a leading, synthesizing role in natural resource issues has created a policy vacuum that a variety of policy entrepreneurs have attempted to fill. For example, the energy policy vacuum has been filled by the Rocky Mountain Institute. But one can ask whether a small think tank in western Colorado compensates for the fact that the region does not have one or more institutions known for their overall expertise in energy matters.

In public land management, various people and institutions are striving to fill the vacuum left by the land grant universities. In rangeland resources, there is the Holistic Resource Management Institute (Bingham and Savory 1990). With regard to National Park Service policies, there is Alston Chase, author of *Playing God in Yellowstone* (1986). In the Yellowstone National Park region, it is the Greater Yellowstone Coalition that has promoted the idea that the park and surrounding lands should be managed as a whole, no matter which agency is managing a particular piece of ground. That concept, which has now spread to the Colorado Plateau, northern California-southern Oregon, and other western areas, was developed initially by F.C. Craighead, Jr. (1979), an independent researcher, and is discussed by Grumbine (1990).

Independent natural resource policy bodies are helpful to those struggling with natural resource policy, but they are no substitute for large, coordinated academic institutions addressing these issues.

Question 3: Why have the West's land grant universities failed to do research that would enable the West to better understand itself and cope with its current transformation.

Answer 3: There is no complete answer, but one can attempt partial answers. The general one is that the land grant universities' sources of funding limit their independence. As documented by Hightower (1973), the land grant universities are funded by the agricultural committees of the U.S. Congress, state legislatures, and user groups.

What does this funding relationship mean in practice? There is no simple answer. Some researchers have the stature and conviction to produce accurate, expert, unbiased work regardless of outside pressures. But there are cases where academic freedom appears to have been compromised; and there are also cases where an entire field appears to see the world as the dominant commodity group sees the world, or at least is fearful of differing publicly with that group. In this section, I give an example of each. They are not meant to provide sweeping proofs. It is unreasonable to expect sweeping, documented proof of how the land grant system works without the resources of—let us say—Congressman John Dingell and his power to investigate universities. These examples are meant to be indicators. They may be isolated and unrepresentative instances; or they may be representative of how the system works.

An example of how politics influences academe is provided by a well-publicized Wyoming case in 1987. The victims were academic freedom and several faculty members at the University of Wyoming, which serves as both a land grant and a state university. A firm with strong political connections, Char-Fuels, had used university and state resources to prepare a study of a coal-conversion process. The firm hoped to convince the state to provide funding for a commercial-level project. When the draft report came back to the university, the vice-president for research, Ralph DeVries, gave it to two professors of engineering for review. They sent back a harsh evaluation of the Char-Fuels report. The university then forwarded the report to the committee in charge of handing out $30 million in state clean coal funds, but suppressed the negative review (Lazarus 1989).

The suppression of this review was part of a pattern. Earlier that year, in January 1987, three chemical engineering professors at the University of Wyoming, including chairman David Cooney, had also been asked to review the Char-Fuels proposal. The three expressed concern about the proposal. The response to their technical evaluation was immediate and intense. They were attacked in print by a high state official who supported the project, and they were then threatened with a lawsuit by an executive of Char-Fuels. When the university attorney, David Baker, indicated to Chairman Cooney that the professors would probably have to handle any lawsuit on their own, the three retracted their criticism, and chose to stay silent on the subject for 22 months. They spoke up only when extensive press coverage and public discussion changed the nature of the politics, and revealed the way in which both the University of Wyoming and state government had been influenced by Char-Fuels. As a result of the exposé, Char-Fuels got only part of the financing it had sought.

This incident, which involves the development of one of Wyoming's major natural resources, illustrates the sensitive political atmosphere in which Wyoming's land grant operates, and its vulnerability to suppression and political pressure. Moreover, the effectiveness of the Char-Fuels process was an objective question of engineering effectiveness, divorced from the more difficult, value-laden issues surrounding clean air and water, threatened wildlife, and the like. When it comes to these more emotional and subjective questions, it seems likely that it would be even more difficult for land grant university faculty to provide balanced and objective information.

One way to address the question of possible bias in land grant faculty members would be to do a survey. So far as I could tell from the literature search, no such survey has been done. However, a survey of sorts was conducted over a period of several years in the early 1970s by Johanna H. Wald, an attorney with the Natural Resources Defense Council, Inc. During that period, she was investigating how the U.S. Bureau of Land Management managed 170 million acres of public land. Ms. Wald recalled, in a January 25, 1991, conversation: "I began trying to find out how rangelands were managed. But I literally could not get one range professor at [the University of California at] Berkeley, or at [the University of California at] Davis, or at any of the other schools in the West to talk to me. Their position basically was: 'There's no problem; there's nothing to talk about.'" Ms. Wald said she finally got help from "former BLM and Forest Service employees. To this day, my help still comes from them. Academics will talk to me now, but they appear only on the other side in court."

Ms. Wald's data, of course, are not the same as an unbiased survey, complete with controls and statistical analysis of the results. But her approach has one advantage over such surveys: respondents could not give theoretical answers. She was asking them to help an environmental group investigate conditions on the range. It was an excellent test of values, especially because the Natural Resources Defense Council is very much an establishment, centrist group.

There are, of course, any number of reasons why all the range professors at all the land grant universities in California and the region I am concerned with refused to talk with or aid Ms. Wald. Their failure to talk with her certainly hints at loyalty to the livestock industry, rather than loyalty to range management independent of a user group, but it does not constitute proof of such loyalty.

However, the important point is that Ms. Wald's lawsuits had an immense, continuing effect on the use of the public range in the western United States, and that the faculty of the land grant universities excluded themselves from that process of change, except indirectly as a result of the support they gave to the livestock industry and the Bureau of Land Management. But those sides lost. The ranchers and BLM were on the defensive, and were unable to stop the forces of change. The NRDC lawsuits, therefore, provide another example of how the land grant faculty and institutions have failed to play a role in the formation of public policy surrounding the West's changing approaches to natural resources.

What policy changes did Ms. Wald and her colleagues put into motion? NRDC won its lawsuit in 1974, forcing the BLM to prepare environmental impact statements on all rangeland under its jurisdiction (Natural Resources

Defense Council, Inc. v. Morton, 388 F. Supp. 829,833 [D.D.C.]). The suit was part of a series of events that led to major changes in federal law, including passage of the BLM's organic act, called the Federal Land Policy and Management Act of 1976 (FLPMA), and the Public Rangelands Improvement Act of 1978 (PRIA). The suit also helped incite the Sagebrush Rebellion, in which public land users throughout the West attempted unsuccessfully to transfer BLM-managed land from federal to state domain. In some circles, Ms. Wald is known as the Mother of the Sagebrush Rebellion.

My purpose is not to recount the judicial and legislative history of range management in the 1970s and 1980s—a good review is contained in the 1985 suit, Natural Resources Defense Council, Inc. v. Hodel, 618 F. Supp. 848 (D.C. California), and in Fradkin (1979)—but rather to point out that one side of this land policy revolution went forward without help from anyone connected with a land grant. Had the NRDC lawsuits lost, and had Congress not passed FLPMA and PRIA, one might argue that the range professors were correct when they told Ms. Wald there was no problem. Or, alternatively, had the land grant faculty members chosen to stay neutral, and not appear as expert witnesses on behalf of the government and the public land ranchers, as they did, one might argue that they were simply maintaining academic neutrality. But in light of the results of her efforts, it appears that the land grant faculty, for whatever reason, missed an opportunity to influence a change in public policy.

Exceptions

Having made a series of very strong statements about the West's public institutions, I now wish to partially eat them. This article would do a disservice to the independent-minded faculty in natural resource schools and departments at the West's land grants and state universities if it did not point out that there are exceptions to the above portrait at each college. However, almost always those exceptions consist of individuals who stand alone or almost alone in their departments or schools, or they are doing natural resource work at western institutions but outside of natural resource departments.

Conclusion

Does the preceding discussion suggest a prescription for change? Is there a model to guide land grant universities along a path that would enable them to better discharge their charter responsibilities, and thereby better serve the West? In answering this question, it is helpful to consider the two separate, although related, challenges that land grant universities face. One of the challenges confronts not just land grant universities, but higher education and scholarly research generally. The other challenge is peculiar to the land grant universities.

The Broad Challenge

In a paper, "The Improvement of Teaching: An Essay to the Stanford Community," released March 3, 1991, Donald Kennedy, a noted biologist and the president of Stanford University, called on his faculty to examine the nature of scholarly research: "The overproduction of routine scholarship is one of the most egregious aspects of contemporary academic life: it tends to conceal really important work by its sheer volume, it wastes time and valuable resources, and it is a major contributor to the inflation of academic library costs."

The breaking down of research into the smallest possible publishable units is not confined to the Stanfords of the world. To the extent that land grant universities are part of the broader academic culture, the drive to publish frequently and narrowly also affects their faculties, and works against the intellectual synthesis and overviews that would be useful to the setting of natural resource policy.

In part, the solution for the land grant universities will come slowly, as part of an overall reform of higher education and research. But land grant universities occupy a special niche in academe. Their charters tie them to the land and to rural peoples. They have both a constituency and a set of challenges posed by natural resource issues that non-land grant universities, with their broader charters and responsibilities, do not have. The term "cow college" can be used in a denigrating way, but the sense of mission and focus imposed by their responsibilities could also help the land grant universities change more quickly than the rest of academe.

A western academic field provides a dramatic example of how change can occur rapidly. The field is that of western history, which over the last decade has attracted a great deal of public attention in the form of media coverage, a surge in student interest, and greatly increased attendance at meetings of the Western History Association (P. N. Limerick, pers. comm., April 25, 1991).

This resurgence has been driven by the relatively new willingness of western historians to grapple with the foundations and underlying assumptions of their field. As mentioned earlier, the father of western history is Frederick Jackson Turner. In his famous paper of 1893, Turner laid out the tenets of western history: that the West was settled by 1890, when an identifiable frontier disappeared; that the settlement of the West had been a key element in the nation's evolution as a democracy; that the West had been "empty" until it was cultivated and civilized by European pioneers moving in from the East; and that the end of this settlement posed a crisis for the United States as a whole (Turner 1894).

Turner's ideas dominated western history for nearly a century. They were quietly questioned at times, but not until the late 1980s did a small group of historians, led by Patricia Nelson Limerick of the University of Colorado, succeed in subduing the Turner thesis. Ms. Limerick did this in her book,

The Legacy of Conquest, which argued that the very notion of a frontier is false; that the land had not been empty before the "pioneers" arrived; and that bloody conquest—of native peoples, wildlife, and the environment—was a more accurate description of what transpired than settlement (Limerick 1987).

Ms. Limerick and her intellectual colleagues—Donald Worster of the University of Kansas, William Cronon of Yale University, Richard White of the University of Washington, and others—have been active within and without academe. (Ms. Limerick, for example, is a regular columnist for *USA Today*.) Moreover, they are writing and speaking just as the region and the nation appear ready to reconsider the traditional views of the West. The success of their and other efforts to reshape the West's view of its past can be seen most clearly in the Academy Award winning movie *Dances With Wolves*. Whatever the movie's artistic value, the *idea* behind it is revolutionary. It turns the John Wayne-John Ford-Louis L'Amour view of the West on its head, as the U.S. cavalry and Anglo settlers become villains and the Indians become heroes. It brings the thesis of Ms. Limerick and the other New Western Historians onto the screen, before millions of people.

The methods and success of the New Western Historians have important implications for the land grant universities. First, the success of their ideas shows the public's willingness, even readiness, to see the West in a different perspective. It is possible that the same openness exists in other fields of western thought, including approaches to natural resources. Their work also illustrates the kind of intellectual synthesis that is needed to bridge the gap between academic research and public understanding and policy. Ms. Limerick's book is a series of essays based on much more detailed and reductionist work done by her fellow western historians. She succeeds in linking together research that the public would never read or care about in order to create a whole that can be grasped by the public.

To sum up, the faculty at land grant universities are caught up in the same "publish or perish" syndrome as academe at large. However, through their ties to the land and to rural peoples, they have the potential for providing a model of reform and renewal.

The Specific Challenge

In addition to the general challenge facing all research and teaching institutions, the land grant universities face a more specific one: they appear to be dominated by certain interests. In *Hard Tomatoes, Hard Times*, Hightower (1973) argued that land grant universities in general are dominated by those who own large farms, by chemical and equipment manufacturers, and by others he identifies as agribusiness.

In the West, the natural resource schools and departments of land grant universities often seem particularly loyal to those who produce commodities: ranchers, loggers, water developers, and so forth. This loyalty is probably a mix of personal conviction on the part of the faculty and the fact that their funding flows from agricultural and natural resource committees of the U.S. Congress, state legislators, and industry.

It is in no one's interest to see the land grant universities go from their present loyalty to commodity producers to loyalty to environmental groups. The Sierra Club and the Natural Resources Defense Council are important institutions, but there is no reason to turn a department of range science at the University of Utah into a branch of the Sierra Club instead of a branch of the National Cattle Association. The goal should be to set the natural resource departments of the land grant universities free to do what they should do best: help the West better understand and manage its public land and the resources on that land. The reform of academic research, discussed above, will help create a climate for such reform. But there is also a need for political reform.

A model for such reform can be found by examining the evolution of the U.S. Forest Service over the past decade. Ten years ago, the agency was dedicated, from top to bottom, to the production of commodities, mainly through logging—to "getting the cut out." Today, the agency is in the midst of a thoroughgoing change in behavior and values. It can be seen in the creation of new programs, such as the New Perspectives Program, which is an attempt to shift emphasis from commodity production behavior to a wider view of the national forests (Franklin, this volume; Lee, this volume). It is also visible in such disciplines within the Forest Service as range management and the Change on the Range Program.

The shift can also be seen in organizational restructuring, as the various national forests hire ecologists and attempt to integrate their scattered approaches to the different resources, and thereby get away from the reductionist approach the agency has long practiced. As a long-time observer of the Forest Service, I have been startled to observe it shifting toward a more holistic view of its resources.

This shift has not been easy, and many would argue that it has not gone nearly far enough. However, the engines of this reform are identifiable. Externally, the Forest Service has been driven by its critics, who have tirelessly beat the agency about the ears in the media, in the courts, in the Congress, and at every step of the Forest Service's own internal appeals process. Although the agency has not lost major battles in the Congress, it has lost many battles in the courts, and those who administer the agency are painfully aware that it now shares management of national forest land and resources with various federal district judges. Internally, the Forest Service has been driven by two major influences. First, its long tradition of independence and of putting the resource first is reasserting itself. Second, it has come under pressure from some of its newer employees, who come to the agency with a different set of land management values than longer-term employees. While I have separated the forces acting on the agency into external and internal ones, they are actually inseparable. They interact with and feed off each other.

The Forest Service, of course, is different from the land grant universities. But there are similarities. Each has a charter responsibility tying it to land and resources in the West; each is a public organization or collection of organizations; and each had grown very close to the West's commodity producing interests. The pressure for change on the Forest Service has not transformed it into a branch of the Sierra Club, and is unlikely to do so, but it has given the agency the incentive and freedom to better carry out its responsibilities to the land it is charged with managing.

Like the Forest Service, the land grant universities have traditions of independence and of caring about the land and rural peoples. What may have been missing—compared with the Forest Service—is external pressure. The land grant universities have not received the attention the Forest Service, Bureau of Land Management, Fish and Wildlife Service, Bureau of Reclamation, and Army Corps of Engineers have received.

Why have those interested in change in the West ignored the land grant universities? For two possible reasons. First, the land grant universities do not directly manage land or have responsibility for wildlife, and therefore are not seen as important in the struggle over land and wildlife. Second, the struggle for the West's public lands has usually been seen as a political issue, rather than one of ideas and information. As a result, land grant universities have not been viewed as important participants in the region.

I believe this has been a serious mistake. The West's natural resource challenges will not yield to politics alone. Even if political conflict were to disappear tomorrow, the region would be at a loss over what to do in a wide variety of natural resource situations. The kind of expertise that land grant universities should be good at providing is badly needed in the West. It is important for those who care about the West's vast, rural landscape and its people to strengthen the land grant universities and their faculty by putting pressure on them to fulfill their charter responsibilities and contribute to public discourse and policy making.

Acknowledgments. I wish to thank Jim Risser and Jim Bettinger of the John S. Knight Fellowship Program in the Department of Communication at Stanford University for providing an atmosphere in which a journalist can attempt a scholarly paper. I also wish to thank Stanford's wonderful corps of reference librarians for their invaluable help.

References

Berry, W. 1977. The unsettling of America. Sierra Club Books, San Francisco, California, USA.

Bingham, S., and A. Savory. 1990. Holistic resource management workbook. Island Press, Covelo, California, USA.

Cass, J., and M. Birnbaum. 1989. Comparative guide to America's colleges. 14th edition. Harper and Row, New York, New York, USA.
Chase, A. 1986. Playing God in Yellowstone. Atlantic Monthly Press, Boston, Massachusetts, USA.
Craighead, F.C., Jr. 1979. Track of the grizzly. Sierra Club Books, San Francisco, California, USA.
Eddy, E.D., Jr. 1956. Colleges for our land and time. Harper and Bros., New York, New York, USA.
Fradkin, P. 1979. The eating of the American West. Audubon **81**:94–121.
Goerold, W.T. 1990. Economic and demographic overview of the Four Corners states. The Wilderness Society, Washington, D.C., USA.
Grumbine, E. 1990. Protecting biological diversity through the greater ecosystem concept. Natural Areas Journal **10**:114–120.
Harris, T. 1989. Feds' toxic coverup is foiled by newspaper. High Country News **21**(22):9.
High Country News Staff. 1987. Western water made simple. Island Press, Covelo, California, USA.
High Country News Staff. 1989. Reopening the Western frontier. Island Press, Covelo, California, USA.
Hightower, J. 1973. Hard tomatoes, hard times. Schenkman Publishing Co., New York, New York, USA.
Hinchman, S., and E. Marston. 1987. The superconducting collider. High Country News **19**(15):1.
Hof, G.H., and R.C. Field. 1987. On the possibility of using joint cost allocation in forest management decision making. Forest Science **33**:1035–1046.
Juday, G.P. 1990. An interview with Jerry F. Franklin, Bloedel professor of forestry, University of Washington. Natural Areas Journal **10**:163–172.
Lazarus, W. 1989. The charring of Wyoming. High Country News **21**(10):1.
Limerick, P.N. 1987. The legacy of conquest: the unbroken past of the American West. Norton, New York, New York, USA.
Martin, P.L., and A.L. Olmstead. 1985. The agricultural mechanization controversy. Science **227**:601–606.
National Association of State Universities and Land-Grant Colleges. 1987. Serving the world. National Association of State Universities and Land-Grant Colleges, Washington, D.C., USA.
O'Laughlin, J., editor. 1986. Below cost timber sales: eight articles. Western Wildlands **12**:2–38.
Popper, D.E., and F.J. Popper. 1987. The Great Plains: from dust to dust. Planning **53**:12–18.
Popper, F.J. 1987. The strange case of the contemporary American frontier. Yale Review **76**:101–121.
Reisner, M.P. 1986. Cadillac desert. Viking, New York, New York, USA.
Shanks, B. 1984. This land is your land. Sierra Club Books, San Francisco, California, USA.
Turner, F.J. 1894. The significance of the frontier in American history. Pages 197–229 *in* Report of the American Historical Association, Washington, D.C., USA. Reprinted in 1963 by Ungar, New York, New York, USA.
Tweeten, L. 1983. The economics of small farms. Science **219**:1037–1041.
Wilkinson, C.F. 1991. A eulogy: prior appropriation (1848–1991). Annual symposium, Water Watch of Oregon, Northwestern School of Law, Lewis and Clark College, Portland, Oregon, USA.

20

Integrating Sustainable Development and Environmental Vitality: A Landscape Ecology Approach

ROBERT G. LEE, RICHARD FLAMM, MONICA G. TURNER, CAROLYN BLEDSOE, PAUL CHANDLER, COLLETTE DEFERRARI, ROBIN GOTTFRIED, ROBERT J. NAIMAN, NATHAN SCHUMAKER, AND DAVID WEAR

Abstract

Opportunities for sustaining humans and their environmental systems can be enhanced by examining how socioeconomic and ecological processes are integrated at the landscape level. Landscape properties—such as fragmentation, connectivity, spatial dynamics, and the degree of dominance by habitat types—are influenced by market processes, human institutions, and landowner knowledge as well as by ecological processes. These same landscape properties affect ecological processes that influence species abundance and distribution, as well as the production of goods and services valued by human society. An approach for understanding these complex interactions includes models that simulate (1) land use changes that alter landscape pattern, (2) effects of landscape pattern on species persistence, invasion of exotics, and resource supplies, and (3) dynamic interactions involving possible feedback processes that can alter land uses or landscape patterns. Adaptive management is recommended for using this approach to attain sustainable development where ecological processes operating at micro, meso, and macro scales are integrated.

Key words. Sustainability, landscape ecology, socioeconomic processes, ecological effects, modeling.

Introduction

The most important challenge facing environmental management is to create and foster a balance between human needs and environmental sustainability (Ruckelshaus 1989, Lubchenco et al. 1991). Sustainability is defined as the process of change in which the continued exploitation or protection of resources, the direction of investment in land, and associated institutional changes are consistent with future as well as present objectives for perpetuating environmental qualities and socioeconomic functions of ecosystems (World Commission on Environment and Development 1987).

One manifestation of the interaction between natural processes and human activities is the pattern of land use (or land cover) observed in a region. Landscapes are dynamic mosaics of natural and human-created patches that vary in size, shape, and arrangement (Burgess and Sharpe 1981, Forman and Godron 1986, Urban et al. 1987, Naiman et al. 1988, Turner 1989). Understanding what went into the creation of these mosaics, how humans depend on landscape functions and products, and how landscape changes affect ecological and socioeconomic processes can provide a sound basis for guiding sustainable development within a region (Naveh and Lieberman 1984, Zonneveld and Forman 1990, Odum and Turner 1990). A broad-based understanding of landscape structure and function is essential for promoting integrated management where human sustenance and environmental integrity are considered part of the same system.

This article summarizes an approach to analyzing how humans affect ecological processes at the landscape scale and how landscape patterns affect environmental conditions and socioeconomic functions essential for sustainability. This analysis is a three-step process involving (1) an examination of the socioeconomic factors influencing landscape change, (2) simulation of landscape change by using economic, social, and environmental variables as drivers, and (3) evaluation of the impacts of landscape change on environmental quality and resource supplies. We conclude with a discussion of a modeling environment designed to integrate the different forms of knowledge acquired from the many disciplines involved.

The approach we present is designed to discover how to enhance opportunities for sustaining human and environmental systems by integrating the socioeconomic and ecological processes influencing landscape patterns and their effects. This objective is addressed through three related questions.

1. How do economic and social factors influence land use practices and thus landscape patterns?
2. What are the impacts of landscape patterns on environmental quality (ecological condition) and resource supply (goods and services)?
3. How can environmental quality and resource supply be managed to foster socioeconomic and ecological sustainability?

The relationship between these questions and the processes they address is displayed in Figure 20.1. The first question links the two driving factors (economic and social) to land use practices and resulting landscape patterns. The second question links landscape patterns to environmental qualities and resource supplies, and the third question evaluates these links in the context

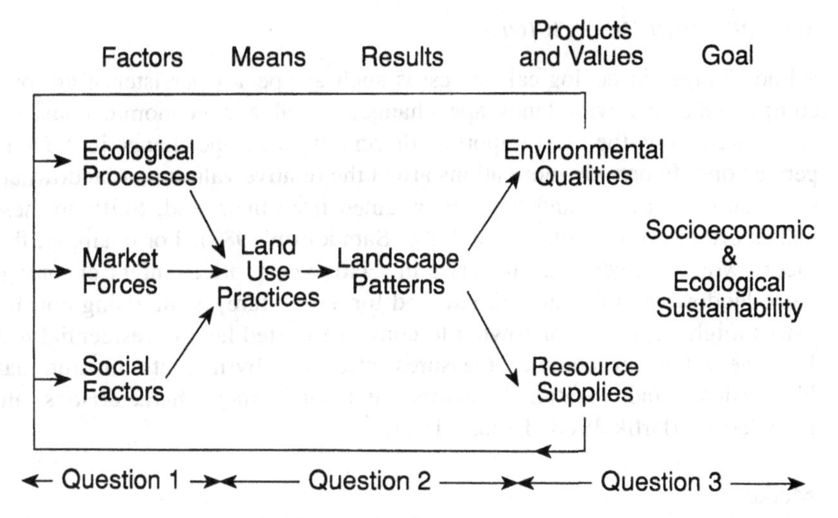

FIGURE 20.1. Processes affecting socioeconomic and ecological sustainability. Sustainability is achieved through the maintenance of acceptable environmental qualities and resource supplies. Ecological processes, market forces, and social factors affecting land use practices are altered by feedback from environmental qualities and resource supplies.

of environmental sustainability. Dynamic processes are represented by feedback loops.

The processes addressed in the first question are modeled separately and are integrated sequentially, beginning with social factors and culminating with a simulation model that uses social, economic, and biophysical information to assign probabilities of transition for land uses. Additional models are used to simulate the effects of changes in landscape patterns on environmental qualities and resource supplies. An adaptive management approach is recommended for answering the third question. Adaptive management involves the use of an experimental or quasi-experimental logic for management problem solving (Holling 1978, Walters 1986).

Influence of Economic and Social Factors on Landscape Structure

Landscape structure is a function of land use, and land use is influenced by social, economic, and environmental factors. We will first examine the economic and social factors driving land use and then discuss their use in models to simulate changes in landscape structure.

Socioeconomic Factors

While changes in ecological processes such as species persistence are outcomes associated with landscape change, social and economic considerations are among the most important drivers of landscape change in the temperate zone. Economic motivations affect the relative values that landowners place on the products and services obtained from their land. Shifts in these values often result in altered land use (Samuelson 1983). For example, this theory would suggest that increases in relative crop prices might encourage some landowners to clear forested land for agriculture, while rising housing prices might prompt a landowner to convert forested land to residential use. Relative values are complex measures influenced by markets; institutional, biophysical, and locational factors; and landowner characteristics and knowledge (Bartik 1988, Brooks 1987).

Markets

Markets are the pricing mechanisms for many important products of land management. The market-determined prices, combined with the costs of management (largely determined by locational factors), define the net returns or rents accruing to the various land uses. In the absence of nonmarket considerations, land will tend to be placed in high rent uses (Clark 1973). In forested landscapes, for example, timber, recreation, and agricultural markets have a critical bearing on the potential return from alternative uses of land (Alig 1986, Parks 1991).

Institutional, Biophysical, and Locational Factors

Institutions are cultural structures such as governmental agencies, interest groups, and the body of laws and policies governing land and resource use. Institutions may influence the discretion of the landowner directly through mechanisms such as zoning and land use regulation. Or they may influence decisions indirectly by altering conditions affecting land rents or, as in the case of public agencies, those affecting budgetary capability (Repetto and Gillis 1988). For example, subsidies for reforestation increase rents accruing to forestry uses, whereas agricultural subsidies may favor planting certain crops. Tax structures that tailor land to specific uses also alter relative land rents. Regulations may protect the habitat of an endangered species or the quality of water entering a stream or aquifer. Institutions affecting land use are a product of processes such as social movements and governmental policy making (which involves interaction among legislative, administrative, and judicial branches of government).

Biophysical attributes determine the potential production from a tract of land (Palmquist 1989). These attributes include soil type and structure, hydrology, vegetative cover, animal habitat, and slope and aspect. For example, slope in conjunction with hydrology defines the stability of the soil, and therefore the sustainability of production under various forms of management. Vegetative cover, such as forest, may tend to remain unchanged because of prohibitive land-clearing costs.

An important attribute for determining land values is the location of a tract relative to cultural and biophysical features (e.g., Thunen, cited in Samuelson 1983). The distance between a tract and the market for its products defines transportation costs. Distances to services (e.g., sewer and water, shopping, a medical center) characterize the relative isolation of a tract, and thus its value for real estate development (Alonzo 1964). Use of adjacent lands also influences management decisions on a particular tract, especially when neighboring lands are developed for residential uses.

Landowner Characteristics and Knowledge

The worth of products and services obtained from the land is influenced by the values, management objectives, and life-styles of landowners (Bartik 1988). Income and budget constraints sway land use decisions as well. For example, compare a landowner engaged in subsistence farming with one who is dependent on a relatively high retirement income. The value to these owners of game products derived from the land might differ considerably. Similarly, one public landowner may be driven by a mandate to generate revenue by cutting timber from state trust lands, while an adjoining public owner may be guided by objectives emphasizing the preservation of original forests, such as a national park.

Land use decisions are also influenced by the extent of the landowner's knowledge about the land's physical capabilities and the relevant product markets. Knowledge about biological systems and markets is distributed unevenly within a society. Landowners and managers may often know more about the biological and management possibilities of their lands than the researchers who are responsible for increasing scientific knowledge (Padoch 1986). Although it is often ignored in economic analysis, landowners behave differently depending on the degree of their knowledge about the land. The length of time that owners or managers have worked with a parcel of land can have a direct bearing on how much they have learned about its characteristics and uses (Chandler 1990); absentee owners are expected to have less knowledge than residents. It is possible to separate landowners or managers into categories of those who are knowledgeable about their lands, its possible uses, and the markets for its products, and those who are relatively uninformed.

Methodology

Calculating Transitional Probabilities

The socioeconomic factors discussed above can be incorporated in models to simulate the propensity of land to change from one use to another. The extent of landowner and manager knowledge can be measured by using methods developed in cognitive anthropology and applied to cultural ecology (Spradley 1979). These methods are derived from systematic interviews of landowners and managers with the purpose of quantifying the amount and accuracy of their knowledge and its influence on their behavior. The remaining economic and biophysical data can be acquired from existing maps and data bases and will involve time series analyses.

The influence of the socioeconomic factors can be simulated by modeling the propensity of a patch of land in a specific use to change to another use (or remain in the same use) as a function of several driving variables. This analysis involves modeling land use shifts between time periods in probabilistic terms by using limited dependent variable approaches (e.g., Parks 1991). These approaches are used to describe the influence of selected explanatory factors on the conditional probability of choices among a limited number of alternatives. The equation representing these conditional probabilities is:

$$P_{ij} = Pr(Y_{ij} = 1/X)$$

where P_{ij} is the probability that land will change from use i to j; Y_{ij} is a binary variable that takes on the value 1 if the tract moves from use i to use j ($Y_{ij} = 0$ otherwise) within the measurement period (Pr is a probability operator); and X represents the vector of socioeconomic factors described previously.

A cumulative distribution function for the change in land use is estimated. The results of these estimations can be evaluated by using standard t-tests to assess the influence of the X-variables on decisions and to provide transition probabilities for land, social, and economic conditions. These transition probabilities are used to drive the landscape-change simulations. Simulations reflecting changes in economic and social conditions are accomplished by altering the values of the driving variables, recalculating transition probabilities, and driving the landscape change simulation model with these new probabilities.

Simulating Landscape Change

Projecting regional patterns of land use and land cover requires integrating the transitional probabilities calculated above with existing land cover patterns. Given a map of existing land cover patterns, the land use transition probabilities are distributed spatially depending on the economic, social, and physical characteristics of each land parcel. The linkage of these probabilities to the spatially explicit data base, such as is stored in a geographic information system (GIS), allows changes in a landscape through time to be simulated. Alternative scenarios can be explored, and the relative importance of different controlling factors evaluated. Thus far, this approach has been applied only to a limited extent but appears promising (Turner 1987, 1988; Naiman et al. 1988, Parks 1991).

Impacts of Landscape Change on Environmental Quality and Resource Supply

Since environmental integrity and resource supply are intimately tied to the structure of landscapes, changes in the landscape inevitably affect the ecological properties of the environment and the abundance and quality of resources produced.

Environmental Quality

The ecological implications of land use change are numerous because landscape patterns influence a variety of ecological processes (Turner 1989, Naiman and Décamps 1990). For example, plant succession, biological diversity, foraging patterns, predator-prey interactions, dispersal, nutrient dynamics, and the spread of disturbance all have important spatial components at broad spatial scales (Huffaker 1958, Holling 1966, May 1975, Peterjohn and Correll 1984, McNaughton 1985, Turner 1987, Senft et al. 1987, Burke 1989, Hardt and Forman 1989, Turner and Gardner 1991). We address one particular ecological response: species persistence as an indicator of environmental quality. We begin with a general discussion, then address indigenous and exotic species.

Species Persistence

The persistence of species is influenced by the number, size, and geographic arrangement of patches across a landscape (Forman and Godron 1986). Thus land management for the maintenance of a particular species may require the protection of habitat patches of a particular size and juxtaposition. An alternative management or conservation goal may be to perpetuate natural fluctuations in the landscape mosaic (e.g., resulting from a natural fire regime), implying that the abundance of certain species will fluctuate as well. The connectivity of habitat is often of particular importance, and several studies suggest that landscapes have critical thresholds in habitat connectivity at which ecological processes will show dramatic qualitative changes (Gardner et al. 1987, Krummel et al. 1987, O'Neill et al. 1988, Turner et al. 1989, Gosz and Sharpe 1989, Rosén 1989, Naiman et al. 1988, Johnston and Naiman 1990). Changes in a landscape that is near a threshold may strongly influence species persistence. For example, habitat fragmentation may progress with little effect on a population until the critical pathways of connectivity are disrupted; then a slight change can have dramatic consequences for the persistence of the population (Gardner et al. 1991, Turner et al. 1991). In addition, changes in habitat connectivity can influence the susceptibility of a landscape to disturbance, such as the spread of an invading organism. Therefore, if the long-term maintenance of biological diversity is a conservation goal, a land management strategy that emphasizes regional biogeography and landscape patterns may be necessary (Noss 1983, Noss and Harris 1986).

Indigenous Species

Concern over the persistence of indigenous species is reflected in the increasing attention given to maintenance of biodiversity (Noss 1989, Lubchenco et al. 1991). Although biodiversity is difficult to define, and even more difficult to measure, diversity of species is conceptually simple and ultimately measurable. Destruction or disturbance of habitat is a major cause of threats to species persistence or species loss (Fahrig and Merriam 1985, Noss 1987, Harris 1988, Yahner 1988, Lord and Norton 1990, Quinn and Harrison 1988, Wilcove 1987). Fragmentation of the landscape has resulted in a loss of communities, the creation of abundant edge habitat at the expense of interior habitat, the alteration of natural disturbance regimes, and the loss of variability of ecological processes over broad spatial and temporal scales.

We now realize, for example, that because habitats for many species occupy areas larger than ecological communities and sites, a landscape approach is necessary to understand how habitat disturbance affects species persistence. The efforts to preserve the northern spotted owl (*Strix occidentalis caurina*) illustrate how some species may require maintenance of suitable habitat conditions over large landscapes (Thomas et al. 1990).

Many species appear to require the relatively undisturbed habitat conditions found in national parks, wilderness areas, old-growth forests, or mature second-growth forests (Carey 1989). These areas often contain landscapes large enough to provide the ecological patterns and processes needed for their survival. Perpetuation of species may require continuity in the natural cycles of disturbance which characterized the development of these landscapes and their dependent species over evolutionary time. These areas constitute ecological refuges within larger landscapes subject to disturbances accompanying the production of timber, agricultural crops, or residential and commercial development.

Suitable habitat conditions can often be maintained on lands used for commodity production or recreation if essential vegetative characteristics and landscape patterns are provided (Franklin and Forman 1987, Naiman et al. 1991). Landscape connectivity is important for species that require advanced seral stages for purposes of migration from one large patch to another; and animals such as deer (*Odocoileus* spp.), elk (*Cervus elaphus*), and anadromous fishes (e.g., *Salmo, Oncorhynchus*) require migration corridors relatively free from residential development or other obstructions to their movement between habitats (Naiman et al., this volume; Stanford and Ward, this volume). Cavity nesting birds can be perpetuated by leaving snags in forests where harvesting would normally require removal of such safety or fire hazards (Oliver et al., this volume; Franklin, this volume).

The impact of organisms on the environment is well known. Vegetative cover and individual species influence soil characteristics by the production of litter and by activities associated with roots; they affect water yield and qualities via evapotranspiration and soil-nutrient interactions; and they even influence climate by their abilities to regulate heat and water fluxes. Similarly, animals exert long-lasting controls on ecological systems by direct habitat alterations and by their foraging strategies (Naiman 1988). Feeding strategies and physical environmental alteration affect plant and animal community composition and biogeochemical cycling and nutrients in soils and water. These changes are important because they reverberate throughout the food web, causing alterations to the ecosystem that cannot necessarily be anticipated with current knowledge and technology. Examples are found in ecosystems undergoing sustained anthropogenic alterations (Zaret and Paine 1973, Vitousek 1986, Thomas et al. 1990).

Exotic Plant Species

The invasion of nonnative species may be the most pervasive influence affecting biodiversity in many systems (di Castri 1990, Coblentz 1990). Exotic species frequently displace native plant or animal species, and thus alter the ecological dynamics of an area. They may even alter ecosystem structure and function. The invasion of the Atlantic shrub *Myrica faya* in Hawaii demonstrates the impact an exotic may have. *Myrica faya* has invaded extremely nitrogen-deficient sites of Hawaii Volcanoes National Park. By actively fixing nitrogen, this plant has altered nutrient cycling, productivity, and primary succession in this region (Vitousek 1986). In California, the exotic ice plant *Mesembryanthemum crystallinum* has altered the physical and chemical properties of the soil by concentrating salt from throughout the rooting zone onto the soil surface (Vivrette and Muller 1977). Numerous other examples of alteration of ecosystem properties by exotics have been documented (Vitousek 1986).

Exotic species typically thrive in highly disturbed or managed environments (Baker 1965, Forcella and Harvey 1983, Fox and Fox 1986, Heywood 1989, Rejmanek 1989), which are often located adjacent to parks and other natural preserves. Hence parks and preserves may be adversely impacted by the invasion of exotic species from adjacent disturbed or managed lands. Once established at a site, an invading plant can disperse to other nearby locations. Thus the distribution, abundance, and juxtaposition of colonized patches and potential sites all influence the movement and establishment of exotics. There are three fundamental questions that concern exotic species from our perspective: (1) What features of the landscape are more susceptible to colonization? (2) What features serve as long-term or ephemeral sites for the establishment of exotics? (3) What features serve as corridors for movement of these species?

A landscape-level perspective addresses source areas (i.e., where exotic species are well established), and potential colonization sites. Source areas must be evaluated for (1) number, diversity, and abundance of exotic species present, (2) the spatial distribution of source patches across a landscape, and (3) the length of time the area serves as a source. Potential colonization sites must be evaluated for (1) susceptibility to invasion, (2) proximity to source areas, and (3) regional and local disturbance regimes that alter susceptibility to colonization or establishment. By considering patches as sources and potential colonization sites, and corridors as paths for movement of exotics, managers can interpret land use practices in terms of increasing or decreasing landscape susceptibility to invasion by exotics with contrasting life history and colonization strategies.

Resource Supply

The design and maintenance of landscape patterns conducive to the perpetuation of desirable ecological processes can significantly affect opportunities for producing resource supplies. For example, the conservation strategy to protect the northern spotted owl may reduce timber harvesting on federal lands by as much as 23% (Thomas et al. 1990, Interagency Economic Effects Review Team 1990, Lippke et al. 1990). Additional regulation of timber harvesting on private land may be necessary to maintain migration corridors for dispersing owls. Reductions of this magnitude over a few years can be socially and economically devastating to people who live in relatively isolated communities dependent for jobs and income on wood resources (Lee 1990). The rate and magnitude of this change may exceed critical thresholds of human adaptability, resulting in a loss of socioeconomic sustainability for people living in local settlements (Fortmann et al. 1990, Lee et al. 1991). Conservation strategies for other species requiring the maintenance of large-scale landscapes could have similar regional or local human impacts. Development activities such as dam construction, rapid timber harvesting, and residential development can also interrupt socioeconomic sustainability by disrupting the habitat upon which people depend for their livelihood (Paine 1982, Muth 1990, Bradley 1984).

These conservation and development activities may simply replace one sort of human habitat with another. Unlike other animals, humans are relatively adaptable to a wide variety of environments and are unique in their ability to consciously transform landscapes into habitats that suit their needs, including the long-term need for a sustainable life support system. The absence of biologically programmed habitat requirements makes it possible for humans to adjust to constraints imposed by the lesser adaptability of other species and natural ecological systems (Berger and Luckmann 1966). However, human adaptability is governed by social and cultural processes that constrain both the rate at which people can accommodate to habitat change or create new habitats and the geographic scale of change that they can accommodate without major social and economic disruptions (Goldschmidt 1990, Lee 1991, Firey 1960). Hence the maintenance and regulation of ecological processes at the landscape scale will involve reinstitutionalizing the temporal and spatial organization of patterns of behavior long geared to the availability of agricultural fields, timber stands, homesites, campsites, or individual plants or animals.

Analysis of human responses to landscape structure is complicated both by the role humans play in structuring their environments to suit their needs and by the fact that their habitat requirements are defined by their culture rather than by genetic programming (Berger and Luckmann 1966). Moreover, humans are capable of satisfying their material needs by exchanging goods and services on a global scale rather than just locally. The challenge for resource managers is to find how the goods and services needed by society can be produced while also progressively building and maintaining the landscape patterns that will ensure perpetuation of essential ecological processes (Firey 1960, Lee 1991, Franklin and Forman 1987, Stanford and Ward, this volume). Land use decisions made with the goal of attaining desired goods and services can be evaluated in terms of their long-term environmental and socioeconomic impacts. For example, individual small and localized land use decisions occurring over several years appear insignificant at the landscape level. However, consideration of the ecological impact that these decisions could have in the aggregate could prevent landscape structure from crossing a threshold that would cause permanent and undesirable change.

Methodology and Feedback

The identification of landscape characteristics (e.g., patch sizes, habitat connectivity) that are particularly important to the persistence of a single species, indigenous species, or supply of a resource is necessary before effects of landscape change can be predicted. These variables can be identified by analyzing field data, or by performing a sensitivity analysis of a spatially explicit model. In the latter case, model simulations are conducted as landscape structure is varied systematically. Landscape characteristics that have a strong influence over the model's projections can be considered as landscape state variables for the species or resource of interest. Changes in the landscape can then be expressed as a trajectory through the state space. At any point, the landscape can be described by the values of each state variable. Figure 20.2 depicts such a diagram for a case in which three landscape state variables have been deemed crucial to the persistence of a species. Land use changes would cause the landscape to evolve through time along a path or trajectory that connects the current landscape to the future landscape.

Consider as an example a spatially explicit demographic model that predicts population size and number of local extinctions for a certain species. Given a particular management regime, one objective is to maximize the species' population size while still being within the state space defined by management options that consider both ecological processes and socioeconomic functions. The criterion used to select a path connecting the present

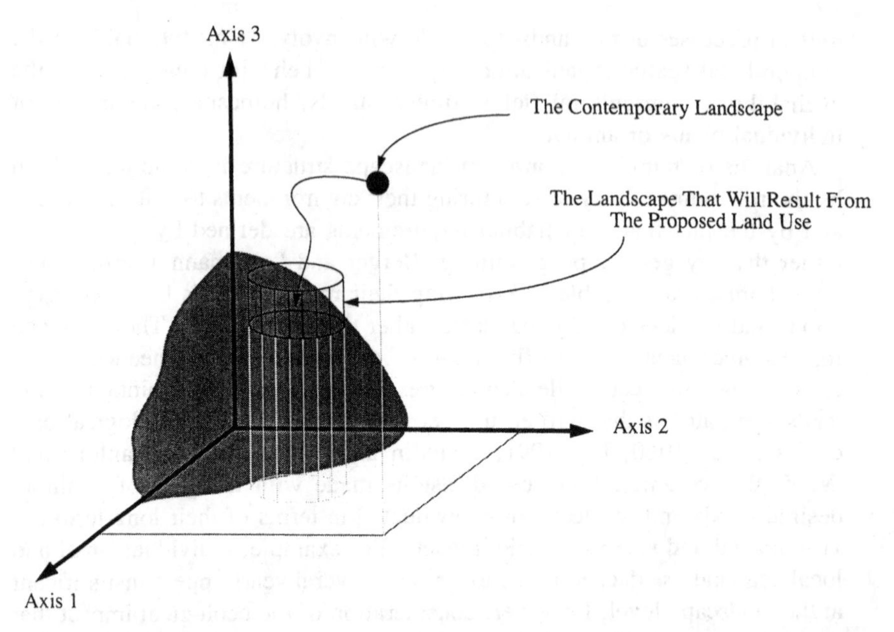

FIGURE 20.2. The contemporary and future landscapes within the landscape state space. The contemporary landscape is represented by a point, whereas the future landscape is represented as a region. Both the endpoint in this region and the path used to get there are variables that can be adjusted by changing the way land is managed. The gray region corresponds to an area of low species persistence. In this example, most of the future landscape would correspond to low species persistence. The path connecting the contemporary and future landscapes ends in the gray region, and would thus not be acceptable. A path connecting the contemporary landscape to the portion of the future landscape outside the gray region could be attained by altering land use pattern.

and future landscapes might be to minimize the number of local extinctions expected during the management period. Using the demographic model, regions of state space could be identified that were unsuitable for the species under study. These regions could then be plotted and, in the case of a three-dimensional state space, the results might look something like the shaded area in Figure 20.2. The results of the optimization could be used in planning land management activities (e.g., timber harvesting patterns). Management regimes that would prevent the landscape from taking excursions into the portion of state space deemed unacceptable for species persistence could be determined. This entire analysis could then be repeated for other species that operate at different scales or trophic levels in the landscape. Landscape trajectories could thus be identified with the highest probabilities for not threatening any of the species under study yet providing resource flows needed to sustain socioeconomic functions. If the species to be studied were chosen so that together they formed a good indicator of the health of the community, conclusions could be drawn about the impact of land use on a community of organisms inhabiting a managed landscape.

The landscape trajectory can be evaluated as a feedback process. Society responds to the state of the landscape and external socioeconomic factors by making land use decisions. These decisions result in the construction of a new landscape, and therefore movement in the state space. This new landscape, with its own set of attributes, affects future land use decision making, just as the original landscape did. The changed landscape may alter future land use decisions by changing socioeconomic functions and causing changes in policies governing landscape management. This inherent feedback process permits the examination of long-term effects of present land use decisions.

A Knowledge System Environment for Integrating Interdisciplinary Modeling

A versatile methodology is essential when seeking to integrate qualitative and quantitative knowledge describing diverse biological, physical, social, and economic processes. One methodology that is designed to integrate knowledge, regardless of its origin or form, is artificial intelligence (Borrow and Collins 1975, Hart 1986, Charniak and McDermott 1987, Luger and Stubblefield 1989). An environment where information is integrated is called the knowledge system environment (KSE) (Coulson et al. 1989).

Modeling Systems

Knowledge integration can be accomplished by using two artificial intelligence techniques: knowledge representation and search (Saarenmaa et al. 1988). Knowledge representation is the form in which each piece of information is stored in the computer. In the KSE, these forms include answers to system queries via a user interface, tabular and spatial information stored in a data base, analytical and simulation models residing in a model base, spatial analysis routines operating within a geographic information system (GIS), and qualitative information organized as a knowledge base. Search is how the information is processed. A search algorithm can be viewed as a path between a problem and a solution and typically is defined in the knowledge base. Along a search path are questions that must be answered before the solution is reached. The specific path taken is defined by the problem posed, the information gathered, and the order in which the information is compiled. Knowledge is ultimately displayed spatially as a map. Pixels or patches on the maps are represented as objects and have assigned to them attributes (such as land cover type, ownership, and transitional probabilities) collected during a search algorithm (Bobrow and Stefik 1986). Search paths conclude with the assignment of attributes to pixels or patches that represent the solution, and the production of a map.

A system to project and evaluate landscape changes is shown in Figure 20.3. Each representation of knowledge is stored in one of the system's components. The general solution algorithm represents the universal search path. A single iteration in a simulation involves two steps: (1) estimation of landscape change, and (2) prediction of the ecological effects of the landscape change. The final product of both analyses is a map which is entered into the data base for subsequent simulations or examinations of feedback processes.

The analysis of landscape change involves two knowledge bases. The first houses the search paths used to evaluate the socioeconomic factors that affect land use decision making. In these paths, transitional probabilities for land use or land cover type are calculated and then assigned as attributes to each pixel or patch in the map. In the second knowledge base reside the paths used to apply the transitional probabilities to the simulation of landscape change. The result here is a map of land cover that includes all the transitions in cover type caused by changes in land use.

The ecological-impacts-of-landscape-change knowledge base contains the search paths used to integrate the simulated landscape produced above with information gathered from the other system components for the purpose of analyzing ecological impacts. The information assembled during a simulation is specific to an ecological classification, such as an endangered species or water quality. Again, the information collected during the search process will be represented as attributes for each pixel or patch. The map produced can illustrate changes in some ecological variable (i.e., increase or decrease in diversity) or show abundances or distributions of the ecological classification.

Only a few modeling systems have been developed that integrate the human reasoning applied to land use decision making with the ecological implications of landscape change (Grainger 1986). The way knowledge is represented and searched for is unique to each problem and is not necessarily superior in any one system. There are, however, distinct advantages in incorporating qualitative knowledge to simulation problems that traditionally were solved using only numerical methods (Raman 1986, Saarenmaa et al. 1988). These advantages are: (1) hypotheses about the dynamics and implications of landscape change can be evaluated by developing mechanistic models of ecological processes and their socioeconomic drivers; (2) event-driven rather than time-driven models can be constructed; and (3) the knowledge base can be easily modified to allow for reasoning about alternate situations (such as predicting the behavior of the system with the introduction of a new landowner type or ecological characteristic, a change in the value of a resource, a change in government regulations, or a change in landowner heuristics).

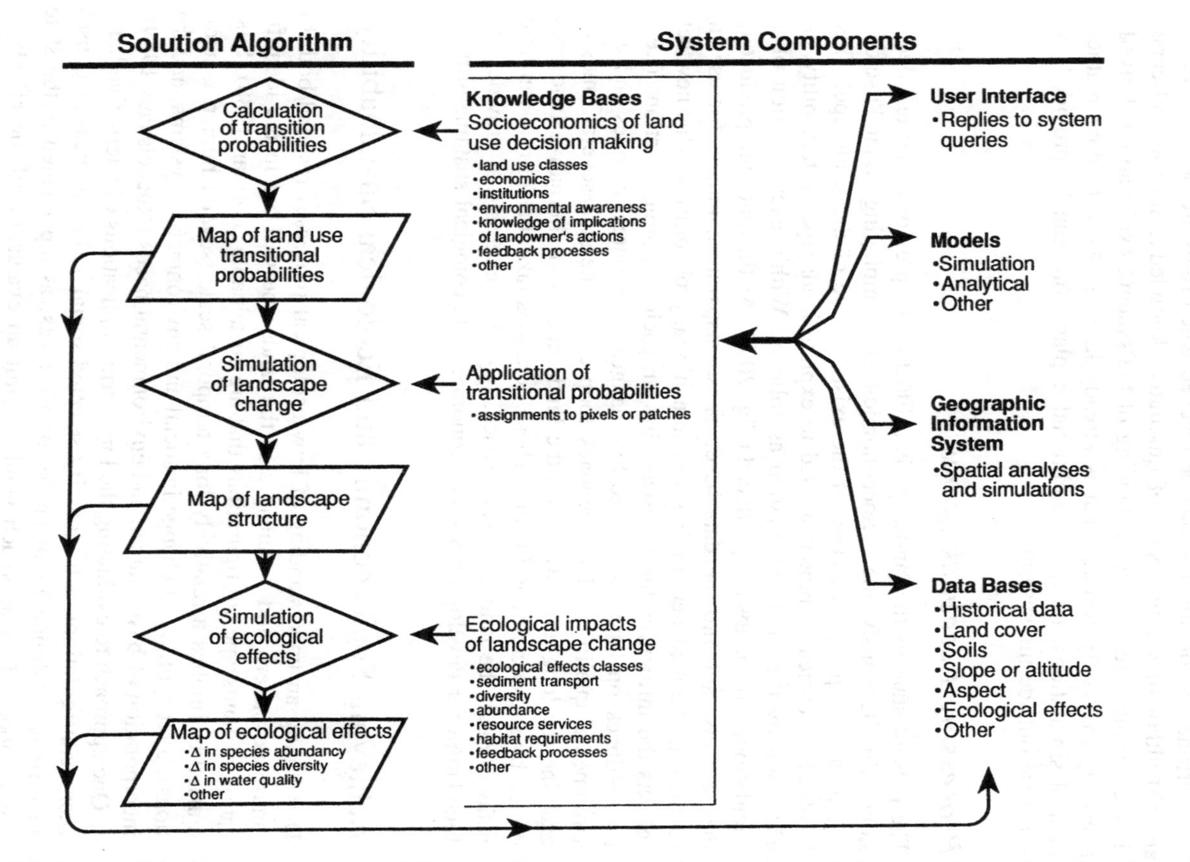

FIGURE 20.3. Schematic of the basic components of the modeling system, their linkages, and the general solution algorithm. Diamonds represent simulations and parallelograms are system outputs.

The disadvantages of including qualitative knowledge are that (1) it is difficult to create a comprehensive set of rules that describe a domain; (2) the rules are not based on an objective and predictable set of assumptions (e.g., erroneous conclusions may be reached when rules are not observed); and (3) although the inclusion of qualitative knowledge into a landscape simulation increases our understanding of the system, explanations derived internally may only restate what is already known. As with pure mathematical simulations, however, additional explanations can be provided by expertise outside the system.

Processing Feedback Loops

The present state of the landscape is a function of previous land use decisions. Simultaneously, the socioeconomic factors impinging on land use decisions are, in part, a function of the existing state of the landscape. This feedback, a critical component used to explain landscape sustainability, is addressed by the modeling system as follows. Within each iteration of a simulation, three maps are produced (Fig. 20.3). As the maps are produced, they are inserted into the data base. These maps are available for use in calculating the next iteration's set of transitional probabilities. This new set reflects the influences the landscape has on socioeconomic functions (i.e., the feedback on land use caused by changes in resource supplies and environmental qualities). This feedback loop can be iterated several times to examine the long-term effects on the landscape from alternative land uses driven by socioeconomic factors. These iterations take the form of multiple paths through the landscape state space. It is this process that provides the foundation for developing a socioeconomic and ecological sustainability.

Achieving Socioeconomic and Ecological Sustainability

Developing an understanding of how ecological processes respond both to landscape pattern and to changes in the landscape will aid in the identification of landscape configurations that are sustainable (Forman 1988). Sustainability implies a relatively long temporal scale (several human generations), a recognition of change in social and environmental systems, and the incorporation of basic human and environmental needs (Ruckelshaus 1989).

One approach to evaluating the long-term implications of alternative landscape configurations is to quantify selected social and ecological responses to various combinations of landscape variables, as represented in the state space model. This approach would permit an evaluation of the effects of landscape patterns, as they change through time, on parameters of social or ecological interest. Given the complexity of the socioeconomic-ecological interactions that we face, the development of aggregated measures of sustainability remains a crucial task.

Moreover, sustainable development requires an approach to science that involves inventing desirable futures (Jacob 1982, Medawar 1982). Science alone cannot tell us what to do to sustain ecological and societal systems, because of the importance of choosing desired conditions. Such choices are inherently social and must involve the people affected by environmental management decisions. Hence the complexity of implementing sustainable development defies the normal scientific process of using hypothesis testing as a primary tool for prescribing actions. Instead, we recommend an interactive approach to goal seeking and monitoring of consequences in which working hypotheses are formulated, tested, reformulated, and retested through an adaptive management strategy (Holling 1978, Lee and Lawrence 1986, Walters 1986). In this way science can serve as an instrument for development without imposing scientific prescriptions in place of social choices.

Conclusions

As this chapter has shown, there is much to be learned about sustaining humans and their environmental systems by focusing on ecological processes that operate at the landscape level. Landscape analysis can facilitate discovery of interactions between socioeconomic and ecological processes that were not clearly discernible when examining individual ecological processes and socioeconomic systems in isolation. Human production of necessary goods and services from lands has readily measurable impacts on ecological processes at the landscape scale that often are not evident when attention is focused on specific sites. Models that estimate land use transition probabilities appear to be effective means for analyzing anthropogenic changes in land cover that affect landscape pattern in ways not detected by focusing on smaller scale ecological processes. Similarly, modeling of the relationships between landscape patterns and environmental quality or resource supplies can show how larger scale ecological processes affect species persistence, the invasion of exotics, and the provision of goods and services used by humans.

However, landscape analysis can be just as limiting as the focus on sites, stands, and ecological communities. Attention also needs to be given to the scale of major biogeographical regions and continents to anticipate and detect acceleration of ecological disruptions in one region or continent when landscape-level regulations constrain land uses or resource production in another country or political jurisdiction (Lippke et al. 1990). For example, given no reduction, or steady increases, in demand for wood products on a global scale, reductions of wood production in the Pacific Northwest may indirectly contribute to accelerated harvesting of forests in Siberia or another region with available wood supplies and less concern for conservation. Regulations designed to conserve landscape-level ecological functions in one region may have the unanticipated consequence of contributing to far greater losses of biological diversity and sustainability in another.

Landscape ecology emerged as a means for understanding the cumulative effect of individual land use decisions on ecological processes operating on a larger scale. Site- and stand-level ecological disturbances accumulated over time and space to have major unanticipated effects on landscape processes. Landscape analysis must develop methods for anticipating the regional and global effects of prescribing landscape patterns that may make ecological sense at the landscape scale. To help attain sustainability of environmental qualities and resource supplies, scientists must learn how to integrate analysis performed at micro, meso, and macro scales. This article has shown how progress can be made in studying sustainability at the meso scale. Further work is needed to develop methods for integrating such landscape analysis with the macro scale. The focus will need to be the global ecological processes (especially atmospheric pollution and biogeochemical cycling), human population growth and distribution, human sustenance activities, and the human ecology of energy and material flows.

Acknowledgments. Support for developing a landscape approach to integrating sustainable development with environmental vitality was provided by the generous contribution of funds or personnel by the following agencies: U.S. Man and the Biosphere Program, U.S. Forest Service (Pacific Northwest Forest Experiment Station and Southeast Forest Experiment Station), SAMAB (Southeast Appalachian Man and the Biosphere), Oak Ridge National Laboratory, University of the South, University of Washington, National Science Foundation, and National Park Service.

References

Alig, R.J. 1986. Econometric analysis of the factors influencing forest acreage trends in the Southeast. Forest Science **32**:119–134.

Alonzo, W. 1964. Location and land use: toward a general theory of land rent. Harvard University Press, Cambridge, Massachusetts, USA.

Baker, H.G. 1965. Characteristics and modes of origin of weeds. Pages 147–172 *in* H. G. Baker and G. L. Stebbins, editors. The genetics of colonizing species. Academic Press, London, England.

Bartik, T.J. 1988. Measuring the benefits of amenity improvements in hedonic price models. Land Economics **64**:171–183.

Berger, P., and T. Luckmann. 1966. The social construction of reality. Doubleday, New York, New York, USA.

Bobrow, D.G., and M.J. Stefik. 1986. Perspectives on artificial intelligence programming. Science **231**:951–957.

Borrow, D.G., and A. Collins, editors. 1975. Representation and understanding: studies in cognitive science. Academic Press, New York, New York, USA.

Bradley, G.A., editor. 1984. Land use and forest resources in a changing environment. University of Washington Press, Seattle, Washington, USA.

Brooks, D.H. 1987. Land use in economic theory. United States Department of Agriculture Economic Research Service Staff Report AGE870806.

Burgess, R.L., and D.M. Sharpe, editors. 1981. Forest island dynamics in man-dominated landscapes. Springer-Verlag, New York, New York, USA.

Burke, I.C. 1989. Control of nitrogen mineralization in a sagebrush steppe landscape. Ecology **70**:1115–1126.

Carey, A.B. 1989. Wildlife associated with old-growth forests in the Pacific Northwest. Natural Areas Journal **9**:151–162.

Chandler, P. 1990. Ecological knowledge in a traditional agroforest management system among peasants in China. Dissertation. University of Washington, Seattle, Washington, USA.

Charniak, E., and D. McDermott. 1987. Artificial intelligence. Addison-Wesley, Reading, Massachusetts, USA.

Clark, D. 1973. The value of agricultural land. Pergamon Press, Oxford, England.

Coblentz, B.E. 1990. Exotic organisms: a dilemma for conservation biology. Conservation Biology **4**:261–265.

Coulson, R.N., M.C. Saunders, D.K. Loh, F.L. Oliveria, D. Drummond, P.J. Barry, and K.M. Swain. 1989. Knowledge system environment for integrated pest management in forest landscapes: the southern pine beetle. Bulletin of the Entomological Society of America **35**:26–32.

di Castri, F. 1990. On invading species and invaded systems: the interplay of historical chance and biological necessity. Pages 3–16 *in* F. di Castri, A.J. Hansen, and M. Debussche, editors. Biological invasions in Europe and the Mediterranean basin. Kluwer Academic Publishers, Dordrecht, The Netherlands.

Fahrig, L., and G. Merriam. 1985. Habitat patch connectivity and population survival. Ecology **66**:1762–1768.

Firey, W. 1960. Man, mind and land: a theory of resource use. Free Press, Glencoe, Illinois, USA.

Forcella, F., and S.J. Harvey. 1983. Eurasian weed infestation in western Montana in relation to vegetation and disturbance. Madrono **30**:102–109.

Forman, R.T.T. 1988. Ecologically sustainable landscapes: the role of spatial configuration. Proceedings of the 1988 World Congress, International Federation of Landscape Architects. American Association of Landscape Architects, Washington, D.C., USA.

Forman, R.T.T., and M. Godron. 1986. Landscape ecology. John Wiley and Sons, New York, New York, USA.

Fortmann, L., K. Kusel, C. Danks, L. Moody, and S. Seshan. 1990. The human costs of the California forestry crisis. Presentation to The Concern for Sustainable Forests, October 3, 1990. Department of Forestry and Resource Management, University of California, Berkeley, California, USA.

Fox, M.D., and B.J. Fox. 1986. The susceptibility of natural communities to invasion. Pages 57–66 *in* R.H. Groves and J. J. Burdon, editors. Ecology of biological invasions: an Australian perspective. Australian Academy of Science, Canberra, Australia.

Franklin, J.F., and R.T.T. Forman. 1987. Creating landscape patterns by forest cutting: ecological consequences and principles. Landscape Ecology **1**:5–18.

Gardner, R.H., B.T. Milne, M.G. Turner, and R.V. O'Neill. 1987. Neutral models for the analysis of broad-scale landscape pattern. Landscape Ecology **1**:19–28.

Gardner, R.H., M.G. Turner, R.V. O'Neill, and S. Lanorel. 1991. Simulation of the scale-dependent effects of landscape boundaries on species persistence and dispersal. Pages 76–89 *in* M.M. Holland, P.G. Risser, and R.J. Naiman, editors. The role of landscape boundaries in the management and restoration of changing environments. Chapman and Hall, New York, New York, USA.

Goldschmidt, W. 1990. The human career: the self in the symbolic world. Basil Blackwell, Cambridge, Massachusetts, USA.

Gosz, J.R., and P.J.H. Sharpe. 1989. Broad-scale concepts for interactions of climate, topography, and biota at biome transitions. Landscape Ecology **3**:229–243.

Grainger, A. 1986. The future role of the tropical rain forests in the world forest economy. Dissertation. Department of Plant Sciences, Oxford University, Oxford, England.

Hardt, R.A., and R.T.T. Forman. 1989. Boundary form effects on woody colonization of reclaimed surface mines. Ecology **70**:1252–1260.

Harris, L.D. 1988. Edge effects and conservation of biotic diversity. Conservation Biology **2**:330–332.

Hart, A. 1986. Knowledge acquisition for expert systems. McGraw-Hill, New York, New York, USA.

Heywood, V.H. 1989. Patterns, extents and modes of invasions by terrestrial plants. Pages 31–60 *in* J.A. Drake and H.A. Mooney, editors. Biological invasions: a global perspective. John Wiley and Sons, Chichester, England.

Holling, C.S. 1966. The functional response of invertebrate predators to prey density. Memoirs of the Entomological Society of Canada **48**:1–85.

Holling, C.S., editor. 1978. Adaptive environmental assessment and management. John Wiley and Sons, Chichester, England.

Huffaker, C.B. 1958. Experimental studies on predation: dispersion factors and predator-prey oscillations. Hilgardia **27**:343–383.

Interagency Economic Effects Review Team. 1990. Economic impacts of implementing the recommendations of the Interagency Scientific Committee appointed to study the northern spotted owl. United States Forest Service, Portland, Oregon, USA.

Jacob, F. 1982. The possible and the actual. Pantheon Books, New York, New York, USA.

Johnston, C.A., and R.J. Naiman. 1990. Aquatic patch creation in relation to beaver population trends. Ecology **71**:1617–1621.

Krummel, J.R., R.H. Gardner, G. Sugihara, R.V. O'Neill, and P.R. Coleman. 1987. Landscape patterns in a disturbed environment. Oikos **48**:321–324.

Lee, K.N., and J. Lawrence. 1986. Adaptive management: learning from the Columbia River Basin fish and wildlife program. Environmental Law **16**:431–460.

Lee, R.G. 1990. Social and cultural implications of implementing "a conservation strategy for the northern spotted owl." Independent report prepared for Mason, Bruce and Girard, Portland, Oregon, USA.

Lee, R.G. 1991. Institutional stability: a requisite for sustainable forestry. *In* 1990–91 Starker Lectures. College of Forestry, Oregon State University, Corvallis, Oregon, USA.

Lee, R.G., M.S. Carroll, and K.K. Warren. 1991. The social impact of timber harvest reductions in Washington State. Chapter 3 *in* P. Sommers and H. Birss, editors. Revitalizing the timber dependent regions of Washington. Final report for the Washington Department of Trade and Economic Development. Northwest Policy Center, Institute for Public Policy and Management, Graduate School of Public Affairs, University of Washington, Seattle, Washington, USA.

Lippke, B., K. Gilless, R.G. Lee, and P. Sommers. 1990. Three-state impact of spotted owl conservation and other timber harvest reductions: cooperative evaluation of the economic and social impacts. Contribution 69, Institute of Forest Resources, University of Washington, Seattle, Washington, USA.

Lord, J.M., and D.A. Norton. 1990. Scale and the spatial concept of fragmentation. Conservation Biology **4**:197–202.

Lubchenco, J., A.M. Olson, L.B. Brubaker, S.R. Carpenter, M.M. Holland, S.P. Hubbell, S.A. Levin, J.A. MacMahon, P.A. Matson, J.M. Melillo, H.A. Mooney, C.H. Peterson, H.R. Pulliam, L.A. Real, P.J. Regal, and P.G. Risser. 1991. The sustainable biosphere initiative: an ecological research agenda. Ecology **72**:371–412.

Luger, G.F., and W.A. Stubblefield. 1989. Artificial intelligence and the design of expert systems. Benjamin/Cummings, Redwood City, California, USA.

May, R.M. 1975. Patterns of species abundance and diversity. Pages 81–120 *in* M.L. Cody and J.M. Diamond, editors. Ecology and the evolution of communities. Belknap Press, Harvard University Press, Cambridge, Massachusetts, USA.

McNaughton, S.J. 1985. Ecology of a grazing ecosystem: the Serengeti. Ecological Monographs **55**:259–94.

Medawar, P. 1982. Pluto's republic. Oxford University Press, New York, New York, USA.

Muth, R.M. 1990. Community stability as social structure: the role of subsistence uses of natural resources in southeast Alaska. Pages 211–228 *in* R.G. Lee, D.R. Field, and W.R. Burch, Jr., editors. Community and forestry: continuities in the sociology of natural resources. Westview Press, Boulder, Colorado, USA.

Naiman, R.J. 1988. Animal influences on ecosystem dynamics. BioScience **38**:750–752.

Naiman, R.J., and H. Décamps, editors. 1990. The ecology and management of aquatic-terrestrial ecotones. UNESCO, Paris, and Parthenon Publishing Group, Carnforth, United Kingdom.

Naiman, R.J., H. Décamps, J. Pastor, and C.A. Johnston. 1988. The potential importance of boundaries to fluvial ecosystems. Journal of the North American Benthological Society **7**:289–306.

Naiman, R.J., D.G. Lonzarich, T.J. Beechie, and S.C. Ralph. 1991. General principles of classification and the assessment of conservation potential in rivers. Pages 93–123 *in* P.J. Boon, P. Calow, and G.E. Petts, editors. River conservation and management. John Wiley and Sons, Chichester, England.

Naveh, Z., and A.S. Lieberman. 1984. Landscape ecology. Springer-Verlag, New York, New York, USA.

Noss, R.F. 1983. A regional landscape approach to maintain diversity. BioScience **33**:700–706.

Noss, R.F. 1987. Protecting natural areas in fragmented landscapes. Natural Areas Journal **7**:2–13.

Noss, R.F. 1989. Who will speak for biodiversity? Conservation Biology **3**:202–203.

Noss R.F., and L.D. Harris. 1986. Nodes, networks, and MUMs: preserving diversity at all scales. Environmental Management **10**:299–309.

Odum, E.P., and M.G. Turner. 1990. The Georgia landscape: a changing resource. Pages 137–164 *in* I.S. Zonneveld and R.T.T. Forman, editors. Changing landscapes: an ecological perspective. Springer-Verlag, New York, New York, USA.

O'Neill R.V., B.T. Milne, M.G. Turner, and R.H. Gardner. 1988. Resource utilization scales and landscape pattern. Landscape Ecology **2**:63–69.

Padoch, C. 1986. Agricultural site selection among permanent field farmers: an example from East Kalimantan, Indonesia. Journal of Ethnobiology **6**:279–288.

Paine, R. 1982. Dam a river, damn a people? Saami (Lapp) Livelihood and Alta/Kautokeino Hydro-electric Project and the Norwegian Parliament. IWGIA (International Working Group for Indigenous Affairs), Document 45, Copenhagen, Denmark.

Palmquist, R.B. 1989. Land as a differentiated factor of production: a hedonic model and its implications for welfare measurement. Land Economics **65**:23–28.

Parks, P.J. 1991. Models of forested and agricultural landscapes: integrating economics. Pages 309–322 *in* M.G. Turner and R.H. Gardner, editors. Quantitative methods in landscape ecology. Springer-Verlag, New York, New York, USA.

Peterjohn, W.T., and D.L. Correll. 1984. Nutrient dynamics in an agricultural watershed: observations on the role of a riparian forest. Ecology **65**:1466–1475.

Quinn, J.F., and S. Harrison. 1988. Effects of habitat fragmentation and isolation on species richness: evidence from biogeographic patterns. Oecologia (Berlin) **75**:132–140.

Raman, R. 1986. Qualitative modeling and simulation: a survey. Pages 9–26 *in* E.J.H. Kerckhoffs, G.C. Vansteenkiste, and B.P. Zeigler, editors. AI applied to simulation. Proceedings of the European Conference at the University of Ghent, Simulation Series, Volume 18, Number 1.

Rejmanek, M. 1989. Invisibility of plant communities. Pages 369–388 *in* J.A. Drake and H.A. Mooney, editors. Biological invasions: a global perspective. John Wiley and Sons, Chichester, England.

Repetto, R., and M. Gillis, editors. 1988. Public policies and the misuse of forest resources. Cambridge University Press, New York, New York, USA.

Rosen, R. 1989. Similitude, similarity, and scaling. Landscape Ecology **3**:207–216.

Ruckelshaus, W.D. 1989. Toward a sustainable world. Scientific American, September: 166–175.

Saarenmaa, H., N.D. Stone, L.J. Folse, J.M. Packard, W.E. Grant, M.E. Makela, and R.N. Coulson. 1988. An artificial intelligence modelling approach to simulating animal/habitat interactions. Ecological Modelling **44**:125–141.

Samuelson, P.A. 1983. Thunen at two hundred. Journal of Economic Literature **21**:1468–1488.

Senft, R.L., M.B. Coughenour, D.W. Bailey, L.R. Rittenhouse, O.E. Salo, and D.M. Swift. 1987. Large herbivore foraging and ecological hierarchies. BioScience **37**:789–799.

Spradley, J.P. 1979. The ethnographic interview. Holt, Rinehart and Winston, New York, New York, USA.

Thomas, J.W., E.D. Forsman, J.B. Lint, E.C. Meslow, B.R. Noon, J. Verner. 1990. A conservation strategy for the northern spotted owl: report of the Interagency Scientific Committee to address conservation of the northern spotted owl.

United States Forest Service, Bureau of Land Management, Fish and Wildlife Service, and National Park Service, Portland, Oregon, USA.

Turner, M.G. 1988. A spatial simulation model of land use changes in a Piedmont county in Georgia. Applied Mathematics and Computation **27**:39–51.

Turner, M.G. 1989. Landscape ecology: the effect of pattern on process. Annual Review of Ecology and Systematics **20**:171–197.

Turner, M.G., editor. 1987. Landscape heterogeneity and disturbance. Springer-Verlag, New York, New York, USA.

Turner, M.G., V.H. Dale, and R.H. Gardner. 1989. Predicting across scales: theory development and testing. Landscape Ecology **3**:245–252.

Turner M.G., and R.H. Gardner, editors. 1991. Quantitative methods in landscape ecology. Springer-Verlag, New York, New York, USA.

Turner, M.G., R.H. Gardner, and R.V. O'Neill. 1991. Potential responses of landscape boundaries to global environmental change. Pages 52–75 *in* M.M. Holland, P.G. Risser, and R.J. Naiman, editors. The role of landscape boundaries in the management and restoration of changing environments. Chapman and Hall, New York, New York, USA.

Urban, D.L., R.V. O'Neill, and H.H. Shugart. 1987. Landscape ecology. BioScience **37**:119–127.

Vitousek, P.M. 1986. Biological invasions and ecosystem properties: can species make a difference? Pages 163–178 *in* H.A. Mooney and J.A. Drake, editors. Ecology of biological invasions of North America and Hawaii. Springer-Verlag, New York, New York, USA.

Vivrette, N.J., and C.H. Muller. 1977. Mechanism of invasion and dominance of coastal grassland by *Mesembryanthemum crystallinum*. Ecological Monographs **47**:301–318.

Walters, C. 1986. Adaptive management of renewable resources. Macmillan, New York, New York, USA.

Wilcove, D.S. 1987. From fragmentation to extinction. Natural Areas Journal **7**:23–29.

World Commission on Environment and Development. 1987. Our common future. Oxford University Press, Oxford, England.

Yahner, R.H. 1988. Changes in wildlife communities near edges. Conservation Biology **2**:333–339.

Zaret, T.M., and R.T. Paine. 1973. Species introduction in a tropical lake. Science **182**:449–455.

Zonneveld, I.S., and R.T.T. Forman, editors. 1990. Changing landscapes: an ecological perspective. Springer-Verlag, New York, New York, USA.

Index